28th Ethylene Producers' Conference 2016

Topical Conference at the 2016 AIChE Spring Meeting and 12th Global Congress on Process Safety

Houston, Texas, USA
10-14 April 2016

ISBN: 978-1-5108-2484-3

Printed from e-media with permission by:

Curran Associates, Inc.
57 Morehouse Lane
Red Hook, NY 12571

Printed by Curran Associates, Inc. (2016)

For permission requests, please contact AIChE
at the address below.

AIChE
120 Wall Street, FL 23
New York, NY 10005-4020

Phone: (800) 242-4363
Fax: (203) 775-5177

www.aiche.org

Additional copies of this publication are available from:

Curran Associates, Inc.
57 Morehouse Lane
Red Hook, NY 12571 USA
Phone: 845-758-0400
Fax: 845-758-2633
Email: curran@proceedings.com
Web: www.proceedings.com

TABLE OF CONTENTS

440798

IMPROVING TEMPERATURE MEASUREMENT AND CONTROL USING THE EXACTUS® OPTICAL THERMOMETERS

Pieter A. Reyniers, Nenad D. Ristic, Marko R. Djokic

PhD researchers

Ghent University, Ghent, Belgium

Kevin M. Van Geem

Full Professor

Ghent University, Ghent, Belgium

Richard Marx

Petrochemical Applications Engineer and Sales Lead

BASF Temperature Sensing Products, Portland, OR, USA

Mark Foerch

Managing Director

Pt Technology GmbH (BASF), Dreieich, Germany

Prepared for Presentation at the 2016 Spring National Meeting

Houston, TX., April 10 – 14, 2016

IMPROVING TEMPERATURE MEASUREMENT AND CONTROL USING THE EXACTUS® OPTICAL THERMOMETERS

Pieter A. Reyniers, Nenad D. Ristic, Marko R. Djokic

PhD researchers

Ghent University, Ghent, Belgium

Kevin M. Van Geem

Full Professor

Ghent University, Ghent, Belgium

Richard Marx

Petrochemical Applications Engineer and Sales Lead

BASF Temperature Sensing Products, Portland, OR, USA

Mark Foerch

Managing Director

Pt Technology GmbH (BASF), Dreieich, Germany

Abstract: Offering excellent accuracy, resolution, repeatability, and stability, EXACTUS® optical thermometry technology provides significant performance advantages in non-contact temperature measurement in a practical, rugged and user-friendly design. To assess the performance of the EXACTUS® optical pyrometer device for controlling the temperature of steam cracking furnaces, ethane steam cracking experiments have been carried out in the Pilot plant set-up of the Laboratory for Chemical Technology (LCT), Gent, Belgium. The control performance of EXACTUS® was compared with that of a weld-on type K thermocouple, the latter being representative of the present state of the art in furnace control via wall temperature measurement. EXACTUS® equipment is found to be suited as a sensor to measure tube metal temperatures of a steam cracking reactor coil, either inside or outside of the furnace. It outperforms a weld-on thermocouple in providing a process value to the controller and hence offers an advantage over the present state of the art. The main benefit of EXACTUS® over thermocouples is when the temperature measurement is used to prevent overshoots of tube metal temperature, which have an adverse effect on the tube longevity and the coke formation.

Introduction

Accurate temperature measurements are of the utmost importance for stable operation and optimal product quality in most, if not all, industrial processes. Several methods for temperature measurement have been developed based on different physical principles [1]. A distinction is made between contact and non-contact temperature measurements, with contact measurement being the industry's favored method. Contact measurements are either based on mechanical effects, e.g. the volume change of gases, liquids or solids, or on electrical effects, e.g. the voltage generated between two dissimilar metals or the change in resistance of a metal. Non-contact measurements are mostly based on the intensity of electromagnetic radiation emitted by a hot surface.

Thermocouples are the most frequently used contact temperature measurement method due to their low cost, wide measurement range and relatively high accuracy. Their major drawback is thermal drift: recurrent recalibration or replacement of thermocouples is needed to maintain their absolute accuracy. Non-contact measurements are almost exclusively based on infrared measurements. These techniques have only recently been applied in industry, after issues of reliability, cost and sensor size were resolved. Research continues improve the reliability and applicability of the method [2,3]. Advantages of infrared temperature measurements are the higher information rate (higher measurement frequency), the high longevity at high temperatures and the fact that no energy is removed from the measured surface. However, as the measurement is based on electromagnetic radiation, the optical path between the object and the sensor must be clear. An accurate measurement is inhibited by dust or smoke, as well as by contaminations on the sensor or the measurement surface. Moreover, only surface temperature measurements are possible. Non-contact temperature measurements have been successfully implemented in heavy-duty sectors such as metal production [4], cement and lime production [5], power generation and the glass industry [6]. Other applications benefit from these measurements as well, e.g. the paper industry [7], food and beverage industry [8] and the automotive industry [9].

In this work, the potential of using optical thermometers for control purposes in a steam cracking furnace is assessed by testing the equipment in a Pilot plant set-up for steam cracking. First, the working principle of non-contact infrared thermometers is explained, followed by a description of the EXACTUS® optical thermometer. In a third section, the Pilot plant set-up for steam cracking of hydrocarbons is presented, including a description on how the EXACTUS® equipment was implemented and the procedure followed during the experiments. Subsequently the results for temperature, product yields and coke formation are presented and discussed. At the end, the conclusions of the study are summarized.

Working principle of optical thermometers

An optical thermometer or pyrometer determines the temperature of a radiating surface based on the known emissivity of the surface (by calibration) together with the measured radiation intensity. This discussion on the working principle is based on the work of Brown et al. [10]. The Stefan-Boltzmann law, given in Equation (1), forms the basis for calculating the temperature. It relates the total radiation intensity J, i.e. the power of the radiation from a certain surface received by the detector, to the temperature T via the Stefan-Boltzmann constant σ and the surface emissivity ε.

$$J = \varepsilon\, \sigma\, T^4 \tag{1}$$

The Stefan-Boltzmann law applies to the full electromagnetic spectrum so it accounts for all the radiation emitted by a certain surface. However, detection of the full radiative power of a certain surface requires very tedious measurements. Therefore, the temperature of a surface can also be calculated based on the radiation intensity at a certain wavelength λ. This facilitates the measurement and via careful selection of the wavelength, interference with the environment can be minimized. Planck's Law (Equation (2)) can be used instead of the Stefan-Boltzmann law to calculate the surface temperature based on the emissivity at a certain wavelength.

$$I'(\lambda, T) = \frac{2\,h\,c^2}{\lambda^5} \frac{1}{e^{\frac{h\,c}{\lambda\,k_B\,T}} - 1} \tag{2}$$

In this equation, I' is the radiation intensity, i.e. the energy per unit time radiated per unit area of emitting surface in the normal direction per unit solid angle per unit frequency, h is the Planck constant, c is the speed of light in the medium and k_B is the Boltzmann constant. Planck's Law is visualized in Figure 1 for wavelengths in the interval 0 10^{-6} m – 7 10^{-6} m at temperatures of 800 K, 1000 K, 1200 K, 1400 K, and 1600 K. As the temperature increases, the maximum radiation intensity increases and the location of the peak shifts to shorter wavelengths. Planck's Law can be used to select an appropriate wavelength for the range of temperatures that is to be measured with the pyrometer. The higher the radiation intensity of the surface at the selected wavelength, the higher the precision of the measurement.

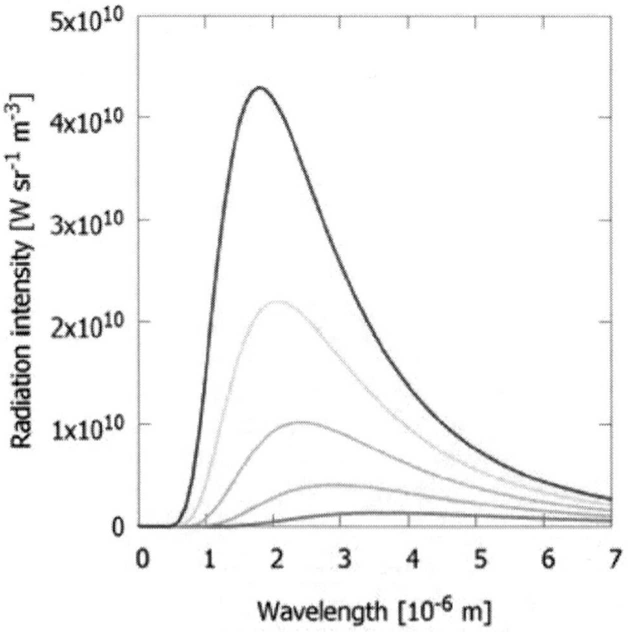

Figure 1. Radiation intensity as function of wavelength according to Planck's Law at different surface temperatures. 800 K; 1000 K; 1200 K; 1400 K; 1600 K

Planck's Law is only valid for black bodies with an emissivity of 1 for all wavelengths. The radiation intensity of a real surface is lower than the radiation intensity of a black body. The ratio of the emitted radiation of a real surface and the radiation intensity of a black body is defined as the emissivity of the real surface, see Equation (3). Emissivity is both temperature and wavelength dependent.

$$\varepsilon(\lambda, T) = \frac{I'_{actual}(\lambda, T)}{I'_{black\ body}(\lambda, T)} \tag{3}$$

Incident radiation on a surface can either be absorbed, reflected or transmitted. These parts are respectively calculated via the reflectance ρ, absorptance α and the transmittance τ. Equation (4) follows directly from the conservation of energy. For opaque bodies, the transmittance is very small so the equation simplifies to equation (5). For black bodies, the reflectance is zero so the absorptance is equal to one.

$$\rho + \alpha + \tau = 1 \tag{4}$$

$$\rho + \alpha = 1 \tag{5}$$

In most applications, the transmittance of the emitting surface can be considered as zero. However, the absorptance of the surface is not equal to one so part of the incident radiation is reflected. To obtain an accurate measurement of the surface temperature of the real surface, the emissivity should be precisely known. As emissivity is both wavelength and temperature dependent, this is not so straightforward. Deviations on the emissivity value can significantly affect the accuracy of the measurement. A 10% error on the emissivity value can result in an absolute error on the temperature measurement of a couple of Kelvin at low surface temperatures but the error could increase to over 100 K at high surface temperatures for measurements at longer wavelengths. An experimental device to measure the temperature and wavelength dependent emissivity of a surface is described in the work of Brown et al. [10].

The issue of not knowing the emissivity can be mitigated by measuring the radiation intensity of the surface at two distinct wavelengths sufficiently close together in the electromagnetic spectrum. In that case, the emissivity can be considered as equal for the two wavelengths. Plank's Law can then be applied to the measurements at the two wavelengths, resulting in a set of two equations with two unknowns: the emissivity ε and the temperature T. So-called multi-wavelength pyrometers implement this strategy for measuring the temperature of a surface with an unknown or varying emissivity [2].

EXACTUS® optical thermometer

The EXACTUS® is an optical thermometer developed by the Temperature Sensing Group of BASF. Figure 2 provides a schematic overview of the equipment. The pyrometer consists of three main parts: a detector, an amplifier circuit and an optical system [10]. Via the collection optics, electromagnetic radiation emitted from the measurement surface is gathered and sent via and optional lightpipe to the photodetector incorporated in the probe head. The filter isolates the required wavelength from the incoming light. The amplified electrical signal from the photodetector is digitized and sent to a computer via a LAN-connection or transformed in an analog signal e.g. a 4-20 mA signal or a 0-10 VDC signal which can be interpreted by a Programmable Logic Controller (PLC).

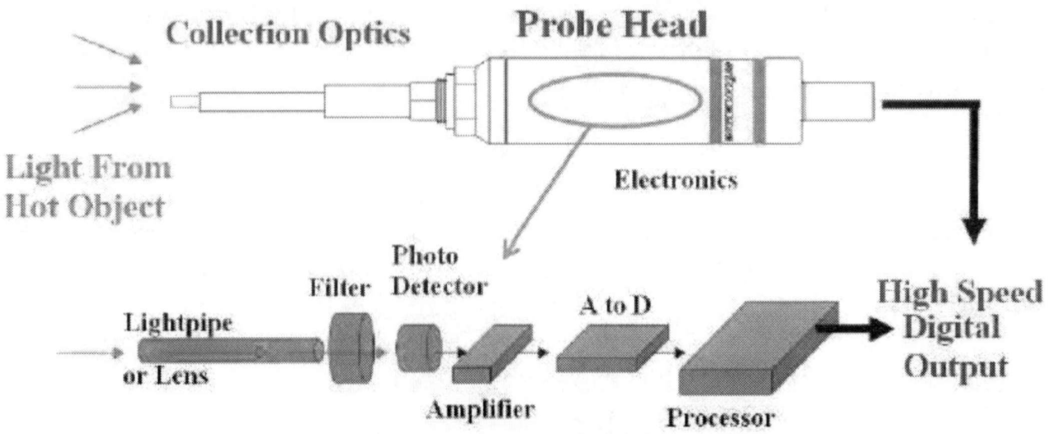

Figure 2. Schematic overview of the EXACTUS® optical thermometer.

The benefits of the EXACTUS® device are a high accuracy down to 1.5 K with a repeatability of 0.10 K, a fast signal gathering up to 1000 Hz and a low drift of less than 0.1 K per year. An optional system with fiber-optic cables allows installing the electronic equipment away from the point where the light is gathered. This is necessary in industrial configurations where strict zoning laws apply. The downside is a slight loss in accuracy as the fiber-optic cable can influence the result. Calibration in place of the hardware can further reduce the measurement error.

Depending on the desired temperature range, light of different wavelengths is analyzed to determine the temperature. A photodetector tuned to radiation with a wavelength of 900 nm allows to measure temperatures in the range of 280 °C to 3000 °C. Radiation at 1600 nm is suited to provide temperature readings in the range of 65 °C to 2800 °C. The wavelengths are selected in such a way to avoid interference with common gaseous components in the air and in combustion gases (H_2O, CO_2, CO, CH_4, etc.).

Initially the EXACTUS® device has been developed for applications in the semiconductor industry where fast temperature readings with a high accuracy are needed in order to have good wafer growth. Optical measurements in such an environment are straightforward as these manufacturing facilities are free of dust. However, when applying EXACTUS® in other industries such as production of bulk chemicals or in the steel industry, it should be guaranteed that there is a clear line of sight between the optics of EXACTUS® and the measuring surface. Additionally, no depositions should be present on the optics and the state of the analyzed surface should not change since these changes will affect the surface emissivity.

Pilot plant set-up for steam cracking of hydrocarbons

The experimental work is carried out in the Pilot plant set-up for steam cracking of hydrocarbons at the Laboratory for Chemical Technology (LCT) of Ghent University, see Figure 3. Since the main parts of this unit, the analytical equipment and the calibration procedure have been described elsewhere [11,12], only a brief description, based on Wang et al. [13], Dhuyvetter et al. [14] and Van Geem et al. [15] is given below.

Figure 3. Picture of the LCT Pilot plant set-up for steam cracking of hydrocarbons.

The process flow diagram of the Pilot plant set-up is given in Figure 4. The feed section controls the supply of different feedstocks to the reactor coil. The measurement of the mass flow rate of liquid hydrocarbons and water is carried out using a CORI-FLOW mass flow controller. The pumping frequency of the CORI-FLOW pump is automatically adjusted by the controller of the CORI-FLOW. Gaseous hydrocarbons are fed to the reactor from pressurized gas cylinders by means of a separate CORI-FLOW mass flow controller.

The furnace, built using silica/alumina bricks (Li23), is about 4.0 m long, 0.7 m wide and 2.6 m high. It is fired by means of ninety premixed gas burners, mounted with automatic fire checks and arranged on the side walls in such a way as to provide a uniform heat distribution. The furnace is divided into seven separate cells, which can be fired independently to set any type of temperature profile.

The reaction section of the reactor coil is about 12 m long, made of high-alloy steel, and has an internal diameter of 9 10^{-3} m. These dimensions are chosen to achieve turbulent flow conditions in the coil with reasonable feed flow rates (Re > 7000). Twenty thermocouples and eight manometers are located along the reactor coil to measure the temperature and pressure of the process gas as indicated on Figure 4. The pressure in the reactor is controlled by a reduction valve. At the reactor outlet, nitrogen is injected as internal standard for the analysis. Before cooling the effluent, a sample is taken for the on-line C_5^+ analysis. After cooling and knock-out of the liquefied products and tar, a second sample is taken for on-line C_4^- analysis. The rest of the effluent is sent to the flare.

For the current campaign, some modifications were made to the set-up in order to accommodate the EXACTUS® equipment. These modifications are discussed in the next paragraph.

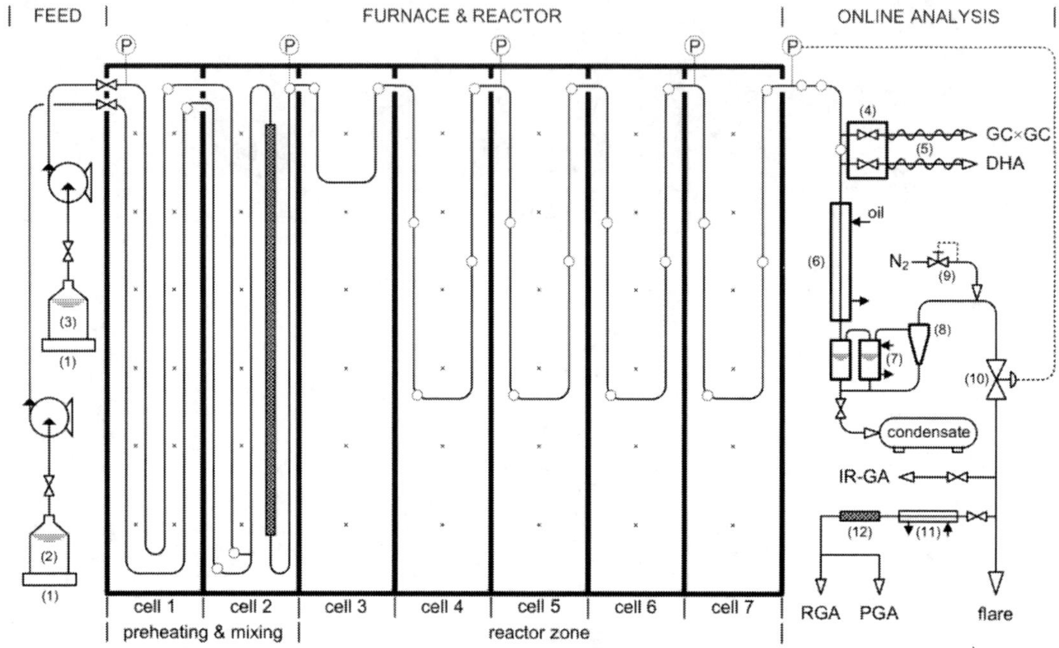

Figure 4: Process flow diagram of the LCT Pilot plant set-up for steam cracking of hydrocarbons. ⍦ : temperature measurements; Ⓟ : pressure measurements.

Installation of the EXACTUS® device in the Pilot plant set-up

A custom collar is fitted around the reactor outlet tube between the insulation box of the transfer line exchanger (TLE) and the flange connecting the reactor to the analysis section, that is in the adiabatic section of the furnace. The custom designed collar displayed in Figure 5 allows to position 3 devices to measure tube metal temperature within a distance of 0.08 m. Two EXACTUS® devices and one spring-loaded thermocouple were positioned in the custom collar in the following order (in the flow direction): EXACTUS® 1, EXACTUS® 2 and the spring loaded thermocouple. In between the insulation box for the TLE and the custom collar with the EXACTUS® equipment, a type K thermocouple was welded to the tube as a reference measurement for the tube metal temperature. This thermocouple is not shown on Figure 5.

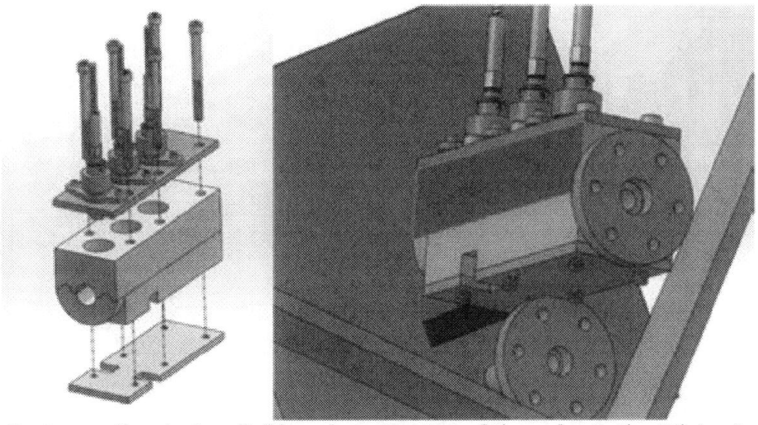

Figure 5. Custom collar design (left) and positioning of the colar in the Pilot setup (right).

Via the viewing ports in the furnace wall, EXACTUS® 3 was installed to measure the tube metal temperature inside the furnace close to the reactor outlet. The lens for this device was positioned by attaching a laser pointer to the fiber-optic cable and manually pointing the lens at the desired point in the reactor. Before each experiment the alignment was checked. The mounted EXACTUS® device and the position of the laser pointer on the coil in the furnace are shown in Figure 6.

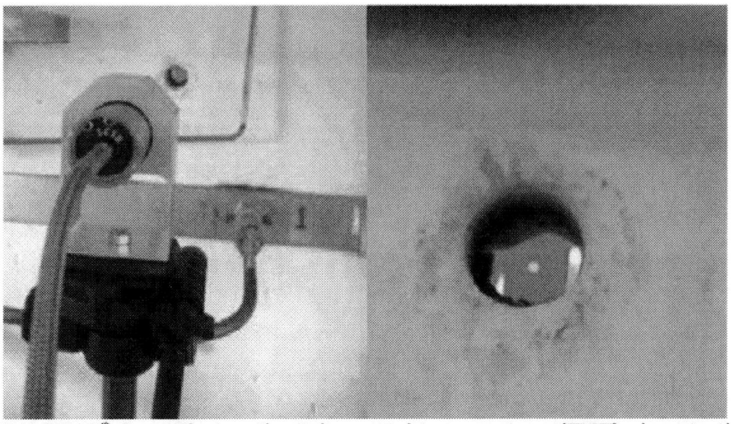

Figure 6. Mounting of EXACTUS® 3, analyzing the tube metal temperature (TMT) close to the reactor outlet inside the furnace (left); Position of the laser pointer on the reactor coil (digitally enhanced) (right).

The newly installed hardware allows to provide a process value for the furnace temperature control in the following three ways:

1. EXACTUS ADIABATIC: EXACTUS® 1, mounted in the custom collar in the adiabatic section
2. TK ADIABATIC: the weld-on type K thermocouple in the adiabatic section
3. EXACTUS FURNACE: EXACTUS® 3, mounted outside of the furnace and viewing the reactor coil via a viewing port in the refractory.

The names EXACTUS ADIABATIC, TK ADIABATIC and EXACTUS FURNACE will be used to refer to these three different control systems.

Experimental procedure

The process value for temperature in the furnace is delivered by either the EXACTUS® optical thermometer in the adiabatic section (EXACTUS ADIABATIC), the weld-on thermocouple in the adiabatic section (TK ADIABATIC) or the EXACTUS® optical thermometer pointed towards the reactor in the furnace (EXACTUS FURNACE). The default control system of the Pilot plant set-up is used during initial stabilization and decoking but it is disabled during the actual experiment (phases 2, 3 and 4). The following experimental procedure is adopted for all experiments.

- Leak test of the hot section using N_2 at a pressure of 4 bar during 5 minutes.
- In case there has been no experiment the previous day: pre-oxidation of the coil with air for 30 minutes.
- Preheating of the reactor using the default control system, reactor: H_2O
- **Phase 1**: Cracking with 3000 g/h ethane and 1155 g/h water (dilution: 0.385 kg/kg) using the default control for stabilization. Addition of 100 ppm sulfur/kg hydrocarbons. Duration: 1h08.
- **Phase 2**: Switch furnace temperature control to the relevant alternative control system. The process set point for this phase is determined based on its stable temperature during phase 1. Continue cracking. Duration: 1h50.
- **Phase 3**: Increase process set point of the relevant alternative control system by 10 °C. Continue cracking. Duration: 1h26.
- **Phase 4**: Increase process set point of the relevant alternative control system by 15 °C. Continue cracking. Duration: 1h26.
- Stop cracking, reactor: N_2
- Decoking of the reactor
- Cooling down of the reactor

All experiments are carried out using ethane as feedstock. It is supplied by Praxair Belgium with a specification of 99.999 %. Sulfur is added to the feed in the form of dimethyl disulfide. A solution of 1.13 wt% DMDS in hexane is made and fed to the reactor at a flow rate of 59.3 ml/h, resulting in a sulfur mass based content of 100 ppm per kg hydrocarbons.

Effluent composition analyses are done back-to-back. The C_4^- analysis takes 29 minutes, and a new sample injection is started once the full C_4^- analysis cycle is finished. The focus in the current campaign is on the C_4^- analysis as only a small fraction of the reactor effluent in ethane cracking has a carbon number higher than 4.

Results and discussion

In total two experiments were carried out with each of the three available control systems listed in the section on the hardware modifications, resulting in a grand total of 6 experiments. The results of all experiments are systematically displayed. Three parameters are investigated in the remainder of this paper for their stability, accuracy and reproducibility: temperature, product yields and coke formation.

445172 A Computational Model to Simulate Decoking in Steam-Cracking Furnace Coils

Abstract:

Steam cracking produces ethylene via non-catalytic high temperature pyrolysis of hydrocarbons inside coils in the furnace's radiant zone. During cracking, coke buildup on the coil's internal surfaces inhibits heat transfer. Ultimately, either a metal temperature limit or pressure drop increase necessitates periodic decoking of coils with steam/air mixture. To further understand the effect of process parameters on decoking, a 2D axisymmetric CFD (computational fluid dynamics) model has been developed to simulate decoking in steam-cracking furnace coils. The model accounts for both oxidation and steam gasification reaction kinetics and incorporates key process parameters and heat flux from a furnace. A multi-stage pseudo-transient numerical framework has been developed to simulate the decoke cycle of industrial furnaces in a computationally affordable manner. Furthermore, the numerical technique ensures mass conservation during various stages of decoking.

443955 Rotordynamic Considerations of Large Frame Centrifugal Compressors for Ethylene Service

Abstract:

To achieve improved economies of scale, ethylene plant sizes have increased to as large as 1000 KTA. Building compressors of this scale requires careful rotordynamic design due to the high flow rate requirement and large size. Maintaining good rotordynamic stability and critical speed separation margin is key to reliable equipment operation. This paper summarizes unique considerations in the compressor design to maintain acceptable rotordynamic performance. The influence of labyrinth seals, impeller secondary flow cavities, damper seals, swirl brakes, gas seals, and bearings on the rotordynamic behavior will be investigated. API rotordynamic requirements will also be discussed.

443954 Maximizing Turbomachinery Reliability and Optimizing Plant Performance with Torque Monitoring Couplings

Abstract:

Couplings with integrated torque monitoring devices have been available for over 40 years. However numerous misconceptions remain regarding the technology and effective use of these devices during performance testing and continuous on-site operation of turbomachinery equipment. An overview of current torque monitoring technologies will be reviewed, calibration methods will be discussed, and a comparison against the industry standard practice of heat balance equations will be provided. Several real-world examples will be presented to demonstrate how torque monitoring couplings can accurately measure operating conditions. Torque monitoring couplings verify the power being transmitted, identify decreases in train efficiencies, and detect damaging torsional resonances as part of a system to achieve optimal plant reliability and performance.

443953 Dry Gas Seal System for Large Ethylene Compressors

Abstract:

Dry gas seal support systems must perform in various applications that typically have multiple process conditions. These alternate operating points may be significantly different than guaranteed conditions. Although flow control is common, design implications of this versus differential pressure control will be compared and evaluated. These control methods as well as alarm & trip strategies in the event of a seal failure need to be addressed and mutually agreed upon by the both the equipment manufacturer & owner. Trends for operation and design as well as impacts of API specifications will be discussed.

443950 Abstract for the Design and Implementation of a Reliable and Cost-Effective Surge and Choke Control System on Centrifugal Compressors

Abstract:

The proposed oral presentation will review the steps involved in the design and implementation of a reliable and cost-effective surge control system for centrifugal compressors. Additionally, it will review the requirements for a successful choke control system. The paper will highlight the basics of surge control system design, instrumentation, and guidelines for a controls retrofit and field testing. Surge and choke control as they pertain to the Cracked Gas Compressor, Propylene Compressor, and Ethylene Compressor will be addressed. Special attention will be given to process upsets that can cause surge and choke events in the compressors, and how a properly-designed control system can ameliorate the impacts caused by these upsets. Events such as a furnace shut-down can cause the Crack Gas Compressor to shut-down if the surge and performance control system is not adequately designed. A trip of the Propylene Compressor can cascade to the Ethylene Compressor and shut it down unless its Compressor Control System can use appropriate feed-forward signals to prevent that occurrence.

Actual design and implementation of surge control systems will be illustrated with two successful field installations: 1) A surge control system for a refrigeration application in a gas fractionation plant, and 2) A surge control and loadsharing control system for two compressors in series. Final mention will be made about choke control that was done for a pipeline compressor to illustrate the technology used to identify choke and the different means for protecting a compressor from choke.

This tutorial content is meant to target rotating equipment, reliability and instrumentation engineers, and process engineers working in ethylene plants as well as other industries that make use of centrifugal and axial compressors. The material covered will be technical in nature, and it will be a good source of training material for new engineers, and a refresher for experienced engineers.

443689 The Extraordinary Challenge of Mega-Projects

Abstract:

Shale related developments led to the renaissance of the petrochemical industry in North America. Last 5 plus years have seen an unprecedented pace of investment in the petrochemical industry. Last one year has seen a significant change in oil prices, dropping from nearly $100/bbl to below $50/bbl, resulting in a slow-down in investments. This slow-down is driven by lower margins, due to flattening of cost of production curve, and will prolong the payout of these investments. Ethylene based petrochemical complexes are capital intensive and industry is looking for improving competitiveness of new investments as compared to naphtha based plants in Asia and other regions. This presentation will explore some of the potential opportunities for managing the capital cost, one of the key element for mega project delivery, which will help improving competitiveness of petrochemical investments - particularly in North America.

443958 Compressor for Ethylene Plants - 101 Tutorial Session

Abstract:

Selection and application of large compressors in Ethylene Plant service will be presented. Details will include: Machine overview, operation, materials of construction, water and oil injection, coatings, etc. The presentation will also contrast and compare unique design features for Process Gas Compressors and the refrigeration services most commonly found in Ethylene plants. A Q/A session will be held at the end of the presentation.

443956 The Newly Developed High Pressure Ratio Compressor for the Olefins Market

Abstract:

The current pressure on cost calls for new technologies that can simplify and reduce the number of compressor bodies installed as well as saving space and weight.

High Pressure Ratio Compression is a technology, that building on our compressor heritage, enable a compressor to deliver very high pressure ratio in one single casing.

The paper will illustrate, as HPRC applied to real cases for charge gas compressor studies, deliver huge amount of benefits and also how process can be re-engineered to achieve:

- higher efficiency,
- reliability,
- reduced footprint
- and cost reduction

Using a Large Data Bank to Develop a Robust and Simple Model to Estimate the Saturation Pressure of Oil Reservoirs

By

Hassan Alkandari[1], Adel Elsharkawy[1], Adel Malallah, Osamah Alomair

Petroleum Engineering Department, College of Engineering & Petroleum
Kuwait University
Po Box 5969 Safat 13060-Kuwait

Abstract

Measurements of saturation pressure are crucial for all hydrocarbon reservoir fluids. Below the crude oil saturation pressure gas reaches a critical saturation; then, a two phase flow occurs and results in decreasing oil production and recovery. To maximize oil production and recovery, the reservoir pressure has to be maintained closer to the original saturation pressure. This pressure is normally measured using bottom-hole samples or surface recombined samples of oil and gas. Occasionally, the samples become unavailable and the pressure needs to be estimated using computational methods.

In this study, a large data bank of 231 crude oil compositions and saturation pressure measurements including literature data, unpublished data, and newly measured data were used to develop two empirical models to predict the saturation pressure of a variety of crude oils if the oil sample is unavailable or the experimental measurements are unreliable. The first proposed model utilizes the extended compositions of hydrocarbons up to the heptane plus fraction in addition to non-hydrocarbons. The second model uses the lumped compositions of light, intermediate, and heavy components in addition to non-hydrocarbon components as the input variables. The lumped model has the advantage of using fewer input parameters while maintaining the thermodynamics. The accuracy and validity of both models to calculate the saturation pressure for volatile oils, black oils, and heavy oils are presented using several compositional data. The models performance is also compared to the Peng-Robinson equation-of-state (PR-EOS), and the Soave-Redlich-Kwong equation-of-state (SRK-EOS) as well as all published methods that use compositions as input variables. The comparison indicates that the proposed models are much simpler and more accurate than the other computational methods. The proposed models treat the heptane-plus fraction as a single component, thus eliminating the splitting and characterizing the plus fractions and the binary interaction parameters needed for the EOS's calculations.

Key words: Saturation pressure (Ps), Bubble-point pressure (BPP), Equation of state (EOS), Correlations, and PVT Data

[1] Corresponding Authors: Hassan Alkandari , Hassan@hkandari.com, Adel Elsharkawy, adel.elsharkawy@ku.edu.kw

Introduction:

Reservoir fluid properties are an essential source of information for all calculations on hydrocarbon reserves. Accurate estimation of the pressure-volume-temperature (PVT) measurement is crucial. Of all the PVT properties, saturation pressure (Ps) is considered the most important parameter. It is well known that gas starts to flow once the reservoir pressure drops below the bubble point pressure and the gas saturation reaches a critical value. To minimize the gas flow and maximize the oil recovery, it is very important to maintain the reservoir pressure close to the original bubble point pressure.

Although the PVT properties are experimentally measured in laboratories and provide reliable results, this process is expensive, time consuming, and sometimes unavailable; therefore, measured field data (separator gas oil ratio, separator pressure, stock-tank oil specific gravity, and reservoir temperature) or reservoir fluid compositions are used to estimate the bubble point pressure.

There are two methods in the petroleum industry to calculate the saturation pressure of crude oil when experimental is unavailable; either using production data or compositional data.

Methods Based on Production Data (Composition is unavailable)

Numerous attempts have been made since the 1940s to develop correlations to estimate the saturation pressure from production data, such as producing gas oil ratio, oil gravity, gas gravity, reservoir pressure and temperature. Table 1 presents summary of all the methods that use production data to estimate the saturation pressure.

Methods based on Compositional data

Many authors have reported that the application of any correlation in other regions results in serious errors. These conclusions arise from the fact that regional crude oils represent a physical-chemical trend; therefore, all attempts to develop a general correlation using field measured parameters to estimate the saturation pressure from a large data bank have failed. For this reason, Elsharkawy (2003) made the first attempt to correlate the saturation pressure to the reservoir fluid composition instead of using the production data. In recent years, there have been several attempts to estimate the saturation pressure of oil reservoirs. These methods are reported in Table 1.

Elsharkawy (2003) presented a correlation of the bubble point pressure for the oil reservoir using the same input data needed for the equation-of-state (EOS) calculations, i.e., the compositional data and the reservoir temperature. Elsharkawy's model is based on the saturation pressure measurements of 60 crude oil samples from the Middle East and 75 collected from the literature. He demonstrates that the calculation of the saturation pressure depends on the type of EOS, the number of pseudo components, the method used to characterize the pseudo-components and the binary interaction number. Elsharkawy compared the performance of his model to the SRK-EOS and PR-EOS for black oils, volatile oils, heavy oils, and synthetic oils. He indicated that his model eliminates splitting the heavy plus fraction, characterizing the pseudo-components and inclusion of the binary interaction parameters needed for the EOS calculations.

Later, Tale et al. (2009) made the second attempt to model the saturation pressure of crude oil using 132 data samples published by Elsharkawy (2003). Tale's model, which is based on the alternating conditional expectation (ACE), has a set of 14 equations with 42 constants. The average absolute relative error (AARE) reported by Tale et al. indicates that the model did not perform well with the selected data. The ACE model has an AARE of 12.3% compared with AARE of 6.7% by Elsharkawy (2003) and 8.1% AARE for the SRK-EOS.

AlQuraishi (2009) attempted to model the saturation pressure using Elsharkawy's (2003) data. The model was developed using linear genetic programming. He correlated the saturation pressure to the formation temperature, methane content, and heptanes plus molecular weight. Thus, AlQuraishi's model neglected the effect of the intermediate and heptane plus content on the saturation pressure. The error analysis reported by AlQuraishi indicates that his model has a 5.8% error compared to 7.7% by Elsharkawy, 9.9% using the SRK-EOS, and 10.1% using the PR-EOS.

Bandyopadhyay and Sharma (2011) presented a model to predict the bubble point pressure, using temperature and fluid composition. They used 129 experimental and literature data originally published by Elsharkawy (2003) to develop their model. The model introduced a temperature interaction parameter that depends on compositions. The proposed model has a set of 24 coefficients and the estimated bubble point pressure is calculated as the sum of the series calculations. They reported an average absolute error (AAR) of 7.83% using their model compared to 8.3% using Elsharkawy's simple model (2003), 10.15% using SRK-EOS, and 10.58% using the PR-EOS.

Khamehchi et al. (2011) used an intelligent proxy simulator (IPS), which is based on an adaptive neural fuzzy inference system (ANFIS) and a structured artificial neural network (ANN) to predict the bubble point pressure and asphaltene onset pressure. The model uses a reservoir fluid composition and temperature to predict the Ps. It is not clear how the Ps is calculated from the proposed method, and they did not present the accuracy of the proposed method compared to the other published methods.

Farasat et al. (2013) presented a new mathematical approach to calculate the crude oil saturation pressure as a function of the temperature and the reservoir fluid composition. The model developed by Farasat et al used 130 experimental data, which were originally published by Elsharkawy (2003). Farasat used a support vector machine (SVM) to model the saturation pressure. They did not clearly demonstrate how to calculate the saturation pressure from the SVM model. Moreover, they did not indicate the overall average absolute deviation (AAD) for their model compared to other published models and both the PR-EOS and SRK-EOS for the 130 data.

Ahmadi et al. (2014) also presented an ANN technique called gene expression programming (GEP) to determine the bubble point pressure of the oil samples. Similar to Farasat et al. (2013), Ahmadi's model used most of the data previously published by Elsharkawy (2003) to construct a non-linear equation using reservoir fluid compositions and temperatures as input parameters. The proposed model is highly complicated and consists of 8 equations that have several constants compared to the one simple equation proposed by Elsharkawy (2003). They compared the performance of their proposed model to the SRK-EOS and PR-EOS; however, their paper does not mention how the heptanes plus fraction were treated and characterized for the EOS calculations. They reported an AAD of 4% using the proposed model compared to 8% using Elsharkawy's model (2003), 10.9% by PR-EOS, 10.5% by SRK-EOS, and 15% by Bandyopadhyay and Sharma (2011).

Gholami et al. (2014) estimated the bubble point pressure using a support vector regression SVR and a supervised learning algorithm SLT based on the statistical learning theory, and then the ACE was used. The model used reservoir fluid compositions and temperature to estimate the bubble point pressure. Gholami et al. used the same data originally published by Elsharkawy (2003). They presented three attempts based on genetic programming to correlate the bubble point pressure: SVR, ACE, and power-law committee machine (PLCM). The ACE-based method resulted in a set of three equations with a matrix of [8 by14] constants. Gholami et al. compared the accuracy of their model to the SRK-EOS and PR-EOS predictions reported by Elsharkawy (2003). Based on their AARD analysis, the PLCM model has the lowest errors followed by the SVR model, Elsharkawy's model, ACE model, PR-EOS, and SRK-EOS.

Lately, Jarrahian et al (2015) presented two empirical models to estimate the saturation pressure of black oils. The first model is based on production information. The second model is based on compositional information mostly from Elsharkawy (2003) and Heidaryan and Moghadasi (2011). They have not presented comparison of the accuracy of their model to Farasat (2013) model, Ahmadi (2014) model, and Gholami (2014) model.

This paper is concerned with the estimation of saturation pressure using compositional data. Therefore, the objective is to develop a robust yet simple model to estimate the saturation pressure for various types of reservoir oils using a larger data bank than that used by other published studies. The accuracy of the proposed models is compared to previously published models as well as SRK-EOS and PR-EOS.

Data bank

This study uses a large data set of 231 crude oil compositions and saturation pressure measurements. The data set consists of 130 data sets originally reported by Elsharkawy (2003); in addition to others collected from literature (Jacopy, 1958; Vogel, 1980; Coats, 1986; Droham, 1980; Dansh, 1992; Moharam, 1996; Rosenegger, 1999); 94 new data sets that have not been utilized before; as well as 7 newly measured samples at the Petroleum Fluid Research Center (PFRC) in Kuwait University. These data represents a wide range of compositions, such as volatile oils, black oils, and heavy oils. The data cover a pressure range of 537 to 5065 psia and a temperature range of 58 to 319 °F, crude oil compositions from C_1 to C_{7+}, non-hydrocarbons, such as N_2, H_2S, and CO_2, and the molecular weight and the specific gravity of the heptane plus fraction.

Data screening and classification

The compositions and the saturation pressure measurements in the data bank are subjected to a meticulous screening process to check the validity of the data. Our screening criteria uses the following steps: the sum of the molar compositions of all the components must add up to one, otherwise, the sample is eliminated. Samples with a missing molecular weight or specific gravity of the plus fraction are also eliminated.

After filtering, the data were reduced to 214 and divided into 179 measurements used to develop the new models, and 35 data sets are used for the blind test of the robustness of the proposed models. Statistical analysis of the input variables of the data set used in training and testing the extended and the lumped models are illustrated in Tables 2.

Several of the reported compositional data have extended analysis up to (C_{20+}). The following techniques are used for lumping or regrouping the extended analysis into single pseudo-fractions (C_{7+}). The mole fraction (X_i) of the individual components of the extended analysis is equal to the plus fraction (Ahmad, 2011).

$$X_{C_{7+}} = \sum_{i=7}^{n} X_i \qquad \text{i=7, 8 … n} \qquad (1)$$

The average molecular weight (Mw) of the individual component of the extended analysis is equal to the molecular weight of the plus fraction.

$$Mw_{C_{7+}} = \sum_{i=7}^{n} \frac{X_i Mw_i}{X_{C_{7+}}} \qquad \text{i=7, 8 … n} \qquad (2)$$

The density of the total plus-fraction can be calculated as follows:

$$\gamma_{C_{7+}} = \frac{X_{C_{7+}} Mw_{C_{7+}}}{\sum_{i=7}^{n} \frac{X_i Mw_i}{\gamma_i}} \qquad i=7, 8 \ldots n \qquad (3)$$

It is important to note the significance of the density of plus fraction because its density determines its paraffinic naphthenic aromatic (PNA) content.

Equation-of-state calculations

The EOS calculation of the saturation pressure is compared with the newly developed models as well as the other methods. The two EOS calculations considered in this paper are the ones most considered and widely used in the petroleum industry to evaluate the volumetric and phase behavior of reservoir fluid properties: the Soave-Redlich-Kwong EOS (SRK-EOS) and the Peng-Robinson EOS (PR-EOS). It should be noted that the accuracy of any EOS calculation of the saturation pressure largely depends on the number of pseudo-components and the splitting schemes used to divide the heptane plus fraction; as well as the characterization and calculation of the critical properties of the pseudo-components and the binary interaction parameters. Therefore, the two EOS calculations of the saturation pressure are not the same, even though the number of pseudo-components, characterization of the pseudo-components, and binary interaction parameters were the same.

Model development

As previously stated, the objective of this work is to provide the petroleum reservoir engineer with a simple yet accurate tool to estimate the saturation pressure for a variety of crude oils using compositional data when experimental is unavailable. Two models are presented in this study to calculate the saturation pressure of the crude oil samples. The first model uses all input parameters, such as the molar compositions of the hydrocarbon and non-hydrocarbon components similar to the EOS models. This model is referred to as the extended model. The second model uses fewer input parameters in which some of the hydrocarbon components have been lumped into a single group; this is designated as the lumped model. Both models are simpler than all published correlations, physically correct, and have an acceptable accuracy compared with the well-known EOS.

Figure 1 shows a correlation matrix of all the data available under consideration. This figure also indicates the relative importance of each of the input variables into the first proposed model. It agrees with most published correlations and the EOS that the methane content, heptane plus content and temperatures have the greatest impact on the saturation pressure followed by the intermediate content. The data understudy indicates that the non-hydrocarbon has the smallest impact. This result is simply a reflection of the molar contribution of each component comprising the crude oil composition.

Model I - Extended Model

The extended model uses 13 variables comprising the molar compositions of the pure components, heptane plus properties, and reservoir temperatures, which are similar to the input parameters for the EOS. Figure 2 shows the correlation coefficient graph of all the input parameters in the extended model. Thus, the elimination of the intermediate effect has a negative impact on the accuracy of the predictive model. Our proposed extended model has the following form:

$$Ps = 124.72N_2 + 17.57CO_2 + 22.55H_2S + 64.22C_1 - 9.80C_2 - 52.49C_3 + 6.16C_4 - 19.22C_5 - 23.63C_6 - 21.43C_{7+} + 435.31\gamma_{C7+} + 1.14Mw_{C7+} + 4.29T \quad (4)$$

Where Ps is the saturation pressure in psia, T is the temperature in °F, and the hydrocarbon and non-hydrocarbon compositions are expressed in mole.

Model II - Lumped Model

The second model is much simpler and uses only 7 -variables (the lumped model). Several lumping schemes are adapted in the petroleum industry for phase behavior modeling and reservoir simulation of the petroleum reservoir fluids (Ahmed, 2011). Hydrocarbon components are grouped into a pseudo fraction based on their molecular weight (Whitson, 1980), the similarity of their physical-chemical properties (Lee, 1979), their carbon number, their true boiling point or their number of families based on the chemical structure (Becker et al., 2015). (Emera and Samara, 2005) proposed lumping the hydrocarbon components into volatile (C_1 & N_2), intermediate (CO_2, H_2S, C_2 to C_4) and heavy categories (C_{5+}) to calculate the minimum miscibility pressure. (Duan et al., 2013) proposed lumping several components into a single carbon number (SCN) with predetermined values for properties, such as critical pressure, temperature, and eccentric factor. Others have proposed optimal lumping schemes based on the objectives (Lin et al., 2008). In lumped model, $C_{1+}N_2$, C_2, CO_2, and H_2S, the intermediate (C_3 to C_6), heptanes plus (C_{7+}) and its properties, and temperature are the input variable. The correlation matrixes of the lumped components are provided in Figure 3. Even though the non-hydrocarbon has the least effect their contribution is important, especially the H_2S in the case of sour crudes. Both models presented by AlQuraishi (2009) and Bandyopadhyay and Sharma (2011) have eliminated the effect of H_2S.

The proposed model has the following form:

$$Ps = 82.115*(C_1+N2)-11.635*C_2+39.158*CO_2+ 38.244*H_2S -1.217*(C_3—C_6) - 890.701*(C_{7+}/MW_{7+}*\gamma_{7+}) + 4.217 *T-1042 \quad (5)$$

Where Ps is the saturation pressure in psia; and T is the temperature in °F.

Result and discussion

Model accuracy

The accuracy of the two proposed models used to calculate the saturation pressures of the previously described crude oil samples is shown in Table 3. This table shows reports the average relative error, average absolute relative error standard deviation and the correlation coefficients for all the methods considered in this study. Table 3 also reports the accuracy of the PR-EOS, SRK-EOS, Elsharkawy's model, AlQuraishi's model, and Bandoypandyay and Sharma's model and Jarrahian model. The methods proposed by Gholami et al., Ahmadi et al., and Jarrahian et al are included in Table 3, even though they produce a unreasonable estimate of the saturation pressures, as explained in the appendix.

This table indicates that the extended and the lumped models have an (AARE) in the order of 8% and 8.5%, respectively. It should be noted that the complications in the Bandoypandyay and Sharma model did not improve the accuracy of the saturation pressure, because it had a large error. It is also noted that Jarrahian et al model produces an average error of 35%. Compared with the EOS, the lumped model has a few constants and eliminates the splitting, characterizing the pseudo-components and binary interaction parameters necessary for the EOS calculations. The lumped model as well as the extended model are very simple and require one-step calculations compared with all the other methods reported in Table 3.

A cross-plot of the calculated saturation pressure using the extended model based on the training data versus the measured data is illustrated in Figure 4A. This figure indicates that most of the data falls on the unit slope line, and there is no over-fit. This trend is supported by the high correlation of regression of the model (97.6%) presented in Table 3. Figure 4B shows a cross-plot of the testing data (28 from literature and 7 new measured in PFRC at Kuwait University). The extended model indicates an excellent agreement with all of the 28 data sets chosen at random, with the exception of a few data points at higher pressures. Additionally, the model has a perfect match for the 7 newly measured samples.

Figure 5A shows the performance of the lumped training-model for capturing the measured saturation pressure for most of the 179 samples. A cross-plot of the calculated saturation pressure using the lumped model based on the training data versus measured data is depicted in Figure 5B. This figure indicates that most of the data scatters falls to the unit slope line. The lumped model also has a high coefficient of regression of 97.4%, which is indicated in Table 3.

Models Validity

The accuracy and validity of the models for estimating the saturation pressure of various types of crude oil are reported in Table 4. This table illustrates the capability of the newly proposed models to predict the saturation pressure for volatile oils, black oils, and heavy oils. Table 4; provides the compositions, properties of the heptane plus fraction, temperature, and measured saturation pressures of the samples. Additionally, the calculated saturation pressure using the PR-EOS, SRK-EOS, Elsharkawy's model (2003), AlQuraishi's model (2009), and Bandoypandhyay and Sharma's model (2011), Jarrahian et al (2015); as well as the extended and lumped models are provided in Table 4. It is clear from this table that the proposed models have a wide range of applications regardless of crude oil types. We have also included information regarding the 7 newly measured data; at the PFRC, compositional, and saturation pressure. At the bottom of Table 4, we have provided the unpublished data of the compositions; and the saturation pressures at low temperatures. It is well known that the saturation pressure of a crude sample is measured at low temperatures in addition to the reservoir temperature as a method to check the quality of the sample.

Figure 6A reports the error of all the methods considered in this paper to predict the saturation pressure for volatile oils (heptane plus content below 30%). Both the extended and the lumped models have an AARE of nearly 4% followed by Bandoypandhyay and Sharma's model (2011); and Elsharkawy's model (2003); the SRK-EOS has nearly twice the AARE of the newly proposed model. AlQuraishi's model (2009) has an AARE of 12% because the model is based on methane and neglects the effect of heavy and intermediate components, whose contribution is highly significant in the reservoir and production performance of volatile crudes. Jarrhaine et al (2015) model is not suitable for volatile oils. It shows a 35% error in estimated saturation pressure.

The performance of the extended and lumped models compared to the various methods for the black oil samples ($35 < C_{7+} < 55\%$) is provided in Figure 6B. The newly proposed models have an AARE close to 10%; followed by Bandoypandyay and Sharma's model 14%, the SRK-EOS and PR-EOS 16%, and Elsharkawy 17%. AlQuraishi's model has an AARE of 24%. Jarrhain et al (2015) model shows a 33% error in estimated saturation pressure even though the authors claim that the model is suitable for black oils.

Figure 6C shows the error level of all the methods as well as the proposed models for the newly measured crude oil samples at Kuwait University using the PFRC. Again, Jarrhain et al (2015) model show a 40% error in estimated saturation pressure.

The performance of all published models at low temperature is illustrated in Figure 6D. The low-temperature calculation indicates that Bandoypandhyay and Sharma's model (2011) has accuracy problems and is not valid at low temperatures because the authors proposed a temperature-composition related coefficient. Also, Jarrhain et al (2015) model is not suitable at low temperature. It shows a 40% error in estimated saturation pressure.

The ability of the proposed models to capture the thermodynamic is thoroughly checked against its independent variables. The validity of the model is compared to the experimentally measured data as well as the predictions by other methods used to capture the physics of increasing the saturation pressure as a function of temperature is illustrated in Figure 7. Two main points are observed here; first, the model matches the measured saturation pressures much better than the other methods. Second, it follows the physical trend predicted by the EOS's. Figure 8 shows the model performance for capturing the effect of changing the saturation pressure as a function of the molecular weight of the heptane plus fraction compared with the SRK-EOS and PR-EOS. The extended model follows the trend of the PR-EOS in increasing the saturation pressure. Elsharkawy's model (2003) was unable to predict the correct trend as predicted by the EOS's. The model sensitivity in capturing the change in the saturation pressure as a function of increasing the methane content is shown in Figure 9. The proposed model as well predicts the correct trend of increasing the saturation pressure when increasing the methane content.

Conclusion

In this paper, two models are presented to estimate the saturation pressure of a variety of crude oils when measures are unreliable or samples are unavailable. The two models are developed using compositional data and saturation pressure measurements from a large data bank comprising literature data, unpublished data, and newly measured and reported data of 231 crude oil samples. The input parameters in the first model are the extended hydrocarbon and non-hydrocarbon molar percentages and reservoir temperatures. The second model uses lumped hydrocarbon fractions of intermediate, and heptane plus fractions in addition to its properties; and the reservoir temperature as input parameters. The accuracy of the two models has been tested and compared to that of the SRK-EOS and the PR-EOS; as well as published correlations that use compositional data as the input parameters. The comparison indicates that the two models are much more accurate than all the published methods. The validity of both models has also been tested using various types of crude oils samples and compared to experimentally measured saturation pressures and estimates using the EOS as well as the published correlations. The proposed models are physically correct and much simpler than all the published methods. The models do not require tedious computations and eliminate splitting of the plus fraction, characterization of the pseudo-components; as well as the binary interaction parameters that are required for the EOS calculations.

Acknowledgement

The authors thank the General Facility Lab at Kuwait University for their cooperation and conducting the compositional and saturation pressure measurements of the 7 new samples described in this paper using Petroleum Fluid Research Center (PFRC) project no. GE 01107.

References

Agarwal, R.K., Li, Y.K., Nghiem, L., 1990. A regression technique with dynamic parameter selection for phase – behavior matching. SPE Res. Eng. February, 115 – 120.

Ahmadi, M. A., Zendehboudi, S, James, L., Elkamel, A., Dusseault, M., Chatsiz, I., Lohi, A., 2014, New Tools to determine Bubble Point Pressure of Crude Oils: Experimental and Modeling Study, *Journal of Petroleum Science and Engineering, v.123, 204-216.*

Ahmed, T., 2011, Splitting and lumping schemes of the plus fraction, working guide to vapour-liquid phase equilibria calculations, Chapter 9, Hydrocarbon Phase Behaviour, Gulf Publishing Co. 111-126.

Al-Gathe, A., Abd-el Fattah, Kh., El-Mtwally, K., 2014, New artificial intelligent approach for bubble point pressure, Int. J innovation and Sci. Research, Vol.4, No.2, July, 121-135.

Al-Marhoun, M.A. 1985. Pressure-Volume-Temperature Correlations for Saudi Crude Oils. Presented at the SPE Middle East Oil Technical Conference and Exhibition, Bahrain, 11–14 March. SPE-13718.

Almehaideb, R.A. 1992. Improved PVT Correlations for UAE Crude Oils. Presented at the Middle East Oil Show and Conference, Bahrain, 15-18 March. SPE-37691.

Al-Najjar, H.S., Al-Soof, N.B.A., and Al-Khalisy, K.M. 1988. Correlations for Bubble-Point Pressures, Gas Oil Ratios and Formation Volume Factors for Iraqi Crude Oils. *Journal of Petroleum Research* (June 1988): 13.

AlQuraishi, A. A., 2009, Determination of crude oil saturation pressure using linear genetic programing, Energy & Fuel, 23, 884-887.

Al-Shammasi, A.A. 2001. A Review of Bubblepoint Pressure and Oil Formation Volume Factor Correlations. *SPE Res. Eval. & Eng.* 4 (2): 146-160. SPE71302.

Asgarpour, S., McLauchlin, L., Wong, D. et al. 1989. Pressure-Volume-Temperature Correlations for Western Canadian Gases and Oils. *J Can Pet Tech.* 28 (4): 103.

Asoodeh, M., Bagheripour, P.,2012, Estimation of bubble point pressure from *PVT* data using a power-law committee with intelligent systems, *J. Pet. Sci. and Eng., Vol. 90–91, July 2012, Pages 1-11.*

Bandyopadhyay, P., Sharma, A., 2011, Development of a new semi analytical model for prediction of bubble point pressure of crude oils, *Journal of Petroleum Science and Engineering, Volume 78, Issues 3–4, September 2011, Pages 719-731.*

Becker, P. J., Celse, B., Guillaume, D., Dulot, H., Costa, V., 2015, Hydrotreatment modelling for variety of VGD feedstocks: a continuous lumping approach, Fuel, 139, 133-143.

Coats, K.H., and Smart, G.T., 1986, Application of regression base EOS PVT Program to laboratory data, SPERE, May 277-299.

Dansh, A. Xu, D. H., Todd, A. C., 1992 A grouping method to optimize reservoir description for compositional simulation of gas injection process, SPERE, Aug. 343-348.

De Ghetto, G. and Villa, M. 1994. Reliability Analysis on PVT Correlations. SPE-28904, Presented at the European Petroleum Conference, London, United Kingdom, 25-27 October.

Dindoruk, B. and Christman, P.G. 2001. PVT Properties and Viscosity Correlations for Gulf of Mexico Oils. Presented at the SPE Annual Technical Conference and Exhibition, New Orleans, 30 September-3 October. SPE-71633

Dokla, M.E. and Osman, M.E. 1992. Correlation of PVT Properties for UAE (United Arab Emirates) Crudes. *SPE Form Eval.* (March 1992): 41.

Droham, J.K., 1988, Enhancing the evaluation of PVT data, OSEA paper presented the 7th Offshore South East Asia, Feb. 2-5, Singapore.

Duan, J., Wang, W., Liu, H., Cong, J., 2013, modelling the characterization of the plus fraction by using continuous distribution function, Fluid Phase Equilibria, 345, 1-10.

Elam, F.M. 1957. *Prediction of Bubble Point Pressures and Formation Volume Factors from Field Data.* MS thesis, University of Texas at Austin, Austin, Texas.

Elmabrouk, S., Zekri, A., Shirif E., 2014, The prediction of bubble point pressure and bubble point oil formation volume factor in absence of PVT analysis, PET. Sci. & Tech., 32, 1168-1174.

Elsharkawy, A. M., 2003, An empirical model for estimating the saturation pressures of crude oils, Journal of Petroleum Science and Engineering 38 (2003) 55-77.

Elsharkawy, A.M. and Alikhan, A.A. 1997. Correlations for predicting solution gas/oil ratio, oil formation volume factor, and undersaturated oil compressibility. *J. Pet. Sci. Eng.* 17 (3–4): 291-302.

Emera, M., Samara, H., 2005, Use of genetic algorithm to estimate CO2-oil minimum miscibility pressure- a key parameter in design of CO2 miscible flood, J. Pet. Sci. Eng., Vol. 46, issue 1-2, 37-52.

Farasat, A., Shokrollahi, A., Arabloo, M., Gharagheizi, F., Mohammadi, A., 2013, toward an intelligent approach for determination of saturation pressure of crude oil, Fuel Processing Technology 115 (2013) 201-214.

Frashad, F., LeBlanc, J.L., Garber, J.D. et al. 1996. Empirical PVT Correlations for Colombian Crude Oils. SPE-36105 Presented at the SPE Latin American and Caribbean Petroleum Engineering Conference, Port of Spain, Trinidad and Tobago, 23–26 April. -MS.

Gharbi, R.B.C., Elsharkawy, A.M., 1999, neural network model for estimating the PVT properties of Meddle East crude oils, SPE Reservoir Eval. Eng. 2 (3), 255-265.

Gholami, A., Asoodeh, M., Bagheripour, P., 2014 How committee machine with SVR and ACE estimates bubble point pressure of crudes, *Fluid Phase Equilibria, Volume 382, 25 November 2014, Pages 139-149.*

Glasø, Ø. 1980, Generalized Pressure-Volume-Temperature Correlations. *J Pet Tech.* 32 (5): 785-795.

Heidaryan, E., Moghadasi, J., 2011, Crude saturation pressure through imperial correlation, Thermodynamics 2011, Athens, 31 August- 3 September 2011.

Hasan, M., 1993, PVT Correlations for Indonesian Crude Oils. *Proc.* 22nd Annual Convention of the Indonesian Petroleum Association, Indonesia, Vol.1, IPA93-2.1-070.

Jarrhian, A., Moghadsi, J., Heidaryan, E., 2015, Empirical estimating of black oil bubblepoint (saturation) pressure, J. Petroleum Science and engineering, (126), 69-77.

Jacopy R.H., Berry V.J.,1958, A method for predicting the pressure maintenance for reservoir producing volatile oil, AIME 213, 59-65.

Kartoatmodjo, T., Schmidt, Z. 1994, Large Data Bank Improves Crude Physical Property Correlations. *Oil Gas J.* 92 (27): 51–55.

Kartoatmodjo, T.R.S. and Schmidt, Z. 1991, *New Correlations for Crude Oil Physical Properties*, SPE 23556.

Khairy, M., El-Tayeb, S., Hamdallah, M., 1998. PVT Correlations Developed for Egyptian Crudes. *Oil Gas J.* 96 (18): 114.

Khamekhcki, E., Behvandi, R., Rashid, F., 2011, Prediction of bubble point pressure & asphaltene onset pressure during CO2 injection using ANN & ANFIS models, Pet. Sci. & Tech. Vol. 1 No.2, 35-45.

Labedi, R.M. 1982, *PVT Correlations of the African Crudes*, PhD thesis, CSM.

Labedi, R.M. 1990, Use of Production Data to Estimate the Saturation Pressure, Solution GOR, and Chemical Composition of Reservoir Fluids, SPE-21164-MS. Presented at the SPE Latin America Petroleum Engineering Conference, Rio de Janeiro, Brazil, 14-19 October,

Lasater, J.A. 1958, Bubble Point Pressure Correlations. *J Pet. Tech.*10 (5): 65–67. SPE-957-G.

Levitan, L.L. and Murtha, M. 1999. New Correlations Estimate Pb, FVF. *Oil Gas J.* 97 (10): 70.

Lee, S., 1979, Experimental and theoretical studies on the fluid properties required for simulation of thermal process, Paper SPE 8393.

Li, Y.K., Nghiem, L.X., Siu, A., 1985. Phase behavior computations for reservoir fluids: effect of pseudo-components on the phase diagrams and simulation results. J. Can. Pet. Technol. Nov. – Dec., 29 – 36

Lin, B., Leibovici, C. F., Jorgensen, S. B., 2008, optimal component lumping: problem formulation and solution techniques, Computer & Chem. Eng., 32, 1167-2272.

Macary, S.M. and El-Batanoney, M.H. 1992, Derivation of PVT Correlations for the Gulf of Suez Crude Oils. *Proc. of the* 11th EGPC Pet. Exp. & Prod. Conference, Vol. 1, 374, Cairo, Egypt,

Moharam, H.M., 1995. Prediction of viscosity of heavy petroleum fractions and crude oil using a corresponding states method. Ind. Eng. Chem. Res. 34, 4140 – 4144.

Moradi, B., Malekzadeh, E., Amani, M., Boukadi, F.H., Kharat, R., 2101, Bubblepoint pressure imperial correlation, SPE 132756.

Obomanu, D.A. and Okpobiri, G.A. 1987, Correlating the PVT Properties of Nigerian Crudes. *J. Energy Res. Tech.* 109 (4): 214-217.

Okoduwa, I. G., Ikiensikimama, S.S., 2010, Bubble point pressure correlation for Niger Delta Crude Oils, SPE 136968.

Omar, M.I. and Todd, A.C. 1993. Development of New Modified Black Oil Correlations for Malaysian Crudes. SPE-25338 Presented at the SPE Asia Pacific Oil and Gas Conference, Singapore, 8-10 February.

Owolabi, O.O. 1984. *Reservoir Fluid Properties of Alaskan Crudes*. MS thesis, University of Alaska, Fairbanks, Alaska (May 1984).

Pedersen, K.S., Thomassen, P., Fredenslund, Aa., 1988a. Characterization of gas condensate mixtures. AIChE Spring National Meeting, March 6 – 10, New Orleans, LA.

Peng, D. y., Robinson. D. B., 1976. A new two-constant equation of state , Ind. Eng. Chem. Fund. 15, 59-63.

Petrosky, G.E. Jr. 1990. *PVT Correlations for Gulf of Mexico Crude Oils*. MS thesis, University of Southwestern Louisiana, Lafayette, Louisiana.

Petrosky, G.E. Jr. and Farshad, F. 1998. Pressure-Volume-Temperature Correlations for Gulf of Mexico Crude Oils. *SPE Res Eval & Eng.*1 (5): 416-420. SPE-51395.

Riemens, W.G., Schutle, A.M., Jong, L.N.J., 1988. Birba field PVT variations along the hydrocarbon column and confirmatory field tests. J. Pet. Technol. Jan., 83 – 88

Rosengger, L., Wu., R.1990, Integrated PVT data characterization- lessons from four case histories, J. Can. Per. Tech., 38, (13), (paper 97-05)

Shojaei, M., Bahrami, F., Barati, P., Riahi, S., 2014, Adaptive neuro-fuzzy approach for reservoir oil bubble point pressure estimation, *J. of Natural Gas Sci. and Eng., Volume 20, September 2014, Pages 214-220.*

Soave, G. 1972. Equilibrium constants from a modified Redlich-Kwong equation of state, Chem. Eng. Sci. 27, 1197-1203.

Standing, M.B. 1947. A Pressure-Volume-Temperature Correlation for Mixtures of California Oils and Gases. *API Drilling and Production Practice* (1947): 275-287.

Tale, M., Alquraishi, A., Al Mushaigeh, E. 2009, New model for crude oil saturation pressure prediction using alternation conditional expectation, Offshore Mediterranean Conf. & Exhibit. Italy 25-27, 2009.

Talebi, R., Ghiasi, M. Talebi, H., Mohammadyian, M., Zendehboudi,S., Arabloo, M., Bahadori, A., 2014, Application of soft computing approaches for modelling saturation pressure of reservoir oil, *J. Nat. Gas Sci. and Eng., Vol. 20, Sep. 2014, 8-15.*

Vazquez, M., Beggs, H. D. 1980, Correlations for Fluid Physical Property Prediction. *J Pet. Tech.* 32 (6): 968-970. Vazquez, M.E. 1976. *Correlations for Fluid Physical Property Prediction,* MS thesis, University of Tulsa.

Velarde, J., Blasingame, T.A., and McCain Jr., W.D. 1997. Correlation of Black Oil Properties at Pressures below Bubble Point Pressure - A New Approach. Presented at the Annual Technical Meeting of CIM, Calgary, Alberta, 8–11 June. PETSOC 97-93.

Vogel, J.R., Yarborough, L. 1980, the effect of nitrogen on phase behavior and phase properties of reservoir fluids, SPE 8815.

Whitson, C. H., 1980, Characterizing hydrocarbon plus fraction, SPEJ, 24 (4), 683-694.

Williams, C.A., Zana, E.N., 1980. Use of the Peng-Robinson equation of state to predict hydrocarbon phase behavior and miscibility for fluid displacement. SPE 8817, SPE/DOE on Enhanced Oil Recovery, April 20–23, Tulsa, OK.

Appendix

In this appendix, sample calculations are presented to show the deficiency of some of the published methods.

Input parameters:

N_2	CO_2	H_2S	C_1	C_2	C_3	C_4	C_5	C_6	C_{7+}	GR.C7+	MWC7+	T, °F	Ps, psia
0.65	0.02	0	45.02	12.45	8.93	6.03	3.02	1.44	22.44	0.81	184	140	3002

Ahmadi et al (2014)

Ahmadi et al. proposed an 8 steps calculation (copied from their paper) to estimate the BPP as follows:

$$\mathrm{BPP} = A \times \mathrm{Mw}_{C7+} - \frac{B}{C_{7+}} + C \times \mathrm{Vol./Inter.} + D \times \mathrm{SG}_{C7+} \frac{E}{\mathrm{SG}_{C7+}} + \frac{F}{\mathrm{Mw}_{C7+}} + T \times 505.967 + \frac{2.45789 \times 10^6}{T} + \frac{G}{\mathrm{Vol./Inter.}} \quad (5)$$

$$A = 1488.46 - \mathrm{SG}_{C7+} \times 1087.64 - \frac{601.512}{\mathrm{SG}_{C7+}} + 1670.58 + T \times 0.40771 + \frac{6345.73}{T} + \frac{8.72863}{\mathrm{Vol./Inter.}} \quad (6)$$

$$B = \frac{4132.53}{\mathrm{Vol./Inter.}} + \frac{6.1977 \times 10^7}{\mathrm{Mw}_{C7+}} + T \times 3181.39 + \frac{1.06393 \times 10^8}{T} \quad (7)$$

$$C = \frac{20964.7}{C_{7+}} - \frac{1.76162 \times 10^6}{\mathrm{Mw}_{C7+}} - T \times 19.1963 - \mathrm{Mw}_{C7+} \times 21.5427 + \frac{12851.5}{\mathrm{SG}_{C7+}} - \frac{712613}{T} - 3477.49 + \mathrm{SG}_{C7+} \times 9532.73 \quad (8)$$

$$D = 1.05345 \times 10^{+06} - \frac{1.50748 \times 10^8}{\mathrm{Mw}_{C7+}} - T \times 676.048 - \frac{1.20252 \times 10^7}{T} - \frac{12178.2}{\mathrm{Vol./Inter.}} \quad (9)$$

$$E = 641357 - \frac{1.02891 \times 10^8}{\mathrm{Mw}_{C7+}} - T \times 301.034 - \frac{1.91446 \times 10^6}{T} - \frac{3873.62}{\mathrm{Vol./Inter.}} \quad (10)$$

$$F = 2.51471 \times 10^{+08} + T \times 25332.5 + \frac{3.47893 \times 10^8}{T} - \frac{244573}{\mathrm{Vol./Inter.}} \quad (11)$$

$$G = 20125.6 - T \times 23.9271 - \frac{703193}{T} \quad (12)$$

Ahmadi et al. define the Vol. /Inter. as the mole percentage ratio of the volatile (C_1& N_2) to the intermediate components (C_2 to C_6, CO_2 and H_2S). Application of the above equations from their paper using the input parameters results in the following:

Vol./Inter.	A	B	C	D	E	F	G	BPP
1.43211	1643.9	1545062	-271.127	45122.71	23642.47	2.57E+08	11753	1,794,123

The calculated saturation pressure (BBP) from Eq.5 is 1,794,123 psia which is unreasonable. This indicates a deficiency in the model.

Gholami et al (2014)

Gholami et al. presented 3 steps calculation (copied from their paper) to estimate the BPP as follows:

$$Tr(M) = \sum_{i=0}^{5} B_i M^i \quad (15)$$

Application of Eq. 15 resulted in the following

N_2	0.00584
CO_2	0.060917
H_2S	0.050636
C_1	-0.47658
C_2	-0.41547
C_3	-0.45083
C_4	-0.02003
C_5	0.068769
C_6	0.273921
C_{7+}	1.814523
SG.C7+	-0.0197
MWC7+	37.03525
T., F	-0.11139

Application of Eq. 16

$$Tr(P_b) = Tr(N_2) + Tr(CO_2) + Tr(H_2S) + Tr(C_1) + Tr(C_2) + Tr(C_3) + Tr(C_4) + Tr(C_5) + Tr(C_6) + Tr(C_{7+}) + Tr(SG_{C7+}) + Tr(MW_{C7+}) + Tr(T) \quad (16)$$

Resulted that Tr(Pb) = 37.8159

$$P_b = (57.08403877741 \times (Tr(P_b))^2) + (1166.9102304199 \times (Tr(P_b))) + 2207.3573204358 \quad (17)$$

Application of Eq. 17 resulted in saturation pressure (P_b) = 127,967 psia, which is unreasonable, compared to the measured one is 3002 psia. This indicates that there is a deficiency in the model.

Jarrahian et al (2015)

Jarrahian et al. proposed the following Eq. (copied from their paper) to estimate the saturation pressure:

$$P_{sat} = A_0 \exp\Big(A_1 z_{N_2} + A_2 z_{CO_2} + A_3 z_{H_2S} + A_4 z_{C_1}^{A_5} + A_6 \sum_{2}^{6} z_{C_i} + A_7 z_{C_{7+}}^{A_8} \gamma_{C_{7+}}^{A_9} Mw_{C_{7+}}^{A_{10}} + A_{11}(T - 459.67)\Big) \quad (2)$$

Application of Eq. 2 resulted in saturation pressure (Ps) = 1812 psia, compared to the measured one is 3002 psia. This indicates that model has inaccuracy problem. This has been the pattern in all the crude oil samples shown in Table 4 and figures 6A through 6D.

Table 1- Summary of attempts to calculate the saturation pressure of crude oils

Method based on production information	Method based on compositional data
Standing, 1947; Elam, 1957; Lasater, 1958; Vasquez, 1976; Vasquez and Beggs, 1980; GlasØ, 1980; Labedi, 1982&1992; Owolabi, 1984; Al-Marhoun, 1985; Obomanu and Okpobiri, 1987; Asgarpour et al., 1989; Al-Najar et al., 1988; Dokla and Osman, 1990; Petrosky, 1990; Kartoatmodjo and Schmidt, 1991&1994; Macary and El-Batanoney, 1992; Omar and Todd, 1993; Hasan, 1993; De Ghetto and Villa, 1994; Farashad et al., 1996; Almehaidaeb, 1997; Elsharkawy, 1997; Velarde et al., 1997; Kairy et al., 1998; Petrosky and Farshad, 1988; Al-shammasi, 2001; Gharbi & Elsharkawy 1999 Levitan and Murtha, 1999; Dindrouk and Christman, 2001 Moradi et al., 2010 Okoduwa and Ikiensikimama 2010 Asoodeh and Bagheripour, 2012 Shojaei et al., 2014 Talebi et al., 2014 Elmabrouk et al., 2014 Kanu and Ikiensikimama, 2014 Jarrahian et al 2015	Elsharkawy , 2003 Tale et al., 2009 AlQuraishi, 2009 Bandyopadhyay and Sharma, 2011 Khamehchi et al., 2011 Farasat et al., 2013 Ahmadi et al., 2014 Gholami et al. 2014 Jarrahian et al 2015

Table 2- Description of the data used to develop and test the extended model

Input	Training model (179 Data)			Testing (28 Data)			Testing (7 new experimental)		
	Min	Ave	Max	Min	Ave	Max	Min	Ave	Max
N_2	0.00	0.35	1.67	0.00	0.27	1.67	0.10	0.26	0.35
CO_2	0.00	1.27	11.37	0.00	1.17	11.37	0.14	0.41	0.83
H_2S	0.00	0.15	3.68	0.00	0.15	3.22	0.00	0.00	0.00
C_1	13.16	33.86	74.18	19.50	34.79	73.36	20.64	28.25	42.79
C_2	3.36	8.07	13.71	4.64	8.26	11.63	6.69	7.88	10.82
C_3	0.89	6.59	11.87	2.46	6.76	12.03	6.05	6.59	7.10
C_4	0.95	4.46	8.4	1.66	4.41	6.58	4.03	5.05	5.93
C_5	0.40	3.21	5.95	1.00	3.11	5.23	3.36	4.33	5.52
C_6	0.00	3.14	6.37	0.99	3.15	5.46	2.88	3.74	4.33
C_{7+}	10.72	38.89	57.73	11.18	37.88	51.22	27.61	43.49	51.31
$SG.C_{7+}$	0.74	0.87	0.959	0.70	0.86	0.93	0.85	0.89	0.92
MWC_{7+}	134	240	368.9	105	223	292	216	262	298
T, °F	58	156	319	58	165	261	121	157	230
Ps, psia	537	2198	5065	1025	2307	5000	1227	1779	3344

Table 3: Error analysis of the proposed models in comparison with all the investigated methods

Method	ARE	AARE	SD	R^2%
PR-EOS	6.24%	10.19%	1028	97.9
SRK-EOS	-3.12%	10.36%	1024	96.7
Elsharkawy (2003)	0.85%	8.57%	1015	96.4
Alquraishi (2009)	-6.06%	10.16%	1088	96.7
Bandyopadhyay (2011)	8.05%	12.28%	1054	96.8
Farasat (2013)	NA	NA	NA	NA
Gholami et al. (2014)	?	?	?	?
Ahmadi et al. (2014)	?	?	?	?
Jarrahian et al (2015)	35.52	35.56	983	95.2
New Extended Model	-0.37%	8.20%	1030	97.6
New Lumped Model	-0.50%	8.46%	1036	97.4

ARE = Average Relative Error

AARE = Average Absolute Relative Error

SD = Standard Deviation

R^2 = Correlation Coefficient

NA = Calculation was not possible, because the weight and bias were not provided.

? = The method produces unreasonable results as indicated in the appendix.

Table 4- Comparing model accuracy for calculations of saturation pressures for various crude oils

Volatile-Oil-Samples (C7+<30)																							
Fluid Source	index	N2	CO2	H2S	C1	C2	C3	C4	C5	C6	C7+	GR.C7+	MWC7+	T, °F	Ps, psi	SRK	PR	Elsh	Alq	Bandy	Jarr	Extd	Lump
Rosenegger	7	0.03	8.39	0	47.4	10.3	6.12	4.2	2.89	2.05	18.6	0.83	180	295	4000	3933	3879	4460	4420	4165	3317	4139	4178
Vogel-80	13	0.65	0.02	0	45	12.5	8.93	6.03	3.02	1.44	22.4	0.81	184	140	3002	2624	2548	2742	2895	2643	1812	3019	2996
Dansh-92	18	0	0	0	46.8	8.77	7.44	4.01	2.56	4.02	26.4	0.77	158	212	2941	2550	2594	3248	3617	3269	1944	3244	3374
	31	0.44	0.38	0	49.1	7.6	6.13	3.84	2.52	2	28	0.84	231	199	3739	3522	3507	3418	3951	3508	2309	3625	3644
Middle East	45	0.03	0.97	0	41.6	10.8	7.52	4.5	3	2.43	29.1	0.85	208	241	3069	3109	2979	3080	3513	3138	1962	3127	3140
Coats-86	51	0.55	1.03	0	36.5	9.93	8.85	6	3.78	3.56	30.4	0.84	200	234	2746	2537	2472	2714	2945	2643	1682	2777	2720

Black-Oil -Samples (35<C7+<55)																							
Fluid Source	index	N2	CO2	H2S	C1	C2	C3	C4	C5	C6	C7+	GR.C7+	MWC7+	T, °F	Ps, psi	SRK	PR	Elsh	Alq	Bandy	Jarr	Extd	Lump
Coats-86	73	1.54	0.08	0	28.4	7.16	10.5	8.4	3.82	4.05	36	0.84	252	131	1694	1674	1618	1615	1738	1644	1084	1713	1713
Middle East	88	0.08	1.37	0.69	36	8.67	5.94	2.97	1.5	3.1	39.7	0.9	274	168	2505	2710	2318	2247	2608	2321	1476	2441	2445
Middle East	114	0.29	0.48	0	28.4	8.29	7.38	5.06	3.42	4.41	42.3	0.88	252	133	1632	1672	1477	1544	1748	1517	984	1587	1598
Middle East	122	0	1.25	0	33.4	9.24	6.07	2.4	1.62	2.78	43.3	0.85	252	230	2110	2313	2278	2424	2737	2438	1410	2388	2412
Middle East	134	0	0.17	0	28.6	8.63	7.43	3.81	2.61	4.11	44.7	0.86	249	134	1548	1551	1429	1477	1764	1433	937	1503	1567
Middle East	146	0.44	0.83	0	27.8	7.54	6.9	4.61	3.1	3.11	45.7	0.9	272	133	1565	1799	1505	1558	1730	1528	1002	1602	1590
moharam-96	149	0.21	0.34	0	20	7.93	8	6.6	5.87	5.08	45.9	0.86	230	235	900	1353	1290	1531	1514	1225	773	1303	1294
Middle East	162	0.35	0.47	0	26.5	7.71	6.05	4.03	4.08	4.22	46.6	0.89	253	135	1500	1613	1380	1427	1631	1391	923	1478	1456
Middle East	176	0.12	0.51	0	28.8	7.91	6.46	3.17	2.07	3.25	47.7	0.88	250	134	1615	1665	1476	1574	1783	1502	981	1571	1614
Middle East	194	0.21	0.15	0	27.8	7.68	6.19	3.41	1.81	2.84	49.9	0.86	228	134	1590	1532	1377	1545	1680	1426	923	1466	1493
Rosenegger	207	0.37	0.02	0	17.3	6.11	8.23	5.47	3.76	3.96	54.8	0.85	242	156	537	938	888	914	1064	836	558	661	731

Newly-Measured Kuwait University Laboratory																						
Fluid Source	index	N2	CO2	H2S	C1	C2	C3	C4	C5	C6	C7+	GR.C7+	MWC7+	T, °F	Ps, psi	SRK	PR	Elsh	Alq	Bandy	Jarr	Extd	Lump
KU-1	209	0.33	0.22	0	25.6	6.87	6.39	5.61	4.68	4.33	46	0.879	222	133	1595	1360	1530	1442	1514	1394	860	1357	1337
KU-2	210	0.35	0.47	0	26.5	7.71	6.05	4.03	4.08	4.22	46.6	0.893	253	135	1500	1400	1620	1427	1631	1391	924	1478	1456
KU-3	211	0.27	0.74	0	23.1	7.2	6.45	5.93	4.66	4.16	47.5	0.906	290	138	1283	1140	1360	1284	1448	1296	800	1287	1219
KU-4	212	0.22	0.26	0	20.6	6.69	6.65	5.11	5.52	3.6	51.3	0.919	298	171	1227	1120	1300	1188	1415	1170	742	1176	1131
KU-5	213	0.19	0.14	0	27.8	8.37	6.88	5.28	4.33	3.85	43.2	0.89	278	121	1524	1300	1560	1402	1659	1409	928	1534	1489
KU-6	214	0.33	0.24	0	31.3	7.48	6.58	4.86	3.7	3.17	42.3	0.9	274	171	1978	1800	2200	1975	2242	1985	1224	2060	2026
KU-7	215	0.1	0.83	0	42.8	10.8	7.1	4.51	3.36	2.88	27.6	0.85	216	230	3344	3070	3400	3083	3568	3139	2003	3213	3200

Low-Temperature Oil Samples (59°F<T<71°F)																							
Fluid Source	index	N2	CO2	H2S	C1	C2	C3	C4	C5	C6	C7+	GR.C7+	MWC7+	T, °F	Ps, psi	SRK	PR	Elsh	Alq	Bandy	Jarr	Extd	Lump
Middle East	101	0	1.12	0.21	27	10.8	9.23	4.97	2.62	3.33	40.8	0.88	249	71	1247	1090	1033	1040	1235	564	808	1148	1205
Middle East	117	0.36	0.17	0	27.2	8.93	8.6	6.07	3.71	2.55	42.4	0.88	271	68	1130	1028	1245	1094	1248	766	820	1239	1228
Middle East	135	0	0.17	0	28.6	8.63	7.43	3.81	2.61	4.11	44.7	0.86	249	70	1270	1055	1255	1130	1311	783	826	1229	1297
Middle East	138	0.14	0.54	0	29.4	7.96	6.19	3.96	3.25	3.84	44.7	0.9	290	59	1265	1065	1350	1191	1301	880	894	1408	1390
Middle East	147	0.44	0.83	0	27.8	7.54	6.9	4.61	3.1	3.11	45.7	0.9	272	67	1320	1125	1400	1201	1268	921	879	1319	1312
Middle East	173	0.06	0.81	0	31.3	7.42	5.68	2.82	1.63	2.64	47.6	0.89	270	68	1460	1215	1535	1441	1463	1026	975	1548	1576
Middle East	193	0.16	0.3	0	24.7	7.51	6.92	4.76	2.65	3.43	49.6	0.88	239	71	1050	975	1156	1008	1107	701	714	958	986

Elsh: Elsharkawy (2003)
Extd: Extended model

Alq: AlQuraishi (2009)
Lump: Lumped model

Band: Bandyopadhyay and Sharma (2011)

Jarr: Jarrahian et. al (2015)

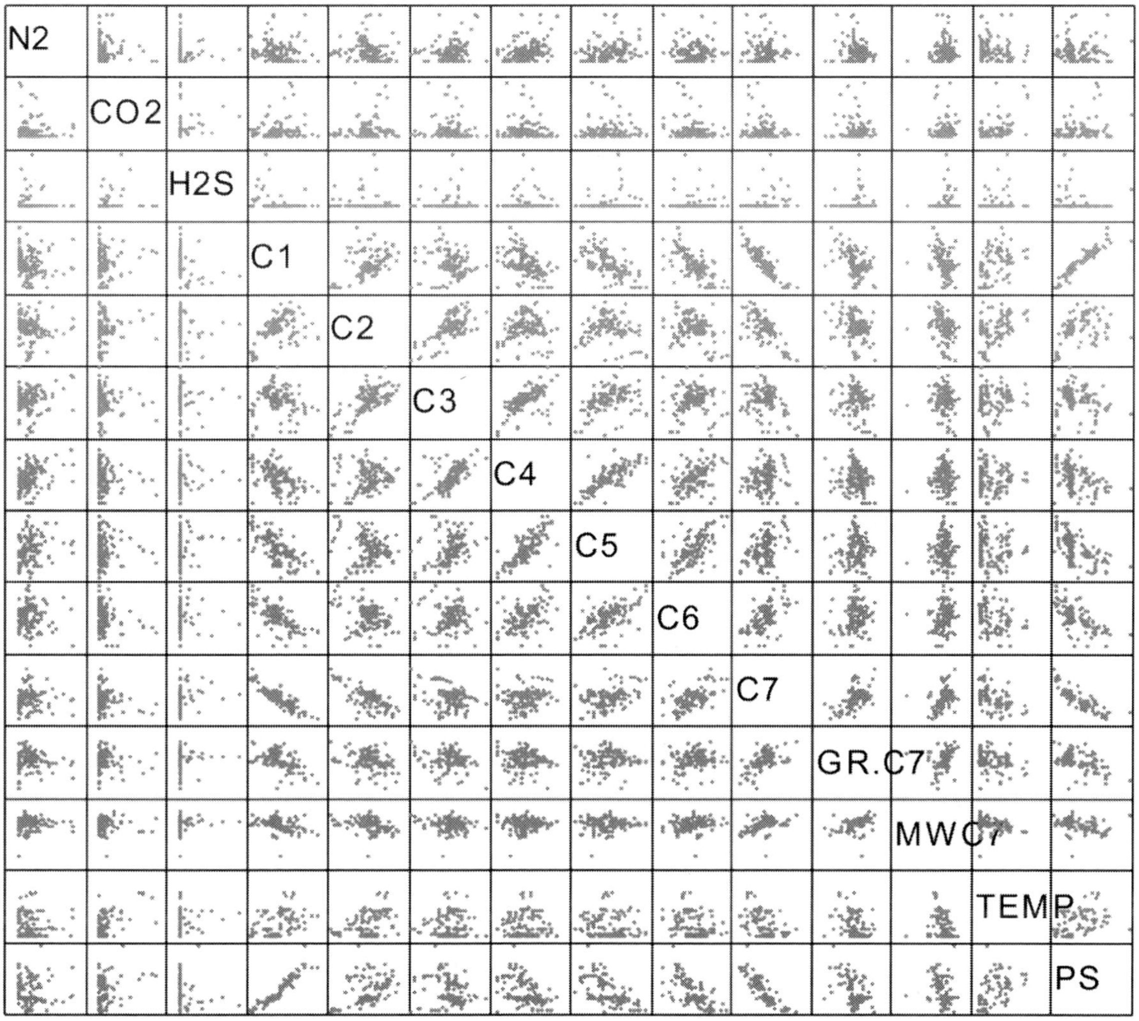

Figure 1: Correlation matrix for all the variables in the oil data bank

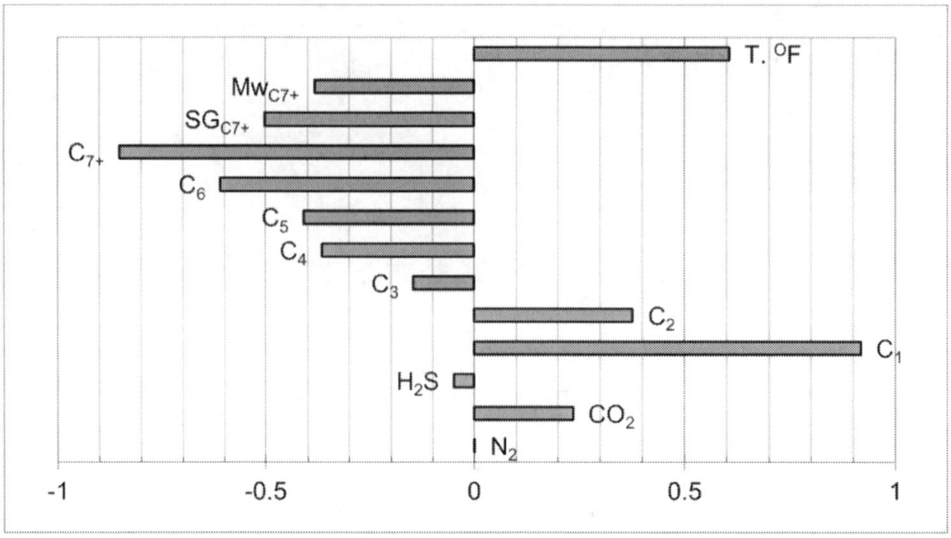

Figure 2: Correlation coefficient for all the variables in the training model (179 data)

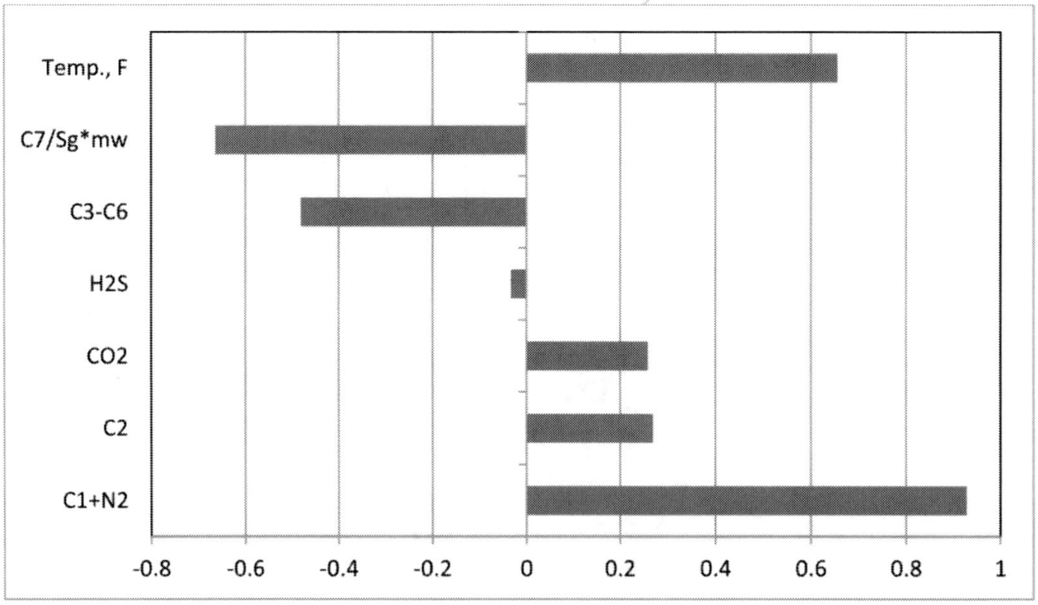

Figure 3: Correlation matrix for the lumped model (179 data)

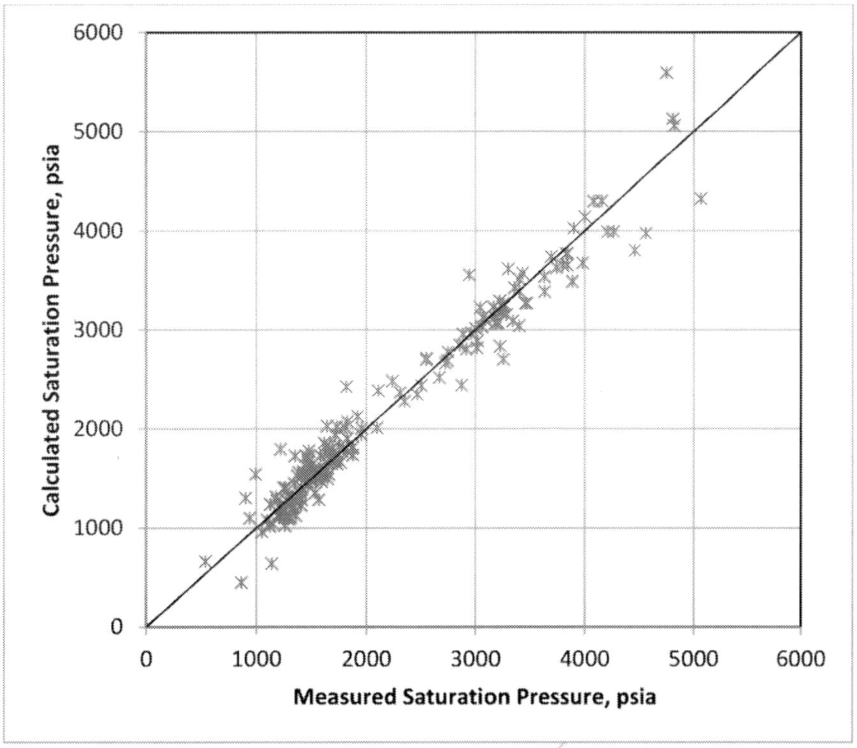

Figure 4A: Cross plot of the extended model using 179 training data

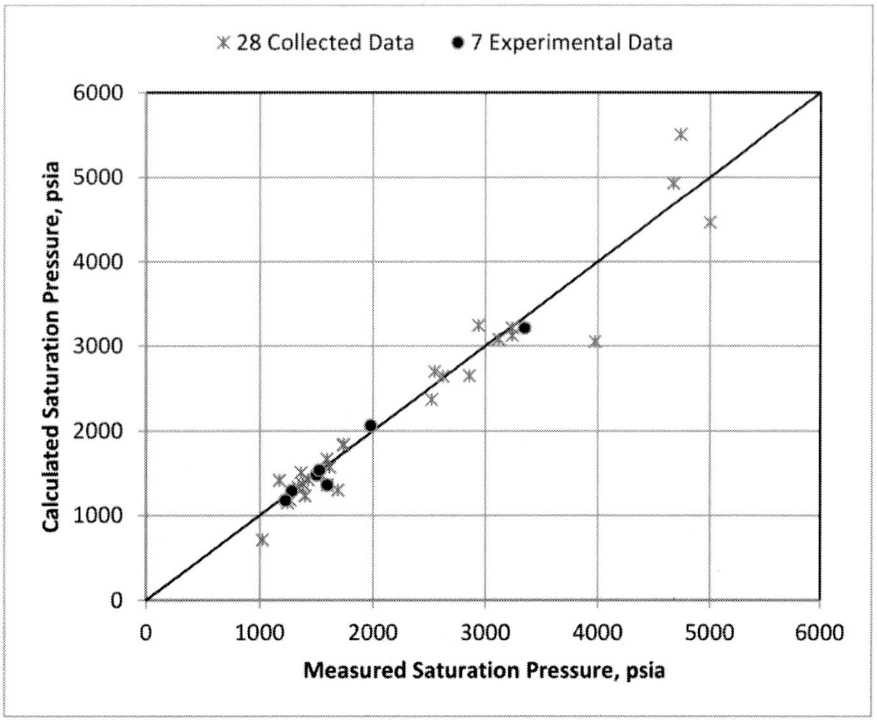

Figure 4B: Cross-plot of the extended model using 35-testing data

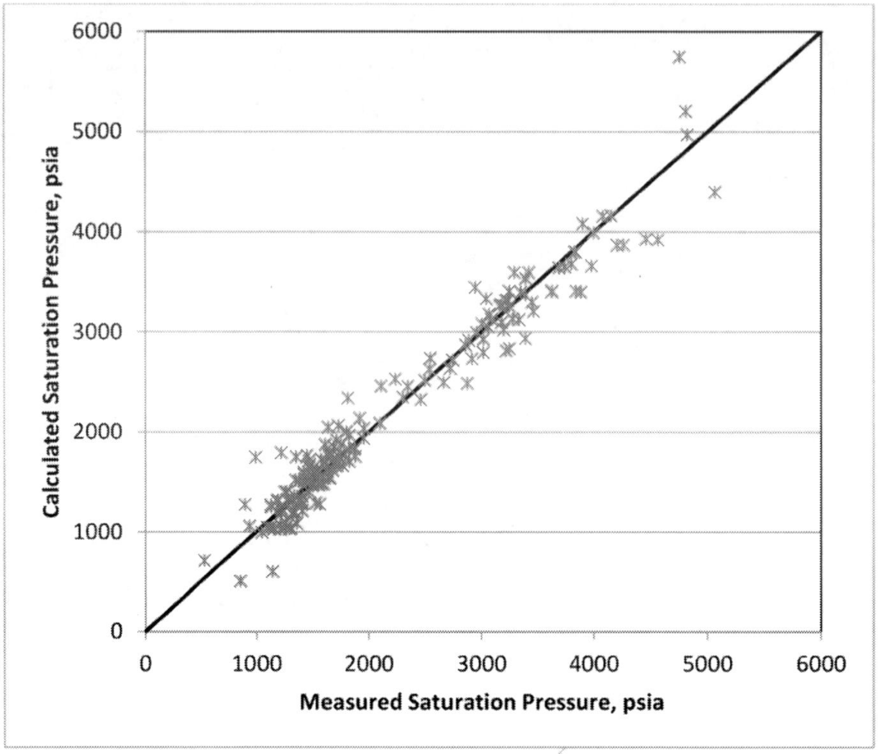

Figure 5A: Cross-plot of the lumped model using 179 training data

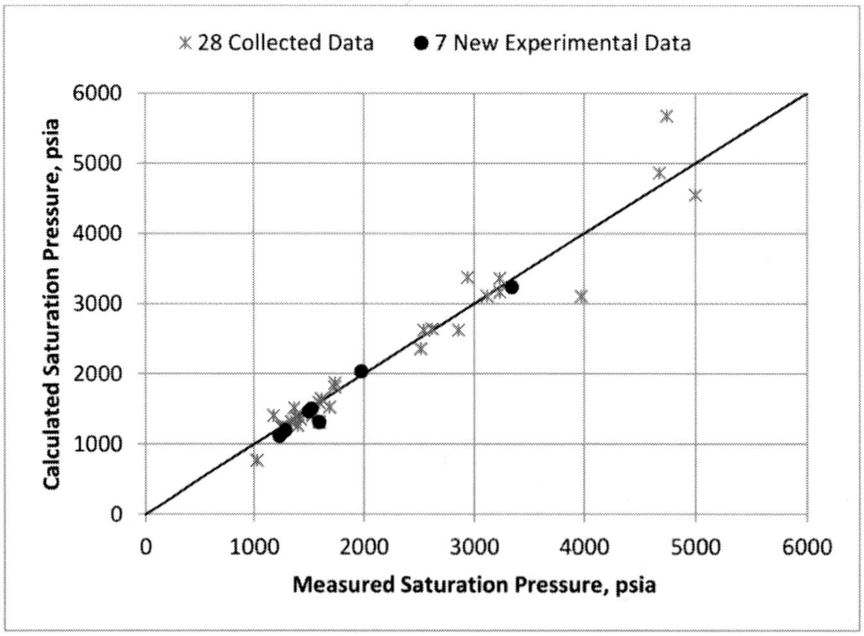

Figure 5B: Cross-plot of testing data for the lumped model

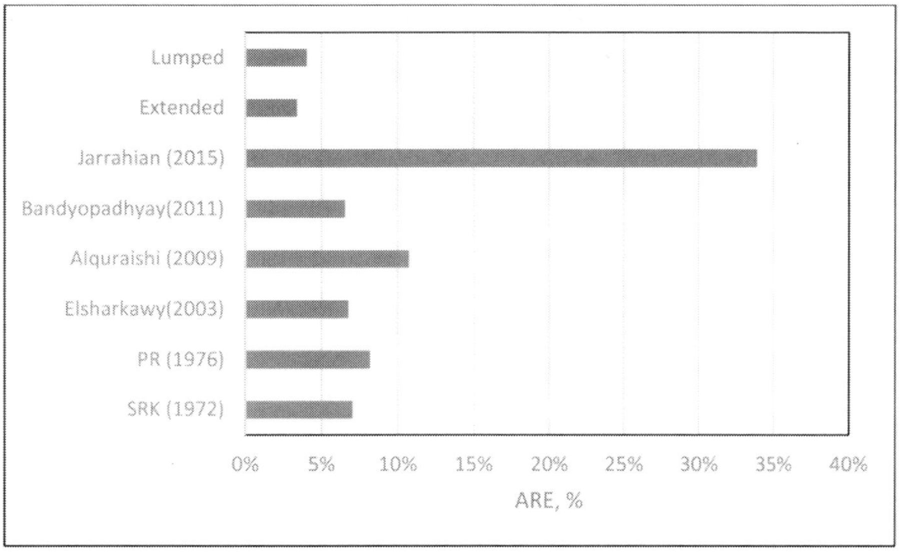

Figure 6A: Comparison of all the models' accuracy for the volatile oil samples

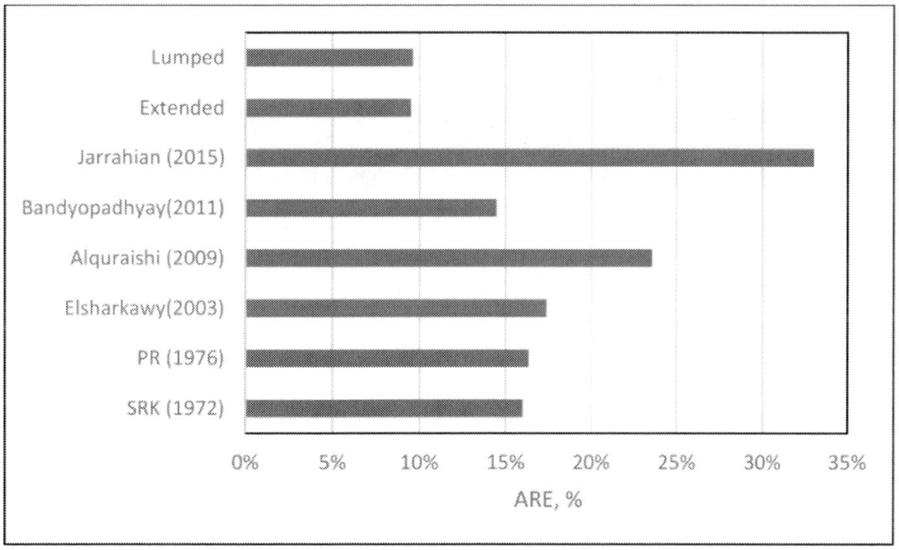

Figure 6B: Comparison of all the models' accuracy for the black oil samples

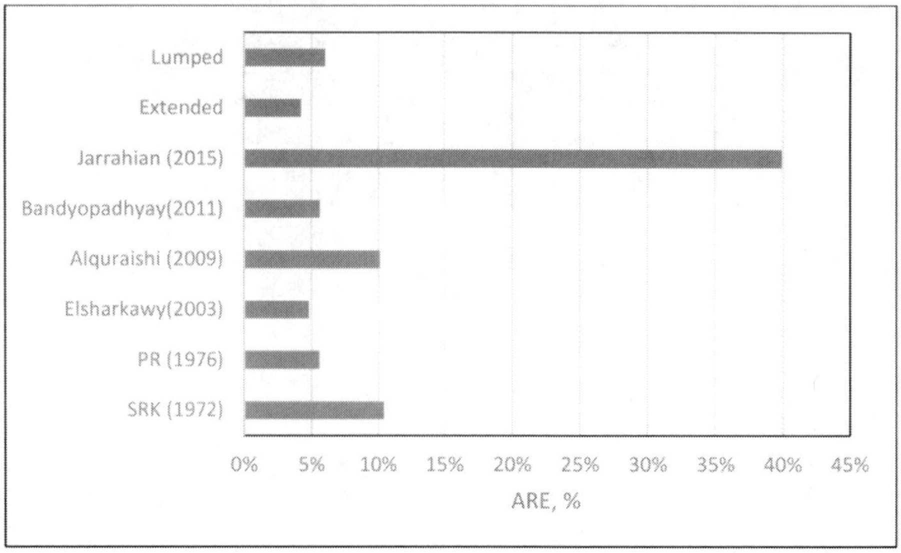

Figure 6C: Comparison of all the models' accuracy for the newly measured samples at KU

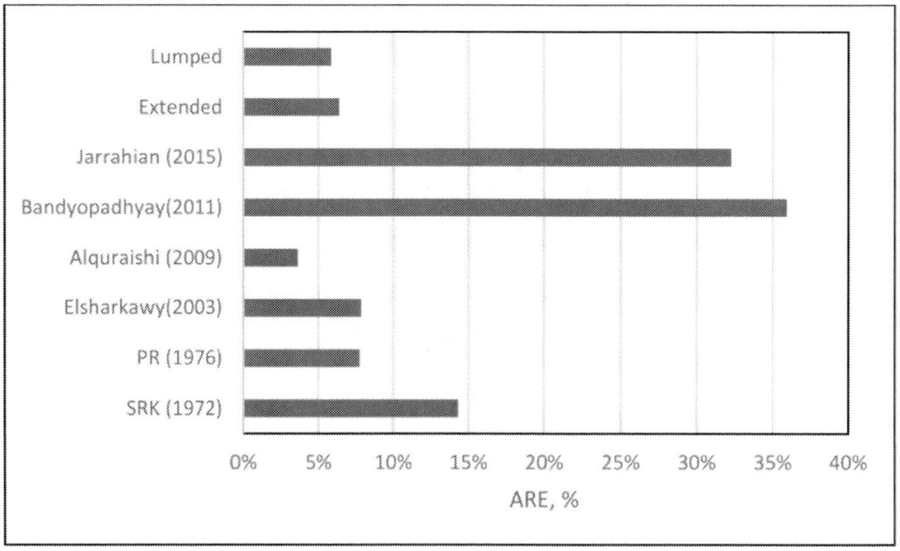

Figure 6D: Comparison of all the models' accuracy for low-temperature oil samples

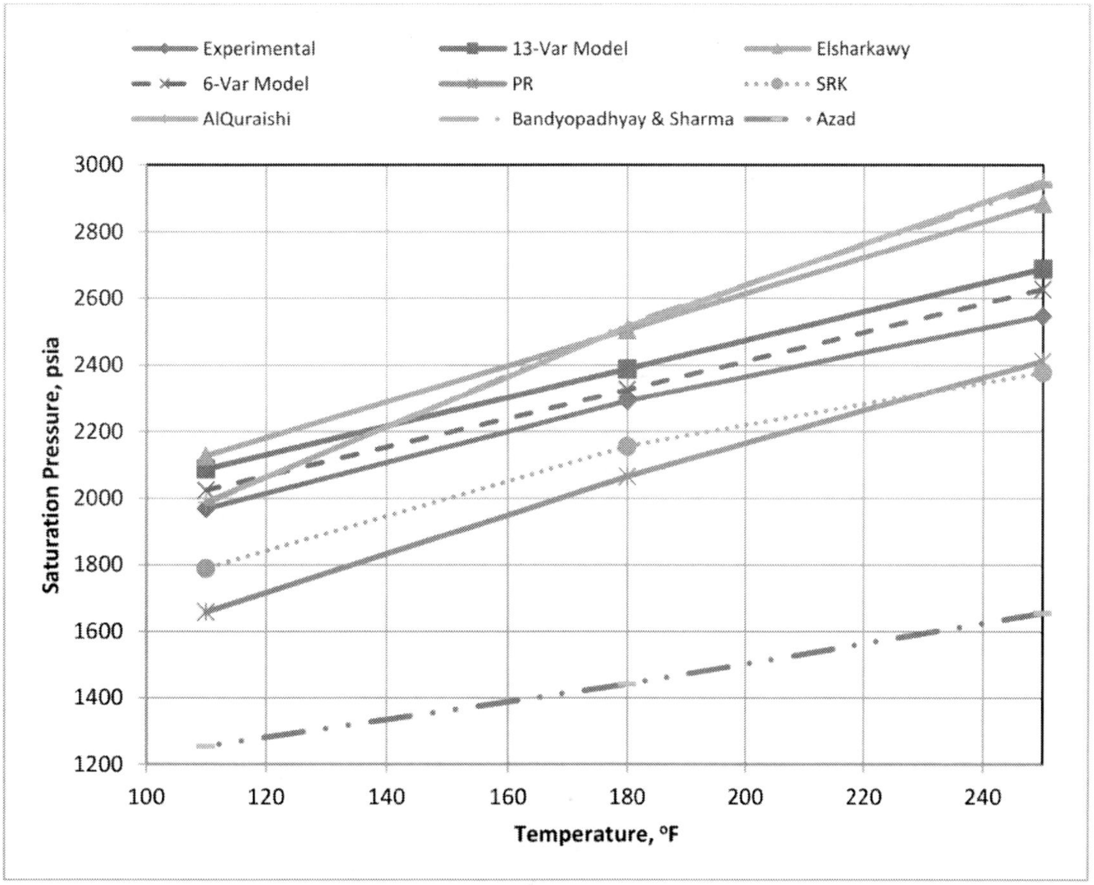

Figure 7: Changes in saturation pressure as a function of temperature for the different models (Sample #70)

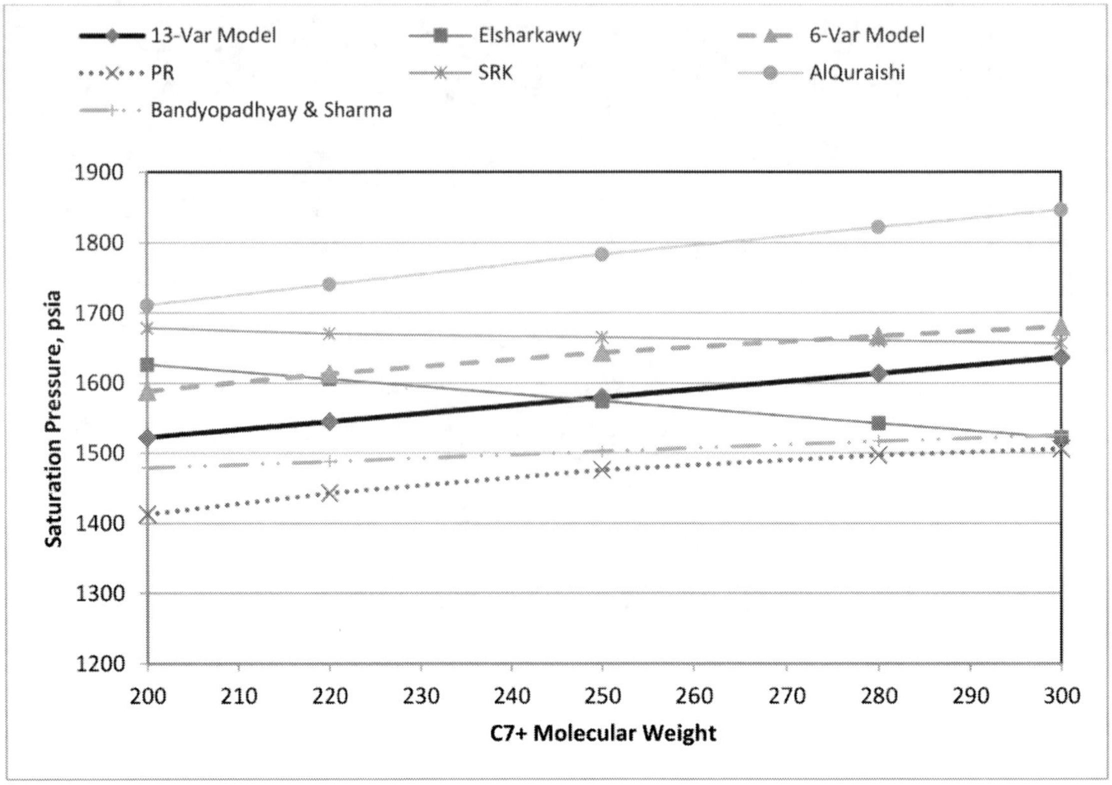

Figure 8: Comparing model performances: Changes in BPP as a function of C_{7+} molecular weight at 134°F (Sample #176)

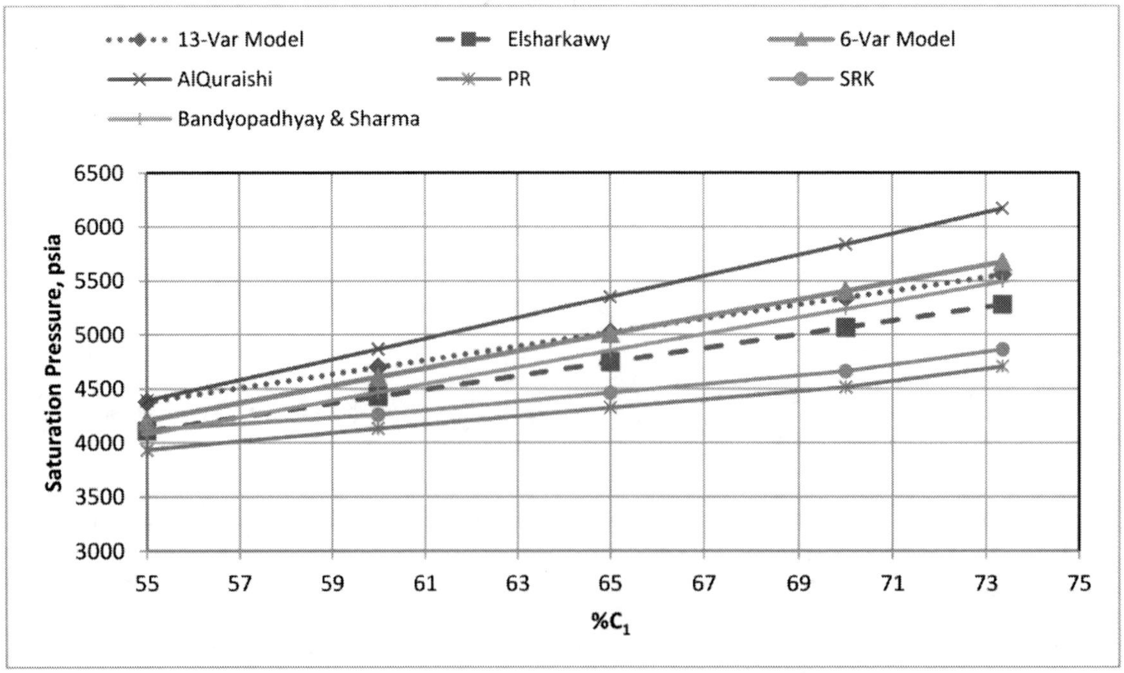

Figure 9: Sensitivity of saturation pressure to methane content at 212°F (sample #2)

439811 Purification of Ibuprofen Enantiomers By a New Technique Combining Distillation and Crystallization

Abstract:

Stripping crystallization (SC) is a new separation technique which combines distillation and crystallization. Basically, SC is operated at a triple-point condition, in which the liquid is simultaneously vaporized and crystallized due to the three-phase equilibrium. In essence, the process is continued until the liquid phase is completely eliminated and only the pure solid crystals remain in the feed. Thus, crystal washing is not required since only pure solid crystals remain in the feed and no impurities are adhered on the crystal surfaces at the end of SC. In the previous research, SC has been successfully applied to separate p-xylene from the mixed xylenes by Shiau et al. (2005, 2008, 2013).

Ibuprofen, also called 2-(isobutylphenyl)-propionic acid, is a non-steroidal anti-inflammatory drug widely used to treat headaches and minor pains. Although it is known that S-ibuprofen is responsible for the anti-inflammatory effects while R-ibuprofen might cause some long-term side effects, ibuprofen is still sold as a racemic mixture in the market due to the complexity involved in the separation of enantiomers. A thermodynamic model is developed in this work to simulate the three-phase equilibrium conditions during the SC operation. The experiments show that seeding with ultrasound mixing or magnetic stirring can be efficiently employed to purify S-ibuprofen from ibuprofen enantiomers. The experimental results, including the final enantiomeric purity and recovery ratio of S-ibuprofen, are consistent with the simulation results predicted by the model.

References:

1. Shiau, L. D.*, Wen, C. C. and Lin, B. S. "Separation and purification of p-xylene from the mixture of m-xylene and p-xylene by distillative freezing", *Ind. Eng. Chem. Res*, **44(7)**, p2258-2265, 2005.

2. Shiau, L. D.*, Wen, C. C. and Lin, B. S. "Separation of p-xylene from the multicomponent xylene system by stripping crystallization", *AIChE Journal*, **54(1)**, p337-342, 2008.

3. Shiau, L. D.*; Liu, K. F. "Investigations into the effects of the cooling rate on stripping crystallization", *Ind. Eng. Chem. Res*, **52**, p1716-1722, 2013.

441535 Hydrogen Disposal Flare System Hazard Identification and Mitigation Analysis

Abstract:

A hazard assessment on a hydrogen disposal flare system was conducted in preparation for a hydrogen turbopump test at the Air Force Research Laboratory (AFRL), Aerospace Systems Directorate at Edwards AFB, CA. The test operations required the disposal of large quantities of liquid and gaseous hydrogen through a commercial flare system to prevent the formation of an explosive environment. As part of the assessment, the primary steps in the test operation process were evaluated for the formation of a potential hazard by looking at each associated failure mode and the resulting affect on personnel and facilities. Evaluation of each operational step was conducted in a Process Failure Modes and Effects Analysis (PFMEA) that identified the probability and severity of the identified effects and associated hazards. If the probability and severity were unacceptable, mitigating steps were applied until the overall hazard risks were reduced to acceptable levels. The recommendations of this assessment were incorporated into both the flare system design and the operational procedures. This paper outlines the results of the PFMEA which include identification of the effects/hazards, available risk-mitigating techniques, and implementation of these techniques used for the mitigation of the identified hazards. Based upon the findings significant design modification to the vender's commercial equipment was needed to comply with recent changes in the code. Most of the identified hydrogen flare system hazards/effects were categorized into four groups: internal explosions, obstructions, external explosions, and maintenance degradation. Many corrective actions were implemented to mitigate both the probability and severity of the identified hazards. In addition, this paper discusses the installation of oxygen monitoring systems on the flare system, determination of appropriate system purge requirements, installation of purge system status alarms, incorporation of a triple redundant flare ignition system with status indicators, and a number of other hazard risk mitigation techniques. This comprehensive description of the hazards and mitigation of a hydrogen disposal system is a plant is vital part to safe operation of both hydrogen commercial and aerospace facilities. <align="center">

442068 Enhanced Furfural Production By an Innovative Bottom Dividing Wall Column with a Decanter Configuration

Abstract:

Enhanced Furfural Production by an Innovative Bottom Dividing Wall Column with a Decanter Configuration

Nguyen Van Duc Long, Le Cao Nhien, Le Quang Minh and Moonyong Lee[†]

School of Chemical Engineering, Yeungnam University, Gyeongsan 712-749, South Korea

Furfural, which are used to derive other industrial chemicals, has been identified as one of the major bio-based platform chemicals that can compete with petroleum-based chemicals. However, the current commercial furfural process has low yield and requires intensive energy. This paper reports the simulation, detailed design and optimization of an industrial process recovering furfural from lignocellulosic biomass, a renewable and inexpensive feedstock. The aqueous feed mixture from the acidic hydrolysis of biomass was fed to the separation process to recover furfural as the main product, along with methanol as valuable byproducts. The purification process was intensified by applying an innovative bottom dividing wall column with a decanter configuration (BDWC-D). The response surface methodology (RSM), which allows the interactions between variables to be identified and quantified, was used to optimize the BDWC-D structure. The predictions by the RSM showed satisfactory agreement with the rigorous simulation results. The heat pump technology and heat integration were then employed to improve the purification process. The results showed that the proposed sequence could save the overall energy requirements and total annual cost significantly as compared to the conventional sequence. This sequence can be employed for both grass-root and retrofit designs.

Keywords: Lignocellulosic Biomass; Furfural; Energy Efficiency Improvement; Dividing Wall Column; Response Surface Methodology

Acknowledgments

The authors wish to acknowledge the financial support from the R&D Convergence Program of NST (National Research Council of Science & Technology) of Republic of Korea and KITECH (Korea Institute of Industrial Technology) (ES150001).

442557 Flow Field Simulation and Distribution Performance Analysis of a Gas Pipe Distributor in Fccu

Abstract:
A fluidized catalytic cracking (FCC) is an important petroleum refinery process, which, aided by highly selective zeolite catalysts, cracks large-molecule low-value heavy oil feedstock into small-molecule highvalue
products. When spend catalyst is fed into the regenerator for reactivating, it is fluidized by the gas distributor at the bottom of regenerator. The gas distributor is designed to uniformly distribute gas to maintain the fluidized state of the fluidized bed and to provide the reaction gas to guarantee the burning coke deposited on the spend catalyst. Therefore, the gas distribution performance of the gas distributor influences the quality of initial gas distribution directly. So the pipe distributor used in a industrial FCC unit, which is composed by main pipes and a number of branch pipes with many nozzle tubes, was investigated by means of flow field calculation. The gas distribution performance of the pipe distributor was analyzed based on the simulated flow field. The simulation result shows that there are a obvious nununiform
gas distribution from the different nozzle gas flow rate. The static pressure increases gradually along the flow direction in the nozzle tube while the axial flow rate decreases, which leads to the nozzle gas flow rate increasing gradually. Furthermore, an obvious bias flow occurs at the inlet of the nozzle tube, which influences the gas flowing into the nozzle. The pressure drops performance of nozzle and pressure varying in the branch pipe are analyzed comprehensively. Due to the nonuniform
gas distribution by the gas pipe distributor in the practical usage, the problems such as overlarge temperature difference among the fluidized bed layers in the regenerator have often happened, which will not only reduce burning coke efficiency and but also decrease activity of the regenerated catalyst. Meanwhile, the erosion of the distributor also occurs, and in consequence the function of the distributor fails, which even causes the nonscheduled shutdown of the FCC unit and leads to tremendous economic loss. The support is provided for optimizations of both structural parameters and operation parameters of the gas pipe distributor and improvement of the gas distribution performance. Furthermore, improvement suggestions were provided based on the above reasoning. The results are expected to be useful for the distributor design.

442925 Identifying Gas Liquid Contactor Performances for Optimized Natural Gas Desacidification Amines Units - from Process Calculations to Experimental Strategy

Abstract:

Amine units are widely used for gas sweetening. The absorber performances of these reactive absorption processes are typically mass transfer limited. The proper selection of gas liquid contactor is thus critical to optimize the size of towers, the solvent flowrate and the duty of the reboiler. This work proposes an approach to identify the most suitable packings for a given application.

The case-study selected here to illustrate the approach deals specifically with a CO_2removal application where the sour gas is at 7 % vol, and the sweet gas specification is 2% vol. The gas flowrate is fixed at 0,18 MMSCFD with a solvent flowrate fixed at the minimum of the thermodynamic constraints (via equilibrium curves). An in-house rate-based simulator with accessible forms of mass transfer correlations, based on a double film model, is used to calculate the absorber height and diameter which are needed to reach the sweet gas specifications. A reference structured packing is compared with standard valve trays.

The calculations show that the bed height required with the first contactor is considerably higher than with the second one while the column diameter is reduced. In this case the use of standard valve trays is more attractive for cost optimization. Interestingly, because mass transfer is accelerated, the absorption rate with the structured packing is found to be thermodynamically limited due to a more severe temperature increase which reduces the driving force in the bottom section of the absorber.

The study determined that the optimum design for the investigated case is a compromise between a full trayed tower, that generally imposes a larger diameter, and a packed tower where the higher mass transfer forces result in a temperature rise in the bottom of the tower which in turn imposes a height increase and a much higher solvent flowrate. . To determine the mass transfer characteristics of a gas-liquid contactor providing this optimum, a sensitivity analysis has been conducted. Mass transfer parameters, i.e., liquid and gas side mass transfer coefficients and interfacial area, were varied by a factor 2 relative to the ones obtained with the reference structured packing. Calculations show that one should use a contactor that reduces liquid phase resistance, specifically, the mass transfer conductance $k_L a_e$ should be increased by at least 50%. This target could be reduced to +30% if one could simultaneously reduce the gas phase conductance $k_G a_e$ by 20%. This result can be explained by the link between mass and heat transfer (Chilton-Colburn analogy) whereby reducing $k_G a_e$decreases the temperature level which in turn thermodynamically increases the mass transfer driving force.

The next step will be to review the literature more extensively to make a short list of pre-selected random and structured packings that best meet the targets. Experimental tests will then be run on IFPEN facilities to measure flooding limits and mass transfer parameters in a small 150 mm diameter column. This step is required to confirm packing's behavior, considering the disparity of the values reported by different authors in the literature. The last step is to consolidate the existing correlations for the most promising packing, and mitigate differences between various authors by performing characterization tests on larger diameter test columns. The final correlations will be implemented in our in-house rate based simulator for industrial use.

The present approach can be applied to determine the required mass transfer characteristics for optimizing other gas treatment applications. However, the conclusions reached here as regards to mass transfer targets and the problematic temperature levels, are valid only to the present case -study, and are not directly applicable to other gas sweetening applications (selectivity, LNG, ...). For example, in a case where H_2S and CO_2 are treated simultaneously the mass transfer limitations could be very different.

443365 Technical Reconstruction of Double Effect Distillation for Methanol Process

Abstract:

Abstract: Methanol is one of important basis raw materials of chemical industry and also an excellent environment-friendly fuel, which make it as a significant commodity in the global market. Distillation is an essential unit operation for purified production of methanol. Product quality, production cost, environmental impact and energy consumption are largely related to distillation process.

Our research effort, under this circumstance, has been devoted since 2002 to energetically optimize 4-column methanol distillation system that is widely adopted in methanol units from coal, with reduced cycling water requirement. This patented process technology saw its industrial application in Chinese largest single unit of 1800Kt/a of Baofeng Group located at North China, with the lowest steam consumption of 0.9/1 t(steam)/t(methanol) compared with normally 25% higher for both simulative case study and on-site operation. Methanol product is in accordance with O-M-232G and the recycling water is reduced significantly from 67 t(CW)/t(methanol) to 52 t(CW)/t(methanol).

The flow diagram of this advanced double effect distillation process is showed in Fig.1. Its advantages arise for limitations found in current units: (1) heat exchange network failed to fully recover energy causing large CW flow to remove extra heat; (2) for pressured column, bubble point reflux need larger reflux ratio than the sub-cooled; (3) the heat of distillate from pressured column was normally ignored; (4) the low concentration of methanol of feed inlet to recovery column leads to higher reflux radio, responsible for unsatisfied unit energy efficiency.

Fig.1 The double effect distillation process.

(Column1.Light ends column; Column 2.Pressured column;

Column 3.Atmospheric column; Column4.Recovery column)

As a solution, a new flow diagram is built up and showed in Fig.2. (1) HEN totally reorganized (not completely show); (2) sub-cooled reflux to pressured column caused 10% reduction to the reflux ratio, economizing steam consumption; (3) top product of pressured column give heat to crude methanol before feeding light ends column; (4) change product and by-product position of recovery column to give efficient unit energy consumption.

Fig.2 The reconstruction of distillation process.

(Column1.Light ends column; Column 2.Pressured column;

Column 3.Atmospheric column; Column4.Recovery column)

It is estimated that the production of methanol will be over 60 million tons in 2014. If this reconstruction could be applied in the whole methanol industry, 1.3 million tons of standard coal would have been saved. This is equivalent to reduce 5 million tons emission of carbon dioxide, so the environmental impact can be largely decreased. Field operation shows reconstruction perfect substitution worth to spread to the whole methanol industry, energetically efficient and low environmentally impacting.

443095 Experimental and Numerical Study of Transport Phenomenon of Falling Film Flow with Different Heating Methods

Abstract:

Falling film flow is a common phenomenon in chemical industry. It is widely used in chemical instruments, such as structured packed columns and falling film evaporators. An experiment was implemented to study the mass transfer phenomenon of binary-mixture (ethanol-water) falling film flows with different heating methods (uniform or concentrated heating). In order to study the detailed flow and mass transfer behaviors, a 2-D multiphase computational fluid dynamics (CFD) model based on volume of fluid (VOF) was developed. This model divides mass transfer into two parts: one is caused by the vapor-liquid equilibrium (VLE) and the other is caused by the heat flux. Results from simulation are inconsistent with the experimental data. Both experiments and simulations show that the ethanol concentration of the vapor with concentrated heating is 1.5% higher than the one with uniform heating.

443101 Thermodynamic Modeling of HNO_3-H_2so_4-H_2o Ternary System with Symmetric Electrolyte NRTL Model

Abstract:

Nitric acid concentration/sulfuric acid concentration (NAC/SAC) process has been widely used for concentrating dilute aqueous nitric acid and recovering spent sulfuric acid. Nitric acid is concentrated by employing sulfuric acid to bind water and break nitric acid-water azeotrope at approximately 68 wt. % nitric acid. To support heat and mass balance calculations and process simulation for NAC/SAC processes, we develop a comprehensive thermodynamic model for nitric acid-sulfuric acid-water ternary system based on previously published thermodynamic models of nitric acid-water [Wang et al., Fluid Phase Equilibria, (2015) (DOI: 10.1016/j.fluid.2015.04.015)] binary and sulfuric acid-water [Que et al., Journal of Chemical & Engineering Data, 56 (2011) 963-9772] binary systems developed with symmetric electrolyte NRTL (eNRTL) activity coefficient model. The present work takes into account one-step hydration of nitric acid molecules, two-step hydration of sulfuric acid molecules, and further hydration of hydronium ions. With up to three temperature coefficients for each eNRTL binary interaction parameter and up to two temperature coefficients for each chemical equilibrium constant, the ternary model is validated with boiling point data from 383.15 to 471.45 K and activities at 273 K. Furthermore, contour plots of boiling points, vapor phase composition, and specific heat capacity of the ternary system are generated for engineering use.

Keywords:

Nitric acid, Sulfuric acid, Electrolyte NRTL model, Vapor-liquid equilibrium, Boiling point, Heat capacity, Nitric acid concentration/sulfuric acid concentration (NAC/SAC)

443503 A Computational Analysis of Fire Scene for Safety in a Compartment

Abstract:

This article details numerical prediction of fire spread carried out to study the distribution of carbon monoxide concentration, smoke dispersion, soot concentration and temperature in laboratory compartment, using a commercial fire modeling software tool. The heat release rate curve predicted from the simulation was compared with literature values carried out in a road tunnel. The time variations of carbon monoxide transportation, smoke concentration and temperature spread in compartment room are reported to evaluate the effect on humans. The soot concentration was described and used throughout this study as an indicator of safety in the compartment room. The effects of vent dimension and ventilation capacity were analyzed by smoke detector response times. This work provides improvements derived from reinterpretation of actual fire scene considering the ability of mechanical exhaust system and ventilation in compartment room to mitigate the hazard of fire.

443640 Preparation and Characterization of (*FURCRAEA spp*) as Green Support of Iron Nanostructured Catalyst

Abstract:

PREPARATION AND CHARACTERIZATION OF (*FURCRAEA spp*) AS "GREEN" SUPPORT OF IRON NANOSTRUCTURED CATALYST

Karen G. Bastidas G[1], Anamaria Barrera B[1], Cesar A. Sierra A[2], Hugo R. Zea R[3]

[1] Department of Chemical and Environmental Engineering. National University of Colombia (Bogota).

[2] Associate professor. Director of the Research Group of Macromolecules. National University of Colombia (Bogota).

[3] Associate professor. Director of the Research Group of Materials, Catalysis and Environment. National University of Colombia (Bogotá).

Abstract.

The availability of natural water resources has declined over time, becoming one of the biggest problems today. Environmental risk in Colombia's rivers has increase due to the widespread use of harmful chemicals in a variety of processes, ranging from illegal mineral extraction to industry wastewater, e.g. Medellin's river, due to its strategic location along a textile industrial corridor might be subject to dye contamination; the gold and platinum mining zones in Choco and Antioquia regions are being heavily affected by the use of mercury in the extraction processes. In 2009, 520 tons of mercury were imported by miners working in this area, amount that exceeds several times the mercury used by regulated industry in previous years (Ministry of Mines and Energy of Colombia, 2007). Not only the environmental risk is a concern but also the economic conditions of many low-income families whose livelihood in many cases depends exclusively of the of the mining.

Due to the issues raised above, the main goal of this study is to research, develop and implement innovative technologies in the synthesis of highly efficient, low cost and environmentally friendly iron based catalysts capable of degradate pollutants, ensuring operational stability. A native fiber (Fique, FURCRAEA spp) has been selected as a solid support for these catalysts due to its abundance, low cost and special physical and chemical characteristics.

The combination of cellulosic fibers with iron oxide nanoparticles could provide exceptional biodegradable composite materials for the treatment of organic dyes and heavy metals present in wastewater. Likewise, the use of iron particles immobilized on fique fibers allows an easy separation catalyst-treated water, turning the biocatalyst in highly recyclable and reusable material.

Fique fiber was initially characterized by determining the percentage of ashes, total lignin, soluble and insoluble lignin, cellulose, hemicellulose, total solids and humidity, obtaining 0.94, 22.58, 0.5, 22.8, 36.3, 27.5 0.701 and 2.1%, respectively. Additionally, the isoelectric point (pzc), acidic and basic sites were measured.

Then, the fiber was subjected to a pretreatment in order to expose more active sites which on time serve as nucleation points for the deposition of nanoparticles. This was done by means of an ultrasound-assisted procedure, varying the sonication time (30, 60 and 90 minutes). The effect of the sonication was followed by measuring the contents of lignin, cellulose and hemicellulose within the fiber after each time. 60 minutes of sonication was found to be most beneficial time since the amount of lignin and hemicellulose present in the fiber was reduced, while conversely the amount of crystalline cellulose exposed increases.

In order to improve the chemical ability of the nucleation sites to interact with iron and then to form stable and size controllable nanoparticles, a cationization procedure was performed on the fiber; this was achieved by the exposure of the fiber to acidic and basic conditions promoting cationization. Three cationization times (1, 2 and 3 hours) were studied and their effect on the fiber chemical surface was evaluated using FTIR-ATR and XRD analysis. The results showed 3 hours as the optimal cationizing time.

Subsequent to the cationization of the fiber surface, a procedure of ionic/cationic exchange was implemented by exposing the fiber to a solution of iron salts in order to replace the cations in the fiber surface and anchoring the iron nanoparticles. Different conditions were studied in order to synthetize zero valent iron nanoparticles (nZVI) and others with different oxidation states. Finally, to track the impregnation process on the fiber, samples from the precursor solution were taken at different times and its iron concentration was determined by atomic absorption spectrophotometry (AAS). Complementary characterization performed by X-ray diffraction, BET surface area and electronic microscopy, among others.

Key words: Natural support (fique), iron nanoparticles, bionanocatalyst.

443764 Optimization of Heat Exchanger Network in Steam Assisted Gravity Drainage Oil Sands Surface Facilities

Abstract:

Steam assisted gravity drainage (SAGD) is the most viable and environmentally safe recovery technology for extracting heavy oil and bitumen. It is a thermal in-situ production process where parallel wells are drilled horizontally into an underground bitumen reservoir. Steam is produced at the central processing facility (CPF) and injected into the reservoir through the wells. The steam heats the bitumen to a point where gravity allows it to flow down to the lower well where the mixture of bitumen and water is then pumped to the surface. The water and bitumen are separated at the CPF. Typically, more than 90% of the energy consumed in producing a barrel of bitumen using SAGD is used to generate the steam that is injected into the bitumen reservoir. Fossil fuels is employed in the process, which causes GHG (greenhouse gas) emissions. The usage of hot and cold utility is reduced by increased heat exchange and optimization of the heat exchanger network, which results in reduction of GHG emissions. In this work, pinch analysis is performed to minimize energy consumption.

443774 Thermodynamic Anaysis of Dehydration of n-Propanol Using Heterogeneous Azeotropic Distillation

Abstract:

Heterogeneous azeotropic distillation is a widely popular process in chemical process industries for the separation of binary azeotropic mixtures into their constituent pure components. It involves the addition of an entrainer which is partially miscible and forms a heterogeneous azeotrope preferably with one of the original components. This is a powerful technique which combines several physical phenomena such as presence of azeotropes and liquid-liquid immiscibilities, and is used widely in industries to enhance difficult or intractable distillation separations. For the design of heterogeneous azeotropic distillation columns, a detailed study of the heterogeneous region is crucial to plan the distillation sequence, which requires experimental vapor-liquid-liquid equilibrium (VLLE) data.

The present study focuses on the investigation of the heterogeneous region of the well known industrially available azeotropic mixture n-Propanol-Water system using Propyl acetate as entrainer. The VLE data of the system was generated by means of modified Othmer still and the still is modified using a probe type ultrasonic homogenizer to account for the heterogeneity. Isobaric vapor- liquid- equilibria and vapor - liquid - liquid - equilibria data at atmospheric pressure have been obtained for the ternary system n-Propanol-water-propyl acetate and the constituent binary systems. The equilibrium data has been correlated using NRTL, UNIQUAC and UNIFAC models and compared with the experimental data.

443776 Study of Ionic Liquids As Entrainers for the Extractive Distillation of *N*-Propanol and *N*-Propyl Acetate Binary Azeotropic Mixture and the Isobaric VLE Studies of the System

Abstract:

Distillation is a widely accepted technique for the separation of binary liquid mixtures. Separation by simple distillation is not possible for binary mixtures that form azeotropes. One of the commonly used methods to separate azeotropes is extractive distillation. It is the process of adding an entrainer that can alter the relative volatility of the mixture and thereby breaking the azeotropic point. Ionic liquids on account of their physiochemical properties as well as the recyclability have been used as a preferred replacement for the conventional entrainers. Extractive distillation with ionic liquids has the advantages of both liquid solvent (easy operation) and solid salt(high separation ability). The corrosion and dissolution problems by using inorganic salts can be rectified by these non-volatile, chemically stable, high electrical conductive molten salts.

The azeotropic system under consideration in the present study is *n*-propanol - *n*-propyl acetate mixture. These compounds which constitute the products/reactants of esterification reaction of n-propanol form a minimum boiling azeotrope at 94.7°C. *n*-propyl acetate has been included in a list of volatile acetates as an ecologically benign CFC-free solvent for aerosol sprays.

Vapour liquid equilibrium(VLE) measurements of *n*-propanol and *n*-propyl acetate are performed at atmospheric pressure in a modified Othmer still which is a continuous distillation still in which vapour phase is recirculated. The most suitable ionic liquids for shifting of azeotropic composition of the system are selected using COSMO-RS model with suitable combinations of anions and cations. The isobaric VLE of the system with [EMIM][HSO_4] is obtained and compared with the existing data. The influence of alkyl chain length on the performance of ionic liquids is also analysed with regard to the previous studies. The effect of polarity is the most important factor on the behaviour of ionic liquid solvation in alcohols. Experimental VLE data were correlated using modified UNIFAC model.

443784 Measurement and Correlation of the Mutual Solubility of Diisopropylamine + Water and Triethylamine + Water Systems at High Pressure

Abstract:

The mutual solubility data of diisopropylamine + water and triethylamine + water systems were measured at the pressure up to 100 MPa and the temperature range was 280.2~316.2 K. Experiments were performed in a high-pressure autoclave with a moveable piston. The experimental data were correlated satisfactorily with NRTL activity coefficient model and the binary parameters were obtained.

443819 Improving High Shear Rate Friction Reduction Behavior of Polyacrylamide-Based Fracturing Fluids Using Xanthan Gum

Abstract:

The main component of slickwater in fracturing low permeability formations is a high molar mass polymer such as polyacrylamide (PAM), which acts as a drag reduction agent in turbulent flow. Polyacrylamide does not degrade in the environment and also suffers from mechanical degradation at high turbulent flow rates. Five different concentrations of PAM as well as mixtures of polyacrylamide/xanthan were prepared to investigate the possibility of improving shear stability of PAM. Drag reduction measurements were performed in a flow loop and friction factors at different flow rates were calculated using rheological parameters. Based on the observations, Xanthan (XG) is not as good a friction reducer as PAM, even at three-fold higher concentration. XG slightly improved the shear stability in mixed polymer solutions at low concentrations. But at high concentrations, XG significantly improves the friction reduction behavior of PAM at nonionic and ionic environments.

443836 Oxidative Dehydrogenation of n-Butane on Carbon Catalysts

Abstract:

Butenes and butadiene, which are one of the important industrial chemical products to produce synthetic rubber , are sythesized traditionally by oxidative dehydrogenation (ODH) of n-butane over various mixed metal oxides. Metal-free carbon nanomaterials has recently received much attention by researchers as a catalyst for the ODH reaction due to its environmental acceptability, corrosion resistance, and unique surface properties. For the reason, the reaction system was modelled in this study for the sythesis of butene and butadiene. For the modelling of the reaction system, the catalytic performance of several carbon catalysts, such as graphite and activated carbon, were initially tested. Then, the reaction mechanisms and kinetics were developed based on experimental results. Lastly, the mathematical modeling of the reaction system was optimized to maximize an overall yield of the main products. This approach allowed us to develop the optimized reaction system more efficiently.

This research was respectfully supported by Engineering Development Research Center (EDRC) funded by the Ministry of Trade, Industry & Energy (MOTIE). (No.N0000990)

443855 Modeling the Thermal Behaviors of an Ultracapacitor Module

Abstract:

Ultracapacitors, also known as supercapacitors, have the potential to meet the increasing power requirements of the sustainable energy systems for the grid-scale energy storage systems and electrified vehicles. As compared to batteries, ultracapacitors offer a higher power density, higher efficiency, and longer shelf and cycle life. Because the performance of an ultracapacitor depends on temperature, it is important to calculate accurately the thermal behavior of a single ultracapacitor cell and the module composed of multiple ultracapacitor cells for the efficient and reliable systems integration of an ultracapacitor in grid and transport applications.

In this work, modeling is performed to study the effect of the layout configugation of the ultracapacitor cells on the thermal behaviors of an ultracapacitor module. The ultracapacitor module is subject to the charge/discharge cycling with constant-current between the nominal voltage of the ultracapacitor module and the half of this voltage. The charge/discharge current values examined are 50, 100, 150, and 200 A. The validation of the modeling approach is provided through the comparison of the modeling results with the experimental measurements.

443856 Modeling the Electrolyte Flow Behaviors of a Zinc/Bromine Flow Battery Module

Abstract:

The zinc/bromine (Zn/Br_2) flow battery is a promising rechargeable power source for energy storage systems because of its inherent chemical simplicity, high degree of reversibility at the electrodes, good energy density, and abundant low-cost materials. The Zn/Br_2 flow battery system, intended for the grid-scale energy storage system, is to be made in modular form. A Zn/Br_2 flow battery module is composed of the multiple stacks consisting of bipolar carbon electrodes. Since the stacks require electrolyte circulation on both the negative and positive sides, two plumbing loops are provided. The flow loops contain an electrolyte storage tank and a circulation pump. A heat exchanger may be included for temperature control in the flow loop. The design of the electrolyte circulation plumbing determines the distributions of electrolyte flow and pressure within the stacks and the efficiency of the stack is dependent on the flow distribution throughout the stack. The modeling of the distributions of electrolyte flow and pressure within the module is needed to optimize the plumbing for electrolyte circulation and thereby to improve the performance of the module. In this work, modeling is performed to study the effect of the layout configuration of the plumbing for electrolyte circulation on the distributions of electrolyte flow and pressure within a 50 kWh Zn/Br_2 flow battery module composed of four stacks. It is seen that the plumbing layout for electrolyte circulation can have a significant effect on the flow rate distribution of electrolyte through the stacks.

443837 Blowdown Dynamic Modeling and Analysis of Pressure and Temperature

Abstract:

Vessel blowdown is one of the safety operation procedures that discharges flammable gas or liquid to flare system during emergency situation such as fire. When blowdown occurs, the liquefaction and freezing effect cause a pressurized vessel to be brittle, exposing it to possible hazard situation. Also the blowdown systems are recommended to be accordance with API(American Petroleum Institute) 521 code. API 521 requires pressure in vessel should be half of the initial pressure or be lower than 100 psig in 15 minutes after commencing blowdown in fire case. Therefore, accurate estimation over the change of temperature and pressure of the vessel during blowdown period is crucial for the design of blowdown system. For the reason, researchers have developed a few dynamic blowdown models so far. However, the model was found to be rather inaccurate under certain conditions. Therefore a new dynamic blowdown model was developed in this work. For the accurate analysis, the model algorithm considers both vapour and liquid phase in non-equilibrium condition that the temperatures of vapour and liquid phases are different. Futhermore, heat transfer from the vessel wall to discharge gas at either laminal or turbulent flow in the vessel was considered in the model to increase the accuracy of blowdown estimation. For thermodynamics, both Peng-Robinson and Soave-Redlich-Kwong equations of state were applied to the model, and the results were compared each other. Finally, the simulation results were compared with Haque et al's experimental and steady-state simulation results.

443974 Observer-Based Model Predictive Control for Time-Delay Systems

Abstract:

Model predictive control is one of the most widely used advanced process control technologies in industry.[1] It has many capabilities such as dealing with input/state constrains and uncertainties. Observer-based model predictive control was developed to estimate states information when the states are not measurable directly. Many chemical processes have time delays in either states or measurements, such as polymer production and some bioprocesses. [2] In this presentation, we are proposing an unknown input observer based model predictive control algorithm for time-delay systems. Unknown input observer is an observer design to eliminate certain types of unknown inputs in the system. Stability of unknown input observer-based model predictive control without time delay was provided by our previous study. Terminal cost and terminal region are designed for a class of non-linear time delay systems. The close-loop stability of the controller is provided. [3][4]

An example of bioreactor is provided which has uncertain time delays in measurements but with a known upper bound. Two measurements are provided which are online biomass and offline ethanol measurements. Model-plant mismatch is considered as unknown inputs in the system. State estimation and control are calculated using the observer and an optimization algorithm.

Reference:

[1] Marcus Reble et al., Model predictive control of constrained non-linear time-delay systems. IMA Journal of Mathematical Control and Information, 2010, 1-19

[2] Seung Cheol Jeong and PooGyeon Park, Constrained MPC algorithm for uncertainty time-varying systems with state delay, IEEE TRANSACTIONS ON AUTOMATIC CONTROL, 50, 2005, 257-262

[3] Mayuresh Kothare et al., Robust constrained model predictive control using linear matrix inequalities. Automatica, 32, 1996, 1361-1379

[4] Wook Hyun Kwon et al., A simple receding horizon control for state delayed systems and its stability criterion, Journal of Process Control, 13, 2003, 539-551

444020 Thermodynamic Modeling of Aqueous Lithium Sulfate and Its Solubility in the Presence of Na+, K+ and Mg2+ Sulfates

Abstract:

This work is an extension of the previous works done by our group towards development of a rigorous and comprehensive thermodynamic model for high salinity produced water that is essential for mass and heat balance calculations and simulation and design of novel treatment processes. Specifically this work presents a thermodynamic model that accurately represents all thermodynamic properties of aqueous lithium sulfate solution including the solubility of Li_2SO_4 in water and also in the presence of Na^+–K^+–Mg^{2+} sulfates. By explicit consideration of solution chemistry of each subsystem, symmetric electrolyte NRTL model is used to correlate available experimental data of all pertinent thermodynamic properties including activity coefficient, osmotic coefficient, excess enthalpy, and liquid heat capacity over the entire concentration range. This facilitates quantification of eNRTL binary molecule–molecule, molecule–electrolyte, and electrolyte–electrolyte interaction parameters. Also, solubility data are used to regress the thermodynamic constants required for modeling the liquid–solid equilibrium of lithium sulfate in water and in the presence of Na^+–K^+–Mg^{2+} sulfates. Displaying a significant prediction ability, the model accurately represents the thermodynamic behavior of the system from infinite dilution up to saturation and temperatures from 298.15 K to 573.15 K.

444044 Organic Solar Cells

Abstract:

This paper describes highlights the need of next-generation "organic solar cells". There is a continuing need for organic solar cells and arrays with very high specific power densities (1000-5000 kW/kg) for generating power in a new generation of ultra light space payloads example in NASA. An emerging technology with promise to meet these ambitious goals are organic solar cells based on very thin films of discrete layers. Theoretical efficiencies for multi junction solar cells are more ~70% as compared to ~30% for single junction devices. There is continuous development going on high mobility, band gap engineered organic donors and acceptors, as well as development of three junction devices and scaled up devices on lightweight flexible polymer substrates. The main interest in this paper is organic semi-conducting crystal: Pentacene, which has photo-conducting states. Organic solar cells manufactured today have low efficiency other than their stability issues. The efficiency of organic solar cells can be improved by either designing polymer materials that absorbs more of solar spectrum or by minimizing the charge collection efficiency by controlling the morphology of polymer films. Once fully developed, organic solar cells would revolutionary power the industry due to their flexibility and minimal weight

444055 A Refined Non-Random Two-Liquid Segment Activity Coefficient Model for Solubility Modeling

Abstract:

Accurate correlation and robust prediction for complex molecule solubility or solid-liquid equilibrium (SLE) in solvents and solvent mixtures have drawn much interest recently [Tanveer et al., *Chem. Eng. Prog.* **2014**, 110 (9), 37-47]. Solubility models such as the Non-Random Two Liquid Segment Activity Coefficient (NRTL-SAC) model [Chen and Song, *Ind. Eng. Chem. Res.* **2004**, 43, 8354-8362] are now routinely used in the industries. However, an important unfilled need is to model the solubility in oligomers/polymers/surfactants such as asphaltene nanoaggregates in petroleum fluids and drug molecules in lipid-based or solid dispersion formulation [Tung, *Org. Proc. Res. Dev.* **2013**, 17, 445-454]. This paper presents a refinement to the very successful NRTL-SAC model to better account for the contribution due to size differences of molecules. Specifically, the Flory-Huggins combinatorial entropy expression for conceptual segments in the model is substituted with the Staverman-Guggenheim expression for real molecules. We present the rationale for the refined model, the updated conceptual segment numbers for common solvents, and the newly identified conceptual segment numbers for common drug excipients. The model performance is illustrated by prediction of drug molecule solubility in various lipid solvents and solvent mixtures, compared to experimental data and previous work by other models. The refined NRTL-SAC model shows enhanced capability to correlate and predict drug molecule solubility in common solvents and excipients used in drug formulation.

444057 Explaining the Paradox of Asphaltene Precipitation with n-Alkanes

Abstract:

Explaining the paradox of asphaltene precipitation with n-alkanes
Md Rashedul Islam and Chau-Chyun Chen
Department of Chemical Engineering, Texas Tech University

Abstract
Paraffins, specifically alkanes, promotes asphaltene precipitation when they are blended with asphaltene-containing petroleum fluids. The alkane/oil volume ratio required for onset of asphaltene precipitation increases with the carbon number and yields a maxima. Experimental study indicates that the maxima lays at C7[1] or between C8 -- C10 [2]. However, precipitation yield decreases asymptotically with increase of the carbon number even if excessively high alkane/oil volume ratio is used [3]. For example, n-pentane be a better solvent for asphaltene than n-hexane in regard of both onset of precipitation and precipitation yield. However, high carbon number alkane beyond the maxima, n-hexadecane for instance, can be a better solvent at onset of precipitation point and a poorer solvent in regard of precipitation yield than n-pentadecane. A thermodynamic explanation for this paradox is necessarily important to find the compatibility of bitumen or crude oil with paraffinic hydrocarbons. In this work, this paradox is explained by the aggregation thermodynamics of asphaltene precipitation [4]. Both the enthalpy and entropy of mixing are investigated to explain the effect of paraffin carbon number on asphaltene precipitation. Athabasca and cold lake bitumen samples are studied.

References
[1] G. Hotier and M. Robin, "Action de divers diluants sur les produits pÃ©troliers lourds: mesure, interprÃ©tation et prÃ©vision de la floculation des asphaltènes," Oil & Gas Science and Technology, vol. 38, pp. 101-120, 1983.
[2] I. A. Wiehe, H. W. Yarranton, K. Akbarzadeh, P. M. Rahimi, and A. Teclemariam, "The paradox of asphaltene precipitation with normal paraffins," Energy & fuels, vol. 19, pp. 1261-1267, 2005.
[3] D. L. Mitchell and J. G. Speight, "The solubility of asphaltenes in hydrocarbon solvents," Fuel, vol. 52, pp. 149-152, 1973.
[4] M. Wang, Y. Hao, M. R. Islam, and C. C. Chen, "Aggregation Thermodynamics for Asphaltene Precipitation," AIChE Journal, Submitted.

444086 Optimization of Layered Graphene Oxide Nanostructures through Controlling the Membrane Deposition Rate Towards Size-Sieving Separation

Abstract:

Single-layer graphene oxide (GO) flakes were coated on anodized aluminum oxide (AAO) substrate through a simple vacuum filtration process, the effect of the deposition rate on the nanostructures of obtained membrane was investigated though pure water flux and organic vapor diffusion, which indicated GO membrane assembled with slow deposition rate might have more ordered lamellar structure and, correspondingly, have smaller effective pore size. In contrast, under fast deposition rate, GO flakes assembled in a more random way. Therefore, through controlling the membrane deposition rate, the nanostructure of layered GO membrane could be optimized for size dependent separation. We found after thermal reduction, 100 nm GO membrane (prepared @ 0.02 nm/s) has effective pore size smaller than 6.1 Å, which exhibit great potential in n-hexane/2,2-dimethybutane (DMB) separation.

444142 Robust Design of a Radiant Wall Burner for a High Temperature Environment Utilizing Computational Fluid Dynamics Simulations

Abstract:

Radiant Wall Burners are often utilized in Ethylene Cracking Furnaces. The conditions and fuel gases for a specific unit can vary considerably within the unit and across similar units. With burner air preheated to high temperatures, premixing the fuel can result in dangerous and destructive flashback, while complete nozzle-mixing of the fuel can result in burner instability at startup or turn-down conditions. These challenges can be solved through the generation of a robust radiant wall burner design, which is facilitated by executing Computational Fluid Dynamics simulations of many different geometric configurations.

444161 Novel Microwave Synthesis Carbon Dots from PINE Bark and Its Application for Heavy Metal Sensing

Monday, April 11, 2016
Exhibit Hall E (George R. Brown)

Saheed Sanni, Chemistry, Vaal University of Technology, Vanderbijlpark, South Africa, Puthalapattu ReddyPrasad, Chemistry, Vaal University of Technology, Vanderbiljpark, South Africa and Augustine E. Ofomaja, Chemistry, Vaal university of Technology, Vanderbiljpark, WV, South Africa

NOVEL MICROWAVE SYNTHESIS CARBON DOTS FROM PINE BARK AND ITS APPLICATION FOR HEAVY METAL SENSING

Sanni S.O., ReddyPrasad P., Ofomaja A.E.

Biosorption and Waste water Treatment Research Laboratory, *Department of Chemistry, Faculty of Applied and Computer Sciences, Vaal University of Technology, P. Bag X021, Vanderbijlpark-1900, South Africa.*

INTRODUCTION

Microwave pyrolysis for the production of carbon has the advantages of being cheap, produces high yield of carbon and is absolutely a green synthetic method [1]. In recent times, the use of microwave pyrolysis have been extended to the preparation of carbon dots. Carbon nanodots (C-dots) falls under the class of carbon nanoparticles with size below 10 nm[2]. C-dots are carbon materials which processes high water solubility and strong fluorescence depending on their size, edge, shape, surface ligands, and defects. Since C-dots preparation can be achieved from difference carbonaceous sources and by varying a range of parameters during their synthesis, the properties of the final product may vary depending on these factors. C-dots have sp^2 characteristics and possess several functional groups on their surfaces that makes them highly hydrophilic and easily functionalized with various organic, polymeric, inorganic, or biological spices[3]. Several synthesis methods that have been explored includes electrochemical synthesis, combustion/thermal/hydrothermal/acidic oxidation, supported synthesis, microwave/ultrasonic, arc discharge, laser ablation/passivation, and plasma treatment[4]. C-dots processes excellent such as water dispersibility, chemical stability and photostability, ease of modification, excitation-dependent multicolor emission, low toxicity, and good cell permeability, great efforts have been paid to their potential applications in biosensing, bioimaging, drug delivery, and other biological related aspects.

METHODS

The carbon dots (C-dots) with green luminescence was synthesized from microwave assisted pyrolysis of pine bark waste our carbon source at 800, 1000 1200 W for 1 Hr. The dots produced were characterized using FTIR, TGA, XRD, UV-Vis and Raman spectroscopy. The dots were applied for heavy metal sensing.

RESULTS AND DISCUSSION

The functional groups present on the C-dots (**Fig.1**) includes -OH stretching band mode at 3413 cm^{-1}, C-H stretching at 2934 and 2845 cm^{-1}, stretching peaks of C=O, C=C at 1731, 1604 and 1437 cm^{-1}. The XRD pattern (**Fig 2a &b**) of produced C-dots depicts sharp peak at 20.9° and 26.8°, which is highly crystalline cores and possess amorphous nature of C-dots. This peculiar peak is ascribed to 002 of the graphitic carbon, which in line with inter-planar spacing of 3.77 Å with higher spacing than bulk graphite at 3.34 Å. The thermogravimetric and differential thermal analysis display (**Fig 3**) that the C-dots possess thermal stability more than 100 °C and mass losses occurs afterwards. The thermal analysis depicts 3 endothermic peaks at 331, 542 and 735 °C, which are ascribed to oxygenated groups such as carboxyl, hydroxyl, carbonyl and phenols. The optical properties of the synthesize C-dots were explored using UV-vis absorption and photoluminescence (PL) spectroscopy. The UV-vis spectra display (**Fig 4**)intense absorption peak in the range of 260 to 290 nm, which is ascribed to π-π^* transition of the aromatic sp2 domain and also presence of multiple polyaromatic chromophores. A shoulder peak at 330 nm which is ascribed to n-π^* transitions of C=O and structural defects of the C-dots. PL spectrum of the C-dots (**Fig 5a**)exhibits an intense maximum emission peak at 432 nm upon excitation at 330 nm, which also shows characteristic yellowish green color under sunlight and intense blue coloration UV lamp (365 nm). The PL properties of as prepared C-dots was varied at (**Fig5b**) different excitation wavelength from 300 to 405 nm. There was red-shifting in emission peak as the excitation peak was increased from 300 to 405 nm and also simultaneously decrease in fluorescent intensity, which is ascribed to quantum effect and surface defects. This behavior of the prepared C-dots is dependent on excitation-fluorescence, which is accordance with other report [13]. The quantum yield of synthesized C-dots was measured to be 17% at maximum excitation at 330 nm

Fig.6 shows the Raman spectrum of the PB-GQDs, which has been usually used to confirm the quality of the prepared PB-GQDs. There is two major Raman peaks at 1370 cm^{-1} (D band), 1580 cm-1 (G band). The peak of G band corresponds to the Sp^2 bonded carbon atoms in a two-dimensional (2D) hexagonal lattice and the peak of G band is associated with the vibrations of carbon atoms with dangling bonds in the termination plane of tangled graphite or glassy carbon. The D/G ratio band ratio is ID/IG: 0.68 is confirm the formation of graphene like structure from synthesized pine bark.

Fluorescence detection of Cu^{2+} ion

The detection of Cu^{2+} was performed at room temperature in phosphate buffer (PBS) solution pH 4.3. 1.0 mL of PB-CQDs solution was added into 0.5 mL of PBS buffer, followed by the addition of 1.0 mL different concentrations of Cu^{2+}. The fluorescence emission spectra were recorded after reaction for 5 min at room temperature (~28 °C).

Selectivity and sensitivity studies

Moreover sensitivity and selectivity is additional significant parameter to assess the performance of the sensing system. Consequently, we examined the fluorescence intensity changes in the presence of representative metal ions such as including Cu^{2+}, Hg^{2+}, Ca^{2+}, Al^{3+}, Ba^{2+}, Mn^{2+}, Ni^{2+}, Pb^{2+}, Co^{2+}, Zn^{2+}, Fe^{3+}, $Cs^{2+}$and Mg^{2+}, of each at a concentration of 15.0 μg mL^{-1} were reacted with a PB-CQDs solution. Fig. 7 A displays the relative change in PL intensity of PB-CQDs in the presence of various metal ions. Fe^{2+}, Mn^{2+} and Hg^{2+} ions cause the slight PL changes (defined as the relative change of PL intensity in 60–80% as compared to blank) which can be attributed to the nonspecific interactions between the carboxylic and/or amine groups and the metal ions. Among these metal ions, Cu^{2+} displays the strongest PL quenching effect on PB-CQDs, assigning to the special coordination between the Cu^{2+} ions and the phenolic hydroxyl and/or amine groups of PB-CQDs which has been widely used for the detection of Cu^{2+} ions or colored reactions in traditional organic chemistry. Such a specific fluorescence quenching effect may originate from the strong interactions between Cu^{2+} ions and the surface groups of PB-CQDs which transfer the photoelectrons from PB-CQDs to Cu^{2+} ions. These results clearly demonstrate that the PB-CQDs-based Cu^{2+} sensor is highly selective to Cu^{2+} over the other metal ions. Fig.8. Fluorescence responses of CCDs to the addition of different metal ions under acetate buffer (pH 4.3). The concentration of each metal ion is 15 μg $mL^{-1.}$ F_0 and F correspond to the fluorescence intensities of PB-CQDs at emission wavelength of 440 nm excited at 320 nm in the absence and presence of metal ions, respectively.

As shown in Fig. 3, the fluorescence intensity of PB-CQDs gradually decreased with the Cu^{2+} concentration. The fluorescence intensity displayed liner responses with the Cu^{2+} concentrations, with two linear ranges of 0.0□14.0 μg mL^{-1} and 15.0□25.0 μg mL^{-1} (Fig.3B). The linear equations are $F/F0$ =0.9827□0.8485C_{Cu2+} (R = 0.993) and $F/F0$ = 0.7565–0.0033 C_{u2+} (R = 0.996), respectively.

REFERENCES

[1] W. B. Lu, X. Y. Qin, S. Liu, G. H. Chang, Y. W. Zhang, Y. L. Luo, A. M. Asiri, A. O. Al-Youbi, X. P. Sun, Economical, green synthesis of fluorescent carbon nanoparticles and their use as probes for sensitive and selective detection of mercury(II) ions, Anal. Chem. **2012**, 84, 5351-5357.

[2] J. Jiang, Y. He, S. Li, H. Cui, Amino acids as the source for producing carbon nanodots: microwave assisted one-step synthesis, intrinsic photoluminescence property and intense chemiluminescence enhancement. Chem Commun **2012**, 48, 9634–6.

[3] Y. Wang, A. Hu, Carbon quantum dots: synthesis, properties and applications, J. Mater. Chem. C **2014,** 2, 6921–6939.

[4] C. Liu, P. Zhang, X. Zhai, F. Tian, W. Li, J. Yang, et al., Nano-carrier for gene delivery and bioimaging based on carbon dots with PEI-passivation enhanced fluorescence, Biomater. **2012,** 33, 3604-13.

Figures:

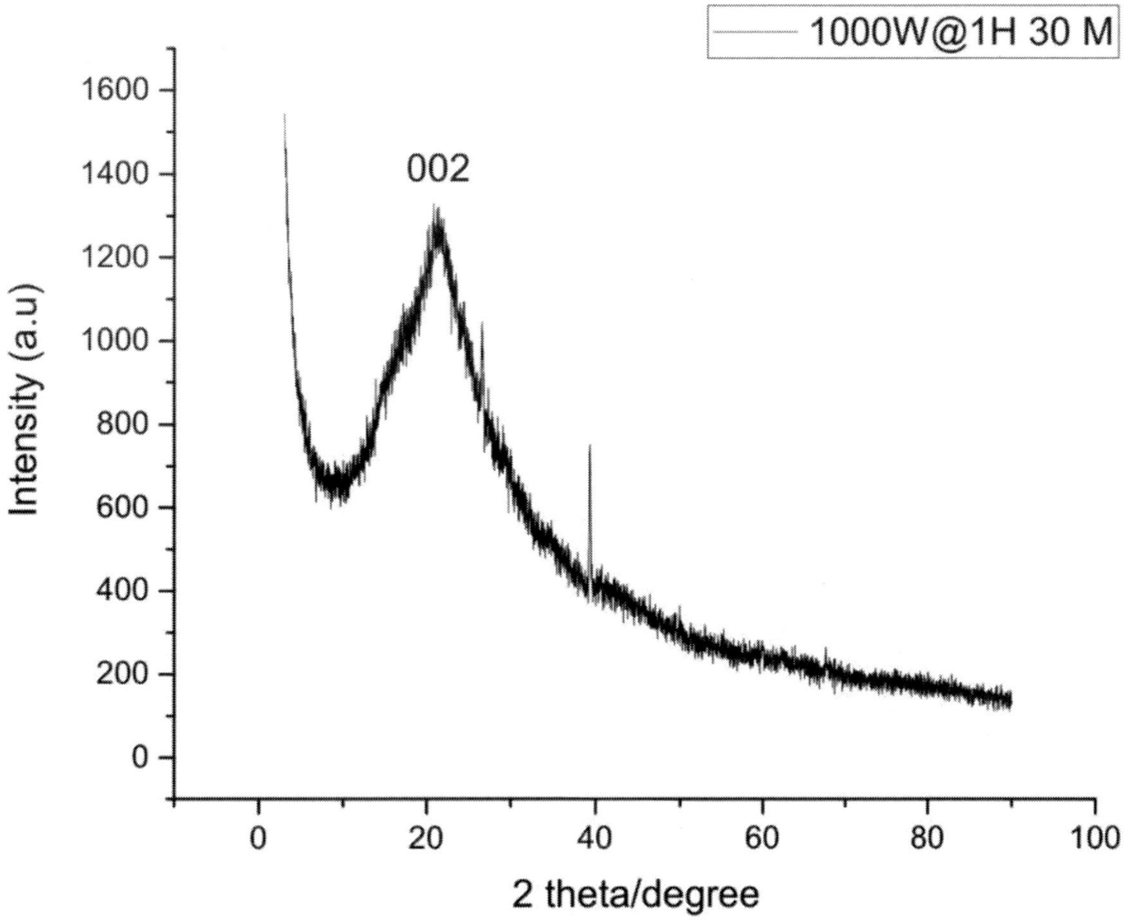
1000W@1H 30 M
002
Intensity (a.u)
2 theta/degree
1600
1400
1200
1000
800
600
400
200
0
0
20
40
60
80
100

Fig 1. FTIR spectra of as-prepared C-dots from Pine Bark

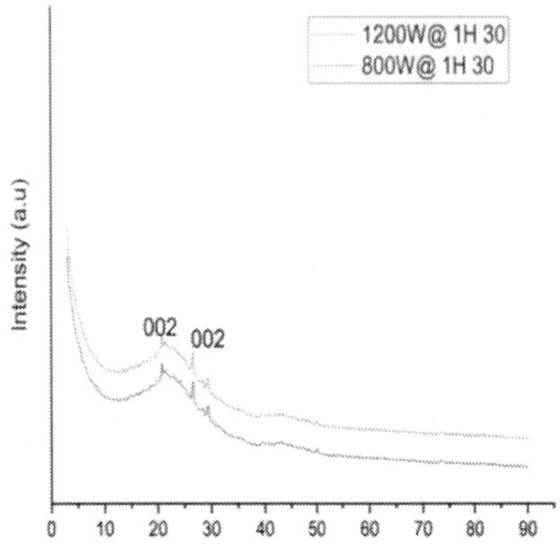

(a)

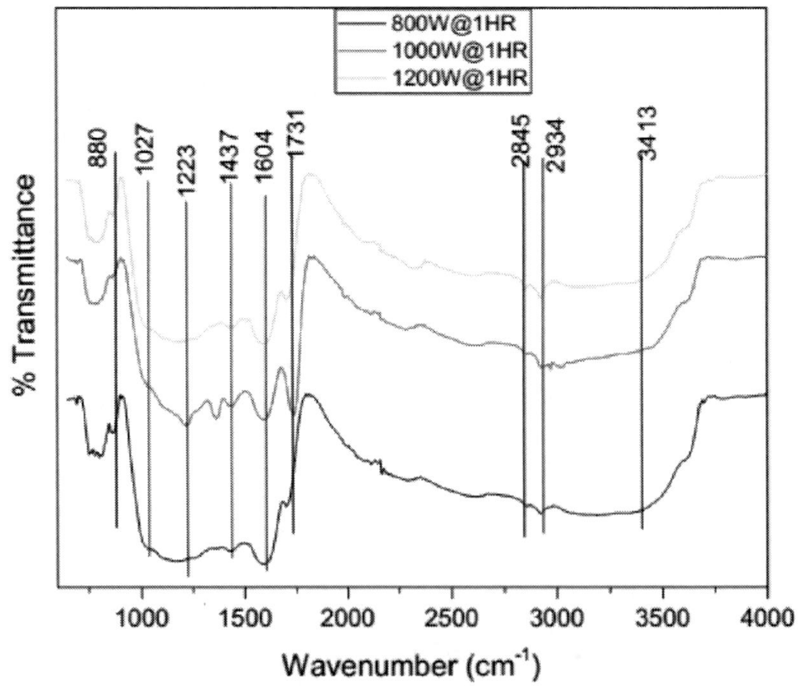

(b)

Fig. 2a & b. XRD pattern of C-dots.

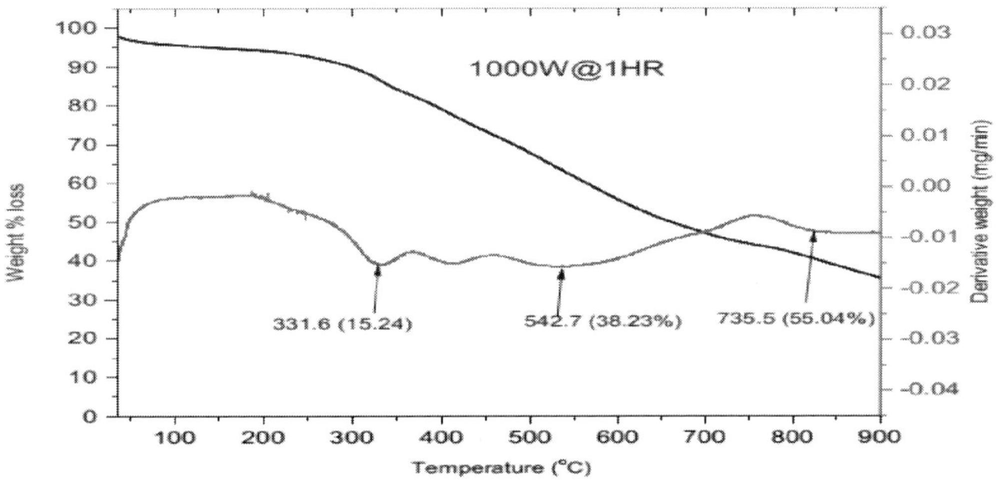

Fig 3. Thermal curves for C-dots under N_2 atmosphere.

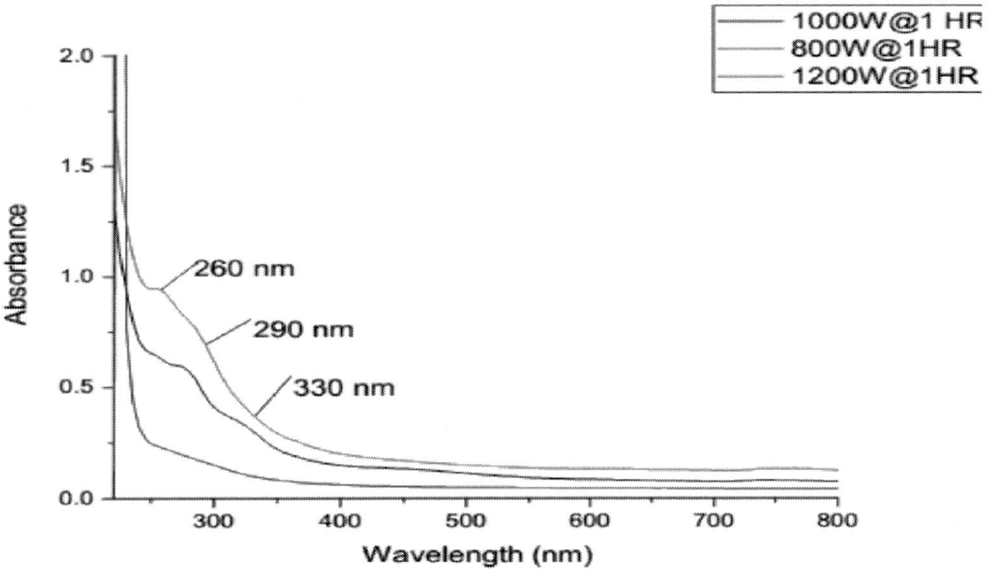

Fig 4. UV-vis absorption of C-dots.

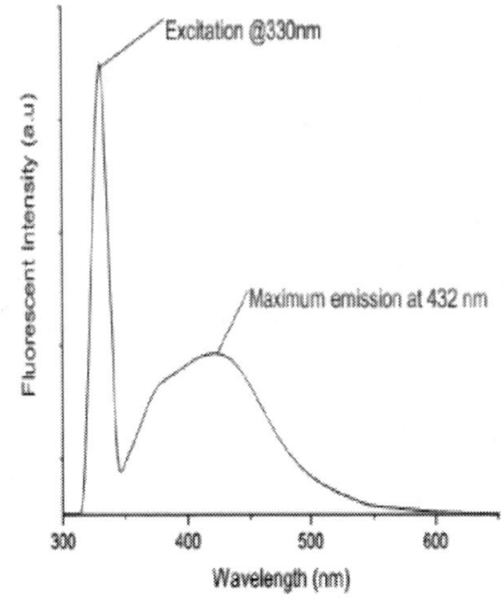

(a)

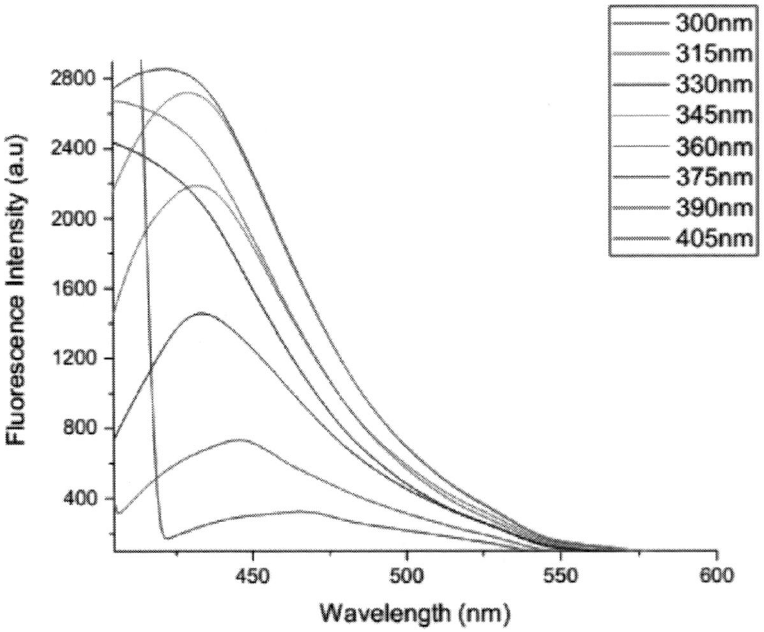

(b)

Fig 5a. PL excitation and emission wavelength (b) PL emission spectra at different excitation wavelength from 300 to 405 nm in 15 nm increments.

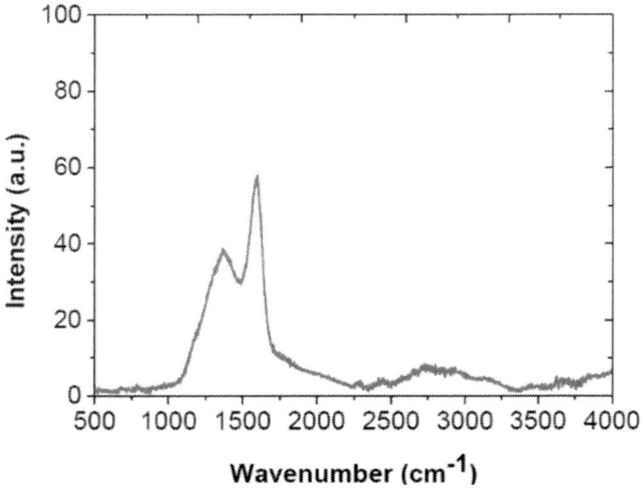

Fig. 6. Raman Spectra of PB-CQDs

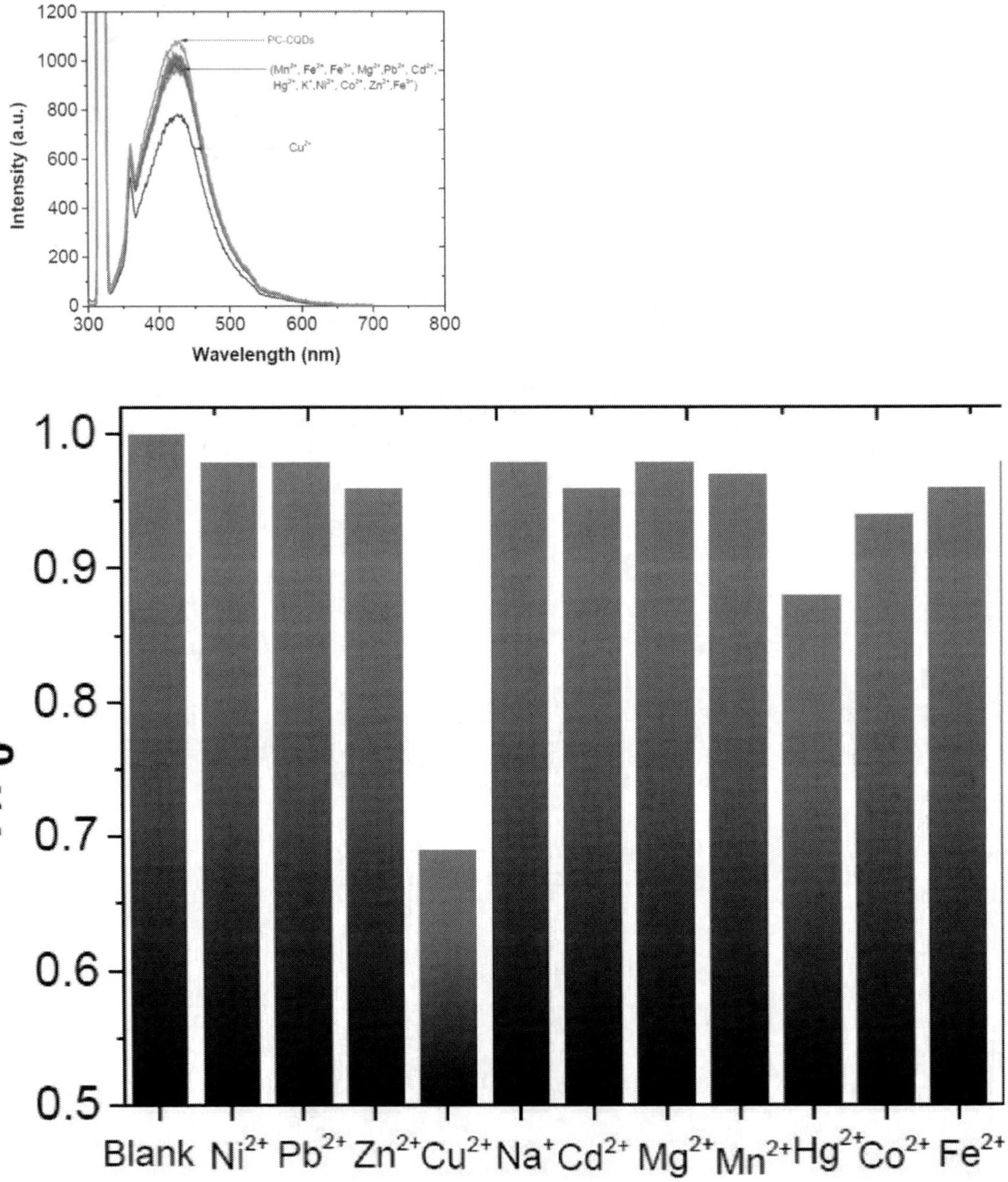

Fig.7. Fluorescence responses of CCDs to the addition of different metal ions under acetate buffer (pH 4.3).

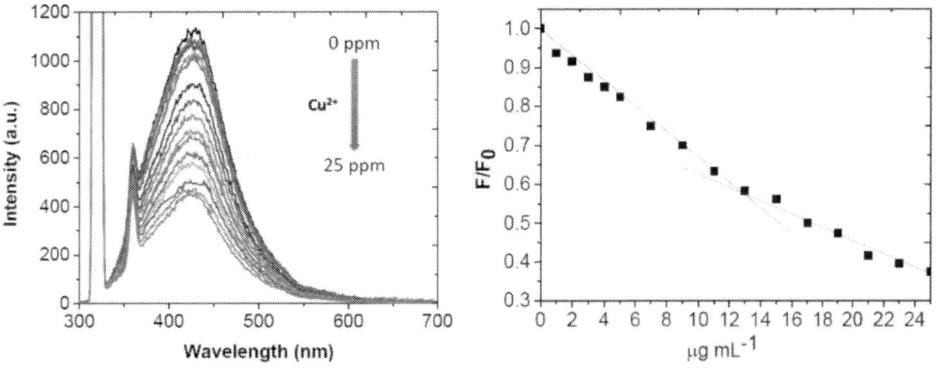

Figure 8. Fluorescence titrations of the y-CDs with Cu^{2+} from 0 to 25 μg mL^{-1}. (Influence of Cu^{2+} concentration on the fluorescence intensity of PB-CQDs (top to bottom: 0, 1, 2, 3, 4, 5, 7, 9, 11, 13, 15, 17, 19, 21, 23 and 25.0 μg mL^{-1})

444269 Functionalization of OH-Terminated Nanoparticles By Cyclic Azasilanes

Abstract:

Colloidal suspensions of functionalized nanoparticles have been suggested for enhanced oil recovery, formation damage remediation, flow assurance, and water treatment operations. Surface modified nanoparticles can also be used as tracers in reservoir studies, active media in thermal management fluids, and fillers in high performance elastomers. Conventional functionalization of nanoparticles with alkoxysilanes and chlorosilanes is limited by needs for catalytic amount of water for the reaction to proceed, requirement of multiple hours to fully saturate the surface, and release of by-products upon reaction completion. Monolayer deposition with a high density of functional groups represents a particularly difficult challenge.

Our ongoing studies indicate that application of cyclic azasilanes that have a strained ring structure with adjacent Si and N centers provides highly efficient functionalization of hydroxylated silica, alumina, magnetite, and diamond nanoparticles which is catalyst-free and by-product-free. The data on synthesis and characterization of functionalized nanoparticles as well as possible reaction mechanism of functionalization will be discussed.

444093 New Development on Hoist Scheduling with the Relaxation of Cyclic Precondition

Abstract:

Multi-Stage Material Handling (MSMH) processes are designed to manufacture materials requiring a series of treatments and in industry, they are broadly used due to their capability to handle large work orders and compatibility with materials with different recipes. A typical example is electroplating process. A hoist handles the transportation of each job among units (stages) according to its processing recipe. The hoist movement plan is predetermined through optimization and this plan is repeated to perform continuity. Obviously, the ultimate goal is to maximize the productivity within each movement plan, which is equivalent to minimize the cycle time to finish one work order. The optimization process is generally called Cyclic Hoist Scheduling (CHS), since every movement plan is a cycle due to its repeatability. This precondition in term of hoist operation and unit usage is general for most cases. However, it could block the really optimal results.

In this study, this precondition has been relaxed for multi-capacity units, thus the unit usage can be more flexible. Thus, the real optimality can be found through CHS. To deal with the cycle connection issue from the relaxation of the precondition, a new module Cycle Connection Identification (CCI) is proposed. Obtained hoist schedule will be checked. Once a schedule with cycle connection problems is identified, a new cycle will be scheduled forcing the smooth connection with the first cycle. The objective of scheduling a new cycle is to match the beginning of the first cycle and the ending of the second cycle. If the two-cycle group is still with cycle connection issue, more cycles will be scheduled till the cycle group is free from cycle connection problems. Each cycle is called a sub-cycle and the constructed cycle group without cycle connection problems is called a full cycle. Note that in each sub-cycle, the work order and all transportation operations are completed.

In this paper, hoist scheduling models with/without such a precondition have been developed and proposed as mixed integer linear programming. The performance of the two models is tested through case studies and the difference is compared and demonstrated in detail.

444359 Development of Novel Emulsified Acid Using Waste Oil

Abstract:

In carbonate reservoirs, hydrochloric acid (HCl) is commonly used for stimulation purposes. The main disadvantage for using HCl is its high and rapid reactivity allowing for less control on acid reaction with formation. In addition, HCl may react with the tubing causing severe corrosion and damages. Several slow-reacting acid systems, though HCl-based, have been developed to retard the HCl reaction and control its reactivity to allow deeper penetration in carbonate reservoirs before it breaks out and reacts with formation rock. However, the selection of the hydrocarbon phase of emulsified acid is critical to the creation and stability of the emulsified acid system. Diesel oil, crude oil and solvents such as xylene have been reported to be successful in the field.

In this work, alternative and less expensive source of oil is developed to replace the current practices. Emulsified acid is a water-in-oil emulsion where oil is the continuous phase and acid is the dispersed one. Industrial and filtered waste oil was used to prepare the emulsified acid of this work. Extensive work has been carried out to study the chemical composition of the waste oil, namely SARA and to select the proper emulsifier. Emulsions were prepared using 15% HCl with emulsifier concentration ranging from 0.5 to 2.0 vol% and 70:30 of acid-to-oil ratio. Droplet size distributions were measured and used to benchmark with other emulsions. Emphasis was given to the stability of the emulsion especially at high temperatures. In order to apprehend the effect of corrosion inhibitor on the emulsion stability, its concentration was varied (up to 0.3 vol%). Later, rheological analyses of the waste oil emulsion were compared to the commonly used diesel-based emulsified acid.

Results showed the potential use of waste oil to prepare high temperature emulsified acid (up to 100°C). The addition of corrosion inhibitor to the emulsified acid caused a negative effect on stability resulting in a less stable emulsion. On the other hand, addition of cationic emulsifier helped to formulate a more stable emulsion. Apart from that, the optimum concentration of the emulsifier was at 2.0 vol%. The rheological data confirms the stability of the new emulsified formulation at higher temperatures. This paper summarizes the findings of using waste oil emulsion and recommends it for field applications. The newly proposed emulsified acid introduces waste oil as a replacement of diesel.

The novelty of using waste oil will help to significantly reduce the expenses of formulating emulsified acid. Since waste oil is basically waste it has a good impact on the environment, that being said, instead of throwing it, the industry will be able to recycle it.

444364 Molecular Simulation of Controlling Ions and Their Role in Low Salinity IOR for Carbonate Reservoirs

Abstract:

Carbonate geology represents around 50% of hydrocarbon-bearing reservoirs in the world and more than 70% in some areas such as the Middle East. Water flooding has not been very efficient in carbonate reservoirs due mainly to the low water wetness. Modifying the ionic composition and salinity of the injected seawater demonstrated that the oil recovery from these carbonate reservoirs could be improved. Although the underlying mechanisms are still a subject of debate, the improved oil recovery has been mostly attributed to the increased water-wetness of the rock through the interactions of specific ions like Ca^{2+}, Mg^{2+}, SO_4^{2-}, and Cl^- , which are present in seawater with the rock surface under favorable conditions

The study of hydrated gas-phase metal ions provides a link between intrinsic chemistry of the isolated ion and its chemistry in solution. The structures and binding enthalpies of a variety of gas-phase and solution mono- and divalent ion hydrates containing up to 12 water molecules have been studied computationally. A DFT and molecular dynamics simulations were employed to study hydrated structures of the form $[\mathbf{M}(H_2O)_n]^{2+}.mH_2O$ for calcium and magnesium and $[\mathbf{M}(H_2O)_n]^{+}.mH_2O$ for sodium, in which there are *n* water molecules in the first coordination shell and *m* water molecules in the second coordination shell. The hydration, dehydration, and interactions of metal ions with Cl^{-1} and SO^{-2}_4 have been studied to better understand the mechanism of low salinity IOR. Density functional (DFT) theory using the B3LYP functional, and 6-311++g** basis set has been used. The charge of the metal ion decreases monotonically as the number of water molecules increases in the complex. Extensive calculations of the energetic contribution of individual structural characteristics as well as the energetic trends in binding energy are developed, which will lead to better understanding of how low salinity water IOR works.

Keywords: Hydration, carbonate reservoir, DFT, binding energy, Low Salinity

*** Formatting Issue - Best Available Paper/Graphic ***

444278 Insight into the Mechanism of Toxic Chromium (VI) from Wastewater Using Magnetite Coated PINE Powder

Abstract:

INSIGHT INTO THE MECHANISM OF TOXIC CHROMIUM (VI) FROM WASTEWATER USING MAGNETITE COATED PINE POWDER

Pholosi A., Ofomaja A., Naidoo E.B.

Biosorption and Wastewater Treatment Research Laboratory, Department of Chemistry,

Faculty of Applied and Computer Sciences, Vaal University of Technology, P. Bag X021, Vanderbiljpark, 1900 South Africa.

INTRODUCTION

A knowledge of the interaction between pollutant molecules or ions and active sites on an adsorbent surface is an important phenomenon in many industrial processes such as catalysis, adsorption and other separation techniques. The use of lignocellulosic materials for Cr(VI) adsorption has gained interest over the years **(Miretzky and Cirelli, 2010)**. The adsorption properties of lignocellulosic materials has been attributed to acidic functional groups on their surfaces which can take part in ion-exchange, complexation, chelation and hydrogen bonding with pollutant species in solution **(Vithanage et al., 2015)**. Two mechanisms for lignocellulosic adsorption of Cr(VI) have been proposed; (i) an electrostatic mechanism in which at low pH, the protonated acidic functional groups on the biomaterial carrying a positive charge attracts the negatively charged Cr(VI) species **(Demirbas, 2005)** and (ii) the adsorption-coupled reduction mechanism in which Cr(VI) is reduced to Cr(III) either in solution (Direct reduction) or after being adsorbed (Indirect reduction) by electron-donor groups of the biomaterial that have lower reduction potential values than that of Cr(VI) **(Gao et al., 2008)**. The Cr(III) ions formed in solution may then form complexes with Cr(III)-binding groups on the biomaterial surface or released into bulk solution by repulsion between Cr(III) and other positively charged groups on the biomaterial **(Miretzky and Cirelli, 2010)**.

METHODS

Sodium hydroxide treated pine coated with magnetite (NTP-NC) was prepared by co-precipitation method. The effect of solution pH was performed in a batch system in which 0.5 g of the adsorbent was contacted in six 250 cm^3 beakers containing 75 cm^3 of 75 mg/dm^3 Cr(VI) solutions set at pH 1, 2, 4, 6, 8 and 10 at 26 °C and agitated at 200 rpm for 2 h and the clear solution analysed for Cr(V/I) left using UV-Vis spectrophotometry at 460 nm

RESULTS

The results of the effect of initial solution pH on the adsorption of Cr(VI) onto NTP-NC is shown in Fig. 1. To confirm the operating mechanism, the change in $[H^+]$ concentration at each initial pH, the oxidation/reduction potential before and after adsorption and the amounts of Total Cr, Cr(VI)and Cr(III) present in solution at each pH was monitored for NTP-NC and the results displayed in Fig1 . The change in $[H^+]$ concentration ($\Delta[H^+] = [H^+]_{final} - [H^+]_{initial}$) at the end of the adsorption process for each initial solution pH was recorded and plotted against the initial solution pH. The results showed that the change in $[H^+]$ concentrations at the end of the adsorption process where higher for lower initial solution pH's (where lower amounts of Cr was found in solution) than for higher initial solution pH's (where higher amounts of Cr was found in solution) and the change in $[H^+]$ concentration became small and

*** Formatting Issue - Best Available Paper/Graphic ***

almost constant after initial solution pH 4. This is consistent with the assumption that the removal of Cr from solution is associated with $[H^+]$ ion consumption and that as the concentration of $[H^+]$ ions reduces in solution, the Cr removal reduces. Cr(VI) ion in acidic solution has a positive redox potential value (above 1.3 V at standard conditions) and in the presence of an electron donor becomes unstable and can be reduced to Cr(III) **(Park et al., 2008)**. The results show that the initial ORP values (before adsorption) for the Cr solutions at pH's of 1, 2, 4, 6, 8, and 10 were 339, 280, 165, 55, -53 and -172 mV for NTP-NC respectively. The positive values of ORP at lower solution pH's indicates higher possibility for reducing Cr(VI), while the lower and negative ORP values indicates lower possibility for reducing Cr(VI). On addition of the adsorbents into the solutions, a change in ORP were observed . The values of ORP at the end of the adsorption process became 330, 268, 30, 9 -29 and -61 for NTP-NC respectively. A reduction in ORP at the end of the adsorption stage signifies that reduction has occurred in the system. **Yu et al. (2014)** showed the amount of Cr(VI) removed from aqueous solution in contact with zero-valent iron (nZVI) could be related to the change in ORP and pH of the solution. The authors observed a decrease in ORP from 400 to 100 mV and a corresponding 10 unit pH increase as Cr(VI) was removed from solution.

Finally, analysis of the prepared Cr solutions at the various solution pH's showed that the concentration of Total Cr obtained from ICP and that of Cr(VI) using UV-vis measurements were very similar, suggesting that Total Cr was almost equal to Cr(VI) in solution. After contacting the various solutions with the adsorbents, the concentration of Total Cr in the pH 1 solution was observed to have reduced to 21.56 mg/dm^3 for NTP-NC, while the Cr(VI) concentrations was 10.66 mg/dm^3 for NTP-NC. With solution pH 2, the concentration Cr(III) left in solution was lower than that of solution pH 1 for both samples. The reason is that at pH 1 there is an abundance of H^+ ions in solution which accumulates on the functional groups on the adsorbent making the surfaces of the adsorbent positively charged and repelling away the positively charged Cr(III) ions from the adsorbent surfaces **(Suksabye et al., 2009)**. At solution pH 2, it is believed that there is less accumulation of H^+ ions at the adsorbent surfaces which led to reduced electrostatic repulsion and increased adsorption of Cr(III) from solution. This increase in Cr(III) adsorption is also confirmed by the reduction in Total Cr in solution between solution pH 1 and 2. The concentrations of Cr(VI) left in solution at pH 2 was higher than at pH 1 for NTP while for NTP-NC the Cr(VI) concentrations were lower at pH 2. For solution pH's higher than 2 and up to 10, the residual Cr(III) in solution becomes very low and almost constant. This is consistent with the small change in $[H^+]$ in solution which is almost constant at this pH range.

FTIR Evidence for the adsorption of Cr(III)

When Cr(III) is formed in solution via adsorption-coupled reduction mechanism it may undergo hydrolysis and/or complexation depending on the solution pH and the total Cr(III) concentration. Hydrolysis may lead to the formation of various Cr(III) species with an increase in solution pH which gives rise to a reduction in its solubility and increasing the chances of precipitation of the Cr(III) species. At solution pH 1 – 4, Cr^{3+} and $CrOH^{2+}$ are the major species which can be adsorbed by negatively charged ligands on the biomaterial surface. On the other hand, at solution pH range of 4 – 10, Cr(III) species (hydroxo complex), such as $CrOH^{2+}$, and $Cr(OH)_3$, are the dominant forms in the aqueous system and at high pH is readily transformed into soluble complex, . When NTP-NC was contacted with Cr(VI) solution for 2 hr, the final solution pH was increased to pH 2.53. The FTIR spectra for NTP-NC (Fig. 2) shows the presence of characteristic Fe-O peaks of bulk Fe_3O_4 at 553.26 cm^{-1}, 564.36 cm^{-1}, 570.21 cm^{-1} and 576 21 cm^{-1}. Functional groups associated with the pine includes the broad band at 3250.99 cm^{-1} representing –OH of cellulose, 2905.17 cm^{-1} representing C-H stretching, while the peaks at 1627.69 and 1602.87 cm^{-1} shows the COO^- and C=O bonds respectively. After 2 hr contact with Cr(VI) solution, the peak at 564.34 cm^{-1} was observed to have split into several new peaks (Fig. 3b (ii)). These emerging peaks are characteristic of Fe-O bonds of maghemite (γ-Fe_3O_4), for example the characteristics peaks of maghemite was observed at 595.21, 685.65 cm^{-1} and 582.70, 634.91 cm^{-1} suggesting a change in the iron oxide phase (Fe_3O_4 to γ-Fe_2O_3) due to the conversion of Cr(VI) to Cr(III) by magnetite nanoparticles. According to **Peterson et al. (1997)**, Cr(III)(hydr)oxide produced is precipitated on the magnetite surface. In this study, α-Cr_2O_3 peak were observed at 604.63 cm^{-1} indicating that Cr(III) was formed during the adsorption process. Finally it was also observed that the carboxylate group at 1635.55 cm^{-1} was shifted to 1637.73 cm^{-1} while a new peak was observed at 1737.72 cm^{-1} suggesting that complexation between Cr(III) and Carboxylate ions also occurred.

REFERENCE

Demirbas, A. Energy Source 27 (2005) 1449–1455.

Miretzky, P., Cerelli, A.F. A review. J. Hard. Mater. 180 (2010) 1-9.

Vithanage, M., Rajapaksha, A.U., Ahmad, M., Uchimiya, M., Dou, X., Alessi, D.S., Ok, Y.S. J. Environ. Manage. 151 (2015) 443-449.

D. Park, C.K. Ahn, Y.M. Kim, Y.-S. Yun, J.M. Park, J. Hazard. Mater. 160 (2008) 422–427.

P. Suksabye, A. Nakajima, P. Thiravetyan, Y. Baba, W. Nakbanpote.. J. Hazar. Mater. 161 (2009) 1103–1108.

R.F., Yu, F.H., Chi, W.P., Cheng, J.C. Chem. Eng. J. 255 (2014) 568–576

M.L. Peterson, A.F. White, G.E. Brown, G.A. Parks Environ. Sci. Technol. 31 (1997) 1573–1576.

Fig. 1 Effect of solution pH on the adsorption of Cr(VI)

***** Formatting Issue - Best Available Paper/Graphic *****

Fig, 2: FTIR of NTP-NC

444280 Copper-Phyllosilicate Core-Shell Nanocatalysts with Balanced Active Sites for Carbon-Oxygen Hydrogenolysis Reactions

Abstract:

Copper-Phyllosilicate Core-shell Nanocatalysts with Balanced Active Sites for Carbon-Oxygen Hydrogenolysis Reactions

Hairong Yue, Weimao Fang, Changmin Liu, Jiankang Ying

College of Chemical Engineering, Sichuan University, Chengdu, China

Email: hryue@scu.edu.cn

Abstract

Hydrogenolysis of carbon-oxygen (C"CO) bonds (e.g., esters, ethers, furfural, and CO_2) has emerged as a versatile synthetic tool in organic methodology, as it could produce a variety of products (e.g., chemicals, fuels, and polymers). Copper-based catalysts have been intensively explored for hydrogenation reactions as the copper sites account for the selective hydrogenation of carbon-oxygen bonds and relatively inactive for the hydrogenolysis of carbon-carbon bonds. Work on understanding the active sites of copper catalysts for hydrogenation reactions indicted that both Cu^0 and Cu^+ species were crucial to the activity of Cu-based catalysts. However, researchers were not able to establish a relationship between activity and Cu^0/Cu^+ active species since the conventional Cu-based catalysts is the deactivation by metal particle growth and unstable surface Cu^0 and Cu^+ active species in the strong reductive H_2 and oxidizing carbon-oxygen atmosphere.

We renctly present on efficient approaches for fabrication of a series of copper-phyllosilicate core-shell nanocatalysts and nanoreactors with balanced and stable Cu^0 and Cu^+ active species. We chose the hydrogenation of dimethyl oxalate, ethylene cyclicarbonate and as probe reactions for a better understanding of properties of active sites and the structure-activity relationship. The results indicated a synergy between copper and the oxide components and the balanced surface Cu^0 and Cu^+ species can greatly improve the catalytic C-O hydrogenolysis performance of the catalysts. These nanocatalysts with balanced and stable Cu^0 and Cu^+ active species, confinement effect, intrinsic high surface of Cu^0 and Cu^+ and unique tunable tubular morphology, has potential applications in the high-temperature hydrogenation reactions.

Keywords: core-shell nanocatalysts, Hydrogenation reaction; copper catalyst

References

1. Yue, H.; Zhao, Y.; Ma, X.; Gong, J.: Ethylene glycol: properties, synthesis, and applications. *Chem Soc Rev* 2012, 41, 4218-4244.

2. Gong, J.; Yue, H.; Zhao, Y.; Zhao, S.; Zhao, L.; Lv, J.; Wang, S.; Ma, X.: Synthesis of Ethanol via Syngas on Cu/SiO2 Catalysts with Balanced Cu^0-Cu^+ Sites. *J Am Chem Soc* 2012, 134, 13922-13925.

3. Zhao, S.; Yue, H.; Zhao, Y.; Wang, B.; Geng, Y.; Lv, J.; Wang, S.; Gong, J.; Ma, X.: Chemoselective synthesis of ethanol via hydrogenation of dimethyl oxalate on Cu/SiO_2: Enhanced stability with boron dopant. *J Catal* 2013, 297, 142-150.

4. Yue, H.; Ma, X.; Gong, J.: An Alternative Synthetic Approach for Efficient Catalytic Conversion of Syngas to Ethanol. *Acc Chem Res* 2014, 47(5):1483-1492

5. Yue, H.; Zhao, Y.; Zhao, L.; Lv, J.; Wang, S.; Gong, J.; Ma, X.: Hydrogenation of dimethyl oxalate to ethylene glycol on a Cu/SiO2/cordierite monolithic catalyst: Enhanced internal mass transfer and stability. *AIChE J* 2012, 58, 2798-2809.

6. Yue, H.; Zhao, Y.; Zhao, S.; Wang, B.; Ma, X.; Gong, J.: A copper-phyllosilicate core-sheath nanoreactor for carbon"Coxygen hydrogenolysis reactions. *Nat Commu*n 2013, 4, 2339.

Stimuli-responsive Rubber in Oil and Gas Applications

Rostyslav Dolog, Sankaran Murugesan, and Valery Khabashesku

Baker Hughes, 14990 Yorktown Plaza Dr, Houston, Texas, 77040, USA

ABSTRACT

A problem that exists today in the oil & gas industry is the loss of mechanical properties by reactive rubber seals due to swelling, and the limited space in downhole well applications to accommodate larger swellable packing elements.

This paper presents cost-effective, reinforcing, stimuli-responsive fillers, such as certain classes of fly ash, for use in rubber composite materials in various oil and gas applications. Apart from cost effectiveness, these fillers can also enable stimuli-responsive effects, as well as improved mechanical properties (e.g. modulus) for rubber composites when swelled in water. Fly ash (Class C) is a common component of Portland cement which was used in the literature to achieve a stimuli-responsive effect when added to Hydrogenated Nitrile Butadiene Rubber (HNBR) as filler. Exposure to water brines can cause a filler to promote an increase in pH which enables the hydrolysis of HNBR nitrile groups to carboxyl groups, thus creating an ionomer. This process is followed by neutralization of the new ionomer with metal ions, and formation of ionic crosslinks, further improving mechanical properties of the rubber.

This paper demonstrates that incorporation of fly ash as a reinforcing filler allows development of swellable seals that can retain, or even increase, their modulus and stiffness upon swelling. These characteristics allow avoiding the loss of mechanical properties by seals due to swelling and also eliminating the need for large swellable packing elements that occupy additional space downhole.

KEYWORDS: Elastomers, stimuli-responsive, fly ash, composites

INTRODUCTION

There is a growing demand for high performance elastomers in the oil & gas industry. Rubber compounds are used for various sealing applications, packer elements, stators for pumps, etc. In general, rubber formulations are comprised of elastomer matrix, curing system, antioxidants, plasticizers, along with reinforcing filler, which provides the compounds with their strength. The reinforcing filler is usually represented by carbon black and, to a lesser extent, silica. Along with other factors, filler mechanical properties and concentration generally define hardness, modulus and tensile strength of the elastomer compound. Throughout the years, many different materials have been tried as alternative fillers for rubber composites, including nanomaterials (CNTs, graphene, nanofibers), recycled rubber particles, and ash/fly ash. In the case of ordinary carbon black, the filler serves not only for reinforcement, but also as a cost reducer, because raw elastomer is often a more expensive component. Mechanical properties of the composites are affected by the aspect ratio of the filler, quality of the dispersion, and interface between blend components. The latter determines the effectiveness of load transfer from polymer matrix to the reinforcing filler. High performance compounds can be designed by incorporating well dispersed fine particles with a high aspect ratio and good compatibility with the polymer matrix.

In this context, the fly ash is a cheaper, low reinforcing filler alternative to carbon black, which will help reduce cost of the compounds for "non-demanding" sealing applications requiring large amounts of raw materials. Fly ash is produced in large quantities as a byproduct of coal burned at power plants, and can be considered as a recycled material. Fly ash costs several cents per pound, while carbon black, depending on the grade, is usually one order of magnitude more expensive. Hemmings et al[1] demonstrated that fly ash can be used as filler for polymeric composites with enhanced mechanical properties. Adoption of this technology in the oil and gas industry could change the design approach for reactive rubber compounds. Further functionalization of fly ash through a cost-efficient process could ensure compatibility with the rubber matrix, while promoting filler dispersion and ensuring an efficient load transfer. Such functionalization's include, but are not limited to, surface modification with fatty acids, dicarboxylic acids or phosphoric acid. The reasoning for surface modification is to attach functional groups with aliphatic chains to fly ash particles to make them compatible with aliphatic rubber chains, or to attach acidic groups on their surface that can interact with nitrile groups of HNBR. Initial attempts towards chemical functionalization of fly ash have been made by Seal et al for treatment of offshore oil spills.[2]

Apart from cost effectiveness, composites with fly ash can show stimuli-responsive properties. It has been shown that adding Portland cement to the HNBR matrix improves its mechanical properties in the swollen state by water uptake.[3] Exposure to water causes Portland cement to grow in volume and mass, and promotes an increase in pH which enables the hydrolysis of HNBR nitrile groups to carboxyl groups, thus creating an ionomer. This process is followed by neutralization of the new ionomer with metal ions and formation of ionic crosslinks, further improving mechanical properties of the rubber. Fly ash (Class C) is a common component of Portland cement, which contains a large variety of metal oxides, and is capable of increasing the

---

[1] Filler Comprising Fly Ash For Use In Polymer Composites. Hemmings et.al., US 6,916,863 B2, July 2005.
[2] Hydrophobic High Surface Area Zeolites Derived from Fly Ash for Oil Spill Remediation, Seal et.al., Environ. Sci. Technol., 2013, 47 (11), pp 5843–5850
[3] Stimuli-Responsive Cement-Reinforced Rubber, Musso et. al., ACS Appl Mater Interfaces. 2014 May 14;6(9):6962-8.

pH of the water solution by hydrolysis, which has a potential to achieve a stimuli-responsive effect when added to HNBR as filler.

Degradation of tensile properties is a common problem for swellable packers and other sealing elements. By selectively using certain classes of fly ash as effective reinforcing filler, one can develop swellable seals that can retain or even increase their modulus and stiffness upon swelling. Therefore, there is a potential to minimize or even avoid the loss of mechanical properties by water swellable reactive packing elements due to swelling. This provides an opportunity to decrease the size of these elements, which are often represented by rather long sections, and hence occupy a significant amount of space downhole. In this paper we have studied the reinforcing effects of fly ash fillers with HNBR compounds as well as stimuli responsive properties of these compounds.

EXPERIMENTAL SECTION

Rubber compounds T1, T2, T3, and T4 are elastomers based on a custom-built peroxide curing system. Formulation F2 was prepared by adding 15 wt. % of fly ash class C to compound F1, which is a commercial swellable NBR sulfur-cured compound.

T series compounds were prepared using a C.W. Brabender® internal mixer operated at 60 RPM for 20 minutes at ~160°F and then molded into sheets and cured at 340°F for 30 minutes in a hot press. Compound F2 was prepared by mixing fly ash with compound F1 in the Brabender mixer for 6 minutes at 60 RPM at ~100°F. Compound F1 is a commercial water swellable formulation that was used as received. F series compounds were cured at 300°F for 60 minutes in a hot press.

Tensile properties of rubber samples were tested using dog bone specimens, according to the ISO37 standard, at room temperature and at 350°F using an MTS® testing system.

Swelling tests in 3% KCl brine and LVT 200 oil were performed to assess the effect of fillers on the chemical resistance of elastomers. Standard one-inch rubber buttons were tested at 200°F at ambient pressure for 6 days. Thermal properties were studied using TGA Q50 and DSC Q2000 instruments.

To analyze the stimuli-responsive effect introduced by fly ash fillers, rubber buttons were aged for 10 days in 3% KCl in the oven at 200°F and then removed from the solution, dried for 30 minutes in the oven and then for 3 hours at room temperature. Samples were later tested in the MTS testing system in compression mode.

Viscoelastic properties of stimuli-responsive samples, based on water swellable compounds, were accessed using the RSA G2 dynamic mechanical analysis (DMA) instrument. This experiment was conducted in a temperature ramp mode of 2°C/minute at 1Hz frequency and 0.2% amplitude strain. A Nicolet 6700 Fourier Transform Infrared (FTIR) spectrometer was used to confirm neutralization of carboxylic groups by metal cations.

RESULTS AND DISCUSSION

The reinforcing properties of fly ash were evaluated at room temperature (Table 1) and at 350°F (Table 2) and compared with N550 carbon black. Sample T3, reinforced with fly ash, demonstrated inferior tensile properties and hardness compared to a compound filled with carbon black. However, at room temperature, fly ash was capable of delivering a reinforcing effect to the rubber compound without sacrificing its elasticity, e.g. the ultimate elongation of elastomer compound remained virtually unchanged after incorporation of 50 parts per hundred rubber (PHR) of fly ash, while its modulus and tensile strength have increased. Therefore, potentially, one can substitute carbon black with high loadings of fly ash in certain applications, thus further decreasing the cost of the rubber product. The summary of mechanical properties is provided in Table 1, E_x denotes modulus of a compound at certain strain (x), i.e. stress at strain.

Sample	N550 CB (PHR)	Fly ash Class F (PHR)	Tensile strength (psi)	E_{25} (psi)	E_{50} (psi)	E_{100} (psi)	Elongation (%)	Hardness Shore A
T1	0	0	829	71	97	121	472	48
T2	50	0	3402	165	273	744	290	68
T3	0	50	1176	94	127	166	467	54

Table 1. Room temperature tensile properties and hardness of elastomer samples

The relatively low increase in modulus and hardness of T3 compared to T1 at room temperature can be explained by a low aspect ratio of fly ash and weak interactions between blend components. The latter problem can be solved by functionalization of fly ash particles.

The general data trends from the high temperature tensile tests closely resemble the results of the tensile test conducted at room temperature. However, at high temperature, rubber specimens break at much lower elongation, which leads to a great decrease in the measured tensile strength. This behavior is typical for elastomers and limits their application to a certain maximum service temperature and is also due to the fact that high temperature tests are conducted at conditions that push the limits of the material. The modulus values at 15% (E_{15}) and 20% (E_{20}) strain are not provided herein, because half of the tested specimens were broken before they reached such strains.

The superiority of carbon black as an efficient filler became even more pronounced at high temperature. However, incorporation of fly ash doubled the tensile strength and increased the elongation at break by one-third when compared to the unfilled elastomer.

Sample	E_5 (psi)	E_{10} (psi)	E_{15} (psi)	E_{20} (psi)	Tensile strength (psi)	Elongation (%)
T1	31	46			91	23
T2	61	130	194	261	832	39
T3	42	73	98	119	173	30

Table 2. Summary of high temperature (350˚F) tensile data for elastomer samples

Thermal properties of samples T1-T3 were assessed with Differential Scanning Calorimetry (DSC) and Thermogravimetric analysis (TGA). TGA allows evaluating thermal stability of samples and to verify the actual composition of a sample from the ash content. The results of the DSC and TGA analysis are summarized in Table 3. During TGA, samples were heated to 600°C at 10°C/minute rate in a nitrogen (N_2) environment, and subsequently heated at 10°C/min to 600 °C in air to burn off carbon-based fillers. This allows decomposing of organic matter first by applying heat, and then burn the remaining carbon-based material in air atmosphere.

The initial degradation of the polymer started at 200°C and progressed to complete decomposition of aliphatic chains in the temperature range of 420-500°C with a highest rate of decomposition at ~460°C. Sample T2 incurred the second rapid mass loss after 600°C due to burning of the carbon black, which comprised most of the ash content at 600°C in a N_2 environment. The DSC experiment demonstrated the glass transition temperature of the elastomer remained largely independent of the type of filler - the respective DSC heating scan was conducted with a 10°C/minute temperature ramp after precooling the sample to -50°C.

Sample	W_{600} (%)	W_{800} (%)	$T_{deg}1$, °C	$T_{deg}2$, °C	T_g, °C
T1	2.4	2.4	452.8	-	-29
T2	29.8	2.7	458.7	616.5	-29
T3	32.8	32.7	447.3	-	-28

Table 3. TGA/ DSC results for elastomer samples

Table 3 contains ash contents (W) at 600°C and 800°C, degradation temperatures (T_{deg}) which represent the temperature of the highest rate of decomposition during TGA experiment, and glass transition temperature (T_g) of the samples, as determined by DSC scan.

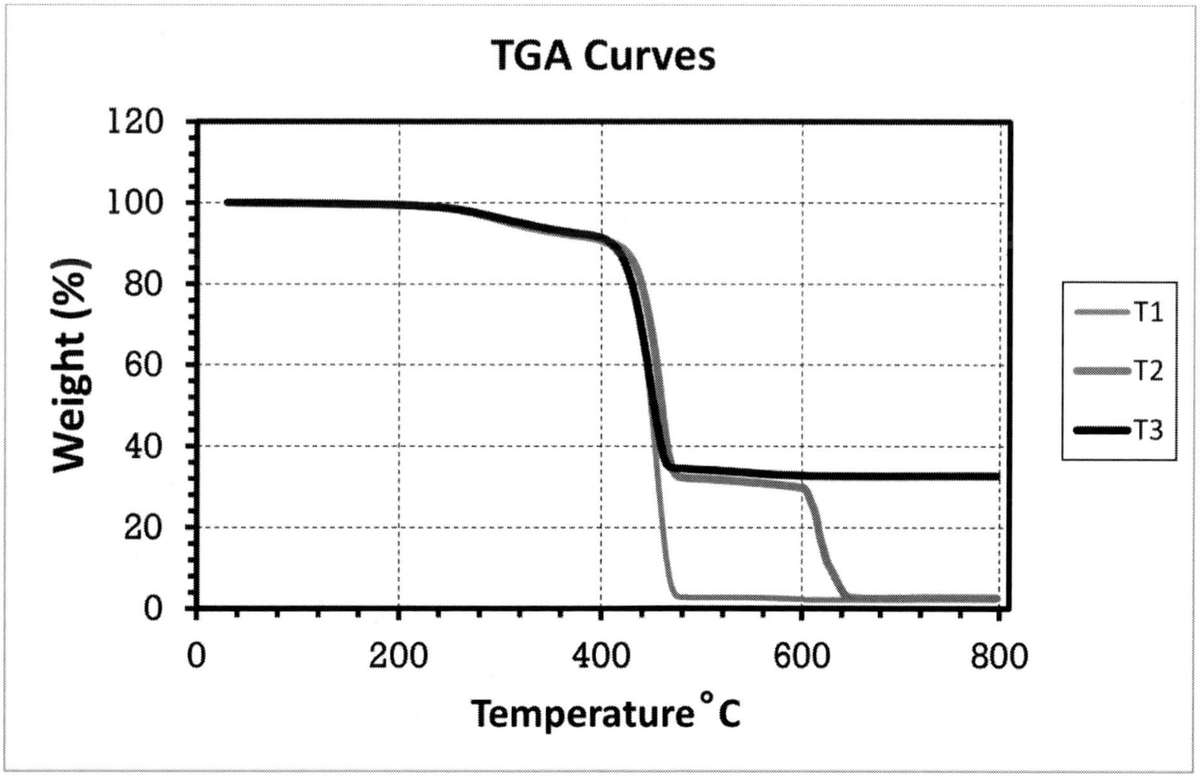

Figure 1. TGA Curves for samples T1, T2, and T3

Testing fluid / Sample	3% KCl brine	LVT 200	Specific gravity
T1	0.6	13.0	0.96
T2	0.3	11.6	1.13
T3	0.8	9.1	1.18

Table 4. Volume increase (%) of elastomer samples due to swelling for 6 days at 200°F

Musso et. al. developed a stimuli-responsive HNBR composite reinforced with Portland cement that was increasing its stiffness and modulus when swelled in water[3]. Fly ash (Class C) is a common component of Portland cement, which contains the components necessary to achieve a similar effect. Therefore, one can avoid the loss of mechanical properties by water swellable packing element seals upon activation. To test this hypothesis, we studied the effect of aging in water brines on the properties of elastomers with fly ash fillers.

Rubber samples were molded and cured into standard size rubber buttons (cylinders of ~0.25" height and ~1" diameter). These buttons were aged for 10 days in 3% KCl brine in the oven at 200°F, removed from the solution, dried for 30 minutes in the oven and for 3 hours at room temperature. The swelling of both compounds was close to ~1% by volume (Table 4). Samples were later tested in an MTS machine in compression mode. Each button was compressed to 70% strain, buttons remained intact during the experiment, and visually all appeared to have a certain level of compression set at the end of the experiment. The results of the experiments are summarized in Table 5 and Figure 2. Only one button was tested for each sample. The provided compression modulus values (C_n) represent values of stress at a percentage strain (n), e.g. C_5 is stress at 5% compression strain.

Compound T4 was reinforced by fly ash Class C, which contains larger weight fraction of CaO than fly ash Class F (~25% vs ~10-15% respectively) and is expected to produce a higher pH and closely resemble cement that was used to prepare stimuli-responsive rubber in the original paper by Musso et.al.[3] The results demonstrate that exposure to water (3% KCl) at high temperatures can improve the modulus for the compounds filled with fly ash Class C (T4), while the control sample (T1) showed the opposite behavior. After aging, the compressive modulus at 50% strain (e.g. stress at 50% strain) of control sample T1 decreased by 19%, as compared to the unaged sample, while compressive strength of sample T4, reinforced by fly ash, increased by 15%. The difference in modulus values is even more drastic at 70% compression strain (-35% and 31%, respectively). This is in agreement with the hypothesis stated earlier.

Sample	Filler	Compressive modulus (psi)				
		C_5	C_{10}	C_{25}	C_{50}	C_{70}
T1	None	19	42	127	468	1810
T1 swollen	None	16	36	109	380	1185
T4	Fly Ash class C (50 PHR)	21	46	134	434	1628
T4 swollen	Fly Ash class C (50 PHR)	29	57	153	501	2125

Table 5. Summary of compressive modulus properties after 3% KCl soaking and 200°F drying

To further investigate the stimuli-responsive property of fly ash filler, the latter was incorporated into a water swellable compound, which provided for significantly more interaction of filler with water to occur. Sample F2 comprises unmodified commercial water swellable NBR formulation F1 mixed with 15 wt.% of fly ash Cass C. Sample F2 was processed into a 0.04" slab, then a rectangular sample was cut from the slab and swollen in 3% KCl solution until 100% volume increase. The F2 sample was then dried overnight in the oven, which lead to de-swelling of the sample due to solvent evaporation. Mechanical properties of the F2 sample were assessed with DMA prior to the swelling test, in a swollen state, and after drying - the data summary is shown on Figure 3.

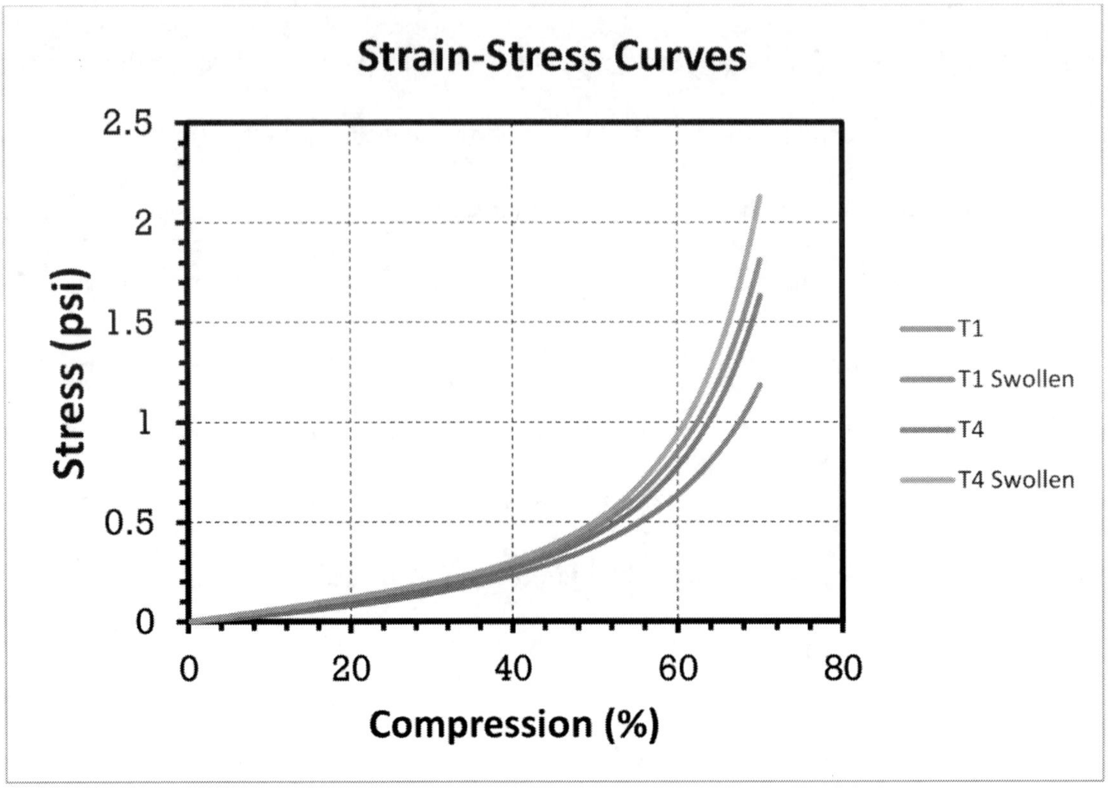

Figure 2. Stress-strain curves

The storage modulus of sample F2 after swelling and subsequent drying was about an order of magnitude higher than that of original sample F2 measured prior to swelling experiment. The storage modulus of the dried F2 sample was also less sensitive to the increasing temperature, contrary to the weakening original sample F2. The increase in the storage modulus of swollen F2 with temperature is simply due to the drying of the sample. The obtained result demonstrated that fly ash can essentially act as a stimuli-responsive material.

The significant improvement in the modulus is believed to be due to the neutralization of the swellable component of the commercial F1, and potential hydrolysis with subsequent neutralization of NBR elastomer. This is confirmed by the shift in T_g to a lower temperature range after drying as illustrated in Figure 3.

Carboxylic groups were detected in the original sample F2 prior to swelling, hence, they cannot be attributed solely to hydrolyzed HNBR, but rather to the super-absorbent polymer present in water swellable rubber compound (see Figure 4). FTIR confirms neutralization of carboxylic groups because a significant increase of the peak near ~1540 cm^{-1} and a lowering of peak intensity at ~1720 cm^{-1} have been observed. This neutralization is believed to promote the formation of ionic clusters in the compound with fly ash considered to be the source of cations. The neutralization of carboxylic groups could as well be attributed to K^+ cations present in the KCl brine solution, however, in that case, the respective peak in FTIR spectra would appear at higher wave numbers, approaching ~1600 cm^{-1}.

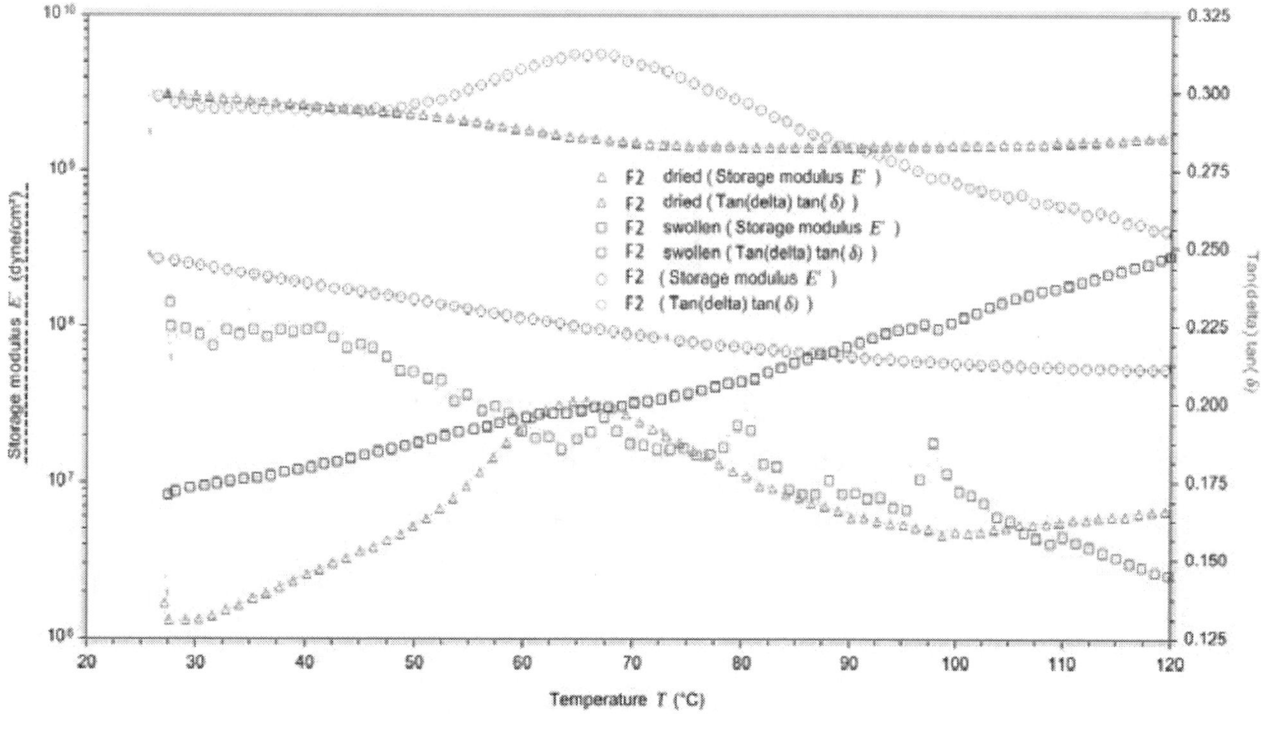

Figure 3. DMA data on F2 sample

It is important to mention that sample F2 exhibited a certain degree of brittleness after drying, which is undesired in rubbers, but this can be managed by incorporation of common ionomer plasticizers.

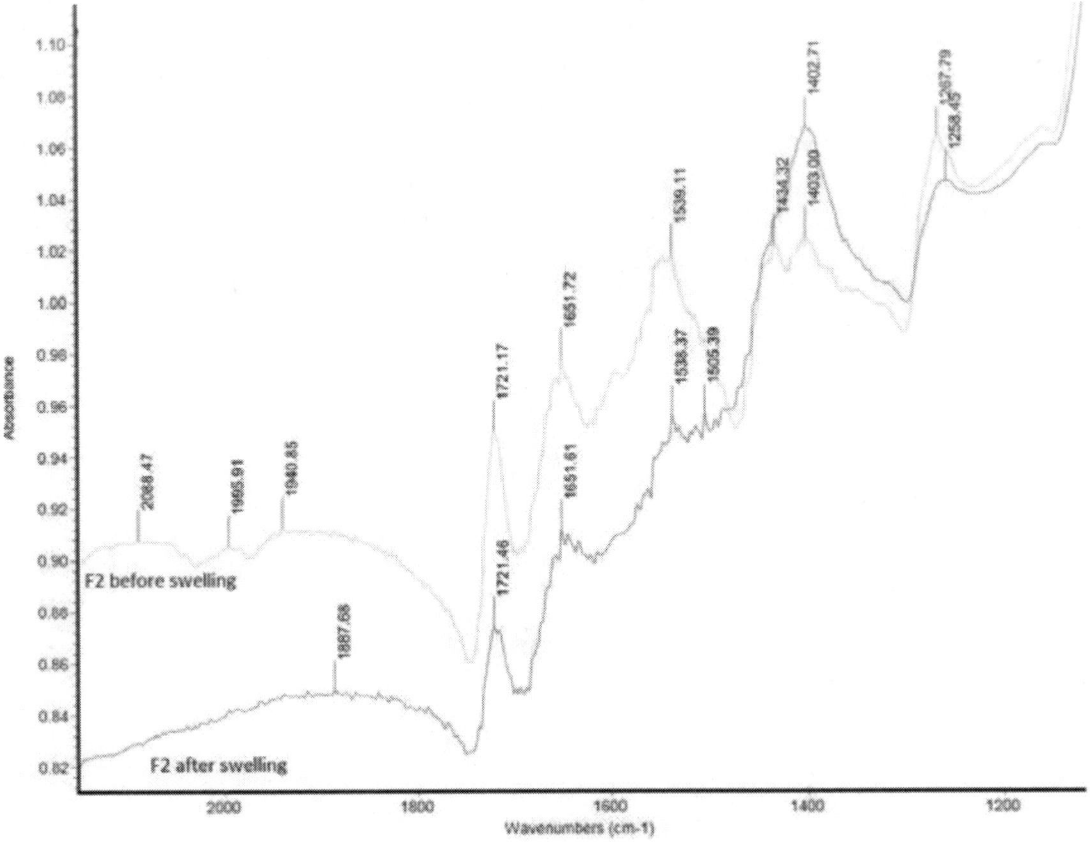

Figure 4. FTIR spectra of F2 sample taken before and after swelling

Conclusions

Fly ash demonstrated a great potential for consideration as cost effective, low reinforcing filler for downhole oil and gas applications. Although the reinforcing properties of fly ash are inferior to carbon black, incorporation of fly ash can improve the mechanical properties of HNBR elastomer compared to the unfilled raw crosslinked elastomer at both room temperature and at 325°F.

Furthermore, composites of HNBR with fly ash Class C demonstrated a stimuli-responsive effect by significantly increasing their modulus upon swelling in water brine. This effect was even more pronounced in composites of fly ash with swellable water compounds, which provided an easy access of filler to water upon swelling. A ten-fold increase in modulus was achieved in such fly ash reinforced sample, after swelling in 3% KCl water brine and drying in the oven overnight. The FTIR confirmed neutralization of carboxylic groups present in the investigated system, which contributed to overall improvement of mechanical properties of the compound upon swelling.

444310 Formal Modeling of Accidents and Automatic Generation of Tutoring Scenarios for Intelligent Safety Training of Process Operators

Abstract:

With the emphasis of accident-free operations and prevention of catastrophic events, safety-training demands for the chemical plants have been on the increase globally. However, the research results on modeling of accidents and involving human operators and the advanced theory and technology for (safety) operator training have been shared limitedly among domain experts and commercial developers. Even with the recent trend of simultaneously delivering the operator training system with the construction of a new process plant, in commercial operator training systems (OTS), safety training used to implement quite limited and obvious scenarios (and the burden was almost up to the trainer), partially due to the difficulty in modeling propagations of probable accidents. By the regulatory requirements or proactive investment by companies, effective scenarios for training of safety and abnormal situation management are necessary to be improved in the aspects of the size of volume as well as the training depth. Thus, considering the emergence of enhanced virtual reality technologies and inexpensive gadgets like HMD, supporting the immersive training environment, the future efficiency of the training heavily relies on the quantity and quality of safety scenarios, which would be essential in guiding the entire training process.

In this research we first overview a government-sponsored research project developing an enhanced safety training system for chemical process operators, using haptic devices in addition to enhanced audio and visual equipment. The initial version of the system runs with a fixed set of training scenarios developed for predetermined accident scenarios. However, with a limited set of training scenarios, the trainees simply memorize and rote learn the steps, required to be mastered to proactively solve the problems with an advent of possible accidents. In addition to that, without formal modeling, the system cannot give timely feedbacks and the performance of the trainee cannot be measured in real-time. Then to enhance the contents and usefulness of the training, the knowledge of process operations is collected and systematically stored as a form of knowledge base, in addition to the accident scenarios obtained by executing Fault Tree Analysis and HAZOP study.

By using a formal modeling language like UML and Petri nets, the accident scenarios and operators' problem-solving tasks are modeled hierarchically and classified into distinguishable subtasks. Then a new training scenario is generated on-line through automatic synthesis based on planning using the knowledge base. Safety training scenarios are generated intelligently for the tutoring of a single operator's job function as well as effective collaboration among a group of operators for real-scale malfunction diagnosis and intervention at the initial propagation stage of an accident. As case study, safety trainings for city gas governing stations are executed based on systematic scenario generation and analyzed for improved effectiveness and field applicability of the system. Reduction of time and cost in the training scenario production and real-time customized training based on new operator's level of knowledge and skill are also discussed in the framework of intelligent tutoring systems.

444376 Ozone Source Apportionments of Local Anthropogenic Emissions

Abstract:

Ozone is one of six principal air pollutants designed by US EPA. NOX and VOC emissions from anthropogenic sources are the most important factors to control ozone pollutant.

In this report, photochemical dispersion models are used to estimate the effect of ozone from local anthropogenic emission sources. EKMA-type diagrams are generated for NOX- and VOC- effects. Furthermore, APCA (Anthropogenic Precursor Culpability Assessment) technology is used to estimate ozone source apportionments.

EKMA-type diagrams show that 10-30% emission reductions could reduce ozone concentrations to meet standard. Further APCA results show the average ozone contributions from regions and emission categories. APCA results explain different effects for different monitoring stations.

444397 Flare Minimization and Emission Characterization during Start-Ups of Chilling Train in Ethylene Plants

Abstract:

Flaring is a crucial safety operation to protect chemical plant personnel and equipment during plant start-up. However, it generates huge amounts of off-spec products, and causes negative environmental and social impacts, as well as results in tremendous raw material and energy losses which could be saved to generate much more needed products. Thus, cost-effective start-up flare minimization strategies which is a double-win practice, simultaneously benefiting both the industrial and environmental sustainability, are becoming more and more important and attractive to the industries.

The chilling train is a very important section for the ethylene plants and accounts for more than 50% of the total start-up time and flare emissions of the whole ethylene plants. Within this section, the charge gas from the upper stream process is refrigerated to a very low temperature (lower than -160 °C) to separate hydrogen and methane. Meanwhile, the left charge gas is liquidized for cryogenic separation. The chilling train system consumes over 40% of the refrigerating capacity of an ethylene plant. During the chilling train start-up, different operational conditions with different compositions and flowrate of precooling process will affect both product separation efficiencies, start-up stability time and flare emission.

Traditional start-up scenarios are developed based on experience and improved by trial and error, which is inefficient and dangerous. In this study, a general methodology on flare minimization for chilling train start-up operations via rigorous dynamic simulation is presented. The methodology starts with setup and validation of chilling train steady-state and dynamic simulation models. The initial state of start-up in the real plant is achieved through systematically transforming from the validated dynamic model, and several operation conditions with different compositions and flowrate of precooling stream are examined to identify an optimal initial state of start-up based on the obtained initial state. Thereafter different flare minimization strategies for the chilling train start-up are conducted on the optimal initial state model. Also, emission source characterization under different flare minimization strategies for chilling train start-up is investigated. Deep insights of the start-up strategies, emission source distribution and dynamic emission profiles are provided. This study provides detailed technical support for both the industry and environmental agencies on evaluating and developing cost-effective flare minimization strategies in the future.

Experimental Study on Najas Minor of Producing Bio-Oil

Abstract: Due to the diminishing of petroleum reserves and the environmental degradation caused by consumption of fossil energy , many attentions have been paid to the application of biomass energy which is acceptable as an abundant renewable energy resource. Najas minor grows in dense clusters with highly branched stems of which a small fragment could propagate into a thrive one.In this study, we researched the properties of Najas minor to make biofuels .Firstly,the najas minor after heavy metal ions degradation were fed into pyrolyser to prepare biofuels. Significant differences of the product was observed compared with only Najas minor pyrolysed.

KEY WORDS: Najas Minor; catalytic pyrolysis; bio-oil

1.Introduction

Energy demands and environmental problems caused by fossil energy have been a matter of growing concern.Bio-oil has been considered to be a promising way for producing renewable bio-fuels or chemicals[1-3].In recent years,many countries pay attention to the importance of biomass that is an abundant renewable energy resource.Biomass can be defined as any organic material mainly consisting of carbon, hydrogen,oxygen, nitrogen, and some other components in small proportions[4].Although the biomass has been studied for nearly a century,most studies mainly concentrated in Lemna minor, Lemna gibba, Spirodela polyrhiza and Landoltia punctata.We still know little about the biological characteristics and potential application value of other varieties of biomass[5].Najas minor,a submersed herb,grows in dense clusters with highly branched stems of which a small fragment could propagate into a thrive one.Najas Minor is native to Europe, Asia and North Africa from the Netherlands to Morocco east to Japan and the Philippines,including China,Siberia, Central Asia, Iran, Turkey, Ukraine, Germany, France Italy and a host of other Countries.

Catalysts used for catalytic cracking are basically divided into the following categories: molecular sieve, alkali metal salt, metal oxide, etc.Different catalysts have different catalytic effects.The use of metal nanoparticles has shown to be very important in recent industrial applications[6].The application of precious metal nanometer materials in catalysis,which has caused the extensive concern of researchers in recent 30 years,has become a hot research topic in the field of catalysis.Gold nanoparticles have received a great deal of attentions due to their special optical and electrical performances, catalytic properties and remarkable quantum size effect, surface effect etc[7]. The synthesis of metal nanoparticles can be classified broadly into wet and dry methods[8]. The cost of production is relatively high both materially and environmentally. There is an unequivocal need to develop more cost-effective and environmental benign methods to alternate these existing methods[9].Researchers in this field turn to use organisms to produce inorganic nanostructures extracellularly[10].

Although there has been a large volume of work on the synthesis of gold nanoparticles[11-13], and obtainning bi-oil by abundant biomass, there are relatively fewer attempts to combine these two methods.In this work, gold nanoparticles were obtained by aqueous chloroauric acid solution with the of Najas Minor at room temperature. The direct pyrolysis and catalytic pyrolysis experiments of Najas Minor were carried out to observe the effects of catalysts on pyrolysis process of Najas Minor.

2. Materials and Methods

2.1. Experimental samples

The Najas Minor collected from Tianjin city and the solutions of Au(III) were used in this study.The stock solutions of Au(III) were prepared by dissolving analytical grade chloroauric acid

($AuCl_3 \cdot HCl \cdot 3H_2O$, Kewei Reagent, China) in deionized water.

2.2.Experimental apparatus and procedure

Pyrolyzer with GCMS,shown in Fig. 1,is mainly used for analyzing the composition of polymer.Pyrolyzer with GCMS includes three parts :quartz pyrolysis tube,pyrolyzer and GCMS.

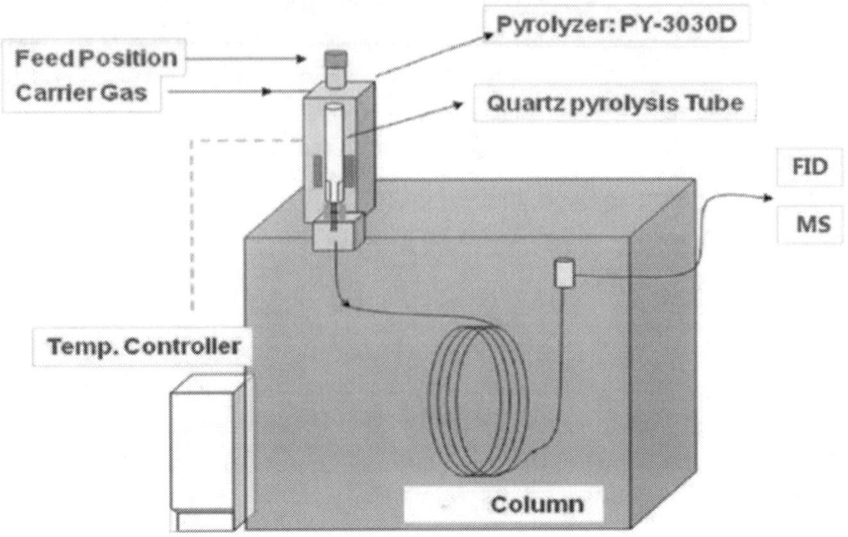

Fig. 1 Diagram of Pyrolyzer with GCMS

The quartz pyrolysis tube (shown in Fig.2) length is 12cm.The sample end diameter is 5 mm,and the pyrolysis products end diameter is 3mm.This special structure is beneficial to the fixation of the reactant in the tube.The top and bottom of the reactants are both supported by quartz wool to prevent reactants blowed by the carrier gas ,or fine particles into the chromatographic column.All pyrolysis experiments were carried out on the EGA/PY-3030D pyrolyzer(Frontier Lab,Japan) .The samples were filled in the special quartz pyrolysis tube to reduce the amount of valuable catalyst and reduce the cost of research work.

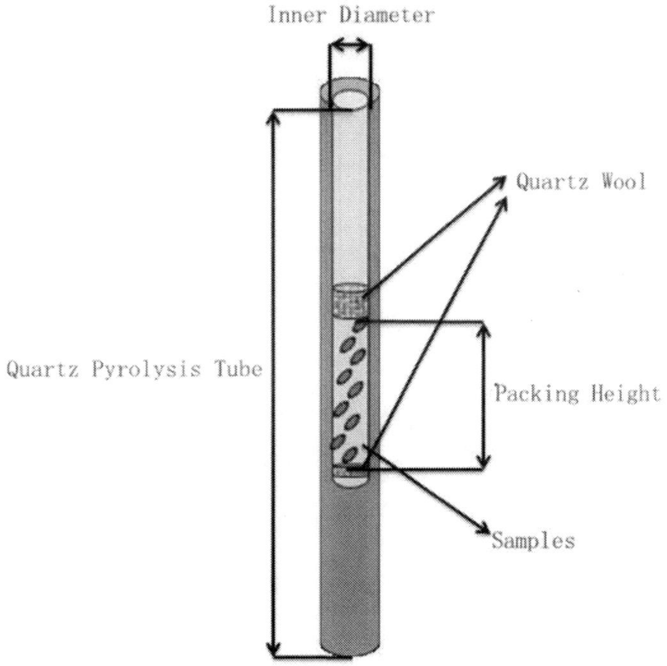

Fig. 2 Diagram of Quartz Pyrolysis Tube

Approximately 0.5g Najas Minors were used in each experiment.The samples were impregnated in 50ml deionized water and 50ml solution of chloroauric acid(0.0793mmol/L) for a night,followed by drying at 80℃ until the weight didn't change.Each experiment used 2.5mg Najas Minors as reactants.Each sample was heated, under the control of heating controller, from ambient temperature to 600 ℃ repeatly and kept at 600 ℃ for 2 h.

3. Results and discussion

3.1.MS Analysis of the Products during the Pyrolysis

Under the same pyrolysis conditions, bio-oil obtained by direct pyrolysis and catalytic pyrolysis were analyzed by GCMS,and the main chemical compounds were shown in Table1 and Table2.

Table 1 Indentification and quantification of main chemical compounds by direct pyrolysis

NO.	peak area%	Formula	Compound
1	12.67	$C_3H_6O_2$	2-Propanone, 1-hydroxy-
2	3.67	C_7H_8	Toluene
3	3.28	C_7H_8O	p-Cresol
4	3.81	$C_5H_{10}O$	Pentanal
5	3.53	C_8H_8O	Benzofuran, 2,3-dihydro-
6	2.17	C_8H_8N	Indole
7	5.93	$C_{20}H_{40}O$	3,7,11,15-Tetramethyl-2-hexadecen-1-ol
8	2.52	$C_{22}H_{42}O_2$	Phytol, acetate
9	5.25	$C_{16}H_{32}O_2$	n-Hexadecanoic acid
10	8.99	$C_{18}H_{30}O_2$	9,12,15-Octadecatrienoic acid, (Z,Z,Z)-

Table 2Indentification and quantification of main chemical compounds by catalytic pyrolysis

NO.	peak area%	Formula	Compound
1	6.01	C_7H_8O	p-Cresol
2	1.75	$C_7H_7NO_2$	3-Hydroxypyridine monoacetate
3	4.58	C_8H_7N	Indole
4	3.05	$C_{10}H9N$	1-Phenyl-1-cyclopropanecarbonitrile
5	3.90	$C_{12}H_{22}O_{11}$	.alpha.-D-Glucopyranose, 4-O-.beta.-D-galactopyranosyl-
6	8.04	$C_{20}H_{40}0$	3,7,11,15-Tetramethyl-2-hexadecen-1-ol
7	2.82	$C_{20}H_{40}$	2-Hexadecene, 3,7,11,15-tetramethyl-, [R-[R*,R*-(E)]]-
8	3.99	$C_{15}H_{28}$	1-Pentadecyne
9	2.12	$C_{15}H_{28}$	3,7,11-Trimethyl-2,4-dodecadiene
10	8.75	$C_{16}H_{32}\ 0_2$	n-Hexadecanoic acid

3.2 discussing

According to the chemical category and common characteristics, the products were divided into hydrocarbons, alcohols, acids, ketones, nitrogen compounds and other compounds.Nitrogen compounds detected by GC/MS analysis ,mainly come from the pyrolysis of amino acids in Najas Minor.Table 3 showed the distribution of pyrolysis products.As shown in the table,the hydrocarbons ,alcohols and nitrogen compounds in biological oil of catalytic pyrolysis were significantly increased compared with direct pyrolysis.The content of hydrocarbons increased from 12.4% to 34.6%.The content of alcohols increased from 9.1% to 16.6%.The gold nanoparticles loaded in surface of Najas Minor, was conducive to the generation of hydrocarbon and alcohol in pyrolysis.By analyzing the GC/MS results, the main hydrocarbons in bio-oil by catalytic pyrolysis were long chain alkanes,and olefins.The main components in gasoline and diesel is hydrocarbons. As a result, whether the biomass can be effective as future power fuel oil, depends largely on the hydrocarbon content of bio-oil after further treatment.

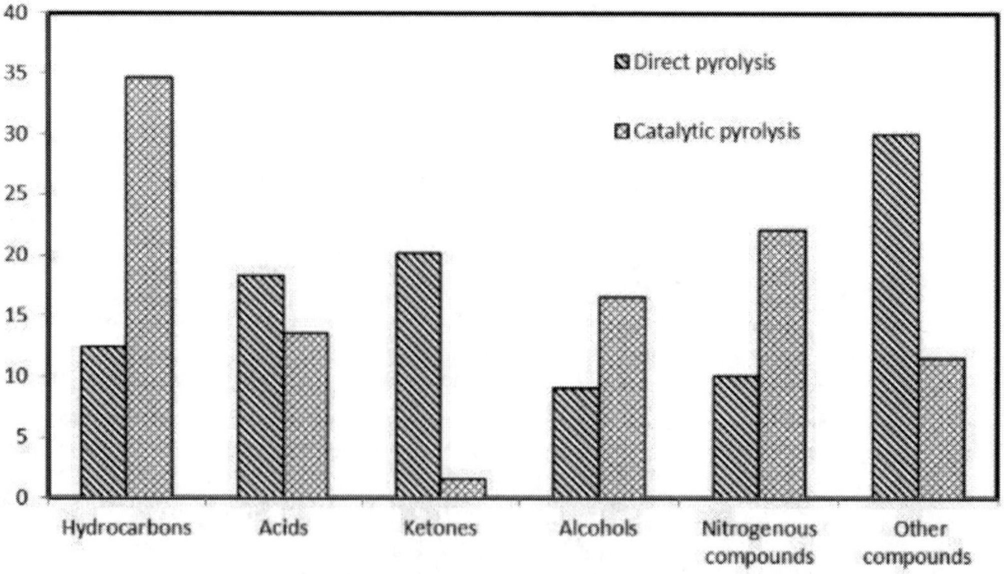

Fig.3 Distribution of pyrolysis products

4.Conclusion

Biomass was explored as a potential biomass energy resource. The experimental study on the direct and catalytic pyrolysis of Najas Minor had been presented, using Pyrolyzer with GCMS,to observe the catalytic effects of gold nanoparticles.In this study,remarkable catalytic effects for pyrolysis were observed.The pyrolysis of Najas Minor with gold nanoparticles,provided more hydrocarbons, alcohols and nitrogen compounds.

参考文献：

[1]Gong F, Yang Z, Hong C, et al. Selective conversion of bio-oil to light olefins: Controlling catalytic cracking for maximum olefins[J]. Bioresource Technology, 2011, 102(19):9247-54.

[2]Zhang Z, Bi P, Jiang P, et al. Production of gasoline fraction from bio-oil under atmospheric conditions by an integrated catalytic transformation process[J]. Energy, 2015, 90:1922-1930.

[3]Hui S, Yuqi D, Jinzhao D, et al. Transesterification of sunflower oil to biodiesel on ZrO2 supported La2O3 catalyst.[J]. Bioresource Technology, 2010, 101(3):953–958.

[4]Yaman S. Pyrolysis of biomass to produce fuels and chemical feedstocks[J]. Cheminform, 2004, 35(31):651-671.

[5]Hejný S. Ellias Landolt The family of Lemnaceae—A monographic study—Vol.1 biosystematic investigations in the family of Duckweeds (Lemnaceae) (vol. 2)[J]. Folia Geobotanica Et Phytotaxonomica, 1993, 28(1):50-50.

[6]Gardea-Torresdey J L, Tiemann K J, Gamez G, et al. Gold Nanoparticles Obtained by Bio-precipitation from Gold(III) Solutions[J]. Journal of Nanoparticle Research, 1999, 1(3):397-404.

[7]Marie-Christine D, Didier A. Gold nanoparticles: assembly, supramolecular chemistry, quantum-size-related properties, and applications toward biology, catalysis, and nanotechnology.[J]. Chemical Reviews, 2004, 35(16):293-346.

[8]Liu B ,, Xie J ,, Lee J Y, et al. Optimization of high-yield biological synthesis of single-crystalline gold nanoplates.[J]. Journal of Physical Chemistry B, 2005, 109(45):15256-63.

[9]Raveendran P, Fu J, Wallen S L. Completely "Green" Synthesis and Stabilization of Metal Nanoparticles[J]. Journal of the American Chemical Society Jacs, 2003, 125(46):págs. 13940-13941.

[10]Absar Ahmad , Satyajyoti Senapati , MIK, et al. Extracellular Biosynthesis of Monodisperse Gold Nanoparticles by a Novel Extremophilic Actinomycete, Thermomonospora sp.[J]. Langmuir, 2003, 19(8):3550-3553.

[11]Selvakannan P, Mandal S, Phadtare S, et al. Water-dispersible tryptophan-protected gold nanoparticles prepared by the spontaneous reduction of aqueous chloroaurate ions by the amino acid[J]. Journal of Colloid & Interface Science, 2004, 269(1):97-102.

[12]Gittins D I, Caruso F. Biological and Physical Applications of Water-Based Metal Nanoparticles Synthesised in Organic Solution[J]. Chemphyschem, 2002, 3(1):110－113.

[13]And A G, Murphy C J. Seed-Mediated Synthesis of Gold Nanorods: Role of the Size and Nature of the Seed[J]. Chemistry of Materials, 2004, 16(19):3633-3640.

444353 A Combined Rheology and Stability Study on Emulsified Acid for High Temperature Applications

Abstract:

HCl emulsion has been a popular formulation to stimulate carbonate reservoirs due to the various advantages that emulsified HCl has over ordinary HCl; including but not limited to deeper penetration and less corrosion damages. Emulsified HCl has even more potential advantages if utilized properly. The reaction kinetics, stability and rheology of emulsified HCl plays a significant role on achieving successful acidizing treatment.

The objective of this paper is to experimentally study the factors affecting the stability of emulsified HCl. The emulsion was prepared using 15wt% HCl and diesel with cationic emulsifier. The acid solution-to-diesel ratio was 70:30. The effect of stirring intensity, concentration of the emulsifier, addition of corrosion inhibitor, temperature, addition rate of acid phase to oil phase and mixing time of the emulsion have been studied. Stirring intensity was varied from 600 to 1500 rpm. Emulsifier dosage ranged between 5 to 20 gpt. Temperature used for heating, to mimic reservoir condition, was as high as 120°C.

The results showed that there is an optimum condition for emulsifier dosage and for mixing time of the emulsion. The optimum emulsifier dosage was found to be 10 gpt with 5 minutes of mixing time. As for the addition rate of acid phase to oil phase, 1.4 ml/min of drop-wise addition was found to give a very stable emulsion. The effect of stirring intensity on the emulsion stability was proportional, the higher the rpm speed the better the stability. Corrosion inhibitor, on the other hand, played a role as a destabilizer. With regards to heating temperature, it was found out that the higher the temperature, the faster the emulsion breaks.

The novelty of this study is the comprehensive approach in combining stablity and rheology of emulsified acid for high temperature applications. The particle size analysis and solution morphology help in better understanding HCl emulsion and how it works.

COMPARATIVE ANALYSIS OF MOLECULAR STRUCTURE IDENTIFIABILTY BASED ON SIGNATURES AND DESCRIPTORS

Zelimir F. Kurtanjek
University of Zagreb, Croatia

Abstract

This work is focused on molecule computer representation aimed for molecule predictive model (QSAR) and inverse modelling. Two different approaches are compared, i.e. the molecule structure coding based on graph theory (Faulon et al.[1] extended valence) and the chemical based descriptors[2]. Applied are 1-3rd level molecule graph fragments and 630 2D molecular descriptors. The molecule test sets are alkanes, alkenes, acetones, aromatics, organic acids and halogenated hydrocarbons in the range of C_1-C_{12}, selection of biomolecules (including aminoacids) and a set of binary ionic liquids (cations: imidazole, pyridinium, quinolinium, ammonium, phosphonium). The comparative comparison of molecule description based prediction potential are statistically evaluated by octanol-water (PWA) partition coefficients (for hydrocarbons), viscosity of ionic liquids and biological impact on acetycholinesterase inhibition. The evaluation is based on two distinct levels. On the first level the comparative analysis based on internal and external correlation between "scores" in multidimensional spaces of the separate first, second and third order valance spaces, and the molecular signature space. Firstly, the potential of information content applicable for inverse modeling is tested on ability for molecule classification by use of linear discrimination (LDA) based on the singular value decomposition (SVD) of corresponding molecular fragments and descriptors. For chemical-biological property inverse modelling proposed is application of constrained genetic algorithm (GA). Tested are linear structure multivariate (chemometric) models such as principal component regression and partial least squares and nonlinear decision trees (random forest). Applied are singular value decompositions for reduction of the molecule signature and the descriptor spaces up to the significance level of p=0.05. The selected level of significance is presumed to be in coherence with experimental data of molecule physical and chemical properties. The main focus is on extraction of the main molecular fragments (partial graphs extracted from molecular signatures) and specific molecule descriptors. Importance of application of open source software modules provided by Bioclipse[3] and R software packeges[4] for computational-statistical numerical algorithms is emphasized.

Key words: Molecule signatures, descriptors, inverse modelling, decision tree

Introduction

Advancement of high-throughput experiments have lead to „big data" databases applied in pharmaceuticals, medicine, new material science, catalysis, and new industrial processes. The role of chemoinformatics as a science for computer encoding of molecule properties and algorithms for data mining and inference are the essential steps in the development. Computer encoding of molecular properties has been a part of chemical engineering from the beginning of process design. It is mostly based on molecule group contributions, but has been extended by application of molecular descriptors (1-2-3 D), molecule signatures, and application of molecular quantum mechanics. Modeling is faced with the problem of „high dimension curse". For example, about 5 000 various molecular descriptors can be determined for each molecule, although most of them are highly correlated. It results in an ill-conditioned problem of modelling with large uncertainties in parameter and poor model prediction accuracy outside modelling training sets. Great improvement has been achieved by application by use of model ensembles, like random forest of many individual decision trees. These models are highly flexible, can be trained on discontinuous and nonlinear functional relations, can encompass various data formats like numerical and categorical variables, and due to large number of randomly selected individual predictors in principle do not have tendency to overfitting. However, when the object of modelling is extraction of important molecular features, due to autocorrelation between descriptors random forest, produced lists of importance are large sets of poorly distinguishable variables. The focus of this work is to apply feature regularization for optimization and reduction of list of variable importance. Compared are two methods for structure encoding, molecular descriptors and molecular signatures of first and second order.

Modeling and Factor Regularization

In this work are selected two distinct data sets of organic molecules from open source data bases (J. Sangster, 1989) and ionic liquids ILs (The UFT/ Merck and NIST, 2015), and lists of the selected molecules are given in Table 1. As target properties for organic molecules are selected the corresponding molecular classes and partition octanol-water (POW) coefficients, and dynamic viscosity in the range of 30-80 ^{0}C of ILs is selected The molecules are coded into SMILES and used as input lists into software for evaluation of molecular signatures. Most popular open source software tools are DRAGON (R. Todeschini, et al., 2009) and PADEL (Chun Wei Yap, 2010). Here are used 1-2 D descriptors generated by PADEL. For the same list of molecules are determined first and second order molecular signatures or extended valances (J.P. Faulon, 2003). The signatures are two dimensional graph based structural molecular fragments. The vertices are atoms and the edges are corresponding bonds (lengths or heights), and are here only used of the first and second level. A molecule is characterized by a complete set of signatures as input data for modelling. Since the models are based on large number of variables (molecular features) possibly lager than number of samples (molecules), usually denoted as p>>n models. They are intrinsically „ill conditioned" leading to good predictions on train data but very poor predictions on test data. Also, produced list of variable importance are very dense "clusters are not distinguished" with large uncertainties (broad confidence intervals). This makes the problem of inverse modelling aimed to predict a set of candidate molecules for given set of target properties, such as in a case of high dimension combinatorial task for selection of ILs cations and anions, very unreliable. In order to reduce model dimensionality and improve robustness and parsimonious of important molecular features here is applied modelling with feature regularization. The model is based on ensemble of random trees (CART models, L. Breiman et al. 1998) with included feature regularization.

Data sets of organic molecules		**Range of number of atoms**			
Molecular classes	C	H	O	Cl	F – I - Br
Alkanes	5-12	12-30			
Alkenes	4-9	6-18			
Acetones	6-24	6-26			
Halogenated hydrocarbons	3-10	6-22	1		
Organic acids	1-16	2-32	2		
Esters	4-11	6-22	2		
Aromatics	1-12	1-21		3	2 – 2 - 1
Data sets of ionic liquids (ILs)					
Cations		**Anions**			
1,3-Dimethylimidazolium (+)		PF_6 (-) Cl (-) CH_3COO (-)			
1-Methylpyridinium (+)		BF_4 (-) $MeSO_4$ (-) $EtSO_4$ (-)			
1,1-Dimethylpyrrolidinium (+)		Tf_2N (-) CF_3SO_3 (-)			

Table 1. The molecular groups of the organic and ionic liquid molecules selected for random forest models and extraction of the most important molecular features. The properties for the organics are obtained from J. Sangster (1989) and for the ionic liquids from Merck and NIST data bases (2015). Molecular descriptors and signatures of ILs ions are recorded separately and adjoined as a single record.

The modelling is formulated as an optimization (functional and multivariable minimization) problem with two parts of the objective function (T. Hastie et al., 2009).

$$\min F_{object.}(X,T,\omega) = SSE(y,\hat{y}) + \phi(T,\omega) \tag{1}$$

The first part is the sum of square errors (SSE) based on difference between observed y and prediction $\hat{y}$ based on input data by an ensemble T of interrelated CART models (L. Breiman et al. 1998), The second part is a selected metrics ϕ for the model complexity depending the sum of leaf probabilities for population of individual CART models and their total number N_T.

$$\phi(T,\omega) = \gamma \times N_T + \frac{1}{2} \times \lambda \times \sum_{i}^{N_T} \omega_i^2 \tag{2}$$

The ensemble output prediction is a sum of predictions of individual T_i tree CART models:

$$\hat{y} = \sum_{i}^{N_T} T_i(X) \tag{3}$$

A very efficient open source code Extreme Gradient Boosting XGB (T. Chen, 2016) is available in R language (R Developm-ent Core Team, v. 3.2.3, 2015). The optimization algorithm is consecutive, starting with first CART model which is appended with a next tree boosted by errors in predictions of previous tree. At each step optimization is solved by linearization of the objective function to quadratic terms. Number and structure of trees as well as the parameters for regularization need to be found by trial and error and minimal number of trees by folding and cross validation error on the training set. The critical step is validation on untrained set to have acceptable model error and avoid model overfitting with minimal model complexity.

Results and Discussion

Tested are models for prediction of the seven molecular classes (Table 1.) based on molecular descriptors and molecular signatures (first and second level). Results for molecular descriptor feature importance are shown on Fig. 1. The effect of regularization is very pronounced. Out of hundreds of descriptors for the molecular classification needed are only few, the most important being BCUTw.1h. On contrary, when an ensemble of independent CART trees is applied, i.e. random forest model, level of accuracy of prediction is the same, but the list of variable importance is densely populated with numerous descriptors.

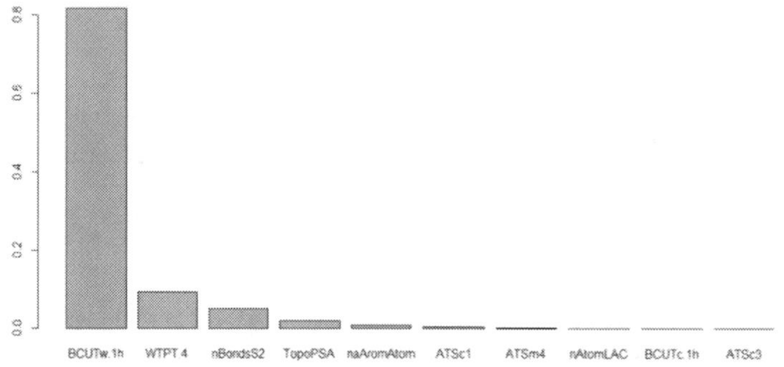

Fig 1. Variable importance molecular descriptor factors for the inverse modeling of the selected molecular classes. The first five descriptors are: BCUTw.1h (based on a weighted version of the Burden matrix which takes into account both the connectivity as well as atomic properties of a molecule), WTPT.4 (Randić's weighted path), nBondS2 (bond count), and naAromAtom (number of aromatic atoms).

The same classification problem was tested by the first and second order molecular signatures. The important signatures are depicted on Fig. 2-3. Accuracy of prediction on trained sets of molecules compared to descriptors is about the same, though the importance factors are not as distinctive as in case of the descriptors.

significantly better when molecular descriptors are applied. The

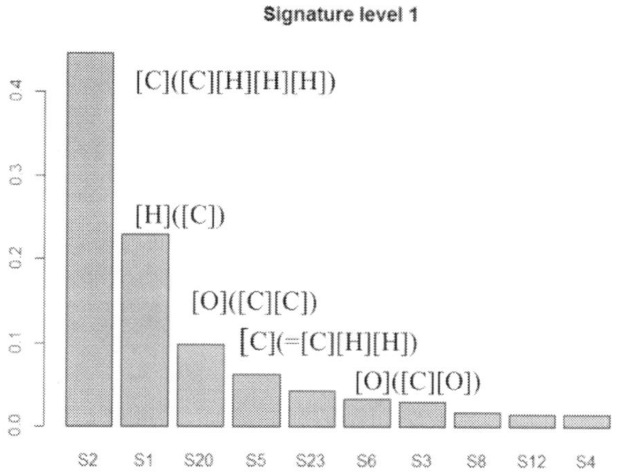

Fig 2. Relative importance of the individual first level signatures for inverse prediction of the selected molecular classes. The first five most important signatures are coded according to SMILES (Simplified Molecular Input Line Entry System).

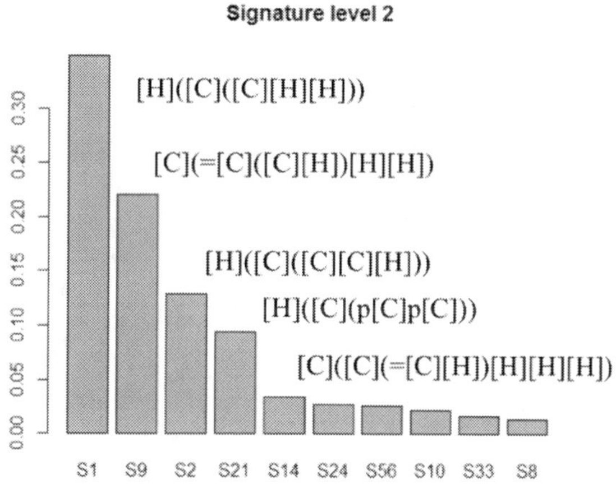

Fig 3. Relative importance of the individual second level signatures for inverse prediction for the selected molecular classes. The first five most important signatures are coded according to SMILES (Simplified Molecular Input Line Entry System).

However, model accuracy on untrained set of molecules, depicted in Fig. 4, is significantly lower for the model with molecular descriptors compared to signatures. The average error with descriptors is about 5% while for signatures about 11%. The "box and whiskers" plots are obtained by error evaluation of ten randomly generated untrained sets.

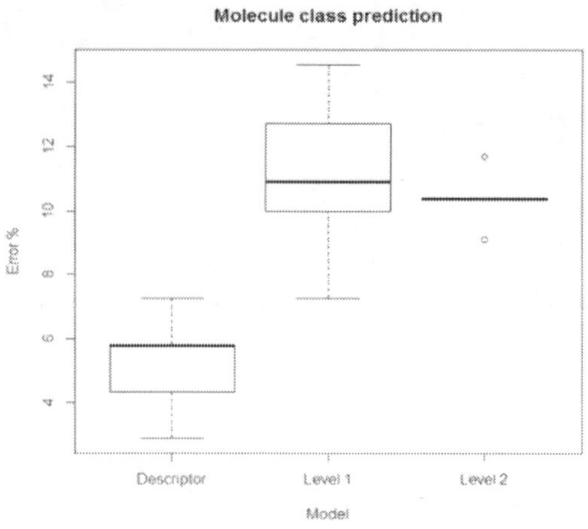

Fig 4. "Box and whiskers" plot for the model validation (relative errors) for inverse prediction of the molecular classes based on the set of untrained molecular descriptors and the first and second level of molecular signatures.

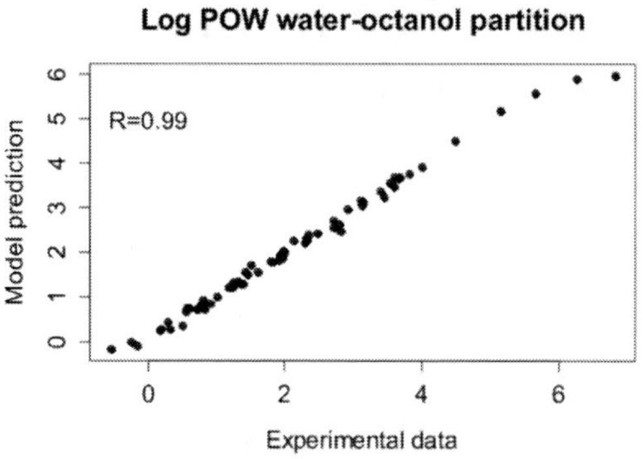

Fig 5. Predictions of the partition octanol-water coefficients POW, for the specified molecular classes of organic molecules, are based on the random forest model with the molecular descriptors.

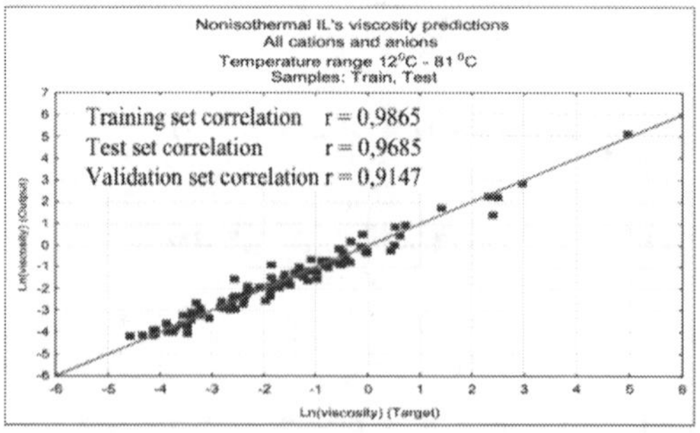

Fig 6. Predictions of dynamic viscosities for the specified groups of ionic liquids (ILs) based on the random forest model with the molecular descriptors. The viscosities cover the temperature range from 12 – 81 ^{0}C and five orders of magnitude.

In Fig. 4 can be observed that dispersion of errors (quantiles) for the second level molecular signatures is negligible, though two outliers are present.

For prediction of molecular properties applied are random forest models based on molecular descriptors. For all of the organic molecules, included in the seven classes, evaluated are octanol-water partition coefficients POW, shown in Fig. 5. The range of POW values is very high, on log scale fro 0 to 6, but the obtained accuracy of the model predictions is high (Pearson correlation R=0.99). The same approach is applied for prediction of ionic liquid viscosity data in temperature range from 12 to 81 ^{0}C. Since temperature effects on viscosity are intrinsically nonlinear, they represent a modelling challenge. However, the decision trees are very efficient for accounting nonlinearities, though nonlinearity is reflected in non-orthogonal decision planes, rendering trees with significantly increased number of branches. For the range of 6 order of magnitude the accuracy of prediction on trained set is R = 0.98 and on untrained set R = 0.91.

Conclusions

Compared are ensemble decision trees models for molecular feature class and property prediction based on extensive list of molecular descriptors and first and second level of extended valances or signatures.

The aim was to test effects of regularization on ensemble of CART decision tree models. Proven is effectiveness of the extreme gradient boosting algorithm in cases with extensive list of (molecular) features.

Obtained is increased accuracy on untrained sets of data, and more importantly, obtained are parsimonious lists for ranking of variable importance lists. This enables further significant reduction in development of robust and simple decision tree models.

The reduction in number of molecular features applied for decision tree models may greatly simplify the challenging problem of inverse modeling based on molecular descriptors and/or signatures.

References

1. J. Sangster, J. Phys. Chem. Ref. Data, **18**(3)(1989) 1111-1227
2. The UFT/ Merck Ionic Liquids Biological Effects Database, www.il-eco.uft.uni-bremen.de
3. NIST, Ionic Liquids Database Standard Reference Database #147, <ilthermo.bouder.-nist.gov /ILThermo/mainmenu.uix>
4. R. Todeschini and V.Consonni: "Molecular Descriptors for Chemoinformatics", (2 volumes), WILEY-VCH,Weinheim (Germany) 2009.
5. Chun Wei Yap, PaDEL-Descriptor: An Open Source Software to Calculate Molecular Descriptors and Fingerprints, J. Comp. Chem., (2010):**32**: 1466-1474.

6. J.-L. Faulon, D.P. Visco Jr., R.S. Pophale, The signature molecular descriptor. 1. Extended valence sequences vs. topological indices in QSAR and QSPR studies, J. Chem. Inf. Comput. Sci. **43** (2003) 707–720.
7. J.-L. Faulon, C.J. Churchwell, J.D.P. Visco, The signature molecular descriptor. 2. Enumerating molecules from their extended valence sequence, J. Chem. Inf. Comput. Sci. **43** (2003) 721–734.
8. L. Breiman, J. H. Friedman, R. A. Olshen, C.J. Stone, „Classification and Regression Trees", CRC, Boca Raton, 1998.
9. T. Hastie, R. Tibshirani, J. Friedman, The Elements of Statistical Learning, Data Mining, Inference and Prediction, Springer Verlag, Second Edition, New York, 2009.
10. T. Chen, T. He and M. Benesty (2016). xgboost: Extreme Gradient Boosting. R package version 0.4-3. https://CRAN.R-project.org/package=xgboost
11. R Development Core Team, , R: A language and environment for statistical computing. R, Vienna, Austria,www.R-project.org 2015.

444791 Dissolution and Precipitation of Minerals Under Different Alkali Concentrations in a Reactive CO_2 Batch System

Abstract:

The primary focus of this research is to study the dissolution and precipitation of minerals and salts from Berea sandstone and brine samples taken from the Great Salt Lake in a reactive CO_2 batch system. Various experiments were constructed to measure dissolution and precipitation reaction rates under different alkali concentrations. Changing the alkali concentrations of the Great Salt Lake brine samples produced changes in dissolution and precipitation of minerals from the Berea sandstone and the Great Salt Lake brine samples. Data from ICP-MS was used to quantify mineralogical changes in the samples and Micro-CT was used to measure the porosity change in the Berea sandstone samples. The results from these analytical instruments showed patterns of mineralogical dissolution and precipitation in the post-experimental fluid that correlate to the change in alkali concentrations.

444825 Significance of Energy Intensity and GHG Emission Reduction Towards Sustainability of a Petrochemical Plant

Abstract:

Sustainability is an important pillar in today's petrochemical industry for improving process efficacy by reducing specific energy consumption, greenhouse gas (GHG) emissions, and wastewater. SABIC (Saudi Basic Industries Corporation) like other giant petrochemical companies has set stringent sustainability targets and investing huge resources to transform its plants from the present position to the first quartile in each category.

SABIC is the world largest producer of ethylene oxide and ethylene glycol. For each ton of glycol produced, 0.2 to 0.5 tons of CO_2 is emitted as GHG. The significance of sustainability "in general" while the methodology adopted to reduce energy intensity, carbon footprint, and wastewater "in particular", will be discussed thoroughly in our talk.

444827 Syngas to Ethylene Glycol: A Technology Overview

Abstract:

SABIC is the world largest MEG producer petrochemical company and most of its MEG is being produced through the hydrolysis of EO. Ethylene oxide (EO) is a simplest form of epoxides and due to its molecular structure, acts as a versatile intermediate for a variety of chemicals such as; monoethylene glycol (MEG), di- and triethylene glycols, ethoxylates, ethanol amines, and glycol ethers. The growing demand of MEG owing to its downstream utilization in polyester industry has prompted the research of an efficient, cost effective, and environmentally benign EO production process.

The major feedstock for MEG are ethylene and syngas. The heavy dependence of MEG producers on ethylene and its scarcity in some parts of the world promoted the need of alternate feedstock for highly demanded MEG. The technological developments utilizing syngas as a feedstock will be discussed along with the commercial analysis and future growth in EO/EG business.

444834 Stability and Rheology of Emulsified Acid in Waste Oil: Effect of Emulsifier Concentration, Mixing Speed, Temperature and Droplet Size

Abstract:

Acid in oil emulsion is used in the industry for stimulation purposes. In particular, HCl as the acid phase and diesel as the oil. Emulsifying HCl has numerous benefits over regular HCl. Perhaps the main reason is due to deeper penetration. Apart from that, less corrosion damages are caused since the external phase is hydrocarbon. Several studies showed the success of replacing diesel as an external phase with other hydrocarbon oil, such as crude oil and xylene.

This work utilizes the extra hydrocarbon left unused –or sometimes dumped- from refineries. This unused hydrocarbon is referred to as waste oil. First the chemical composition of waste oil was studied using HPLC, SARA fractionation as well as FTIR. This helped in selecting the proper emulsifier agent for the waste oil. The HCl-waste-oil emulsion was prepared using 15wt% HCl with a ratio of 70:30 of acid-to-oil. Extensive work was done to ensure reproducibility of data. This study was done to experimentally investigate the effects of distinct important variables on thermal stability and apparent viscosity of the HCl-waste-oil emulsion. These variables are; emulsifier concentration, mixing speed and the addition of corrosion inhibitor. All experiments were conducted at a high temperature 120°C. To assess the quality of the HCl-waste-oil emulsion, droplet size was measured as a function of temperature (up to 100 °C).

The results showed that the HCl-waste-oil emulsion is a shear-thinning fluid. Power-law model was applicable seamlessly to all of the apparent viscosity data for all of measured temperatures. The optimum conditions were found to be; 0.5vol% and 600 rpm for emulsifier concentration and mixing speed, respectively. Apart from that, the emulsions prepared with corrosion inhibitor were less stable compared to the ones without it. However, the addition of corrosion inhibitor tended to generate monodispersed emulsion but bigger size droplets for temperatures 25°C and 50°C and polydispersed droplets for higher temperatures. Droplet size measurements showed an average size of 1.47 to 3.09 μm. Overall, the lab results showed a promising potential for the HCl-waste-oil emulsion to be used in the field.

This work takes into account; reducing the budget of acidizing job and at the same time environmental concerns. It makes use of the unwanted hydrocarbon from refiners. And, it saves the environment by using the waste oil instead of dumping it.

444994 Multiscale Simulation of Proteins Motion in the Microchannel

Abstract:

Protein adsorption is an important phenomenon, which plays a prominent role in chemical, industrial and medical areas. In this work, the dynamic behavior of the egg-white lysozyme has been investigated in a micro-channel with a multiscale modeling approach. The Molecular Dynamic simulation, free energy computation and Langevion Dynamics (LD) simulations have been combined to study lysozyme adsorption process on Au (111) surface. The gold surface has strong attraction to proteins. The protein-surface adsorption or desorption free energies is calculated. From the case studies of LD simulations, the surface has effects on the protein dynamic adsorption process. By the comparison of LD, MD and the experiments, the time step is concluded as an important factor to study protein adsorption behavior at the water-surface interface. Fine time step at the scale of 1 picosecond is necessary to provide enough accuracy to resolve proteins' dynamics trajectories.

445004 Grand Challenges in Future Energy and Chemicals: Climate Change and Sustainability As Drivers for New Technology and Innovation

Abstract:

Global demand for energy will double by 2050, driven by increasing population and affluence especially in emerging economies. While fossil fuels will continue to underpin the energy supply for the near future, concern over impact on climate change continues to grow, driving interest in expanded use of renewables and development of carbon capture and sequestration or re-use solutions. In addition to CO_2, water use and air emissions are critical issues to be addressed in future and current energy supply, as is energy storage for effective use of renewables. This presentation will provide an overview of what industry is doing today to address these issues, including the factors and metrics used in decision making.

445029 Electrospun Nanocomposite Fibers for the Removal of Heavy Metals in Water

Abstract:

Uniform electrospun nanofibers with small diameters were successfully prepared by systematically optimizing the electrospinning parameters including polymer concentration, salt loading, applied voltage, tip-to-collector distance, and flow rate. These parameters are shown to affect the fiber morphology and the addition of iron (III) nitrate salt was found to be able to reduce the fiber size. Due to the conductive nature of iron salt in polymer solution, even a relative lower voltage of 6 kV could generate a stable fluid jet and yield the nanofibers. The thermal properties of the composite fibers were investigated by thermal gravimetric analysis (TGA) and differential scanning calorimetry (DSC). Rheological studies were also conducted. The composite nanofibers were further carbonized to carbon nanofibers. Applications of the nanofibers in environment remediation have shown promise.

445030 Heavy Metal Detection in Treated Water Via a Fabric Nanocomposite Sensor

Abstract:

We present the conception of an electrochemical sensor for on-site/portable water analysis. This flexible sensor was developed to quantify the amount of heavy metals (in real-time), alleviating the wait time, large sample size, possible contamination, and expensive analytical equipment associated with current procedures for water analysis. This sensor is composed of a multiwalled carbon nanotube (MWNT) functionalized polymer mat, for a flexible yet conductive sensor. The MWNTs are then functionalized with a cyclo-oligomeric calixarene, which is shown to selectively form a supramolecular complex with metal ions. Upon complex formation, the charge carriers are drawn away from the carbon layer and the detection is prevalent at appropriate levels. The optimization of MWNT and calixarene functionalization, as well as the selectivity and sensitivity of the sensor will be discussed.

447119 Development R&D of Manoir Industries and Emphasis on Coating

Abstract:

The Senior VP Technical Support & Projects will present the development R&D of Manoir Industries and especially the coating (it is our lattest development) as well the services we can provide such as metallurgy expertise, advice, convector cleaning, site welding, welding supervision in order to increase the furnace productivity. So it is 60% R&D and 40% services. Manoir Industries manufactures cast components for the petrochemical market and develops high temperature alloys such as the Manaurites.

444478 Biogas Reforming with Addition of Carbon Dioxide to Syngas over Ni-Based Bimodal Pore Catalyst

Abstract:

Methane dry reforming of (MDR) has been widely studied due to its important applications in producing of syngas ($CO + H_2$), removal of two greenhouse gases, and upgrading of biogas (mainly composed of CH_2 and CO_2) into value added chemicals. Compared to the noble metal catalysts, nickel-based catalysts are suitable for industrial scaled-up MDR due to their high activity and low cost. However, nickel-based catalysts have been encountering a significant challenge that is the active metal sintering and coke deposition, leading to the catalyst deactivation and then the poor stability. Nevertheless, it has been reported that the deactivation of nickel catalysts can be suppressed by adding promoters like strong Lewis bases with the enhancement of chemisorb CO_2, such as MgO and CaO, and like lanthanide elements with the capacity of oxygen storage and release, such as CeO_2 and La_2O_3. Al_2O_3 is a commonly applied catalyst carrier own to its good pore size dispersion, high specific surface area and high mechanic strength.

Herein, we report that the bimodal NiCeMgAl catalysts with different NiO-loading were synthesized by the refluxed co-precipitation method and were evaluated for MDR performance by using real biogas from anaerobic digestion of food wastes. The sample containing 15wt% NiO-loading (Ni_{15}CeMgAl) was found to be active enough at 750 °C with a high CH_4 conversion of around 90%. Ni_{15}CeMgAl catalysts kept their bimodal porosity after reduction and MDR process. The evolution of the Ni_{15}CeMgAl catalyst before and after MDR process was investigated by SEM, TEM, XRD, and TGA techniques. The correlation between the structure evolution and catalytic performance change was also discussed.

444551 Continuously Adjustable, Molecular-Sieving "Gate" on 5A Zeolite for Distinguishing Small Organic Molecules By Size

Abstract:

Zeolites/molecular sieves with uniform, molecular-sized pores are important for many adsorption-based separation processes. Pore size gaps, however, exist in the current zeolite family. This leads to a great challenge of separating molecules with size differences at ~0.01 nm level. Here, we report a novel concept, pore misalignment, to form a continuously adjustable, molecular-sieving "gate" at the 5A zeolite pore entrance without sacrificing the internal capacity. Misalignment of the micropores of the alumina coating with the 5A zeolite pores was related with and facilely adjusted by the coating thickness. For the first time, organic molecules with sub-0.01 nm size differences were effectively distinguished via appropriate misalignment. This novel concept may have great potential to fill the pore size gaps of the zeolite family and realize size-selective adsorption separation.

16 AIChE
Spring Meeting, Houston, TX

The Principle of Intrinsic Reality
Solving the World's Woes through Chemical Engineering and Economics

Jed Moyano Bellen, SCPM, PMP
Hyundai Engineering & Construction Co. Ltd.
9 Temasek Blvd., #27-02 Suntec Tower 2, Singapore 038989
jed.bellen@hdec.co.kr

Prepared for Presentation at
American Institute of Chemical Engineers
2016 Spring Meeting
11th Global Congress on Process Safety
Houston, Texas
April 10-14, 2016

UNPUBLISHED

The Principle of Intrinsic Reality Solving the World's Woes through Chemical Engineering and Economics

Jed Moyano Bellen, SCPM, PMP
Hyundai Engineering & Construction Co. Ltd.
9 Temasek Blvd., #27-02 Suntec Tower 2, Singapore 038989
jed.bellen@hdec.co.kr

Keywords: Principle of Intrinsic Reality, Chemical Engineering, Economics, Sustainability

Abstract

Human beings in profound rationality always seek for explanations of the phenomena around them. To understand these phenomena, human beings have developed systematized body of knowledge as a practical way of dealing with life's challenges. Among the systematized body of knowledge developed are chemical engineering and economics.

An avant-garde way of looking at each of these disciplines leads to the Principle of Intrinsic Reality. This principle relates the fundamental laws of the two disciplines to glean insights on the constraints imposed by conservation laws on economic systems at both the macro and micro levels.

It can be argued that if mass and energy in the universe are constant, then logically diminishing returns and marginal utility are bound to be observed. This logic is quite evident in a microorganism culture which follows the Monod growth model.

It can also be thought that conservation laws imply scarcity of resources. As such, it enjoins everyone to find optimal and efficient use of resources in the context of trade-offs, production decisions and sustainability. It even calls for the development of new technologies to produce more outputs from less inputs (i.e. shifting of production possibility frontier to the right). In this way, the Principle of Intrinsic Reality leads to the imperative for sustainable optimization.

Furthermore, it is deemed that the conservation laws can explain the phenomenon called "polarity of wealth." As mass and energy when harnessed create wealth, then the constancy of mass and energy implies boundary for wealth creation. In effect, wealth can neither be created nor destroyed but can only shift from one entity to another thereby creating polarity in wealth distribution. This assertion can be the key towards more equitable restructuring of economic systems to close income gap and thereby result to stable and resilient populace.

In conclusion, this paper envisions for a paradigm shift from traditional money-centric view of chemical engineering and economics to a synergistic convergence of disciplines which can powerfully tackle the world's woes and thereby effect sustainability.

1. The World in which We Live Today

The past few years saw unprecedented global challenges which have put into question existing economic models and scientific paradigms. Among these challenges are the dystopian global economy, climate change, environmental degradation and terrorism. The magnitude of these challenges has grown to a profound level that newer ways of thinking and perspectives become necessary. With that end in mind, this paper attempts to look at the fundamental laws of chemical engineering and economics in new ways to recognize our intrinsic reality from which these pressing challenges can be tackled at large.

2. The Principle of Intrinsic Reality

It is an observable fact that advancement in science increasingly blurs boundaries of various disciplines in tackling proactively the pressing global issues or problems. Within this purview, two seemingly diverging disciplines namely chemical engineering and economics can be taken together to glean insights on our intrinsic reality and what it means for our economic and ecological systems on a larger scale.

In the author's contention, intrinsic reality is grounded on the fundamental laws of chemical engineering and economics. These laws are the Conservation Laws (Chemical Engineering) and Law of Diminishing Returns (Economics). It can be argued that the Law of Diminishing Returns is observable because of the constraints imposed by the Conservation Laws. As such, this leads to the Principle of Intrinsic Reality which can be stated thus;

> *In any system, both at the macro- and micro-levels, the limiting factor for growth is the constancy of potential inputs.*

Potential inputs mean not the actual inputs to processes but the constant material or energy available which can be harnessed to create wealth. The reasoning or proof of this principle is presented in subsequent discussions.

2.1 Proof Based on the Monod Growth Model

The best illustration to understand the Principle of Intrinsic Reality is the Monod Growth Model. Monod in his 1949 seminal paper presented an empirical relation between exponential growth rate and concentration of essential nutrient. The data was fitted into hyperbolic equation:

$$R = R_K\left(\frac{C}{C_1 + C}\right)$$

In this equation, R is the growth rate during the exponential phase; R_K is the rate limit for increasing concentrations of C; C is the concentration of the nutrient; and C_1 is the concentration of nutrient at which the rate is half the maximum.

Figure 1: Graphical Representation of Monod Growth Model

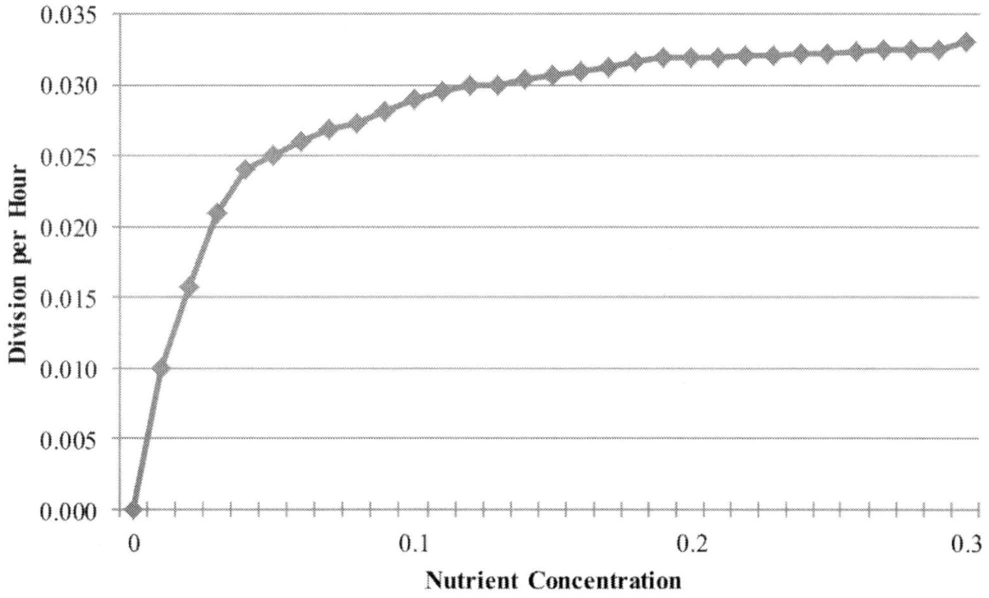

Monod (1949) further pointed out that the limiting factors commonly found are exhaustion of nutrients, accumulation of toxic metabolic products and changes in ion equilibrium.

The foregoing discussion has been observed at the micro-level with the bacterial culture. At the macro-level (i.e. the world at large), analogous phenomena are also observed to occur because of the same constraint imposed by the constant quantity of resources (i.e. material and energy). One example of these phenomena is that human beings cannot grow indefinitely. If they do, then how can the world sustain that growth with limited available resources (e.g. foods) to support it?

Another analogous phenomenon is in population growth. Thomas Malthus (1789) predicted that when population growth exceeds agricultural production, the population level is forced to return to subsistence-level conditions. This phenomenon as Malthus (1789) believed is due to the restriction of available resources supporting population growth, which others have claimed to be exploding as the earth becomes over populated.

In a surprising manner, a similar phenomenon can be observed in production processes in which the increase in a single factor of production assuming all other factors constant, results not to more marginal output but to a reduced one. This was attributed by classical economists, Thomas Malthus and David Ricardo, to the decreasing quality of inputs while neoclassical economists attributed this to the disruption in the entire production process as additional units of input are added to a fixed amount of capital. But in view of the Principle of Intrinsic Reality, the phenomenon is viewed intrinsic to the production system in which case infinite amounts of mass or energy cannot be processed with perfect efficiency. Hence, the law of conservation of mass and energy comes into the fore.

It can be deduced from the discussion that there is an underpinning intrinsic reality to the more obvious phenomena in the world from the micro-scale (bacterial growth) to the macro-scale (human growth and production processes).

2.2 Proof Based on Diminishing Marginal Utility

Another way to prove the Principle of Intrinsic Reality is through the diminishing marginal utility observable among humans and perhaps other organisms as well. Marginal utility, as defined in economics, is the gain from an increase or the loss from a decrease in the consumption of goods or services. The diminishing marginal utility is an observable phenomenon in which the utility or satisfaction an individual gains diminishes as he or she consumes more of the same unit of goods or services. Graphically, this can be represented as in Figure 2.

Figure 2: Graphical Representation of Diminishing Marginal Utility

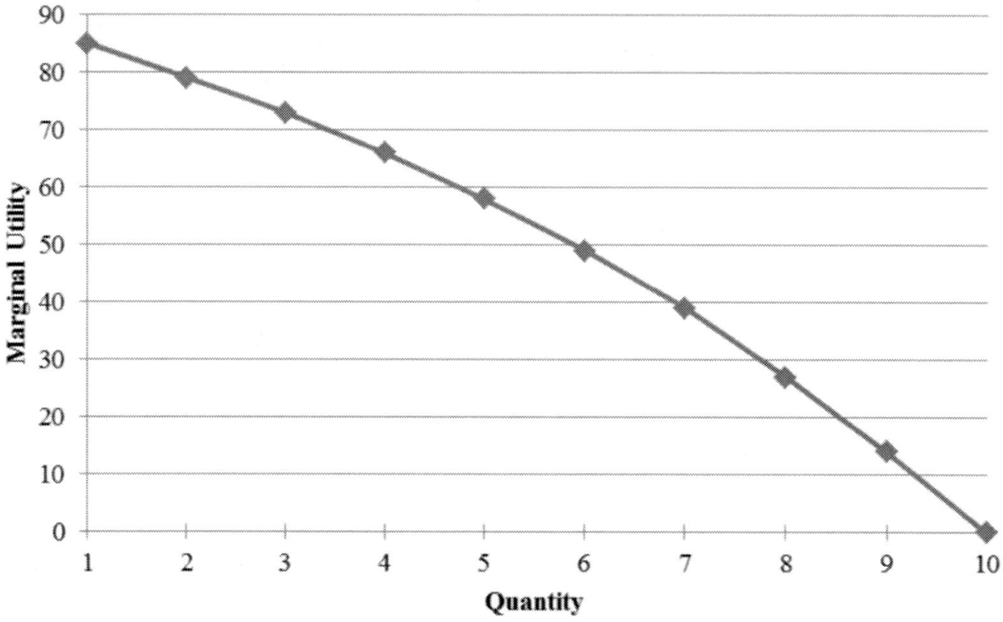

It can be deduced from this phenomenon and in support of the discussion in section 2.1 that human beings cannot grow indefinitely due to the fact that in our intrinsic reality, there is always a limit so as the constancy of available resources (i.e. mass and energy) can support our existence.

This means that in view of our intrinsic reality, we cannot have infinite greed otherwise an upsurge of unintended consequences are bound to happen (i.e. environmental degradation, terrorism and eventually demise of humanity).

Therefore, recognizing our intrinsic reality is the first step towards solving the world's woes and thereby resulting to sustainability and security.

2.3 Expanding the Production Possibility Curve through R&D

The caveat here is that the constant mass or energy (i.e. the limits) leads to trade-offs. Trade-offs is encountered when personal or business decisions are made. In making these decisions, one has to ensure that he or she becomes better off with his or her chosen option. We also realize in our life that we cannot get all that we want and we have to make trade-offs so long as it places us in a better or preferred position. This is more of a philosophical discourse, but it goes to show that the concepts of chemical engineering can be as useful in life and in business as it is in chemical processing.

To tackle these trade-offs, the Production Possibility Frontier in economics can be used. In view of the Principle of Intrinsic Reality where growth is limited by the fixed amount of material and energy, the Production Possibility Frontier needs to shift to the right as shown in Figure 3.

Figure 3. Graphical Representation of Expanding the Production Possibility Curve

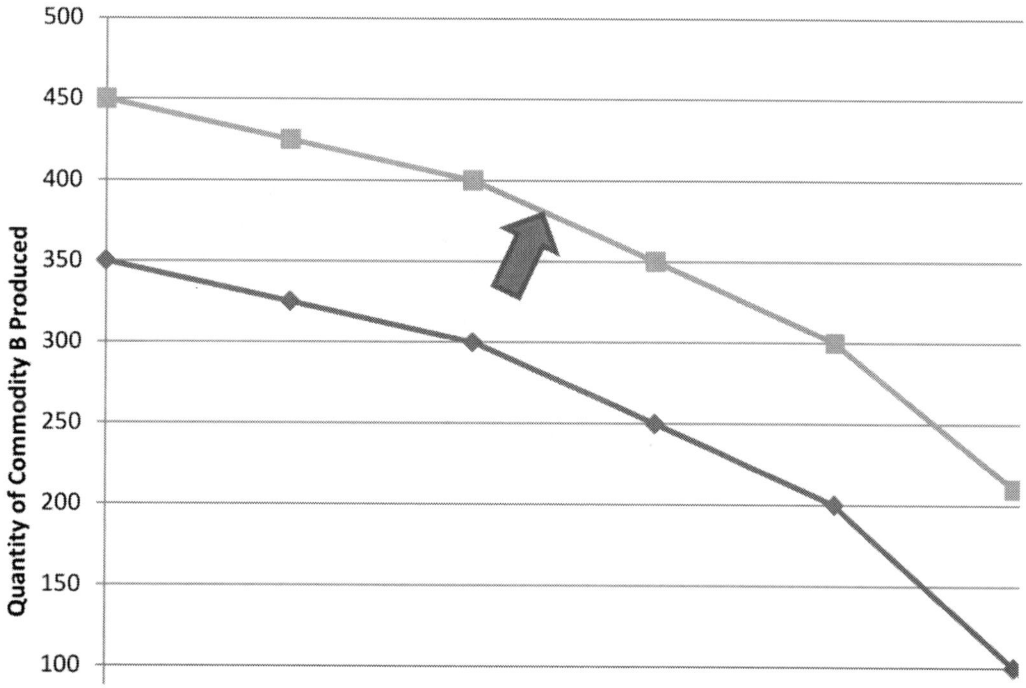

This shift has an implication on the practice of chemical engineering as Research and Development (R&D) is needed to achieve the said shift. R&D enables the economy to produce more of the two commodities (i.e. Commodity A and Commodity B) resulting in economic growth. The key challenge here to Chemical Engineering, particularly on Chemical Engineering R&D, is to look at the bigger picture where economic growth emanating from R&D output is pivoted on sustainability; considering its impact to the environment and the society at large.

3. Paradigm Shift in the Practice of Chemical Engineering

"Money" or "Profit Maximization" has been at the center of the practice of chemical engineering (Smith 2005). Various mathematical optimization methods are used to design systems, processes or components of processes with the end in mind of optimizing profits. This mental model however, will no longer work in our situation today as we face economic, ecological, social and spiritual crisis (Scharmer 2010).

To address this, Scharmer (2010) pointed out seven acupuncture points to shifting capitalism into a regenerative ecosystem economy. Within this purview, the practice of chemical engineering has a greater role to assume. For example, Scharmer (2010) suggested that the design of all production and consumption cycle be completely earth-to-earth, i.e. without necessitating for landfill and that it has to co-evolve with the natural ecosystem. In this regards, the way we do optimization should be revamped into sustainable optimization. By sustainable optimization, it means optimization of the entire system and adding in environmental as well as societal objectives to the equation.

In this definition, two things need to be considered. The first one is systems thinking which emphasizes the understanding of the entire system so the interaction among components will be known and unintended consequences mitigated (Senge 2006). This is a holistic approach to problem solving and finds enormous use in process design.

It should be noted however that the construction of processing plant, whether it is a greenfield or a brownfield project, is rooted on projected market demand. In some instances, economic volatility necessitates companies to experiment with multiple pathways to survive. In this circumstance and within the purview of systems thinking, design thinking comes into the fore. Design thinking is a collective principle emphasizing empathy for users, doing prototyping and tolerance for failure (Kolko 2015). This principle finds enormous use in chemical engineering designs such as in product development, process development, systems improvement and the like. This is also useful for simplifying or humanizing chemical engineering designs with the end in mind of contributing to sustainability. It also has to be noted that of paramount importance in using this tool is imagining the future with the view in mind of status quo and the likely outcome for sustainable development.

The second consideration is the inclusion of environmental and societal objective functions to the objective functions (economic criteria) of profit, cost, energy and yield (Edgar, et al. 2001) in optimization. The environmental objectives can be the impact or risk to environment and the societal objectives can be the impact or risk to society or common good.

These are just few examples of the needed paradigm shift in chemical engineering practice in consideration of the principle of intrinsic reality. But there can be more. A picture of the envisioned new paradigm for the practice of chemical engineering is shown in Figure 4.

Figure 4. New Paradigm in the Practice of Chemical Engineering

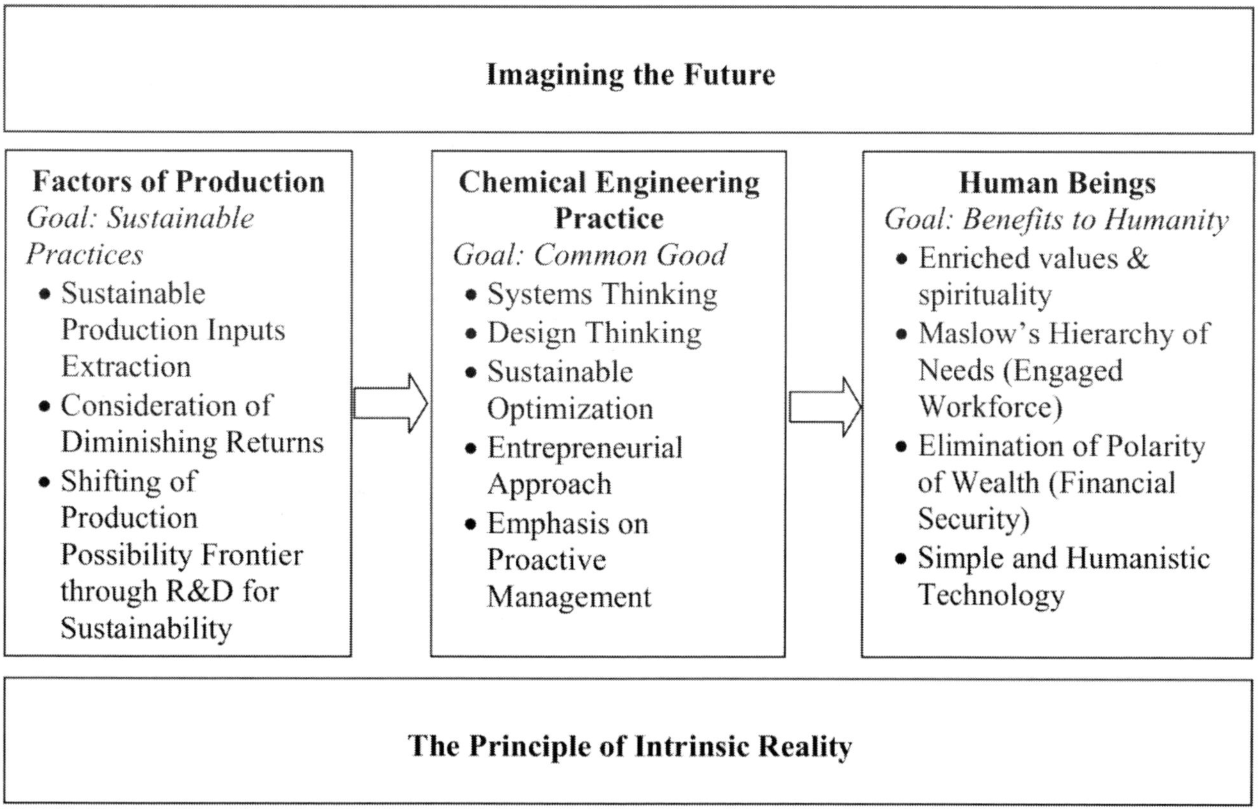

It should be realized that at the core of our economic systems is chemical engineering which can drive the economy into positive direction with the least unintended consequences. Related to that is capitalism in which upgrading of its operating systems can affect the practice of chemical engineering to a substantial degree. A discussion of this is provided in the following section.

4. Wealth Creation and Its Imperative for Sustainability

The development of this paper was inspired by the book "Freakonomics" in which the authors Steven D. Levitt and Steven J. Dubner (2005) asserted that economics can explain almost everything. It came to mind that if economics as a basic science can explain almost everything, then why not chemical engineering which was founded on chemistry, the central science?

Reflecting on it deeper, while explanations clear our doubts, we need more than that so we can tackle the world's woes. As such, this author thought that perhaps combining the two disciplines of chemical engineering and economics can be a potent force towards this end. How this can possibly be done is discussed in this section.

The starting point here is a review of the economic history of the world. In his book "A Farewell to Alms", Gregory Clark (2007) reviewed the economic history of the world in which he divided it into two periods; the period before 1800 and the period after. This is graphically represented in Figure 5.

Figure 5. Shares of Major Economies in World GDP through Time

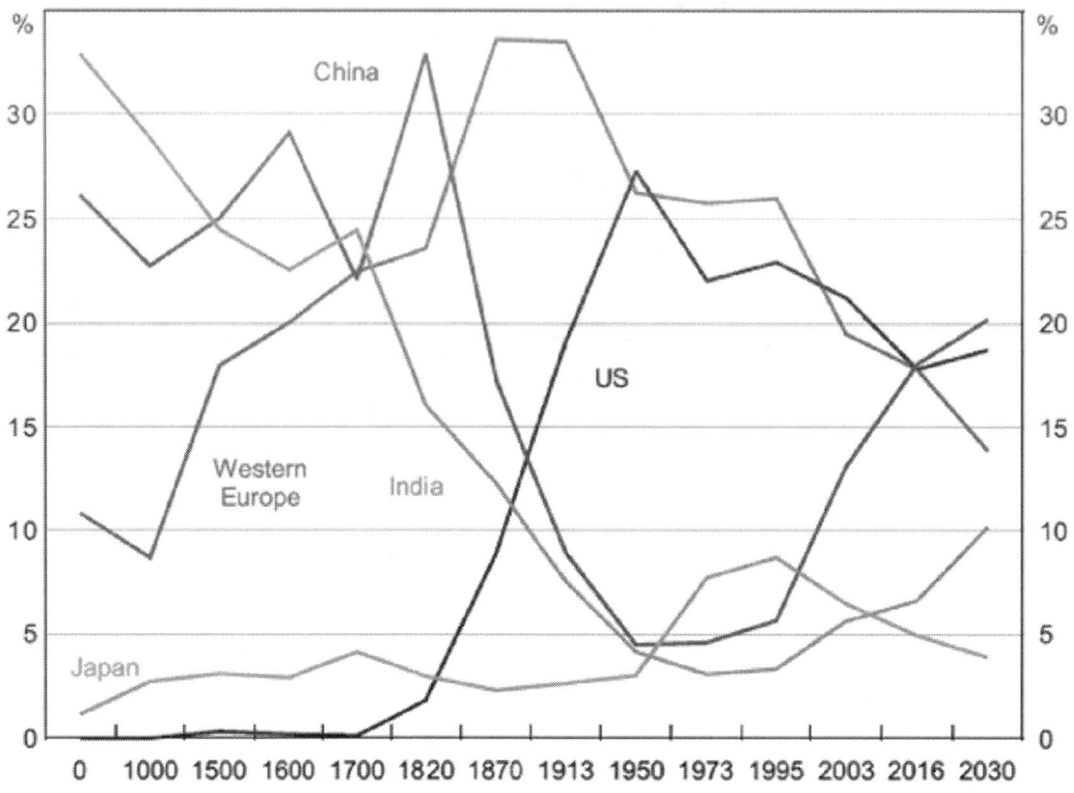

Note: Based on purchasing power parity (PPP) valuation of country GDP

Sources: International Monetary Fund (IMF), *World Economic Outlook* database, April 2011; Maddison (2006)

Thompson (2012) interpreted this graph by stating that "everything to the left of 1800 is an approximation of population distribution around the world and everything to the right of 1800 is a demonstration of productivity divergences around the world." In other words, everything to the left before Industrial Revolution depicts economic power by the sheer size of population and everything to the right after the Industrial Revolution shows the effect of mass production on country's economic output.

Clark (2007) on the other hand called the era before the industrial revolution, i.e. ca. 1800, as Malthusian era in which income per person varied across societies and epochs but growth is not observed. As Clark (2007) explained, this was due to a mechanism called Malthusian trap in which short-term increase in income through advancement in technology is dragged down by population growth. He further explicitly stated:

> "Economics as a discipline arose in the dying decades of the Malthusian era. Classical economics was a brilliantly successful description of this world. But the torrent of goods unleashed by the Industrial Revolution not only created extremes of wealth and poverty across nations, it also undermined the ability of economic theory to explain these differences."

In view of this statement, Figure 5 is reinterpreted in this paper. The contention of this author is that *polarity of wealth* can be gleaned from the graph. *Polarity of wealth* means the uneven concentration of wealth in the global economic system as some parts of the world are richer than others. The polarity can also be observed to shift as some richer parts of the world decline in economic power and some poorer parts of the world rise in economic wealth.

A historical example can be gleaned from Figure 5. India and China up until 1500 CE had 50 and 60 percent share of the world's economy, respectively. In 1 CE, China had a population of almost 60 million, while the area of the present day United States had a population of 680,000. In the late 18th century, the Industrial Revolution rendered countries with large population, just countries with the largest populations.

Looking at North America, it took them 1800 years to overtake China's economic output. However, by 1950, the US economic output was three times as great as China, even though its population is three times less. Comparing China with UK in 1913, the UK had a population of 45 million and China had 437 million but the economic output was almost the same. Analyzing what happened in Europe and later in the United States, Europe was on the lead even before 1800 (Nye 2012).

In 2010 according to data, China overtook Japan as the second biggest economy, though the numbers may not be strictly comparable. Japan's data are seasonally adjusted while those for China were not. Angus Maddison suggested that China and India were the biggest economies in the world for most of the past 2000 years. That is why economists find it a mystery why they fell so far behind and now they are flourishing ('China's economy' 2010). This may not be mysterious considering the *polarity of wealth* as was mentioned earlier and that this polarity can be argued to emanate from the Principle of Intrinsic Reality.

The reasoning here is that the world has constant mass and energy. Assuming that all these mass and energy can be harnessed to create wealth, it could mean that there is a limit for wealth creation. Theoretically, the limit can be expanded indefinitely through R&D but experience demonstrates that this was not achieved yet or getting anywhere near being achieved. As such, the constant mass and energy vis-à-vis the limit for wealth creation can be thought of as the reason for the polarity of wealth, i.e. wealth can only shift from one entity to another.

The implication of this assertion can be demonstrated by the fact that the 1% richest have accumulated more wealth than the rest of the world combined (Oxfam International 2016). The world has never seen 100% of its population to be wealthy. As such, time and again, equitable distribution of wealth is called for and mechanisms to achieve such_should be made a priority in government policies.

In a more concrete way, Scharmer (2010) suggested measures to address this growing inequality which is consistent with the mindset espoused by the Principle of Intrinsic Reality. These measures are as follows:

- Creation of a New Class of Common Property Rights

The creation of common property rights is intended to address the cost to society of private economic undertakings. For example, a private company is extracting natural resources from a certain place. While the private company can profit from that endeavor, the community in that place might suffer the undesirable consequences (e.g. flooding due to forest denudation). To ensure sustainability and institutionalize the voice of future generations in this case, a trust can be established which will own the rights and distribute funds (e.g. from cap-and-trade system for carbon emissions) to citizens (one share per person), communities and the government for investment in public infrastructure.

The trust is an effective means because it does not function like a company (driven by profits) or government (driven by special interest groups) but that it is accountable for long-term sustainability of the specific commons it manage for future generations.

- Creation of a Human Right to Basic Income

One way to ensure equitable distribution of wealth is to create a human right to basic income for every human being. This right when combined with free or affordable access to healthcare and education would create a playing field where everyone has a chance to pursue aspirations and dreams. In other words, it enables individuals to put their real creativity for the benefit of larger community. Another intention of this measure is to free up paid labor (a form of dependence) so that human beings can enhance their creativity and entrepreneurial skills.

- Inclusive Capitalism

To eradicate the *polarity of wealth*, capitalism should be made inclusive so that money and capital flows are directed to all sectors of the economic system. Currently, our money system leads to accumulation of oversupply of money and capital in high financial but low environmental and social gain areas. At the same time, there is an undersupply of money and capital in areas that address societal and community needs. It implies that there is too much money in areas where it is not needed but not enough on areas where it is needed. To address this imbalance, it is recommended to redesign the system so that the current extraction-based financial system is replaced by institutions which can support the real economy and thereby serve the ecology and community.

It should be noted that shifting of wealth is also observed in various geographical regions. When this occurs, geo-political instability can result. This goes to show that if we want lasting stability and security within the limits of our intrinsic reality, then we should not tolerate unabated greed. Otherwise, *polarity of wealth* will continue to persist and we cannot realize our full potential as individuals and collaborators for common good.

5. Way Forward...

Recognizing our intrinsic reality leads us to greater understanding of our very nature bounded by the constancy of mass and energy. It shed light on why growth stabilizes after addition of more inputs. Within this purview, trade-offs are also encountered which necessitates for R&D. R&D initiatives however should be underpinned on sustainability.

Looking at the practice of chemical engineering from the lens of intrinsic reality, the approach should be futuristic while considering relevant aspects related to factors of production and human beings.

The bigger picture shows that Engineers particularly Chemical Engineers are major players in driving the real economy. As Scharmer (2010) aptly puts it, "The essence of technology and knowledge creation, and the ultimate driving force of the economy, is the same: it is the human capacity to create." As such, if sustainability is to be achieved, Chemical Engineers have a greater responsibility to innovate for regenerative and social technologies. In this way, we can stem the polarity of wealth by providing affordable technology to enable disadvantaged individuals enhance their creativity, innovation and entrepreneurship. This can help them partake in the benefits of inclusive capitalism.

As demonstrated in this paper, the synergy emanating from the convergence of Chemical Engineering and Economics can powerfully tackle the world's woes. As such, there is a good reason for chemical engineers and economists alike to collaborate particularly on sustainability initiatives.

6. References

[1] 'China's economy overtakes Japan's in real terms' 2010, *The Economist*, 16 August, viewed 5 March 2016, http://www.economist.com/node/16834943

[2] Clark, G 2007, *A Farewell to Alms: A Brief Economic History of the World,* Princeton University Press, Princeton, NJ.

[3] Edgar, T, Himmelblau D & Lasdon, L 2001, *Optimization of Chemical Processes*, McGraw-Hill, New York.

[4] Huang, Y & Wang, B 2011, 'From the Asian Miracle to an Asian Century? Economic Transformation in the 2000s and Prospects for the 2010s', in H Gerard & J Kearns (eds), *Proceedings of the Reserve Bank of Australia annual conference*, Reserve Bank of Australia, Sydney, viewed 30 January 2016,

http://www.rba.gov.au/publications/confs/2011/huang-wang.html

[5] Kolko, J 2015, 'Design Thinking Comes of Age', *Harvard Business Review*, pp. 66-71, viewed 24 February 2016, https://hbr.org/2015/09/design-thinking-comes-of-age

[6] K.N.C. 2012, 'Mis-charting economic history: more 2,000 years in a single graphic', *The Economist blog*, web log post, 20 June, viewed 29 February 2016,

http://www.economist.com/blogs/graphicdetail/2012/06/mis-charting-economic-history

[7] Levitt, S & Dubner, S 2005, *Freakonomics: A Rogue Economist Explores the Hidden Side of Everything*, William Morrow, New York, NY.

[8] Malthus, T 1798, *An Essay on the Principle of Population*, J. Johnson in St. Paul's Church-yard, London.

[9] Maslow, A 1943, 'A Theory of Human Motivation', *Psychological Review*, vol. 50, pp. 370-396, viewed 24 February 2016, http://psychclassics.yorku.ca/Maslow/motivation.htm

[10] Monod, J 1949, 'The Growth of Bacterial Cultures', *Annual Review of Microbiology*, vol. 3, pp. 371-394, viewed 2 February 2016,

http://www.annualreviews.org/doi/abs/10.1146/annurev.mi.03.100149.002103

[11] Nye, J 2012, 'Fascinating new graph shows the 'economic history of the world since Jesus'', *MailOnline*, 23 June, viewed 15 November 2015,

http://www.dailymail.co.uk/news/article-2163610/Fascinating-new-graph-shows-economic-history-world-Jesus.html

[12] Oxfam International 2016, *An economy for the 1%, how privilege and power in the economy drive extreme inequality and how this can be stopped*, Oxford, viewed 5 March 2016,

https://www.oxfam.org/sites/www.oxfam.org/files/file_attachments/bp210-economy-one-percent-tax-havens-180116-en_0.pdf

[13] Samuelson, P & Nordhaus W 2009, *Economics*, McGraw-Hill Higher Education, New York, NY.

[14] Scharmer, O 2010, 'Seven Acupuncture Points for Shifting Capitalism to Create a Regenerative Ecosystem Economy', *Oxford Leadership Journal*, vol. 1, pp. 1-21, viewed 24 February 2016, http://www.oxfordleadership.com/journal/vol1_issue3/scharmer.pdf

[15] Senge, P 2006, *The Fifth Discipline: The Art & Practice of the Learning Organization*, Doubleday/Currency, New York.

[16] Smith, R 2005, *Chemical Process Design and Integration*, John Wiley & Sons Ltd., West Sussex, England.

[17] Thompson, D 2012, 'The economic history of the last 2,000 years in 1 little graph', *The Atlantic blog*, web log post, 19 June, viewed 29 February 2016, http://www.theatlantic.com/business/archive/2012/06/the-economic-history-of-the-last-2-000-years-in-1-little-graph/258676/

Acknowledgment

The author would like to thank Hyundai Engineering and Construction Co. Ltd. for the support in this endeavor. The author is also thankful to Ms. Sofia Ayu for meticulously editing this manuscript and Mr. Marcial N. Barroga for the poster design.

444116 Impacts of the New National Ozone Standard

Abstract:

Now that EPA has announced a new lower national ozone standard, there will be potential impacts to all sectors of the economy. This includes small industry, large industry, small businesses, large businesses, and potentially every individual consumer in the United States. This presentation will discuss the current state of the ozone air quality and the challenges to meet the new lower ozone standard. Challenges for industry located in ozone non-attainment areas include lower major source thresholds and higher NSR offset ratios for permitting. Challenges for state regulatory agencies include resources needed to address more non-attainment areas and addressing background ozone and international emissions contributions.

LUMMUS CONSULTANTS INTERNATIONAL

Towards a Level Playing Field

Permitting and Regulatory Compliance Around the World

Gordon Combs
Lummus Consultants International
(a CB&I Company)

LUMMUS CONSULTANTS INTERNATIONAL

Environmental Regulations in the USA

- In the beginning …
  - ✓ 1948 – Federal Water Pollution Control Act ("FWPCA")
    - Established vague and often contradictory framework for national water pollution control policy
    - Established the Water Pollution Control Advisory Board
      - Surgeon General or his designee
      - Department of the Army representative
      - Federal Works Agency representative
      - Department of Agriculture representative
      - Six more Presidential appointees
        - Engineer with expertise in sewage and industrial waste disposal
        - Person with interest in wildlife conservation
        - Representative of municipal government
        - Representative of State government
        - Representative of an affected industry
        - One more person
    - Supported technological research into industrial waste treatment
    - Provided financial support to State and local entities for construction of treatment facilities

LUMMUS CONSULTANTS INTERNATIONAL

Environmental Regulations in the USA

- The Modern Era
  - ✓ 1970 – Environmental Protection Agency ("USEPA")
  - ✓ 1971 – Federal Clean Air Act ("CAA")
  - ✓ 1972 – Federal Clean Water Act ("CWA")
  - ✓ 1974 – Federal Safe Drinking Water Act ("SDWA")
    - Established underground injection control ("UIC") program
    - Does not apply to injection of brine or other fluids associated with the production of oil or natural gas
  - ✓ 1976 – Federal Resource Conservation and Recovery Act ("RCRA")
    - Established definitions of "solid waste" and "hazardous waste"
    - Completed the circle of regulations
      - Take it out of the air and put it into the water
      - Take it out of the water and put it into the ground
      - Take it out of the ground and put it into the air
- Regulatory system based on enforcement ("Command and Control")

A World of **Solutions** April 2016 2

LUMMUS CONSULTANTS INTERNATIONAL

International Regulations

- Europe and other "Developed" Countries
  - ✓ Comparable laws and regulations to the USA
  - ✓ Comparable promulgation schedule to the USA
  - ✓ Comparable enforcement policies to the USA
- Developing Countries
  - ✓ Comparable laws and regulations to the USA (today)
  - ✓ Promulgation schedules
    - Vary widely by country
    - Some nations still developing first regulations in the 1990s
    - Virtually all nations have comparable regulatory system today
  - ✓ Enforcement
    - Varies by country
    - More emphasis on working with industry to reach "win-win" solution

A World of **Solutions** April 2016 3

LUMMUS CONSULTANTS INTERNATIONAL

Project Finance

- Definition
  - ✓ Method of financing in which the lender looks primarily to the revenues generated by a single Project, both as the source of repayment and as security for the exposure, rather than to the parent company; i.e., "non-recourse" financing
  - ✓ Usually for large, complex and expensive Projects such as power plants, petroleum refineries, chemical processing plants, mines, oil and gas production, transportation, telecommunications and environmental treatment facilities
  - ✓ Financing of new Projects and refinancing of existing Projects
  - ✓ Borrower usually a Special Purpose Entity that is not permitted to perform any function other than developing, owning and operating the Project
  - ✓ Loan repayment depends primarily on the Project's cash flow and on the collateral value of the Project's assets
- Reserve-based financing in extractive sectors which is non-recourse and where the proceeds are used to develop one particular reserve (e.g., oil field) is considered to be a Project Finance transaction

A World of Solutions April 2016 4

LUMMUS CONSULTANTS INTERNATIONAL

Financial Institutions

- Multilateral Financial Institutions (Public Sector)
  - ✓ World Bank Group
    - International Bank for Reconstruction and Development ("IBRD") - government Projects
    - International Development Association ("IDA") – government Projects in poor countries
    - International Finance Corporation ("IFC") – loans for private Projects
    - Multilateral Investment Guarantee Agency ("MIGA") – political risk insurance
  - ✓ Regional Institutions
    - InterAmerican Development Bank ("IDB")
    - European Bank for Reconstruction and Development ("EBRD")
    - Asian Development Bank ("ADB")
  - ✓ Export Credit Associations ("ECAs")
    - United States Export-Import Bank ("US Ex-Im")
    - Export Development Canada ("EDC")
    - Japan Bank for International Cooperation ("JBIC")
- Development Banks (Private Sector)
    - Citigroup (USA)
    - Credit Agricole Corporate and Investment Bank (France)
    - Sumitomo Mitsui Banking Corporation (Japan)

A World of Solutions April 2016 5

World Bank Standards

- History
  - ✓ 1987 – Pollution Prevention and Abatement Handbook ("PPAH")
  - ✓ 1990s – Safeguard Policies
  - ✓ 2006 – Performance Standards
  - ✓ 2012 – Performance Standards (current revision)
- Today
  - ✓ Performance Standards – define borrower's responsibilities for managing the Project's environmental and social risks
  - ✓ Environmental, Health and Safety ("EHS") Guidelines – successor to PPAH
  - ✓ Common reference for Projects worldwide
  - ✓ All standards available via the Internet (for free!)
  - ✓ "Does the Project Comply with the World Bank Standards?"

IFC Safeguard Policies – 1990s

OP 4.01 Environmental Assessment

OP 4.04 Natural Habitats

OP 4.09 Pest Management

OP 4.36 Forestry

OP 4.37 Safety of Dams

OP 7.50 International Waterways

OP 4.20 Indigenous Peoples

OD 4.30 Involuntary Resettlement

OPN 11.03 Cultural Property

Child and Forced Labor Policy Statement

LUMMUS CONSULTANTS INTERNATIONAL

IFC Performance Standards - 2006

1. Social and Environmental Assessment and Management Systems
2. Labor and Working Conditions
3. Pollution Prevention and Abatement
4. Community Health, Safety and Security
5. Land Acquisition and Involuntary Resettlement
6. Biodiversity Conservation and Sustainable Natural Resource Management
7. Indigenous Peoples
8. Cultural Heritage

A World of Solutions April 2016 8

LUMMUS CONSULTANTS INTERNATIONAL

IFC Performance Standards – 2012

1. Assessment and Management of Environmental and Social Risks and Impacts
2. Labor and Working Conditions
3. Resource Efficiency and Pollution Prevention (reference the EHS Guidelines)
4. Community Health, Safety and Security
5. Land Acquisition and Involuntary Resettlement
6. Biodiversity Conservation and Sustainable Management of Living Natural Resources
7. Indigenous Peoples
8. Cultural Heritage

A World of Solutions April 2016 9

IFC EHS Guidelines

- Technical reference documents
  - ✓ Performance levels normally acceptable to IFC
  - ✓ Achievable in new facilities at reasonable cost by existing technology
  - ✓ Reference several globally accepted standards (e.g., WHO, USEPA, API, IMO)
- Project required to meet host country or IFC requirements
  - ✓ Whichever is more strict
  - ✓ Site-specific requirements if justified by Environmental Impact Assessment

General EHS Guidelines

- Cross-cutting EHS issues potentially applicable to all industry sectors
- Categories
  - ✓ Environmental
  - ✓ Occupational Health and Safety
  - ✓ Community Health and Safety
  - ✓ Construction and Decommissioning

LUMMUS CONSULTANTS INTERNATIONAL

Sector-Specific EHS Guidelines

- Industry-specific examples of Good International Industry Practice ("GIIP")
- Categories
  - ✓ Agribusiness / Food Production
  - ✓ Chemicals, including
    - Large Volume Petroleum-based Organic Chemicals Manufacturing
    - Natural Gas Processing
    - Petroleum Refining
  - ✓ Forestry
  - ✓ General Manufacturing
  - ✓ Infrastructure, including
    - Crude Oil and Petroleum Product Terminals
    - Ports, Harbors and Terminals
  - ✓ Mining
  - ✓ Oil and Gas
    - LNG liquefaction and regasification
    - Oil and gas exploration, development and production (onshore and offshore)
  - ✓ Power Generation, Transmission and Distribution

LUMMUS CONSULTANTS INTERNATIONAL

Equator Principles

- Purpose
  - ✓ Designed for use by private sector financial institutions – *particularly when one or more multi-lateral institutions is not involved*
  - ✓ Based on policies and guidelines of the World Bank Group
- History
  - ✓ 2003 – 10 banks from 7 countries created the Equator Principles
  - ✓ 2006 – First general revision
  - ✓ 2013 – Current revision (available via the Internet – for free!)
  - ✓ Today
    - 83 banks in 36 countries worldwide
    - 70 percent of international Project Finance debt in emerging markets

LUMMUS CONSULTANTS INTERNATIONAL

Equator Principles Categories and Requirements

1. Review and Categorization
   - ✓ Category A – potentially significant adverse environmental and social impacts
   - ✓ Category B – potentially limited adverse environmental and social impacts
   - ✓ Category C – minimal or no adverse environmental and social impacts
2. Environmental and Social Assessment
   - ✓ ESIA required for Category A and (some) Category B Projects
   - ✓ Alternatives analysis required if GHG emissions exceed 100,000 tonnes CO2-e per year
3. Applicable Environmental and Social Standards
   - ✓ Designated (developed) countries – host country standards
   - ✓ Non-designated countries – IFC Performance Standards and EHS Guidelines
4. Environmental and Social Management System and Action Plan
5. Stakeholder Engagement
6. Grievance Mechanism
7. Independent Review – pre-financial close
8. Covenants – Borrower agrees to comply (in writing)
9. Independent Monitoring and Reporting – construction and operations
10. Reporting and Transparency

A World of **Solutions** April 2016 14

LUMMUS CONSULTANTS INTERNATIONAL

Conclusions

- Today, industry cannot "export pollution" by building facilities in developing countries with lax environmental and social requirements
- Some international requirements (e.g., offsite noise control) are stricter for Projects in developing countries than in the USA
- Mitigation of potential social impacts just as important (if not more important) as mitigation of potential environmental impacts
- Financial institutions are a primary driving force behind international environmental and social progress associated with public and private industrial development
- Golden Rule of Business: "Whoever has the gold makes the rules!"

A World of **Solutions** 15

LUMMUS CONSULTANTS
INTERNATIONAL

CB&I

LUMMUS CONSULTANTS
INTERNATIONAL

A World of **Solutions**

BRACEWELL

AIChE Ethylene Producers' Conference
Ethylene Plant Environmental Panel

Proposed Amendments to EPA's Risk Management Program Regulations

April 12, 2016

Matt Paulson
Bracewell LLP

BRACEWELL

Overview

- Background on EPA's Risk Management Program
- Overview of Existing RMP Regulations
- Review of Proposed Rule
  - Prevention Program Requirements
  - Emergency Response Preparedness Requirements
  - Information Availability Requirements
- Specific Requests for Comment

2

BRACEWELL

Background on Risk Management Program

- 1984, Union Carbide, Bhopal India
- 1985, Waxman (D. CA) introduces first proposed legislation
- 1986 EPCRA
- 1990 CAA Amendments

3

BRACEWELL

Overview of Existing RMP Regulations

- Applicability
  - More than a threshold quantity
  - Of a regulated substance
  - In a process
  - At a stationary source
- Three Main Components
  - Hazard assessment
  - Prevention program
  - Emergency response program

4

BRACEWELL

Overview of Existing RMP Regulations (cont.)

- Three-Tiered Program
  - **Program 1**: No affect to public of worst case release; no accidents with offsite consequence in five years; limited to hazard assessment, coordination, and plan submission
  - **Program 2**: Not eligible for Program 1/not subject to Program 3; streamlined prevention program requirements; additional hazard assessment, management and emergency response requirements
  - **Program 3**: Not eligible for Program 1 and either subject to OSHA PSM standard or listed industry sector; accident prevention requirements (14 PSM elements); Program 2 requirements

5

BRACEWELL

Background on Proposed Rule

- Why is EPA revising the RMP regulations now?
  - 2013 West Fertilizer incident
- The following incidents are also highlighted by EPA in the preamble:
  - BP Texas City
  - Tesoro Anacortes
  - Chevron Richmond
  - Williams Geismar

6

BRACEWELL

Background on Proposed Rule (cont.)

- August 2013 – Executive Order 13650
- December 2013 – OSHA PSM RFI
- May 2014 – Working Group Report
- July 2014 – EPA RMP RFI

7

BRACEWELL

Background on Proposed Rule (cont.)

- July 2014 – EPA RMP RFI (cont.)
  - Areas addressed in RFI but not covered by proposed rule:
    - Regulated Substances List
    - RAGAGEP
    - Safety Critical Equipment
    - Safety Case
    - Facility Siting
    - Automatic Detection and Monitoring
    - Offsite Consequence Data
    - Emergency Shutdown

8

BRACEWELL

Background on Proposed Rule (cont.)

- Dec., 21, 2015 – EPA proposed rule sent to OMB
- February 25, 2016 – Pre-publication version of proposal released
- *Federal Register* publication – March 14, 2016
- Comment deadline – May 13, 2016
    - Unusual, shorter "soft" deadline for information availability-related comments – April 13, 2016
- Final rule issued – Fall 2016?

9

BRACEWELL

Overview of Proposed Rule

- Prevention Program Requirements
    - Incident Investigation
    - Third Party Compliance Audits
    - Safer Technology and Alternatives Analysis
- Emergency Response Preparedness Requirements
    - Coordination with Local Responders
    - Facility Exercises
- Information Availability Requirements
    - LEPC Disclosure Requirements
    - Public Availability of Information
- Specific Requests for Comment

10

BRACEWELL

Prevention Program Requirements

- Incident Investigation
  - Revisions to Definition of "Catastrophic Release"
    - "The owner or operator shall investigate each incident that ~~which~~ (1) resulted in~~, or could reasonably have resulted in~~ a catastrophic release ~~of a regulated substance~~ . . . or (2) Could reasonably have resulted in a catastrophic release (i.e., was a near miss)."
    - "Catastrophic release means a major uncontrolled emission, fire, or explosion, involving one or more regulated substances that ~~presents imminent and substantial endangerment to public health and the environment~~ results in deaths, injuries, or significant damage on-site, or known offsite deaths, injuries, evacuations, sheltering in place, property damage, or environmental damage.

11

BRACEWELL

Prevention Program Requirements (cont.)

- Incident Investigation (cont.)
  - New definition – "root cause"
    - The report shall include . . . "factors that contributed to the incident including the initiating event, direct and indirect contributing factors, and root causes."
    - "Root cause means a fundamental, underlying, system-related reason why an incident occurred that identifies correctable failure(s) in management systems."
  - Use of undefined term – "near miss"
    - Preamble discussion: "any unanticipated release of a regulated substance, and some process upsets could potentially lead to a catastrophic release."

12

BRACEWELL

Prevention Program Requirements (cont.)

- Third Party Compliance Audits
  - When is it required?
    - RMP-reportable accident, or
    - Regulatory authority requires one based on any noncompliance
  - Auditor Qualifications
    - RMP program knowledge, certification, training and PE
    - Source type and process unit experience
    - Experience with applicable RAGAGEP
    - Receive no financial benefit tied to outcome
    - No consulting services for owner/operator in past three years or for three years post-audit

13

BRACEWELL

Prevention Program Requirements (cont.)

- Third Party Compliance Audits (cont.)
  - Audit Report
    - Must provide to EPA
    - Include drafts
    - No legal privileges – becomes publicly accessible
  - Audit Findings
    - 90 days after completion of audit owner/operator must submit response to each finding with a schedule for "promptly" addressing deficiencies
  - Submission to Audit Committee of Board of Directors

14

BRACEWELL

Prevention Program Requirements (cont.)

- Safer Technology and Alternatives Analysis
  - Applies to Program 3 facilities in three industries:
    - paper manufacturing
    - coal and petroleum products manufacturing
    - chemical manufacturing
  - Facilities are required to evaluate the "feasibility" of safer technology and alternatives when conducting PHAs

15

BRACEWELL

Prevention Program Requirements (cont.)

- Safer Technology and Alternatives Analysis (cont.)
  - PHA must include analysis of potential safer technology and alternatives in the following order of preference: IST or ISD, passive measures, active measures, procedural measures
    - IST/ISD: measures that would replace or reduce the use of regulated substances or make operating conditions less hazardous or complex
    - Passive measures: dikes, berms, blast walls
    - Active measures: alarms, safety instrumented systems
    - Procedural measures: policies, operating procedures, training

16

BRACEWELL

Prevention Program Requirements (cont.)

- Safer Technology and Alternatives Analysis (cont.)
  - Definition of "feasibility": "capable of being successfully accomplished within a reasonable time, accounting for economic, environmental, legal, social and technological factors."
  - EPA rejected this same concept in 1996 rulemaking – concluded that the existing PHA requirement was sufficient.
  - What has changed?
    - Safer technologies are "becoming more widely available and being adopted by some companies"

17

BRACEWELL

Prevention Program Requirements (cont.)

- Prevention Program Takeaways:
  - Potential for significant increase in number of formal investigations
  - Mandates identification and correction of management system failure even where no such failures exist
  - Releases with only on-site impacts can trigger formal investigation
  - Third party audits could be required in response to any violation
  - Disclosure of audits could fundamentally undermine audit process
  - Rigid auditor qualifications could exclude many auditors
  - STAA would require burdensome justifications for technology and design choices

18

BRACEWELL

Emergency Response Preparedness

- Coordination with Local Responders
  - Documentation of coordination efforts:
    - Names and contact information of individuals involved
    - Dates of coordination
    - Conclusions
    - Next steps

19

BRACEWELL

Emergency Response Preparedness (cont.)

- Facility Exercises
  - Annual tabletop emergency response exercises
  - Field exercise every five years
  - Annual notification exercises

20

BRACEWELL

Emergency Response Preparedness (cont.)

- Emergency Response Takeaways:
  - Potential to overwhelm local responders
  - Owner/operator would bear enforcement risk – "ultimate burden rests with the owner/operator"
  - Security concerns

21

BRACEWELL

Information Availability Requirements

- LEPC Disclosure Requirements
  - Regulated substances on site
  - Compliance audit summary reports
  - Summaries of incident investigation reports
  - Summary if IST implemented or planned to be implemented
  - Reports on emergency response exercises/drills

22

BRACEWELL

Information Availability Requirements (cont.)

- Public Availability of Information
  - Names of hazardous chemicals
  - MSDSs for all regulated substances at facility
  - Five-year accident history
  - Summary of emergency response program
  - Summary of emergency response exercises
  - LEPC contact information

23

BRACEWELL

Information Availability Requirements (cont.)

- Information Availability Takeaways:
  - Public already has access to much of this information
  - Public release of sensitive security information
  - Unclear how much detail is required in "summaries"

24

BRACEWELL

Specific Requests for Comment

- Increase scope of RCA
- Require incident investigation report completed before restart
- Expand third party audit trigger to include conditions "suggesting a concern"
- Require third party auditor to submit draft report to regulatory authority before provided to owner/operator
- Require implementation of STAA, rather than feasibility analysis
- Create "score card" for third party auditors to grade facilities
- New facility siting requirements (extensive discussion)
- Emergency shutdown (extensive discussion)

25

BRACEWELL

Specific Requests for Comment (cont.)

- Request for Comment Takeaways:
  - Could provide EPA with argument against claim that adding new requirements to final rule not in proposal violates scope of proposal/reasonable outgrowth limitations
  - Facility siting and emergency response most likely to result in new requirements in final rule that were not in proposal

26

BRACEWELL

Matt Paulson
Bracewell LLP

111 Congress Avenue, Suite 2300
Austin, Texas 78701-4061
512.494.3659

711 Louisiana Street, Suite 2300
Houston, Texas 77002-2770
713.221.4959

matthew.paulson@bracewelllaw.com

Texas
New York
Washington, D.C.
Connecticut
Seattle
Dubai
London

bracewelllaw.com

27

Application of Dynamic Simulation in Improving Plant Efficiency

By

Tim Song, Maryn Pols, Prakash Ramdoss

CB&I
2 Riverway, Suite 1300
Houston, TX 77056
Tel: +1 713 375 8711

Prakash.ramdoss@cbi.com

Abstract

In ethylene plants, considerate amount of CO_2 and NOx emissions come from gas turbine exhaust gas and furnace flue gas. The hot gas turbine exhaust gas may be reused as oxygen source in the thermal cracking furnaces in place of cold ambient air, thus increasing the overall thermal efficiency, potentially reducing CO_2 and NOx emissions from gas turbine combustion chamber and furnace combustion chamber combined. However the integration between gas turbine operation and the furnace operations raises operating flexibility and plant safety concerns, as the combined process needs to be able to operate in both ambient air mode and turbine exhaust gas mode and be able to transit from one mode to another in a safe and robust manner.

This work is part of a revamping project at an ethylene plant, where a new gas turbine is to be installed and integrated with the existing furnaces to improve plant efficiency. The hot gas turbine exhaust gas is routed to the furnaces, instead of the cold ambient air there by reducing fuel need for the furnaces. The work demonstrates that the dynamic process simulation models can be utilized to study the transition between modes of operation, revealing critical process insights. The transient analysis can also be used to determine key operating parameters such as damper opening and closing time parameters to ensure safe and smooth transitions and operations. Consequently, the analysis assures the plant operators with high confidence the robustness of integration between the gas turbine and furnace processes which increases the plant fuel efficiency thereby reducing emissions.

Introduction

The flue gas from the new gas turbine is reused as oxygen source for the thermal cracking furnaces, therefore reducing CO_2 and NOx emissions from gas turbine exhaust gas and furnace flue gas combined.

A transient analysis for the flue gas system is required to ensure that sufficient combustion air flow to the cracking heaters is maintained at all times during the system transient conditions to prevent sub-stoichiometric combustion and positive heater firebox pressure leading to personnel safety concerns.

The purpose of the computer modeling is to dynamically represent the combustion air supply system and induced draft flue gas system to allow simulation of the transient response of these systems during upset conditions.

The computer model determines the pressure, temperature, mass flow rate and composition at critical points in the system as a function of time. The interaction of the combustion air, the louvers and dampers and the controllers is very important to an accurate calculation of the dynamic response of the integrated combustion air system. The time delays associated with the transient phenomena in the ducting, the sensor response and the valve motions are modeled to determine if the control systems can respond fast enough to prevent an unstable or dangerous condition. The residual heat in the refractory-lined duct system is taken into account for an accurate representation of the flows. Boundary conditions for the computer analysis are provided by models representing the fans, gas turbine, louvers and dampers, controllers and cracking heaters.

The transients of the upset scenarios described below are analyzed for the new gas turbine duct system.

1. The new gas turbine trip.
2. Trip of trim fan associated with the new gas turbine.
3. Trip of the induced draft fan on a thermal cracking furnace.

The primary purpose of the transient analysis is to ensure that the post-trip interlock logic and control strategy:

- Provides sufficient combustion air flow to the cracking heaters at all times during the system transient conditions avoiding shutting down heaters.
- Prevents sub-stoichiometric combustion.
- Prevents a severe pressure transient which may temporarily lead to positive heater pressures causing concerns regarding personnel safety.
- Avoids potential damage to the combustion air and gas turbine exhaust ducts due to excessive draft pressure.
- Prevents the trip of the GTG as a result of a single heater upset.
- Prevents potentially destabilizing pressure oscillations in the system.

The transients were performed using a computer dynamic simulation model of the new gas turbine combustion air system and induced draft flue gas system which includes all principal components (gas turbine, trim fan, combustion air supply ducting and stack, heater radiant sections, heater convection sections, ID fans and FD fans), control dampers and associated control systems.

The results have been utilized to implement appropriate post-trip interlock logic and control strategies for the revamping project.

System Description

The existing heater island consists of a number of thermal cracking furnaces, with half of them are already integrated with an existing gas turbine, supplemented with ambient air. In order to improve overall plant energy efficiency, a new gas turbine exhaust gas will be reused for the rest of thermal cracking heaters currently using ambient air for combustion.

The existing turbine exhaust gas duct system will be modified and expanded to accommodate the integration of the new machine with the thermal cracking heaters currently operating on 100% ambient air. An overflow connection between the new duct system and the existing duct system will be installed to maximize the turbine exhaust gas utilization in case a heater on the new duct system is shut down. The new duct system will have one operating trim fan without spare.

All thermal cracking furnaces have induced draft fans at the furnace top, while some of them have forced draft fans at air inlet and some do not. Upon a new gas turbine trip, some thermal cracking heaters will trip to forced draft operation while others will trip to natural draft operation. The duct branch to each of these heaters will be equipped with an ambient air intake system.

The ID fans control the pressure at the firebox arch. The furnaces that do not have forced draft fans will each be equipped with a dedicated air intake duct with control damper, which will allow ambient air operation using the ID fan as driving force to pull in combustion air when the gas turbine is down.

The furnaces require additional oxygen for stoichiometric combustion and the trim fan by side of the turbine exhaust duct provides the additional oxygen by mixing ambient air with turbine exhaust gas in the mixer. The turbine exhaust stack provides relief during upset events leading to high pressure in the duct. The main duct is pressure controlled by manipulating trim fan inlet damper, while the trim fan operates at constant speed. All furnaces are lined up along the main exhaust duct. The cross flow duct with a cross damper connects the new duct system to the existing GTG system. Normally the cross flow duct is closed. When one or more furnaces in the new duct system are down, the surplus exhaust gas will be routed to the existing duct for better plant efficiency. The furnaces' FD fans and FD fan dampers during normal turbine exhaust gas operation are closed and the fans are not running. Upon a trip of the new turbine, the furnaces will trip to forced draft operation if there is a forced draft fan associated with it, otherwise natural draft ambient air operation.

Within each heater, the simulation model takes into account the buoyancy effect, burner combustion air side pressure loss, and the convection section hydraulics loss at flue gas side in addition to the induced draft fan and the fan damper characteristics.

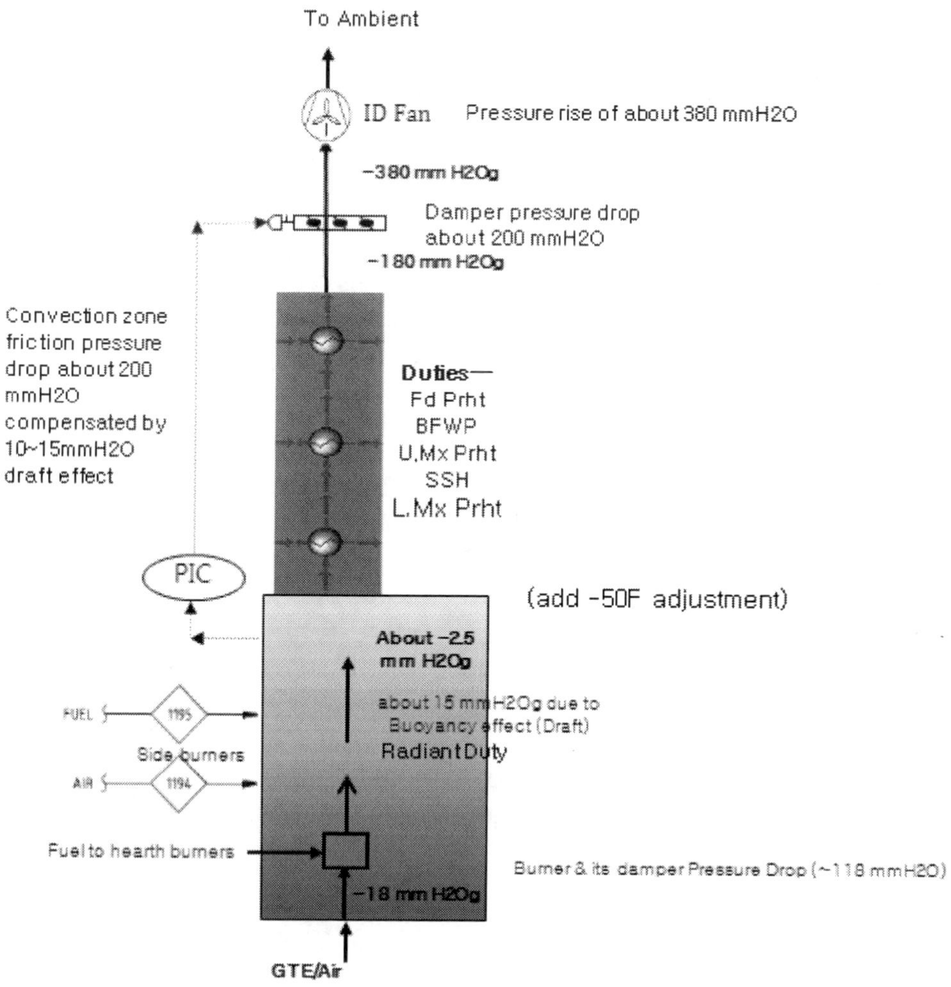

When gas turbine is tripped, both of the flow and temperature decay over time. In order to predict the pressure loss and temperature effect during transition from turbine gas operation to ambient operation and vise versa, duct sizes and duct metal wall thickness, as well as insulation layer are considered in simulation model.

All fans and dampers are modeled with performance curves defined by flow vs pressure head/loss.

Within each furnace, the process side duty demand is correlated to flue gas side by cross limiting control mechanism, such that the coil outlet temperature is properly controlled. The process side feed rate to each furnace is assumed to remain constant during the transients.

On the flue gas side of each furnace, the arch is pressure controlled by the overhead ID fan inlet damper while the ID fans operate at constant speed.

Buoyancy is rigorously considered in the firebox as well as in the flue gas stack according to past engineering practices.

Combustion is taking place in the firebox. The combustion is rigorously calculated thermodynamically, that is, the combustion reaction ensures atomic balance. If there is shortage of oxygen, carbon monoxide will be generated.

The furnace convection section is rigorously modeled at the flue gas side with actual flue gas path volume and measured heat transfer coefficients.

Process side modeling is mostly simplified with duties proportional to the feed rate and inlet temperature. The purpose of modeling feed temperature effect is such that the coil outlet temperature can eventually be adequately predicted and used as input signal to cascade control the firing via cross limiting control mechanism.

Results and Discussion

Gas Turbine Trip

The gas turbine trip is imposed 10 seconds into the simulation. At 10 seconds, the turbine is tripped and the corresponding events are carried out by the simulation model, primarily closing flue gas duct dampers to each furnace while jump open induced draft fan damper opening of each furnace momentarily. The ambient air entry to the furnaces without forced draft fans will be active when duct pressure becomes negative.

During gas turbine trip the concern is the shortage of flue gas/air, therefore oxygen rate, going into each furnace. The simulation predicts the flow rates, as well as oxygen rates, going into a typical furnace are as follows -

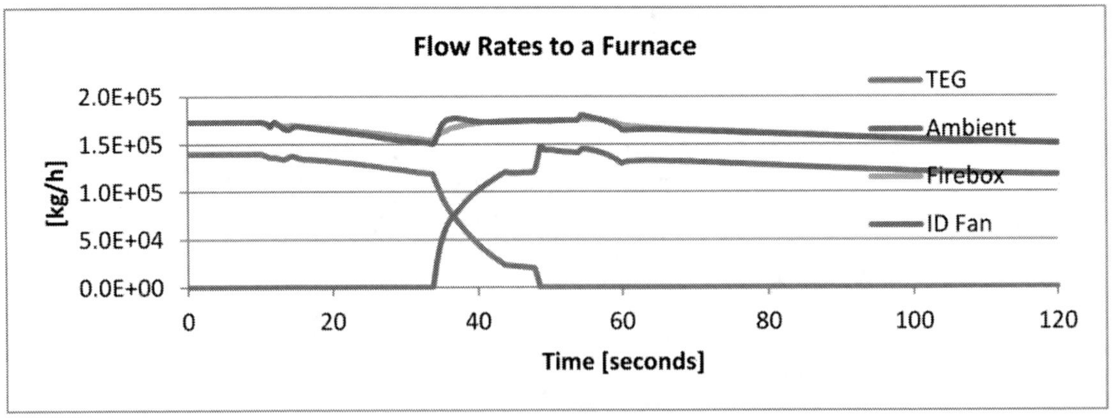

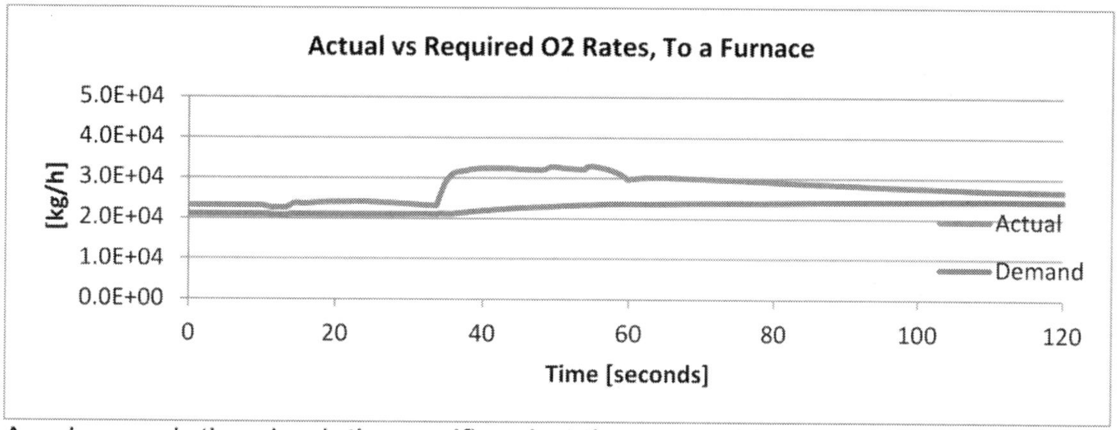

As observed, the simulation verifies that the gas turbine tripping interlocking actions are adequate to ensure sufficient air/oxygen supply to furnaces. As such, sub stoichiometric combustion has been avoided, while production rates remain unchanged.

Trim Fan Trip

The trim fan trip is imposed 10 seconds into the simulation. The sequence of events is implemented according to definition, primarily starting up the forced draft fans for those furnaces with one while leaving the rest of the furnaces intact.

The concern is the potential shortage of oxygen in the furnaces' firebox and positive pressure in the furnaces with forced draft fans.

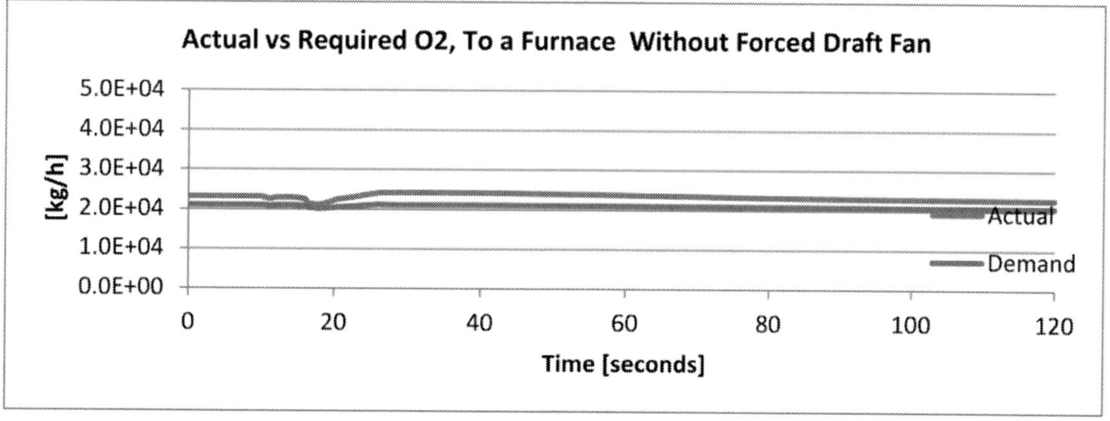

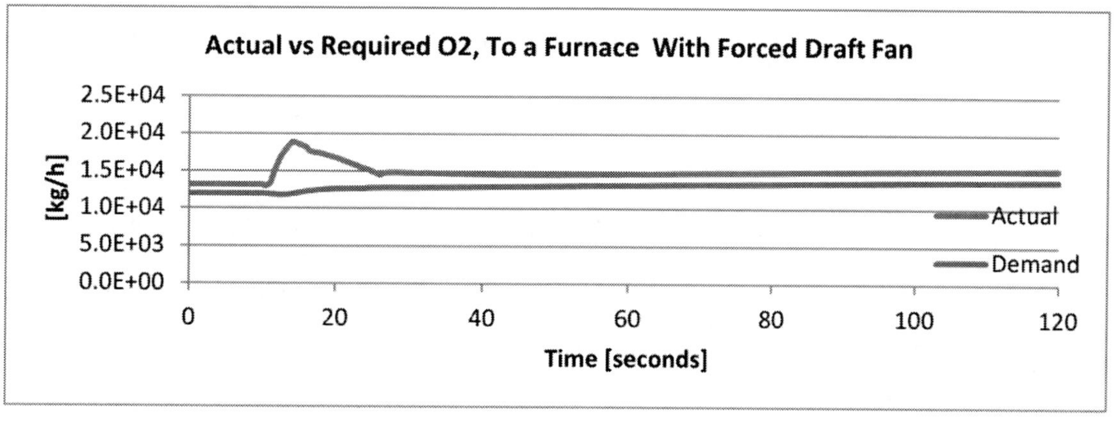

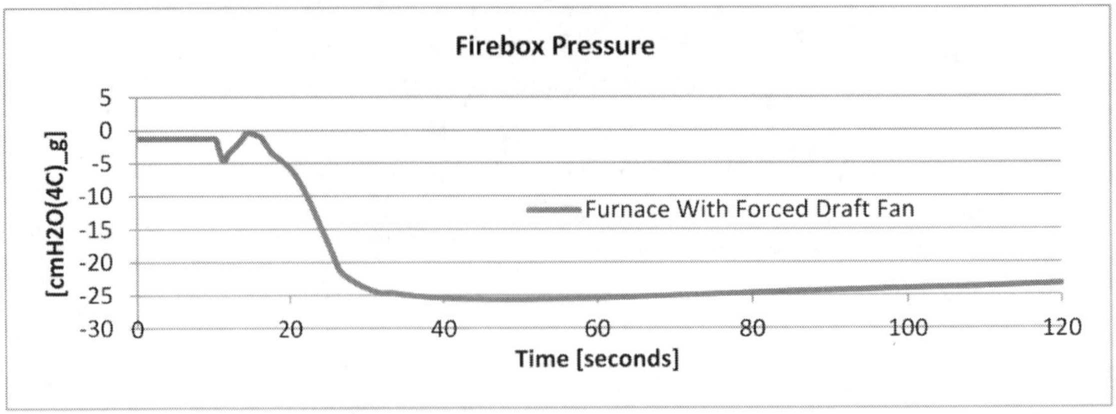

As seen, for a furnace without a forced draft fan, the oxygen supply reaches minimal level 5 seconds into the trip, yet still remains over the oxygen demand at the time. The oxygen supply is ample for furnaces with forced draft fans and the fire box pressure is also below ambient, with the highest pressure at about 5 seconds into the trip.

Induced Fan Trip

The induce fan trip is imposed 10 seconds into the simulation. Some details actions of the trip are listed below:

1. The induced draft fan is tripped and the fan is spinning down in about 3 minutes.
2. Trip close process side feed valve.
3. Trip close all fuel valves.
4. Opening up process side dilution stream valve by 25%.

Since the induced fan takes quite some time to come down, the flue gas damper will stop the flow from furnace inlet side much faster than the fan spinning down process. The net result is that the firebox pressure is predicted to go sub-ambient for as low as -150 mmH2O(g) for the furnace that is tripped, as consequence of its induced fan trip. However the abrupt stop of turbine exhaust gas consumption by the tripped heater does not disrupt the rest of the running heaters. The reduction in flue gas consumption causes the stack damper to open and vent the surplus flue gas to atmosphere. The tripped heater can be shut down in a controlled manner without leading to positive firebox pressure or severe shortage of oxygen in the firebox. Thus, safe operation is anticipated during the induced draft fan trip.

Conclusions

The transient behaviors of the flue gas system bridging turbine exhaust and furnace burners are examined and the results indicate that the system can safely transit from turbine exhaust gas operation to ambient air operation during gas turbine trip, under the condition of a carefully selected series of mitigation measures. That is, both sub-stoichiometric combustion and positive firebox pressure can be avoided.

The results also indicate that the system can safely transit from the induced draft fan trip of a heater, and trim fan trip as long as proper mitigation measures are implemented.

Highlights of the key mitigation measures during gas turbine trip include:

1. For each furnace, switch arch pressure controller and TEG/air intake flow controller to manual mode and open ID fan dampers and TEG dampers on heaters without FD fans to a certain degree while holding TEG dampers opening for heaters with FD fan.

2. When FD fan speed is up, open FD fan damper to a certain degree while closing the corresponding TEG damper, making sure the net flow to the heater remain largely unchanged.

3. When the main duct pressure drops below ambient, open stack damper fully; shutdown trim fan and close its damper; open up ambient air intake dampers to a certain degree. At same time, close TEG dampers such that the net air flow to heaters largely remains constant.

4. Optionally adjust air dampers opening to manage air flows into heaters.

5. Eventually when the process in each heater is sufficiently stable, switch the arch pressure controller back to auto, restore the air damper controls to auto and conclude the transition.

Highlights of trim fan trip are as follows:

This is an event with lesser impacts than the gas turbine trip event. The key measure is to slowly ramp close the TEG damper in about 20 seconds while allowing the combustion air flow controllers to automatically open the ambient damper after a short delay to allow the FD fan discharge pressure to equal with the duct pressure. As a result of operating on FD fans, the surplus TEG will eventually be vented to atmosphere through the stack damper. The firebox pressures can be adequately maintained below ambient while no oxygen shortages are seen.

Highlights of ID fan trip are as follows:

When the fan of a furnace trips, the TEG damper is tripped closed, stopping the combustion air flow into the tripped heater. The remaining running heaters can effectively reject the surplus TEG causing duct pressure to rise to the point that the stack damper pressure controller set point is reached and venting through stack takes place. Positive pressure is not observed in the tripped heater firebox due to residual capacity of the tripped fan during the long period of fan rotor decay time while the faster TEG damper stroke time stops the TEG flow to the heater in just a few seconds, which causes firebox pressure to go down to as low as -150 mmH2O(g).

References

1. Transient Analysis Duty Specification
2. Basis of Design
3. Material Balance
4. Process Description
5. Equipment List
6. Hydraulics Summary
7. Process Control Description
8. PFD and P&ID
9. Interlock Summary
10. Convection section rating tool reports
11. Damper characteristic definition
12. Plot plan
13. Duct Pressure Loss Calculations
14. Reference projects done in the past.
15. Heater burner data sheets
16. Heater data sheets
17. KOM
18. TRIM FAN DAMPER Perf. Curve
19. Pressure Drop Calculation
20. Damper with Actuator Data Sheet
21. Technical Data Table for Process Damper
22. FDF datasheet & curve
23. MOM Study Review Meeting
24. DCS Data for Analysis
25. Actual ID fan damper opening

AIChE Paper 86A

DESIGN, OPERATION AND STARTUP STRATEGY OF A NEW PARALLEL C2 SPLITTER

André Bernard
Process Design Engineer
NOVA Chemicals (Canada) Ltd.

Prepared for Presentation at the 2016 Spring National Meeting
Houston, TX, April 10th – April 14th, 2016

AIChE Paper

DESIGN, OPERATION AND STARTUP STRATEGY OF A NEW PARALLEL C2 SPLITTER

André Bernard
Process Design Engineer
NOVA Chemicals (Canada) Ltd.

Abstract

Feed transition to larger volumes of ethane is known to increase the operating loads in a C2 Splitter that was originally designed for a plant that processes a mix of NGL and Naphtha feedstocks. Revamp options such as changing to high performance trays may sometimes overcome the shortfall in capacity that can be expected due to the additional ethane feed. However, in some cases upsizing the tower or installing a parallel unit become the only viable options. This paper discusses the implications of processing larger volumes of ethane feed on a C2 Splitter along with the revamp options that could be considered. Operation and startup strategies are also discussed.

Introduction

NOVA Chemicals Corunna Site was originally started up in 1977. The olefins unit was designed and operated as a naphtha cracker. Over the years, through revamp and expansion projects; feed flexibility was introduced such that the plant could process a mix of liquid and natural gas liquid (NGL) feedstocks. The last plant expansion was implemented in 2005 and brought the ethylene production to 180% of the original nameplate capacity[1,2]. At that point the plant could process a feed slate with an NGL contribution varying between 30 to 65% of the total feed. The NGL feedstocks were predominantly propane and butane with a limited ethane contribution varying between 5 to 15% of the total feed.

Market conditions for olefins feedstocks around 2010 resulted in a shift by many plants towards lighter feeds. Availability of ethane from shale gas propelled ethane as the feedstock of choice for ethylene plants. This shift in feedstocks market conditions drove a project for converting the Corunna cracker to 100% NGL feed, primarily ethane.

C2 Splitter System

The Corunna site olefins unit process scheme is a front end Demethanizer with back end Acetylene reactors. The feed to the C2 Splitter system is a mixture of mostly ethane/ ethylene coming from the overhead of the Deethanizer after conversion of acetylene in catalytic reactors. The C2 Splitter system prior to the feed conversion project is illustrated on Figure 1. It is a high pressure system, which uses -40 °C propylene refrigerant (C3R) for condensing overhead vapors. Heat to the bottom reboiler is provided by condensing C3R vapor. A side reboiler provides heat integration with charge gas entering the chilling train. Ethylene product is withdrawn at tray 10 and ethane is recovered at the bottom of the tower and recycle to the cracking furnaces. The top 10 trays are used as a pasteurization section, preventing accumulation of methane and hydrogen coming from the Acetylene reactors. A small rectifier tower and a vent condenser using ethylene refrigerant (C2R) allows minimization of the vent stream, which has a relatively high ethylene content. The vent stream is recycled to the charge gas compressor (CGC).

The tower was originally equipped with 122 conventional two pass trays with V1 valves. The tower was first re-trayed in the late eighties with 153 high capacity trays. A second re-tray was done in 2005 with a different type of high capacity trays[3]. The tray count in the 2005 revamp was reduced to 123.

A dilute ethylene product draw was configured in the tower in the mid-nineties in conjunction with a Styrene plant revamp as a mean of increasing the tower capacity.

The side reboiler tube bundle was upgraded with enhanced surface in the late eighties and the original 3 shell and tube condensers were replaced by a single core in kettle exchanger in the 2005 revamp.

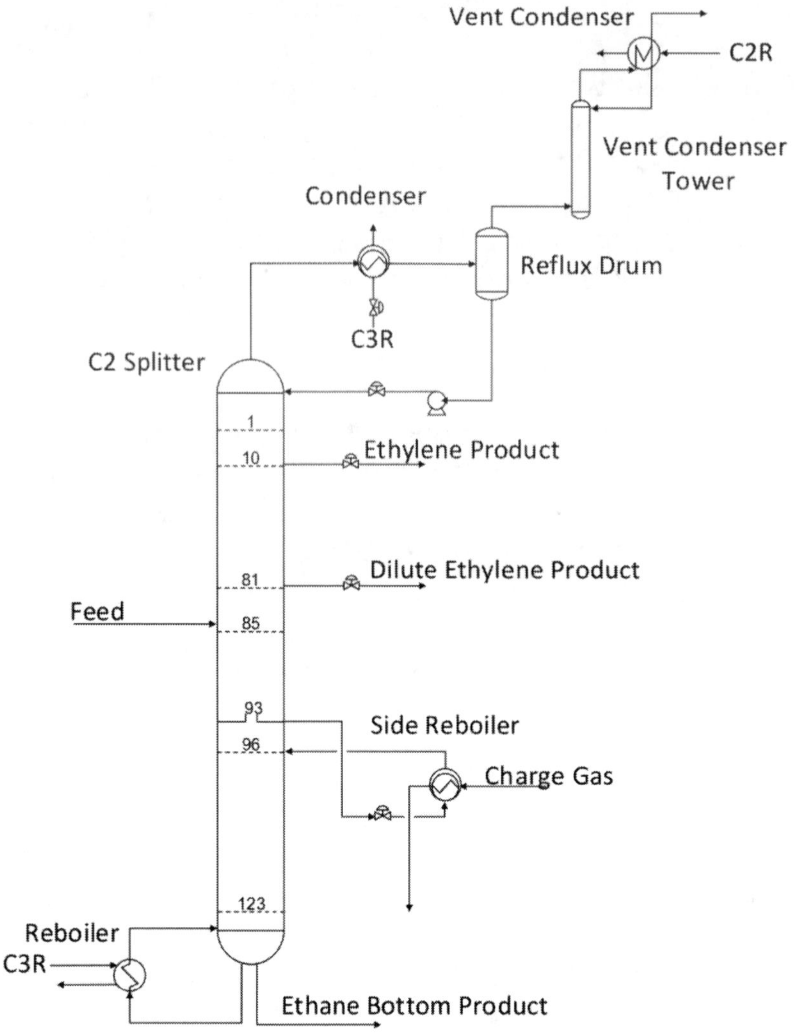

Figure 1: C2 Splitter System

Revamp Project

High crude oil price and availability of ethane from the Marcellus shale in 2010 led to a project for converting the Corunna Ethylene unit to 100% NGL feed. The design feed slate, based on ethane/ butane/ propane; substantially increased the volume of ethane cracking.

The timing of the project was a challenge as the study phase was initiated at the beginning of 2010 with the next plant turnaround schedule for the fall of 2011. Given the short timeline, priorities were given to the identification of tie in's required for executing the project outside the plant turnaround. The 2011 turnaround window was therefore used strictly for providing tie in's required by the project. Identification and design of the tie in's was done as the project was in the basic engineering phase. Detail engineering was started a few months following the turnaround and completed a year later.

Understanding that higher ethane feed would impact the capacity of the C2 Splitter along with a manufacturing desire to eliminate production of dilute ethylene made the existing tower evaluation a high priority. Determination of the tower capacity relative to future operating conditions and modifications required for meeting the project objectives were critical for establishing the list of required tie in's.

Impacts of Feedstocks on C2 Splitter Operation

The plant historical ethane/ ethylene tower feed composition relative to feedstock transition is provided on Table 1. The composition for 100% ethane is also provided as it represent the bookend of the feed spectrum. As shown on Table 1, the ethane fraction in the feed increases as more NGL especially ethane is being cracked.

Table 1: Ethane/ Ethylene Feed Composition Relative to Feedstock

Feedstock	Ethane (mole %)	Ethylene (mole %)
Naphtha	15	84
Mix Naphtha/HAGO/NGL	15 - 17	81 - 85
100% NGL Project (Design)	32	67
100% Ethane Cracking	38	62

Figure 2 shows that although the number of theoretical stages required is not significantly different, the reflux ratio increases by 30 to 40% when comparing ethane to naphtha cracking. It is also observed that slim improvement in reflux ratio are obtained beyond 120 theoretical stages.

Figure 3 shows the implication of the ethane feed composition on the reflux ratio for a tower with 120 theoretical stages and feed at its optimum location. This plot suggest that although the plant had transition from the original 100% naphtha cracking operation to a mix of NGL and liquid feed (Naphtha and HAGO); the ethane feed composition had only increased slightly resulting in a 2.2% increase in required reflux ratio. The ethane feed composition expected by the 100% NGL feed project would result in a reflux ratio increase of 23.6% due to significant increase in ethane cracking. An all-ethane cracking operation would cause the reflux ratio to be 36.6% over what was required on naphtha. For similar ethylene production the internal traffic in the tower would therefore increase with the implementation of the 100% NGL feed project.

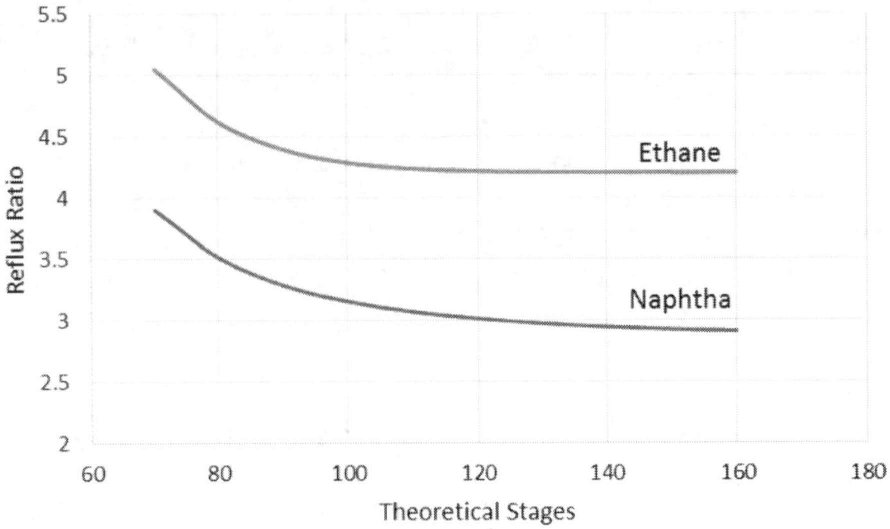

Figure 2: Impact of Theoretical Stages on Reflux Ratio

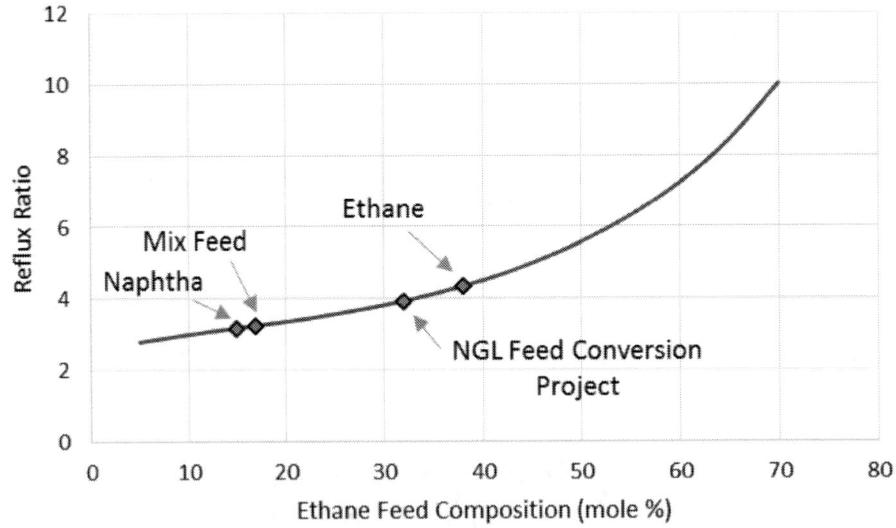

Figure 3: Impact of Ethane Feed Composition on Reflux Ratio

Another interesting observation is the migration of the optimum feed location as the ethane feed composition increases. For the naphtha and mix feed, the optimum feed stage location is located at stage 70 for a tower having 120 stages. The optimum location migrates to stage 65 on a feed equivalent to the 100% NGL feed project and to stage 62 on an all ethane feed. The migration of the optimum feed point is explained by the difference in composition profile, which also causes a change in temperature profile as illustrated on Figure 4a and 4b. For a conventional tower operating with a bottom reboiler and overhead condenser, the change in temperature profile has little significance. However, with side reboiler operation this change may have significant implications as discussed later.

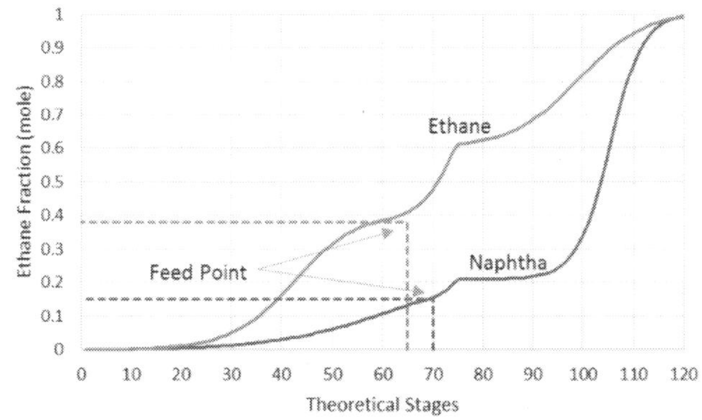

Figure 4a: Ethane Composition Profile

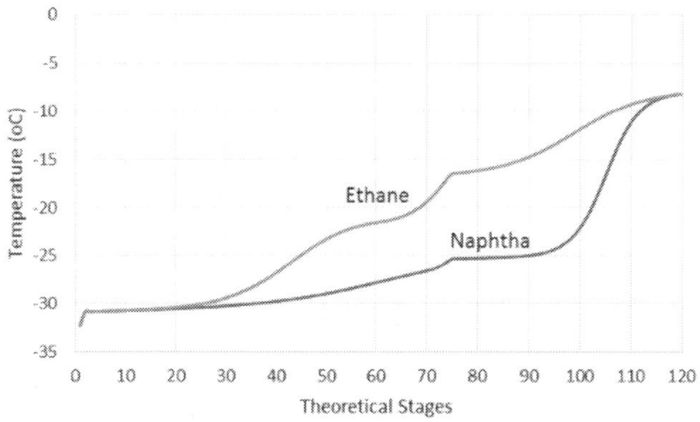

Figure 4b: Temperature Profile

Existing C2 Splitter Evaluation

A performance test was conducted on the C2 Splitter following the 2005 re-tray. The tower was false loaded up to the design load for the re-tray and the overall tray efficiency evaluated at 78%[3]. The tower simulation was adjusted with plant data and provided a relatively accurate prediction of its operation. Using the simulation and tray performance diagram provided by the vendor, the maximum tower capacity without dilute ethylene was evaluated at 110% of the plant required capacity based on the feed slate processed after the 2005 revamp. The maximum capacity was define as the rate where flooding would first occur as predicted by the tray performance diagram.

In the initial stage of the 100% NGL feed project, the tower capacity was re-evaluated based on the new expected feed slate. A first evaluation reveals that the tower would reach the flood limits at 92% of the capacity required by the project; however, to reach this capacity, the heat duty required would exceed the capability of the existing side reboiler by 165%. The warmer temperature profile of the C2 Splitter caused by the higher ethane feed composition as described earlier, would cause the charge gas side of the side reboiler to pinch at a warmer temperature; hence providing less heat duty. Because of the new feed slate, a drop in charge gas flow across the side reboiler is also expected, which would further reduce the duty available for the side reboiler.

Proper balance between the side and bottom reboiler is critical in the tower due to the narrow tray operating window in that section. Tray operating windows for the rectifying section and the tray section located between the side and bottom reboiler are illustrated on Figure 5. Because of the short tray spacing between the side and bottom reboiler (12 in), the trays in that section are limited by downcomer backup. The larger tray spacing in the rectifying section (20 in) causes the jet flood and downcomer backup limits to be more or less equivalent around the expected operating line (left side of Figure 5). The tray section between the side and bottom reboiler (right side of Figure 5) offers a relatively narrow operating range between weeping and downcomer backup, which explain the criticality for proper balance between the side and bottom reboiler. Insufficient side reboiler causes higher duty on the bottom reboiler and would cause the trays to operate above the downcomer backup limit. On the other hand, excessive side reboiler causes lower bottom reboiler duty, which causes the tray to operate below the weep point. Higher ethylene slippage in the tower bottom is typically observed when operating below the weep point.

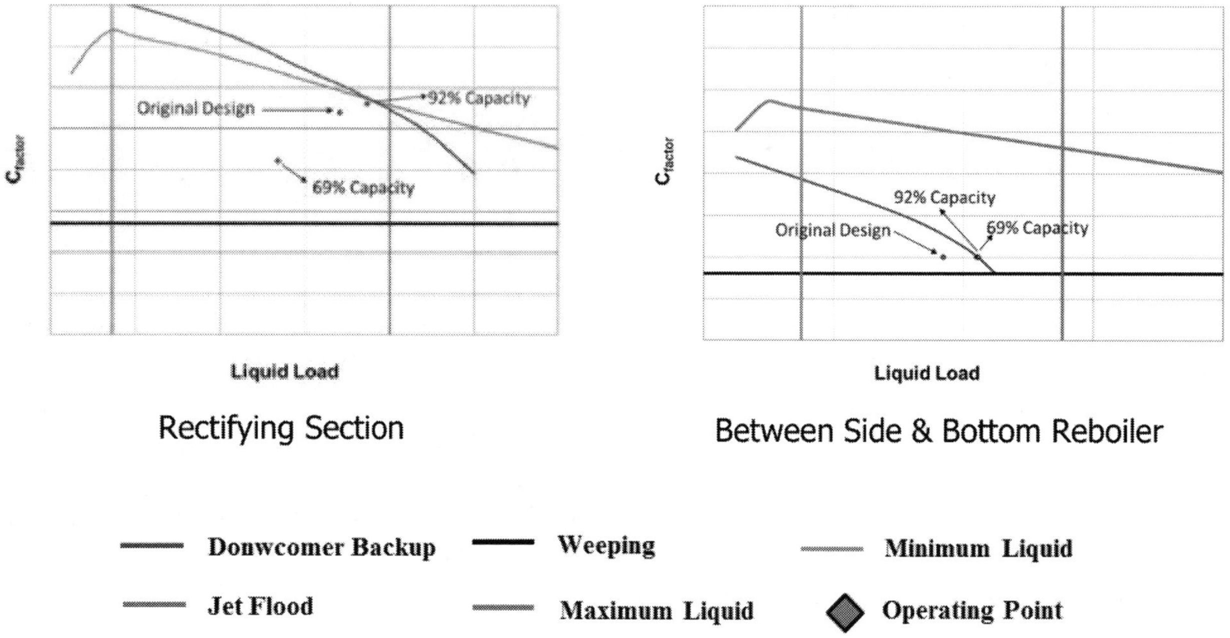

Figure 5: Operating Points vs Tray Operating Limits

Operating the side reboiler with the heat available from the charge gas stream would reduce the capacity of the tower to 69% of the required project capacity. As illustrated on Figure 5, the section between the side and bottom reboiler would operate at the downcomer backup limit, while the trays in the rectifying section would still have significant capacity available. The original design and 92% capacity points show a tower that is more balanced from a hydraulic standpoint when compared to the 69% capacity point.

C2 Splitter Revamp Options

The existing C2 Splitter evaluation concluded that the exiting tower in its current configuration would not support the capacity expected by the project. The following options were considered for meeting the required project capacity.

- Parallel side reboiler with higher cracking severity.
- Tower re-tray
- Parallel side stripper
- Parallel tower

Parallel side reboiler with higher cracking severity

As shown above, without any constraint on the side reboiler the tower capacity increases from 69 to 92% of the project required capacity. Increasing the severity of ethane cracking would reduce the ethane content in the feed thus reducing the tower reflux ratio. A 7% gain in tower capacity was predicted when increasing the ethane conversion by 5% in the furnaces. The combination of a parallel side reboiler using C3R vapor for heating medium along with higher cracking severity on furnaces processing ethane could potentially bring the tower capacity within the target required by the project. Although requiring relatively small capital investments, this option would offer slim capacity upside if any, making the tower sensitive to any transient conditions. Furthermore, the furnaces cycle time would be impacted causing more frequent operation transitions for decoking[4].

Tower Re-Tray

The tower is already equipped with high capacity trays believed to provide the maximum capacity. Change in tray design may provide some upside, for example modifying the trays between the side and bottom reboiler may move the downcomer backup limit, however, any of those changes may provide slim increase in capacity, which would fall well below the capacity required by the project.

Consideration to tray technology operating beyond the system limit by using centrifugal forces[5] or de-entrainment devices[6] has been evaluated. A hybrid design using a combination of high capacity trays along with tray design for operation above the system limits was found capable of increasing the tower capacity to 120% of the project required capacity.

Limited experience with the tray technology brings an element of novelty and risk assessed unacceptable to the project. A tower re-tray would also have to be implemented at a much later plant turnaround which does not align with the project schedule.

Parallel Side Stripper

Building on the idea of a parallel side reboiler, a parallel side stripper may offload the tower and provides the capacity required by the project. The concept as illustrated on Figure 6, involves condensing the existing vapor feed and re-routing it to a new parallel side stripper. The existing side reboiler is maximized with additional material withdrawn to feed to the new side stripper. The overhead vapor from the side stripper feeds the C2 splitter at the current feed point.

This option would increase the capacity to 103% of the project requirements, which is a slim increase in capacity involving significant capital investment. Furthermore, significant energy penalties would be expected due to increases in condenser and reboiler heat duties.

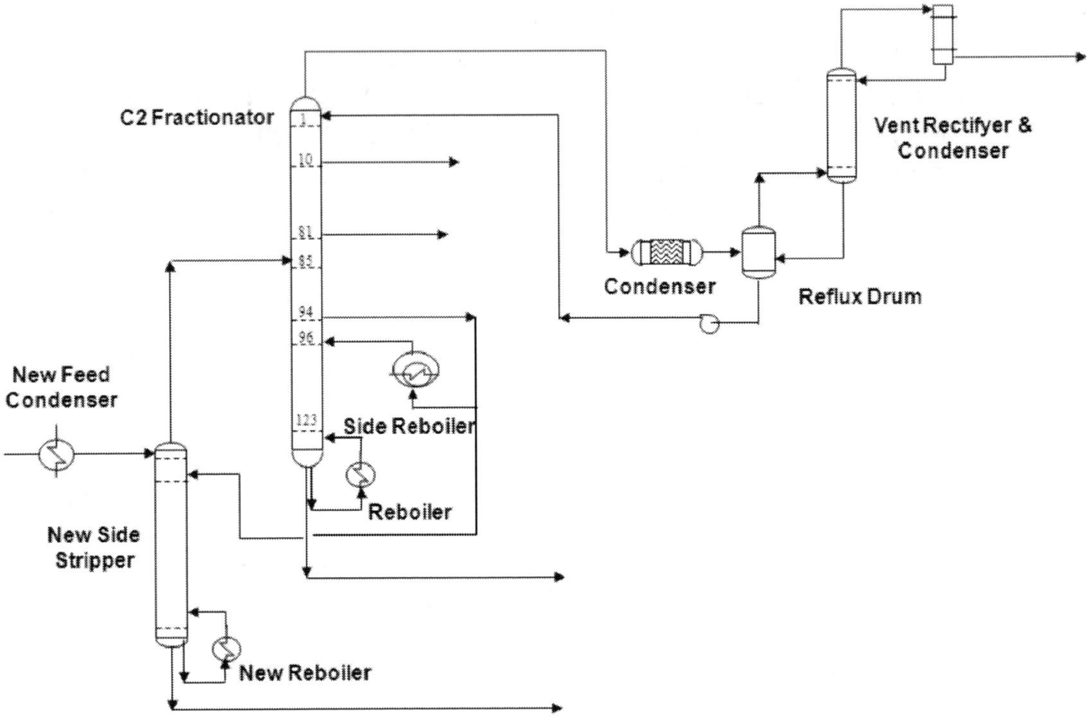

Figure 6: Parallel Side Stripper

Parallel C2 Splitter

A full parallel C2 Splitter system would ensure meeting the production capacity required by the project. With proper tie in's, it can be build and commissioned outside a plant turnaround. Although the most expensive, it was the selected option as it aligns with the project constraints and objectives.

Parallel C2 Splitter Design Considerations

For a given operating pressure, the height of a tower is driven by the number of trays required to achieve the desired fractionation; which is independent of capacity. Capacity will dictate the diameter required for the tower. As described above, the existing C2 Splitter system has a 31% shortfall in capacity in its current configuration. With a parallel side reboiler this gap can be reduced to 8%. Given the magnitude of the investment, designing strictly to meet the project capacity would not make much sense from a long term perspective. Alignment with business plans regarding potential future expansions and feedstocks is imperative for setting up a design strategy for the new tower.

A low pressure C2 Splitter system with a heat pump[7] would reduce the height of the tower while increasing its diameter. However, this option was ruled out as it is more applicable to front end acetylene reactors. Furthermore, it would involve a significant increase in capital given the associated compressor unit and potential operability issues with back end acetylene reactors. The new C2 Splitter was therefore chosen to be a high pressure system similar to the existing tower.

The following criteria were considered for scaling the size of the new tower:

- Future expansion plans.
- Future feedstock strategies.
- Current and future operability.
- Energy efficiency.

Under sizing the tower would limit the ability for expanding the plant capacity while oversizing it would affect energy efficiency. For setting up the size of the new tower a roadmap to expansion was set based on business strategies and other plant constraints. To meet the project capacity, the existing tower would operate at capacity without a parallel side reboiler (69% of the project capacity). The new tower would first be configured with conventional trays. A road map to expansion would be staged by adding a parallel side reboiler on the existing tower followed by a re-tray of the new tower with high capacity trays. Based on those considerations, the new tower diameter was set identical to the existing one. To maximize capacity upside, the initial step was to design conventional trays meeting a turndown rate equivalent to the project capacity (31% of the total required capacity). Further reduction in capacity, if required could be achieved by turning down the old tower.

The capacity of both towers was evaluated for the configuration considered by the project as well as for potential expansion configurations. A capacity range was established as a function of potential feed slate and is summarized on Table 2 relative to the project required capacity.

Further increase in capacity can potentially be accomplished by considering tray technologies that operates beyond the system limit. Although un-acceptable from a schedule and risk standpoint for this project, experience and available data in the future may make this option viable once completing the capacity roadmap described on Table 2.

Table 2: C2 Splitter System Capacity Roadmap

	Old Tower (% of project capacity)	New Tower (% of project capacity)	Total (% project capacity)
Project configuration	60 to 70	70 to 75	130 to 145
Parallel side reboiler on old tower	80 to 92	70 to 75	145 to 165
Parallel side reboiler on old tower & new tower re-tray	80 to 92	85 to 95	165 to 190

Operation Strategy

As illustrated on Figure 7, the effluent from the acetylene reactors splits between the two C2 Splitters. One tower is fed at a fixed rate through a flow controller. Feed to the other tower is regulated by a pressure/ flow cascade control loop with the Deethanizer pressure being the control variable and flow to the C2 Splitter the manipulated variable. The controls are configured such that operation could select which tower is base loaded on flow control with the other tower taking feed variations through the pressure/ flow cascade control loop.

The new tower vent is routed to the existing vent condensing system of the old tower to minimize recycle to the CGC. Provisions were made to vent the new tower directly to the CGC in case of pressure instability. The existing ethylene product drum is shared by both towers.

Reflux, ethylene product draw and reboil rates are control by an advanced control application (APC) as per the existing tower, with the exception that the new tower does not have a side reboiler. The APC application uses analyzers and flows to regulate its control variables for meeting product specifications and keeping the tower in material balance.

For plant startup, a single tower operation will be sustained until feed rates allow a smooth transition to a two tower operation. In a similar fashion for plant shut down, a transition from two to single tower operation will first occur as plant rates are reduced.

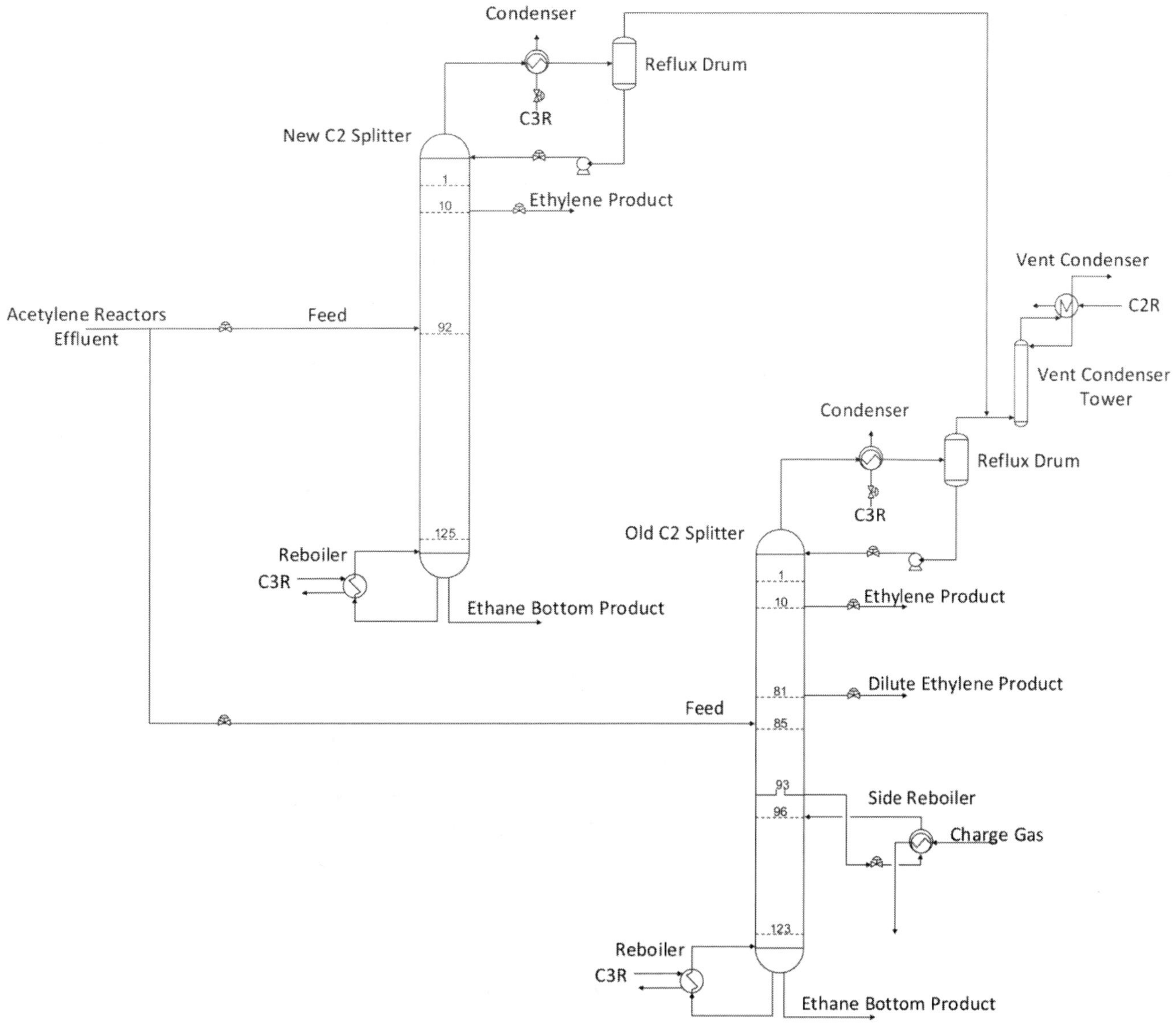

Figure 7: Parallel C2 Splitter System

Fabrication

Process engineering involvement during the release and approval of mechanical drawings of the tower and associated hardware might diminish, however participation remains crucial as other disciplines may not fully understand the implications of certain components in achieving efficient fractionation. Fabrication of a new tower involves different parties such as vessel fabricator, internals (trays, distributors), reboiler and condenser suppliers. Furthermore it is not unusual to see different people involved as work progresses from conceptual engineering to fabrication. Alignment amongst those different players must be scrutinized as drawings are being issued and work progressed. Items such as location of feed nozzles, thermowells, pressure taps, tray rings and manways are to be checked on the vessel drawings. For instance thermowells should be positioned in location exhibiting significant temperature changes and preferably located in the liquid phase near the bottom of downcomers[8]. Tray rings should be consistent with the specified tray spacing. Feed nozzles should be at the expected location with orientation that fits with distributor design. Tray drawings, reboiler and condenser are to be consistent with their associated process datasheets. Any differences should be assessed through revised process datasheets and implications on capacity understood.

It is a good practice to visit the fabrication shop and inspect one tray from each different design. The tray or part of it depending on size, should be laid out on the floor. Downcomer width, flow path length and weir heights are typical measurements that should be done at different places along the length of the tray. On valve trays, a valve count and random measurements of valve lift should be performed. For sieve trays, although tedious, a hole count should be done, which allows verification of the tray hole area. Ensuring that dimensions are within tolerance and consistent with the mechanical drawings will avoid speculation on tray geometry defects in the event of problems once the tower is up and running. When feasible, it is also a good practice to test liquid distributors with water.

Figure 8: Reflux Distributor Water Test

By following the above practices, two design changes were introduced following the inspection at the fabrication shop. Installation of a false downcomer around the reflux distributor was recommended by the design engineer after noticing sideways movement from the liquid jet leaving the nozzles while testing the distributor with water (Figure 8). The false downcomer was to prevent projection of the liquid on the tray deck rather than its intended inlet sump. Projection of the liquid on the tray deck could potentially cause excessive froth on the top tray.

The feed distributor, illustrated on Figure 9; is an H pipe where vapor is projected inwards and outwards from the two laterals. During drawing reviews, it had been pointed out by process engineering that the feed was slightly superheated and that it should not project vapor outwards to the downcomer plates. According to simulation, the vapor feed enters the tower 10 °C above the liquid flowing down the downcomer. During the inspection at the fabrication shop, it was found that holes had been provided on both side of the laterals as shown on Figure 9. Given the risk that the hot vapor may flash the liquid and choke the downcomer, impingement plates were added to the design.

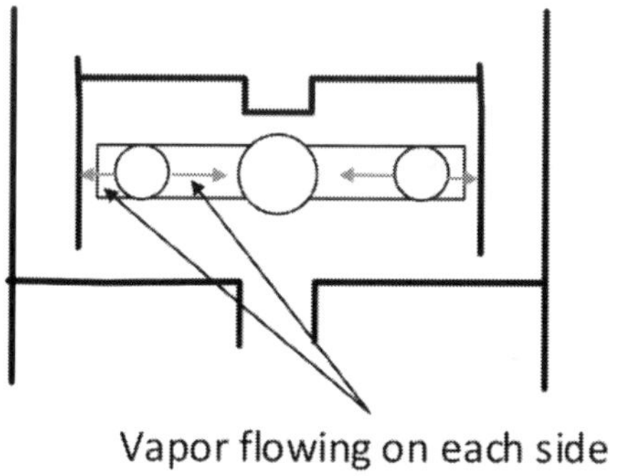

Figure 9: Vapor Feed Distributor

The trays and associated internals were installed at the vessel fabrication shop with the tower on the horizontal position. The vessel fabrication shop was only a few miles from the plant site. However as illustrated on Figure 10 and 11, moving and erecting the tower was a challenging and spectacular event.

Figure 10: New C2 Splitter Transportation from the Fabrication Shop

Figure 11: New C2 splitter Lift

Commissioning

As stated earlier, the existing tower would meet 69% of the project capacity with the design feed slate. This implies that commissioning of the new parallel tower had to occur prior to full feed transition to minimize production impacts. However, a too-soon transition would also involve potential turndown issues.

Commissioning of the new tower was planned with the plant cracking 68% of the expected ethane feed volume. The cracking furnaces feed was broken down into 40% ethane, 10% propane, 20% butane and 30% liquids (naphtha and HAGO). While still cracking liquids, this feed slate lead to a C2 Splitter feed composition relatively close to the one expected by the project but at rates reduced to 78% of target. Dilute ethylene was still produced on the existing tower, providing a 6% boost in capacity. Part of the ethane feed system is integrated with ethylene refrigeration and had not been commissioned at that time. Once fully commissioned, vaporization of the ethane feed would be accomplished by condensing ethylene vapor from the ethylene refrigeration compressor (ERC) discharge. This integration reduces power on the propylene refrigeration compressor (PRC) as condensation of ethylene from the ERC discharge is otherwise accomplished by propylene refrigerant. For commissioning the tower, the ethane feed was vaporized by low pressure steam, compounding the impact of higher PRC power on the steam system.

Until commissioning of the new tower, operation of the existing tower was under the responsibility of the operation group and no significant issues had been reported. Process design got involved in setting up a transition strategy for commissioning the new tower. The plan involved inventorying the tower with ethylene followed by a period of total reflux operation. Beside usual checks for leaks, the total reflux operation also proved the functionality of the overhead condenser and reboiler. It however increased the power required to run the PRC, hence causing a higher steam demand. For feeding the tower, target operating conditions were calculated from simulations with expected outcomes on the towers, the PRC and steam system. The plan was to ramp up feed rates on the new tower to the plan targets while reducing feed on the existing tower. Given a reduced rate operation at 78% of the project target, conditions were calculated to prevent or minimize false load on both towers. Turndown was limited by weeping in the bottom section of each tower. The target operating conditions and the transition plan with expected impacts was presented to operation prior to proceed with the commissioning activities.

As feed was introduced to the new tower and slowly ramped up, operation reported that the steam demand had increased beyond the comfort level of operators and that the propylene refrigeration compressor was operating near its maximum power. Operating conditions were gathered and compared against targets. As shown on Table 3, the feed to the new tower was only at 35% of target and 110% on the old tower. Feed rates to the cracking heaters had also been reduced such that the total feed to both towers was at 88% of target.

Review of the operating conditions listed on Table 3 revealed that the side reboiler duty on the old tower was quite high and causing unnecessary false load. Reducing the side reboiler duty to its target would be associated with a reduction in reflux, hence condenser duty. The PRC power would consequently decrease, which would also help the steam system. Operation was then instructed to slowly ramp down the tower draw rate to the side reboiler. Operation reported early in this transition that the front end of the chilling train was warming up and making the ERC unstable. As discussed in a paper by NOVA Chemicals[9], problems with the C2R system were known prior to commissioning the new tower and were caused by a fouled sub-cooler, which was loading up the ERC 3rd stage; up to that point, those problems should have been independent from the tower operation. Figure 10 illustrates the path of charge gas from the dryers to the front end of the chill train. Charge gas enters the side reboiler at about 15 °C and is first cooled down in the C2 Splitter side reboiler. Heat integration between charge gas and the C2 Splitter reduces the load on the propylene chiller, which uses 1st level propylene refrigeration. Normally reduction of heat duty on the side reboiler should be picked up by the propylene chiller without any consequences on the C2R system. The propylene chiller is a core in kettle exchanger that had been in service for almost 10 years and had always provided relatively tight temperature approach. Simulation work prior to commissioning the tower suggested heat duties within the capability of the exchanger. However, as load was reduced on the side reboiler, the charge gas temperature exiting the propylene chiller warmed up. The additional heat duty was then shifted to the 1st C2R core exchanger in the cold box and contributed to increase the load on the ERC 3rd stage.

Table 3: C2 Splitter Transition Operating Conditions

Old Tower	**% of Recommended Target**
Feed	110
Dilute ethylene draw	58
Reflux	134
Side reboiler draw	267
C3R to reboiler	101
New Tower	**% of Recommended Target**
Feed	35
Reflux	92
C3R to reboiler	90
C2 Splitter System	**% of Recommended Target**
Feed	88

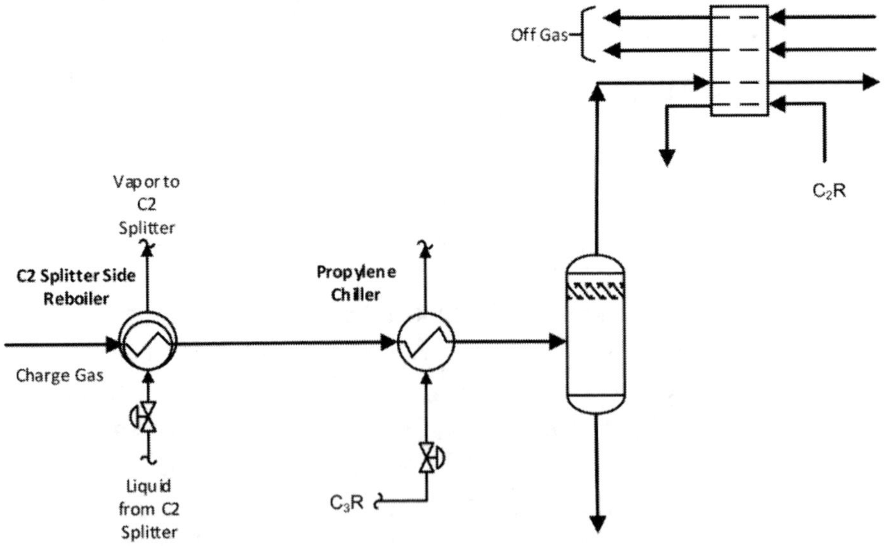

Figure 10: Side Reboiler & Charge Gas Chilling

Accepting that the propylene chiller would not provide the heat duty required, the side reboiler was set back to conditions allowing a stable operation of the ERC. An attempt was made at reducing the load on the bottom reboiler. However, as shown on Figure 11 the operating point on trays located between the bottom and side reboiler were already at the weep point. As a result, ethylene slippage in the bottom of the tower started rising. The bottom reboil rate was therefore brought back to its previous rate.

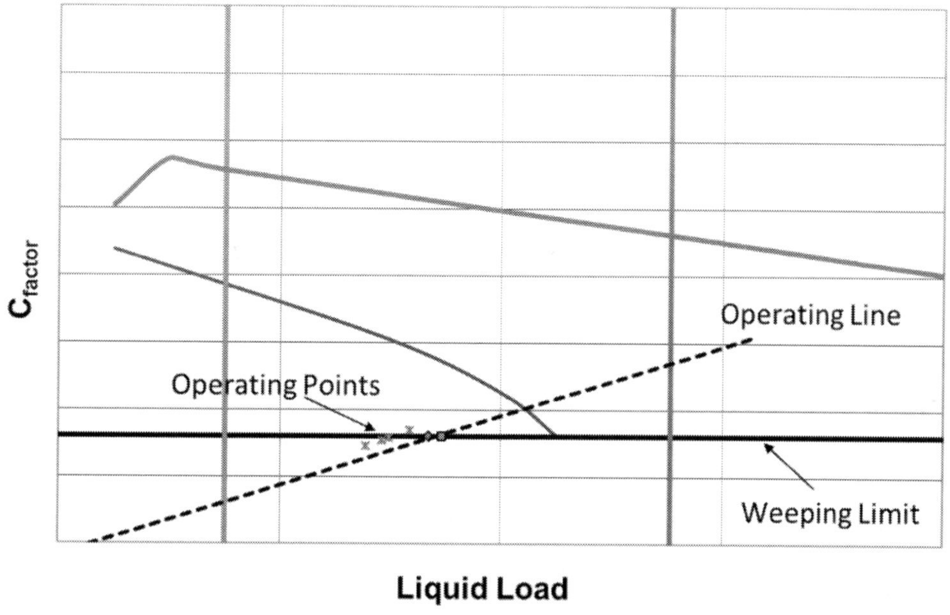

Figure 11: Old Tower Bottom Trays Operating Window

While going over this sequence of events, it became odd that the side reboiler could provide so much heat duty. As previously discussed, limitations in side reboiler duty were expected when cracking larger volume of ethane due to a shift in temperature profile. However, as shown on Figure 12; false loading the tower moved the temperature profile down allowing the tower draw to enter the side reboiler almost 3 °C colder than expected without false load.

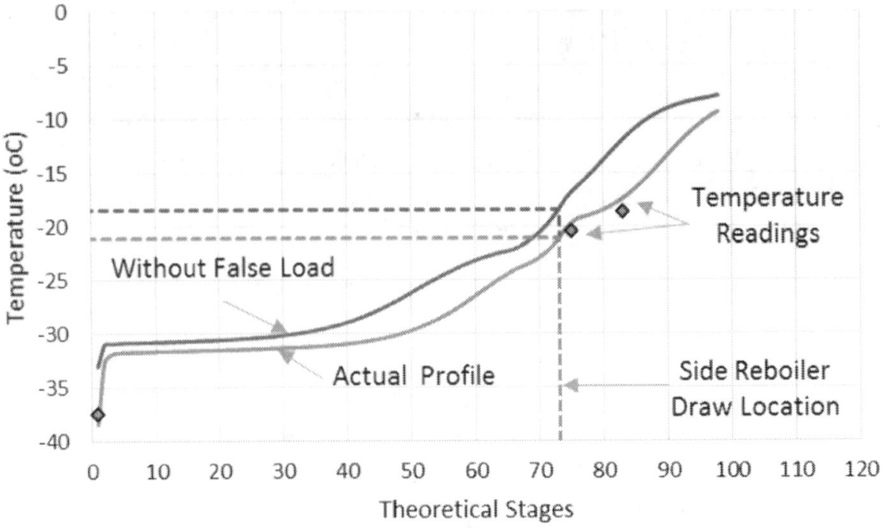

Figure 12: Old Tower Temperature Profile

Recognizing the implication of maximizing the side reboiler heat duty relative to charge gas cooling and stable ERC operation left few options in terms of reducing load on the old tower. Tray weeping limitations in the bottom and the need to satisfy a cooler temperature profile imposed constraints beyond the need of just fractionating ethylene. At that point, it did not make much sense to progress the feed transition to plan targets until gaining a better understanding on constraints encountered. The option of bringing the new tower down was contemplated. However, both towers were fractionating effectively and although high, compressor power and steam loads had dropped to levels acceptable by operation.

While feeding only 35% of the plan target rate, the new tower was stable and provided effective fractionation. The tower was false loaded with the reflux and reboil rates at 90% of targets. As shown on Figure 13, the trays were operated with a good margin from the weeping limit. Venting of non-condensables in the old tower venting system was done smoothly and appeared to be working effectively.

Until resolution of the chill train/ ERC constraints, operation of the towers were left at conditions similar to those reported on Table 3. Increase in plant feed rates would be managed by directing additional feed to the new tower. The old tower would still be operated with conditions ensuring adequate heat exchange with the chill train via the side reboiler.

Operating data on the propylene chiller was gathered for a period covering the previous 600 days of operation. The log mean temperature difference (LMTD) and pressure drop was plotted over time. The results revealed that until about 8 months prior to commissioning the new tower, the propylene chiller LMTD was operating at or below the design value. The pressure drop of the exchanger looked low and relatively steady for the same period. Then a slight increase on the LMTD was observed 6 to 8 months prior to the commissioning activities without any significant increase in pressure drop. The LMTD and pressure drop started showing a constant rise for the following 6 months. Increase LMTD and pressure drop are symptoms that typically correlate with fouling. The timing of the changes appeared to correlate well with the feedstock transition towards larger volume of ethane. The fouling theory was also supported by lab analyses on some of the charge gas streams.

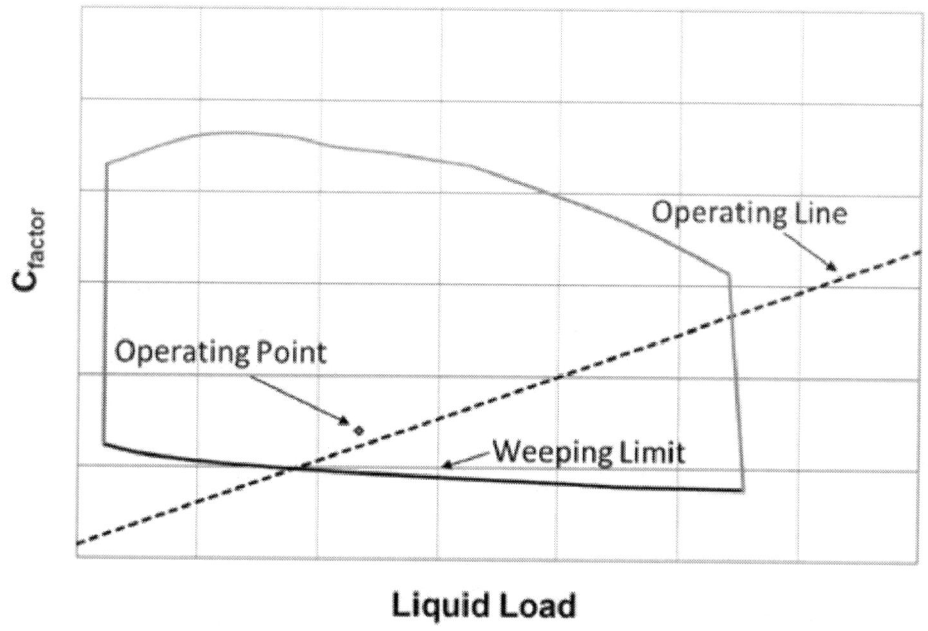

Figure 13: New Tower Stripping Section Tray Window

Implications of the newly discovered propylene chiller constraint were evaluated with simulations relative to the ability of meeting the project feed slate and production targets. Without false loading the old tower, the side reboiler would pinch at warmer temperature and shift the heat load beyond the current capability of the propylene chiller. The residual heat duty would shift to the ERC core exchanger in the cold box. This would not be a problem without limitations on the ERC. However, as experienced during commissioning of the new tower, the ERC was currently experiencing a constraint on the 3rd stage. Although false loading the old C2 Splitter was done during commissioning, false loading for meeting the temperature approach required on the side reboiler would cause a significant increase in condenser duty, which would likely push the PRC power beyond the capability of the machine at the targeted production rate. This evaluation was based on the current state of the propylene chiller; further deterioration in performance would be expected due to the typical nature of exchanger fouling. Given the threat on meeting the project objectives, the project team expedited the design and purchase of a new exchanger. Required tie in's were installed during a short plant outage. The new exchanger was installed and now operates in parallel with the existing units. Since then, the plant has been meeting and exceeding the production targets. The two C2 Splitters have been operating effectively without any problems reported by operation.

Conclusion

Due to the increase in required reflux ratio, higher volume of ethane cracking will reduce the capacity of a C2 Splitter. In this case a reduction from 110 to 92% of the targeted production was evaluated. However, side reboiler limitations and tight tray spacing in the bottom section further reduce the capacity of the tower to 69% of the targeted production. A challenging project schedule imposing construction activities outside normal plant turnaround, limited the number of debottleneck options. Although expensive, a new parallel tower was the only option meeting the project constraints and objectives. To make the investment most profitable, alignment with future business plans and competitive feedstocks was incorporated in the design strategy. Definition of a road map to expansion provided a staged approach for future increase in capacity, ultimately pushing the C2 Splitter system capacity to 163 - 187% of the current project target depending on feedstock.

This work illustrates the importance of thorough process engineering from front end development all the way through commissioning. The use of plant data, simulations and tray operating windows was instrumental in understanding existing equipment limitations, which is imperative for defining debottleneck options. Besides generating process loads for the new tower and tray design, simulation results allowed identifying potential problems with the feed distributor design.

Ethylene plants are highly integrated, and constraints in systems believed non-related at first can shift loads on other systems and limit operation flexibility. During commissioning of the new tower, fouled exchangers increased loads on the ERC and required maximizing heat integration on the old C2 Splitter at the expense of efficient fractionation. Those conditions led to higher power usage on the PRC due to significant false load on both C2 Splitters, which also resulted in higher steam demands. Simulation of the olefins unit ahead of the commissioning activities and definition of process targets along with effective process data gathering tools helped identify the above constraints before they became unmanageable. The use of simulations was also instrumental for establishing new process targets that would aligned with the newly discovered constraints. Although not per original plan, the new tower was commissioned successfully. Further evaluation on the constraints encountered during the commissioning activities provided understanding and early warning, which allowed proper action for removing a constraint that would otherwise be a threat to the success of the project.

Nomenclature

APC:	Advanced Process Control
C2R :	Ethylene Refrigerant
C3R :	Propylene Refrigerant
CGC:	Charge Gas Compressor
ERC:	Ethylene Refrigeration Compressor
HAGO:	Heavy Atmospheric Gas Oil

NGL: Natural Gas Liquid
PRC: Propylene Refrigeration Compressor

References

1. M. de Boer, B. Evans, Ethylene Cracker Capacity Increase And Modernization, Ethylene Producer Conference, 2006 AIChE Spring Meeting, Orlandos, FL
2. A. Bernard, R. Hayden, Planning and Designing the Modernization of the Recovery Area of a Flexible Cracker, Ethylene Producer Conference, 2004 AIChE Spring Meeting, New Orleans, LA
3. A. Bernard, D. Summers, W de Villiers, Improve Product Ethylene Separation, Hydrocarbon Processing, April 2009.
4. S. Katrekar, P. Nair, Technology Leveraged Continuous Process Analysis to Optimize the Performance of Cracking Furnaces, APGC, May 2007, Kuala Lumpur
5. W. de Villiers, P. Wilkinson, D. Summers, Developments in Splitter Revamps, Ethylene Producer Conference, 2004 AIChE Spring Meeting, New Orleans, LA
6. Z.P. Xu, B.J. Nowak, K.J. Richardson, SimulFlow™ Device Capacity Beyond System Limit, Distillation Topical Conference, 2007 AIChE Spring Meeting, Houston, TX
7. D. Leung, J. Hu, W. Lam, Advanced C2 Processing System (ACPS), Ethylene Producer Conference, 1999 AIChE Spring Meeting, Houston, TX
8. H. Z. Kister, "Distillation Design", McGraw-Hill.
9. D. Gent, R. Hayden, Conversion of a Flexible Cracker to a Mostly Ethane Feed, Ethylene Producer Conference, 2016 AIChE Spring Meeting, Houston, TX

AICHE PAPER NUMBER 86B

INTERPRETATION AND APPLICATION OF EN 746-2 REQUIREMENTS TO THE DESIGN AND OPERATION OF BURNER MANAGEMENT SYSTEMS ON CRACKING FURNACES

David J Brown
Senior Engineer – Cracking Furnaces
Total Research & Technology Feluy

Lode De Leye
Senior Engineer – Steam Cracking
Total Research & Technology Feluy

Sandra Gainier
Expert Engineer Furnaces
Total Petrochemicals Gonfreville

Dries De Cooman
Process Engineer Front End Studies
Total Olefins Antwerp

Jim Middleton
Manager of Furnace Technology and Design
Technip Stone & Webster Process Technology

Prepared for Presentation at the 2016 Spring National Meeting
Houston, Texas, April 10-14, 2016

INTERPRETATION AND APPLICATION OF EN 746-2 REQUIREMENTS TO THE DESIGN AND OPERATION OF BURNER MANAGEMENT SYSTEMS ON CRACKING FURNACES

David J Brown
Senior Engineer – Cracking Furnaces
Total Research & Technology Feluy
Lode De Leye
Senior Engineer – Steam Cracking
Total Research & Technology Feluy
Sandra Gainier
Expert Engineer Furnaces
Total Petrochemicals Gonfreville
Dries De Cooman
Process Engineer Front End Studies
Total Olefins Antwerp
Jim Middleton
Manager of Furnace Technology and Design
Technip Stone & Webster Process Technology

Abstract:

In 2010 Total, working closely with its selected licensor and basic engineering contractor, Technip Stone & Webster Process Technology ('TS&W'), developed a philosophy for the burner management system ('BMS') of a pair of new furnaces being designed for Total's Gonfreville site (part of the Normandy Platform.) The key features were operator safety; compliance with the European standard (EN 746-2:1997); and reliability. The idea of having remote ignition (from a local panel) and relying on flame monitoring (no visuals) was rather unusual for a cracking furnace. The operators were especially concerned. In practice the system has proved to be reliable, convenient and, most importantly, safe.

The BMS was adapted by Total and retrofitted to a pair of older furnaces at Gonfreville. Later, after TS&W was awarded a sub-contract for a near-duplicate pair of furnaces for the Antwerp Platform, Total provided detailed feedback from the experience in Gonfreville to allow the design to be fine-tuned. It was also reviewed for compliance with the new edition of the European standard (EN 746-2:2010).

In the paper the authors describe, compare and contrast the design of the BMSs for the three cited references and report on how the systems performed in practice. Finally they consider the implications of the publication of the International Standard, ISO 13577-2:2014, that will eventually supersede EN 746-2 in Europe, and is likely to become the standard adopted by countries which do not currently have a national standard.

Overview

Starting in 2010, TS&W was responsible for licensing and performing the basic engineering of two new pyrolysis furnaces for Total at the Gonfreville site, part of the Normandy Platform. In fall 2012 Total implemented an upgrade of the BMS for two older furnaces, originally started-up in 1988, also at the Gonfreville site. Finally, in mid-2012, TS&W started work on two new furnaces to be installed at the NC-2 Plant, part of the Antwerp Platform. The Antwerp furnaces started up in 2015.

European Standard Requirements

When EN 746-2: 1997 was published it challenged the European petrochemicals industry to a radical rethink of its approach to the design and operation of burner management systems ('BMSs') for cracking furnaces. In 2000 the European Ethylene Producers' Committee ('EEPC') set up a Working Group to consider the implications of the standard (Matthews, 2000). The results were presented at an EEPC HSE ('Health, Safety & Environment') Conference (Overwater, 2001) and then summarized in a final report in 2002. The report confirms that EN 746-2 has the "...status of a national standard in all EU ['European Union'] member state[s] and EFTA ['European Free Trade Association'] countries in September 1997...."; that it "...clearly states its applicability...in petrochemical industrial thermoprocessing...."; and that it covers "...all thermoprocessing equipment installed since September 1997...". Although the standard is not retroactive it could, according to the writers of the report, apply if a furnace undergoes a major revamp. Borealis went further (Enqvist, 2010) and described the development of a strategy to upgrade the BMSs of its old furnaces, some dating back to the early 1970's, and the implementation, starting in 2004.

In late 2009 Total agreed internally some key points on the implementation of EN 746-2 for cracking furnaces as follows:

- It applies to new furnaces and existing furnaces in which burners are replaced.
- It will be applied to floor burners and balcony/terrace-wall burners only and will not be applied to wall burners for the following reasons:
    - Too many small burners
    - Technical issues with pilots and flame detection
    - Has to be covered by procedure
- All floor burners will have continuous pilots.
- A limited amount of floor burners, necessary to reach a minimum flue gas temperature of 750°C, plus a margin for reliability reasons (based upon common sense and operating feedback) will be called 'start-up burners'.
- All start-up burners and their associated pilot burners will have automatic safety shut-off valves ('XVs').
- All start-up burners will be equipped with a remote ignition device, flame detection by means of an ionization rod on their associated pilot burners, and flame detection by means of a UV detector on the main burners.
- Prior to light off, a tightness test to all burners has to be performed. Nitrogen will be used for the test.

- 95% of the burners are either balanced or natural draught: there will be no airflow metering.
- Pilot fuel gas shall be different from the fuel gas to the main burners (natural gas recommended).
- A flue gas temperature measurement at the arch is mandatory.

These general principles were applied in the two projects in Gonfreville. For the Antwerp project we had to consider the changes between the 1997 edition of EN 746-2 and the 2010 edition. The key changes are listed in the table below.

Table 1. EN 746-2 changes.

Clause	EN 746-2:1997	EN 746-2:2010
5.2.2.3 (1997) 5.2.2.3.4 (2010)	Valves controlling capacities greater than 1200 kW shall be equipped with a *leak tightness device.* [And from 3.36 b), definition of leak tightness device, '*...two safety shut-off valves in series, fitted with proof of closure switches to close the fuel line, and a third valve fitted with a switch to prove that it is open, to vent safely the space between them....*' In other words a double-block-and-bleed.]	Automatic shut-off valves controlling capacities greater than 1 200 kW shall be equipped with a *valve proving system.* If this is applicable, then the valve proving system shall comply with EN 1643 or give an equivalent level of safety. The signal of a leaking valve from the *valve proving system* shall stop the current start-up when the test takes place during start up or prevent the next start-up when the test takes place after shut down. This function shall meet the requirements of a protective system according to 5.7.2 and 5.7.3. [3.93 defines a valve proving system as follows, '*...system to check the effective closure of automatic shut-off valves by checking leak tightness NOTE It consists of a programming unit, a measuring device, valves and other functional assemblies....*']
5.2.2.3.1 (2010)	No equivalent	Low cycling applications intended to operate continuously for periods longer than 1 year shall have a redundant safety shut-off system at least comparable with the flow sheet given in Figure C.5 that allows testing the effective closure of the valves at least once a year.
5.2.6.1	For systems where the pilot burner remains in use during main burner operation, separate flame sensors to supervise the pilot and main flames shall be fitted. The main flame sensor shall be so positioned that it cannot in any circumstances detect the pilot flame.	For systems where the pilot burner remains in use during main burner operation, separate flame detector device to supervise the pilot and main flames shall be fitted. The main flame detector device shall be so positioned that it cannot in any circumstances detect the

		pilot flame. *In the case that the pilot burner ignites under all circumstances the main flame, it is sufficient to supervise only the pilot flame providing the flow rate of that pilot burner is checked by a protective System according to 5.7.2 and 5.7.3 (e.g. minimum gas pressure switch).*
Annex C (informative) Typical examples of piping and components	No equivalent	'Figure C.5 — Central pipework for low cycling applications' is the most relevant.

Implementation on new furnace project at Gonfreville

The new furnaces at Gonfreville are back-to-back, single radiant cell USC72SU (symmetrical U) configuration. About 85% of the heat input is provided by 24 x John Zink Ultra-Low NO_x PSMR floor-mounted burners; the remainder of the heat is from 2 x rows of John Zink LPMW wall-mounted burners, 48 in total. The burners are balanced draft. The air dampers on the floor burners are linked with a jackshaft. An end view of one of the furnaces is given in Figure 1.

Figure 1. Furnace overview.

All the floor burners are equipped with continuous pilots. Half of the floor burners are designated as start-up burners. These start-up burners have igniters and ionization flame detectors on the pilots; UV flame detectors to monitor the main flame; and local XVs on the both the pilot fuel supply and the main fuel supply. Typical start-up and non-start-up (a.k.a. 'main') burners are shown in Figure 2.

It is important to note, in passing, that pilots are used as a safe and convenient way to ignite the main flame of the high heat release floor burners, not as a safety device. Total agrees with the philosophy of Platvoet (2013) that "...pilot burners should be considered as ignition devices, not as safety devices..." and follows the requirements of EN 746-2:1997, paragraph 5.2.6.1 to provide "...separate flame sensors to supervise the pilot and main flames..."

The steps up to reaching the prescribed auto-ignition temperature (the so-called "start-up period" defined in paragraph 5.2.6.3

of EN 746-2) are carried out by an operator from a local panel in a safe, remote location. Other operators have to evacuate the area. This is a fundamental part of Total's philosophy. The local panel is shown in Figure 3. Those steps are described in some detail below.

Quite a different approach has to be used with wall-fired furnaces. There are technical reasons why remote ignition and remote flame monitoring are usually not practicable. This is discussed both by Enqvist (2010) and Platvoet (2013). The topic is not covered further in this paper.

Figure 2. Start-up and main burners.

Pre-Ignition Steps

Before igniting the burners a leak test must be performed to confirm the integrity of the fuel gas isolation valves, and the firebox must be purged.

Purge

The furnace purge can be performed concurrently with the leak testing and also after failed or aborted ignition attempts.

The ID fan is running. The arch pressure set point and the floor burner damper positions are set for 25% of the air flow rate at maximum heat input. The purge timer, set for five complete air changes, is started.

After purging, the ID fan continues running to automatically control the arch draft at normal design value.

Leak Test

The system leak tightness test is carried out using nitrogen. It is, in principle, feasible to use the normal fuel gas. There are pros and cons for both approaches. Total's view is that if the leak tightness test fails then there is a risk that fuel gas may leak into the firebox and cause an explosive mixture: in other words a procedure intended to avoid a potentially dangerous situation is responsible for creating a dangerous situation. The disadvantage of residual nitrogen making it

Figure 3. Local panel.

difficult to ignite the first burners can be managed – see discussion below.

The fuel gas is blinded. The ID fan is running. The nitrogen supply is connected to the fuel gas header, but the manual valve on the nitrogen supply remains closed. All valves are closed except the manual valves on the fuel gas to pilots and main fuel gas connections on start-up burners.

A hand switch starts the leak test. In a regular, defined sequence the integrity of the various XVs is tested. This procedure, in its slightly improved version, is described in more detail for the Antwerp project.

A fixed and defined time after the successful completion of the leak test, all of the 'valve test ok' indicators will automatically reset. Therefore ignition of the pilots must proceed within the defined time of completion of the leak test.

If the leak test fails then the DCS operator needs to reinitialize the leak test by pressing again the hand switch.

Significant amounts of nitrogen in the fuel gas header could prevent the successful light-off of pilots and burners. The nitrogen is purged from the pilot fuel gas header, and from the main fuel gas header, with a pre-defined sequence of successive pressurize/depressurize cycles using fuel gas. This is efficient: the only trouble encountered is the lighting of the very first pilot, for which the first attempt to light may fail because of small amounts of residual nitrogen.

Ignition Steps

The floor burner dampers are set to about 15% open using the jackshafts. The original intent was to close the dampers for the main, non-start-up floor burners fully (minimum stop) by disconnecting from the jackshafts. In practice this was found to not be necessary.

The first pilot has to be lit within 10 minutes of validation of the atmospheric analysis or else the process has to be done again: atmospheric analysis, validation, launch of the pilot lighting sequence.

The operator starts the automatic pilot ignition sequence by pressing a local panel push button. All pilot spark igniters are energized and after 3 seconds the pilot XVs to the start-up burners open automatically. The ignition timer stops after 5 seconds. The status of the pilot flame detectors, ionization flame type, start-up burners only, is checked. If all 12 pilot flame detectors are positive then the automatic lighting sequence will move on to the next step. If less than 12 positive flame signals are detected then a second, and finally a third, attempt is made.

If some pilots on start-up burners still do not have positive flame signals, the pilot isolation XVs are closed on these burners.

The originally planned success criteria for lighting the pilots were as follows:

- If fewer than 8 pilot flames are shown as positive, or if any 2 adjacent pilot flames fail to be shown as positive, then the furnace start-up will have failed, a trip is invoked and the leak test and purge sequence will have to be repeated.

These were found, in practice, to be too rigorous. The criteria were modified. The revised method is described below in detail for the Antwerp new furnace project below.

After various adjustments to the process system – not discussed in detail in this paper – the fuel gas header is pressurized. The gas firing is gradually increased by opening the XVs to start-up burners, one at a time, following a burner lighting sequence described in the written procedure (no automatic sequence programmed for the burners). If any start-up burner does not have a positive pilot flame signals, the XV cannot be opened. A positive main flame signal from the main flame UV detector is required within the ignition timer limit. If a positive flame signal is not received, the XV on that burner will close. Three tries are allowed to light the main flame on each start-up burners. A further three attempts are allowed after a timed 20 minute delay – this was a modification to the original plan and described in more details for the Antwerp new furnace project.

When an arch temperature of 180°C is reached dilution steam supply is started.

The next phase of the ignition sequence was originally planned as follows:

- When 8 or more main flames are shown as positive for a period of 15 seconds, and no adjacent start-up burners are unlit, the start-up over-ride on the main furnace trip is removed. Subsequently, if less than 8 main flames are shown as positive, the furnace will trip.
- The main fuel gas pressure control is released to DCS control. If the number of positive pilot flame signals falls below 8 before the arch temperature reaches 750°C an alarm warns the operator; if the number of positive main flame signals falls below 8 the furnace trips; if 2 adjacent start-up burners do not show a positive flame signal the furnace trips.

Again, this was modified based on the experience of the first few cold light-offs. The modified procedure, now used at both sites, is described for the new furnace project in Antwerp. See below.

The firing on the start-up burners is increased until the arch temperature is above 750°C (determined by 3oo3 voting of arch temperature). The dampers are properly adjusted.

Once the arch temperature is above 750°C, the XV for the main burner pilots is permitted to be opened, to allow lighting of main burner pilots.

The main burner pilots are manually lit, one at a time, by opening the manual valve to each burner and with either a portable, manual igniter or cross-ignition. If there are any start-up burner pilots not showing positive flame signals the associated XVs are opened and pilots are lit from the built-in igniter or cross-ignition.

Once the arch temperature is above 750°C, the XV for the main burners is permitted to be opened, to allow lighting of main burners. The main burners are then lit from the pilots using the local, manual isolation valves.

The XV for the wall burner header is also permitted to open. The lighting of the wall burners proceeds in the same manner as the main floor burners, except that the wall burners, having a rather low heat release, do not have pilots and are lit directly with a manual igniter or by cross-ignition.

The furnace start up is now complete and all burners are lit.

Figure 4. Older furnaces at Gonfreville

Implementation on existing furnaces at Gonfreville

The older furnaces at Gonfreville are back-to-back, single radiant cell USC48U configuration. They are 100% floor-fired by 24 x John Zink Low NO_x PSFFG burners. The burners are forced draft, air pre-heated with BFW. The furnaces are shown in Figure 4. Pilots were added to the burners as part of the safety upgrade. Half of the burners (12) were designated as start-up burners. The BMS retrofit was basically identical to what was used for the new furnaces.

Implementation on new furnace project at Antwerp

The pair of new furnaces at Antwerp is very similar to the pair of furnaces at Gonfreville. The return of experience ('REX') from Gonfreville enabled some improvements to be made. In addition the new version of EN 746-2 required some rethinking.

After a detailed analysis it was decided that the double-block-and-bleed system could be considered to be equivalent to the valve proving system ('VPS'), newly mandated in the 2010 Edition of the EN. This was accepted by the Notified Body (accredited third-party inspector in EU, also referred to as 'NoBo'). It is worth noting that ISO 13577-2 also does not require a VPS for most cracking furnace applications (see below). De Boer (2013) describes a different viewpoint.

Natural gas was used as the supply for the pilots (see paragraph 5.2.2.5 of EN 746-2). This was not practicable for the Gonfreville projects.

The new edition of the EN introduced a new requirement that ".......a redundant safety shut-off system at least comparable with the flow sheet given in Figure C.5....." shall be provided. Since the purpose of the parallel system is to allow "...testing the effective closure of the valves at least once a year...." the solution proposed for Antwerp, and accepted by the NoBo, was to install a bypass around the 2 x XVs with a removable spool piece that is not connected during normal operation, and with isolation valves that are car-sealed closed (CSC). If and when the by-pass is connected it requires continuous supervision by the field operator.

Purge
This follows the Gonfreville procedure.

Leak Test

The leak test is carried out using nitrogen. Several of the valve tests are performed simultaneously. The first steps are the same as described for Gonfreville.

The fuel gas system venting is tested by opening all XVs up to the first individual burner isolation and opening all atmospheric vents. The PCV's are opened. The header pressure should fall to atmospheric pressure within 180 seconds.

The upstream, primary XVs (main fuel gas and natural gas to pilots) are closed and confirmed to close within 5 seconds.

The nitrogen fill connection manual valve is opened and pressurizes the fuel gas and pilot gas headers upstream of the primary XVs. The field operator depresses the appropriate hand switches on the local panel to confirm that the nitrogen valve is opened and to start the automated valve leak tests.

Test 1: pilot gas primary XV, and Test 5: burner fuel gas primary XV. The downstream and intermediate 'bleed' XVs are confirmed to close within 15 seconds. The PT's between the closed block valves are confirmed to rise by not more than 0.1 bar in 60 seconds.

Test 2: pilot gas secondary isolation and bleed XVs, and Test 6: burner fuel gas secondary isolation and bleed XVs. Pilot gas primary XV is confirmed open within 15 seconds. Downstream pressure is confirmed to rise to above 2.0 barg within 10 seconds. The upstream valve is confirmed closed within 15 seconds. The downstream pressure is confirmed to hold (falls by less than 0.1 bar in 60 seconds). The same procedure is followed for the main

Figure 5. Furnaces and local panel.

fuel gas supply except that a longer time is allowed for pressurizing (larger volume to fill).

Test 3: start-up pilot gas XVs (12 total) and main pilots XV, and Test 7: start-up burner XVs (12 total), main (non-start-up) burners XV and wall burners XV. The pilot header up to the pilot gas XVs and the main pilot XV is pressurized, the main isolation XVs are closed, and then the header is confirmed to hold pressure using the same time-pressure criteria as above. Next the main fuel gas header is pressurized up to the start-up burner XVs, the main burners XV and the wall burners XV. The main isolation (upstream) XVs are closed, then the main header is confirmed to hold pressure using the same time-pressure criteria as above.

Test 4: main pilot gas manual valves test, Test 8: main burner manual valves test, and Test 9: wall burner manual valves test. The pilot header up to the start-up pilot gas XVs and the main pilot gas manual valves is pressurized, the main isolation XVs are closed, and then the header is confirmed to hold pressure using the same time-pressure criteria as above. Similarly the main header is pressurized up to the start-up burner XVs, the main burner manual isolation valves, and the wall burner manual isolation valves. The main isolation (upstream) XVs are closed. Then the main header is confirmed to hold pressure using the same time-pressure criteria as above.

Nitrogen is vented from the fuel gas headers. The fuel gas and pilot gas valve integrity test is complete.

24 hours after the successful completion of the valve integrity test, all of the 'valve test ok' indicators will automatically reset. Therefore ignition of the pilots must proceed within 24 hours upon completion of this test. This is a change from the Gonfreville protocol based on REX: it is not always practicable to go to the ignition steps as quickly as originally specified.

If the leak test has failed then the control room operator will need to reinitialize leak test by re-pressing the appropriate push button.

After the successful completion of the valve integrity test, the nitrogen supply must be isolated from the fuel gas network. Significant amounts of nitrogen in the fuel and pilot gas headers could prevent the successful light-off of pilots and burners. The nitrogen shall be isolated and purged from the fuel gas header and pilot gas header, prior to the lighting of the pilots.

Ignition Steps

The ignition steps have been described in detail for the Gonfreville new furnace project. Here we simply described the differences.

A series of conditions have to be satisfied before the ignition sequence can start:

- It is less than 24 h since the successful completion of the valve integrity test.
- The fuel gas and pilot gas headers have been purged of nitrogen.
- The pilot gas low pressure trip is overridden and the control system is lined up.
- The field operator has confirmed, by pushing a button on the local panel, that the fuel supply is commissioned and that the firebox atmosphere test is complete and passed.
- The 10 min post-purge timer has not timed out (note that the post-purge timer only starts when the damper position and fan speed are adjusted to ignition conditions, not when the minimum purge time of 20 minutes is reached).

For the start-up burner pilot ignition sequence:

- If less than 6 pilot burners are confirmed lit, the pilot ignition sequence is deemed to have failed and the furnace trips.
- If between 6 and 7 start-up pilots are confirmed lit, the ignition sequence is halted for a period of 20 minutes. After this period the ignition attempt cycle counter is reset, and the operator can initiate another 3 pilot lighting attempts by depressing the hand switch on the local panel.
- If 8 or more start-up pilots are confirmed lit, the distribution of the lit pilots is checked to make sure the pattern of light start-up pilots will provide an even firing pattern. If the lit pilots are in an acceptable pattern, a permissive will be provided to open the start-up burner fuel gas XVs.
- If 8 or more start-up pilots are confirmed lit, but the pattern is unacceptable, the ignition sequence is halted for a period of 20 minutes. After this period the ignition attempt cycle counter is reset, and the operator can initiate another 3 pilot lighting attempts by depressing the hand switch on the local panel.

The REX from Gonfreville suggests that 8 start-up burners can heat the furnace to 750°C (auto-ignition, safe temperature) without adjusting the main, non-start-up burner dampers to limit the overall amount of excess air entering the firebox. This means that the main burner dampers do not need to be disconnected from the common jackshafts.

For the burner ignition and furnace warm up phase:

If a positive main flame signal is not received after 3 attempts, the XV on that burner will be locked closed for 20 minutes; but the operator may choose to continue the warm up sequence by attempting to light a different burner. After the 20 minute period a further three attempts to light the main flame of this burner will be allowed.

If the number of positive flame signals from different burner locations (i.e. either positive pilot flame signal or positive burner flame signal; if both flame signals from the same burner location are positive, this only counts as one positive signal) falls below 6 before the arch temperature reaches 750°C the furnace trips. In order to reach an arch temperature of 750°C at least 8 of the 12 start-up burners should be lit.

International Standard Requirements

ISO 13577-2 was published in 2014. It is clear that the origin and source of the new international standard is EN 746-2. There are, however, important differences, most of which have been described by Van der Bij (2015) and by Clauss (2014).

EN 746-2 requires a VPS*, compliant with EN 1643, for automatic XVs controlling capacities greater than 1 200 kW. ISO 13577-2 only requires an automatic [valve proving] system if the industrial furnaces and associated processing equipment (TPE) is "....intended to be started up more than two times [twice] a year...". Hence an automatic VPS should not be required for most cracking furnaces, although each furnace type has to be evaluated individually.

*__NOTE:__ *this is a new requirement from 2010. The original version of EN 746-2 only required a "leak tightness device", of which a double-block-and-bleed is one permitted variant. See Table 1.*

EN 746-2 mandates "a redundant safety shut-off system" so that annual in situ testing can be made of the automatic shut-off valves if the furnace operates continuously for more than 1 year (see Table 1 of this paper, ref. paragraph 5.2.2.3.1). The equivalent paragraph of ISO 13577-2 (4.2.2.6) states "...NOTE It is commonly agreed that valves are to be tested annually **unless longer testing intervals can be justified by the risk analysis**....[highlighting by the authors]". Van der Bij (2015) claims that the testing interval can be extended to 3-4 years by carrying out a risk analysis according to the so-called Method D of ISO 13577-4.

EN 746-2 mandates a pre-purge rate at least 25% of the maximum combustion air flow rate. In ISO 13577-2 the rate doubles to 50%.

Conclusions

The standards in Europe for burner management systems have become much rigorous in the last 15 years or so. Although such standards are, strictly speaking, only applicable to new projects they can also apply to revamps in which there are major changes to the combustion system. With the publication of the international, ISO standard, this approach is likely to be adopted in countries outside the EU.

Total and TS&W have developed a stringent but flexible design. It has proved, in practice, to be reliable, convenient and safe.

The requirements are still evolving and certainly the BMS will likely be somewhat different for each furnace type considered.

Acknowledgement

Heurtey Petrochem S.A. was the Engineering, Procurement and Construction ('EPC') contractor for the furnace projects at both Gonfreville and Antwerp.

References:

1. "Industrial thermoprocessing equipment – Part 2: Safety requirements for combustion and fuel handling systems", European Standard EN 746-2: 1997.
2. Matthews, R. (2000), "European Ethylene Producers' Committee [EEPC], EN 746-2 (Industrial Thermo-processing Systems), Working Group Agreed Terms of Reference".
3. Overwater, K. (2001), "EN 746-2 Furnace Protection Work Group - final report –", *EEPC HSE Conference,* The Hague.
4. "European Ethylene Producers Committee, Final Report, Issue Group to consider the application of European Standard EN-746/2 to steam cracking plants", 2002.
5. Enqvist, K. (2010), "State of art BMS requirements - Upgrade of old furnaces and implementation in new furnaces", *AIChE Spring National Meeting*, San Antonio, Texas.
6. "Industrial thermoprocessing equipment – Part 2: Safety requirements for combustion and fuel handling systems", European Standard EN 746-2: 2010.

7. Platvoet, E., Strupp, C., & Theis, G. (2013), "Safety considerations for the design and operation of ultra-low NO_x burners in cracking furnaces", *AIChE Spring National Meeting*, San Antonio, Texas.
8. De Boer, M. (2013), "Burner management system upgrade as a result of cracking furnace firing system replacement", *AIChE Spring National Meeting*, San Antonio, Texas.
9. "Industrial furnaces and associated processing equipment — Safety — Part 2: Combustion and fuel handling systems", International Standard ISO 13577-2:2014(E).
10. Van der Bij, M. (2015), "Safety requirements for furnace fuel handling systems", *EEPC Ethylene Seminar*, Prague.
11. Van der Bij, M. (2015), private communication.
12. Clauss, R. (2014), "IG [Issue Group] Cracking Furnace Safety Final Report", *EEPC Ethylene Seminar*, Warsaw.

449973 Ethylene Cracking Furnace Burner Applications Operating with Significantly Less Than 90 mg/nm^3 NO_X Emissions

Abstract:

This abstract will focus on the field results of eight (8) ethylene cracking furnaces, which use Zeeco GLSF Enhanced-Jet burners at Petrochina Sichuan Petrochemical Company Limited (SPC) that achieve field measured NOx emissions significantly less than 100 mg/Nm^3. Using ultra-low emissions burners in ethylene cracking applications have proven to be an economical solution at Zeeco for achieving lower NO_x levels than 100 mg/Nm^3 from existing fired equipment. Providing a next generation low NO_x burner footprint that fits into ethylene cracking applications has proven to be a challenge in many applications. Ultra low NO_x burner flame pattern(s) or profile(s) that can operate properly within a required burner to burner spacing and heat flux limitations have also proven to be a challenge.

This paper will focus on an application, which uses the GLSF Flat Flame Enhanced-Jet Burner model operating in eight (8) ethylene cracking furnaces that achieved less than 100 mg/Nm^3 in the field. For this application, NOx emissions averaged around 50 mg/Nm^3 in the field. The furnace field measured NOx emissions ranged between 35 mg/Nm^3 (17 ppmv) to 65 mg/Nm^3 (32 ppmv) at design heat release.

For applications requiring less than 100 mg/Nm^3 NOx emissions, Zeeco has developed an ultra-low NOx emissions burner technology, "GLSF Enhanced-Jet Flat Flame Burner", which provides a very small mechanical footprint and produces an excellent heat flux profile without flame to flame interaction. The GLSF Enhanced-Jet floor burners utilize the basic "free-jet" method of mixing fuel gas and inert products of combustion to form a new combustible mixture before entering the combustion air stream. The resulting mixture, burns at a lower adiabatic flame temperature, resulting in lower thermal NO_x production. In addition, the GLSF Enhanced-Jet burner is designed to greatly reduce issues of flame interaction and flame rollover experienced with other burner designs. Flame interaction between burners can cause flame impingement upon the process tubes and increased emissions. Flame impingement can result in shorter run lengths and higher tube metal temperatures which are problematic for prolonged ethylene production.

In conclusion, Zeeco and Petrochina Sichuan Petrochemical Company Limited (SPC) will co-author the paper and provide specific application details and the resulting field emission test results. This paper will focus on an application, which uses the GLSF Flat Flame Enhanced-Jet Burner design in eight (8) ethylene cracking furnace that achieved significantly less than 100 mg/Nm^3 in the field. This application will show that the field measured NO_x emissions were between 35 mg/Nm^3 (17 ppmv) to 65 mg/Nm^3 (32 ppmv) at design heat release for this highlighted project.

444925 Compressor Wash Oil Supply in a Light Feedslate Environment

Abstract:

The shale gas development, primarily in the US, has resulted in lower cost feedstocks for gas cracking and thus the addition of new gas cracking capacity. This increased capacity has resulted in increasing demand for Compressor Wash Oil. Furthermore the transition of some naphtha (liquid) crackers to gas feedstocks has further increased demand for Compressor Wash Oil. With this transition fewer internal streams are available for use as Compressor Wash Oil. Current products used for Compressor Wash Oil are generally C10-C12+ aromatic based. This paper will focus on the desirable properties for a Compressor Wash Oil.

442001 WAO Reliability Improvements

Abstract:

The Wet Air Oxidation (WAO) system used for processing spent caustic at an Ethylene plant was plagued with extensive fouling. It required cleaning of the heat exchangers, piping, and the Wet Air Oxidation reactor at a certain frequency.

An analysis of the foulant was conducted, and improvements implemented accordingly to reduce the fouling. These have led to reduced frequency of cleaning, increased run-time, achieve design temperatures, and increased spent caustic through put to the WAO system. It has also led to a reduction in waste water generated from equipment preparation for cleaning.

AIChE Paper 86f

Simulation-Based Assessment and Mitigation of Fouling in Water Quench and Charge Gas Compressor Sections of Gas Crackers

Mathieu Zellhuber
Process Engineer Petrochemical Plants
Linde AG, Engineering Division

Matthias André
Senior Process Lead Engineer Petrochemical Plants
Linde AG, Engineering Division

Thomas Emmert
Manager Process Engineering Petrochemical Plants
Linde Engineering North America

Gunther Kracker
General Manager Process Engineering Petrochemical Plants
Linde AG, Engineering Division

Prepared for Presentation at the 2016 Spring National Meeting
Houston, Texas, April 10-14, 2016

Simulation-Based Assessment and Mitigation of Fouling in Water Quench and Charge Gas Compressor Sections of Gas Crackers

Mathieu Zellhuber

Process Engineer Petrochemical Plants

Linde AG, Engineering Division

Matthias André

Senior Process Lead Engineer Petrochemical Plants

Linde AG, Engineering Division

Thomas Emmert

Manager Process Engineering Petrochemical Plants

Linde Engineering North America

Gunther Kracker

General Manager Process Engineering Petrochemical Plants

Linde AG, Engineering Division

Abstract: Fouling in the hot section of the separation train is widely recognized as a major challenge for successful, long-time operation of gas crackers. Root causes for fouling can be of very different kinds and are consequently hard to determine at first sight. As for the causes, mitigation measures applied in plant operation are also very diverse, ranging from chemical and wash oil injections to targeted management of recycles and small revamps. It is hence of utmost importance to know when to apply which measure.

Process simulation has become an essential tool for the design of new plants, allowing precise prediction of the separation steps. The present paper shows how Linde uses process simulation for understanding root causes of fouling in operating plants and for developing suitable mitigation measures. Although hot section fouling is of high physical and chemical complexity, we have developed a pragmatic approach to evaluate and compare fouling risks and quantities in the water quench and charge gas compressor sections. The methodology has been successfully applied during a recent case study for an existing plant experiencing fouling in charge gas compressor intercoolers and suction drums.

These improved tools can thus be used to solve actual fouling issues, but also to prevent future problems, for instance before proceeding to a feed change. This permits anticipating potential fouling risks and developing mitigation strategies in advance. The developed prediction methods are similarly used for the design of new plants.

1. Introduction

The potential fouling of equipment items and/or piping sections represents a significant concern for the operation of ethylene plants. In many cases, fouling is observed first in terms of decreased plant efficiency, mostly linked to higher pressure drops and/or thermal resistance. This typically results in higher utility consumption and maintenance effort; in more severe cases, equipment damages, unplanned shutdowns and production losses can occur. The potential impact of fouling issues on the overall plant economics is therefore high; this explains the increased attention brought to this topic by licensors and plant operators.

Fouling cannot be handled in a straightforward, singular manner, because the term stands for the panoply of physical and chemical processes that can affect plant operation all along the gas processing route. This high complexity particularly applies to the hot section of the separation train, i.e. to quench and charge gas compressor sections, where stream compositions, pressure and temperature levels are wide-ranging. As a consequence, different fouling mechanisms may come into play, involving various critical cracked gas components. Moreover, these fouling mechanisms will strongly depend on the specific plant setup.

The resolution of past fouling issues and the avoidance of future ones therefore become challenging tasks. The aim of the present paper is to show how the efficient use of process simulation tools can prove to be very useful in this task.

Given the high physical complexity of the problem, it is clear that the employed simulation tools cannot predict fouling mechanisms in all its details, since this would go well beyond industry-standard computational resources. Instead, the developed approach aims at collecting data from standard process simulations and in-house developed post-processing tools in order to evaluate the fouling tendency in a given case.

This approach was applied in the frame of a study made for an existing plant, which experienced fouling issues in the charge gas compressor section after a change of the feed scenario. The root cause analysis was performed by Linde using process simulation that reproduces real operation data. The analysis methods developed allowed to identify the root causes and to evaluate the effect of potential mitigation measures.

Although the study was performed on a gas cracker with mixed ethane/propane feed, the developed methods can be applied to other plants and feed stocks in a similar way.

2. Fouling in water quench and charge gas compressor sections

2.1 Process description

The main task of the water quench section is to ensure sufficient cooling of the charge gas stream down to cooling water temperature level and consequently to condense significant amounts of dilution steam and heavier hydrocarbons. Its main element is the quench tower; in modern designs, it is divided into a lower water saturation section and, separated by a chimney tray, the upper sections, in which the charge gas is cooled by direct contact with circulating quench water. The heavy end of the hydrocarbon spectrum is partially condensed in the lower and upper column stages and retrieved from the system.

The water quench section is followed by the charge gas compressor, whose main purpose is to elevate the charge gas to a pressure level sufficient for the subsequent cryogenic separation into value products. Moreover, with its repeating sequence of compressor stage, intercooler and suction drum, it acts as a continuation of the water quench section, by performing a multistage condensation of water and heavy hydrocarbon components. The molecular weight of the condensed hydrocarbons is decreasing along the compressor stages, resulting in varying compositions of condensates collected in the suction drums.

The similar functions of water quench and charge gas compressor sections in terms of partial charge gas condensation led us to consider them conjointly regarding fouling risk assessment, as will be described in section 3.

2.2 Fouling mechanisms and influence parameters

Previous works and experience indicate that the solidification of heavy hydrocarbons is a potential fouling threat in the water quench section [1, 2]. Indeed, some of the typical charge gas components like naphthalene have a high melting point (80°C at 1 atm) compared to components with similar molecular weight [3]. As a consequence, part of the hydrocarbon components condensed in the quench tower may solidify, precipitate and cause plugging of column internals. This has for example been addressed by Schmigalle and Tolergard [1], who calculated pour point profiles in the quench tower to determine the fouling risk linked to plugging. They could elaborate appropriate fouling mitigation methods for an existing plant to ensure sufficient margin between pour point and process temperatures and avoid the solidification of heavy hydrocarbons.

In the charge gas compressor section, the formation of polymer material is generally referred to as the most common fouling threat. Previous works describe different types of mechanisms and reactions which lead to the formation of macromolecules that can deposit and cause plugging [4-6]. Very relevant is for instance thermal polymerization, in particular of styrene. These reactions are often reported to occur within the compressor casing where temperatures are highest. Indeed, temperature has an exponential impact on reaction rates. Nevertheless, also other parameters need to be taken into consideration to evaluate the risk of polymerization [6, 7]:

- Amount of monomer components in the system, in particular double-bonded, unsaturated hydrocarbons.
- Residence time of monomer components.
- Amount of solvent material, in particular aromatics.
- Presence of oxygenates promoting polymerization reactions.
- High iron contents due to piping and equipment corrosion.

This forms the theoretical background on which the study was performed for an existing plant. The following section will first describe the actual fouling issue and then describe the root cause analysis made.

3. Investigation of fouling in existing plant

3.1 Plant history and problem description

The considered plant has been operated with 10+ years of service with mixed ethane/propane feed. In recent years, propane furnace load and cracking severity have been increased beyond original design values resulting in altered charge gas compositions. Since last major turnaround in 2013, a continuous pressure drop increase could be observed at the 2nd compressor interstage, i.e. between stages II and III. Plugging material was found in both interstage cooler E2 and suction drum D3, see Figure 1; the latter was causing the major part of the pressure drop. Unplanned shutdown for cleaning was needed in 2015.

The foulant material collected at both locations was analyzed and consisted in both cases mainly of organic material, indicating the presence of heavy pyrolysis components as well as polymer materials.

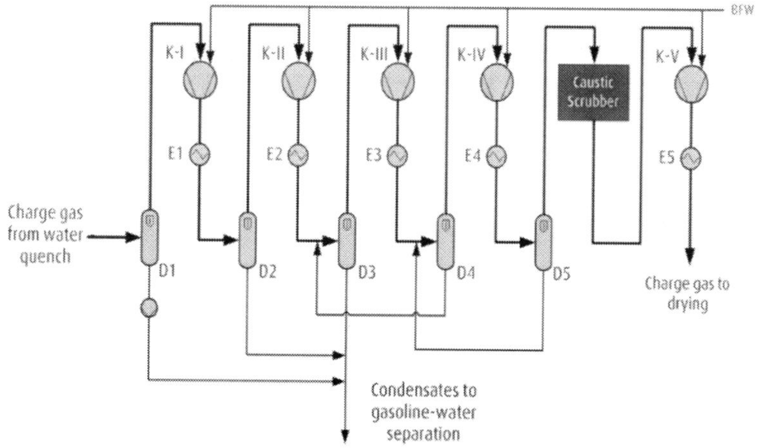

Figure 1: Sketch of considered process topology in charge gas compressor section

3.2 Methods employed for root cause analysis

In order to develop suitable mitigation measures, it is most important to gain a solid understanding of the fouling root causes. Therefore, the investigation was initiated by analyzing operation data provided by the plant owner in order to set up a realistic simulation of current process operation. This included detailed kinetic simulations in ethane and propane furnaces based on actual feed load and conversion data. Furthermore, internal recycle settings and operation set points were adapted in the main process simulation to give a realistic representation of the actual plant operation.

Thereafter, a specific analysis of the process simulation was carried out in order to evaluate which parts of the water quench and charge gas compressor sections are specifically prone to the considered fouling mechanisms.

3.3 Evaluation of fouling risk linked to heavy cracking products

In the present work, the pour point approach described by Schmigalle and Tolergard [1] was employed to evaluate whether solidifying heavy hydrocarbons may contribute to the fouling observed. The pour point calculation is based on proprietary property packages. While in its original application it was applied to the water quench tower only, we applied it now additionally to the charge gas compressor section, more precisely to the intercoolers. Indeed, the foulant material retrieved from the plant was partly composed of heavy components like naphthalene, as mentioned above.

The pour point of the relevant hot section streams is mostly determined by the heavy end of the total cracked gas, which is a very small fraction, in particular for cases of gas cracking. The exact heavy end quantities and compositions are hence difficult to determine with certainty and are moreover sensitive to operating conditions. In order to account for the challenging nature of pour point predictions, an estimation range was determined. This is shown in Figure 2 for intercoolers E1 to E4, in which the process temperature curve is plotted along with the estimated pour point range.

When comparing process and pour point temperatures at the intercoolers, one reckons a potential pour point violation at the second intercooler E2 and, to a slightly minor extent, at the first intercooler E1. It can be followed that, in the first two intercooler stages, fouling can potentially occur due to solidification of heavy cracking products. Indeed, most of the naphthalene is condensed at the second compressor interstage. Further downstream, the hydrocarbons condensed from the cracked gas tend to become lighter, which explains the reduced risk of heavy hydrocarbon solidification at the later compressor stages.

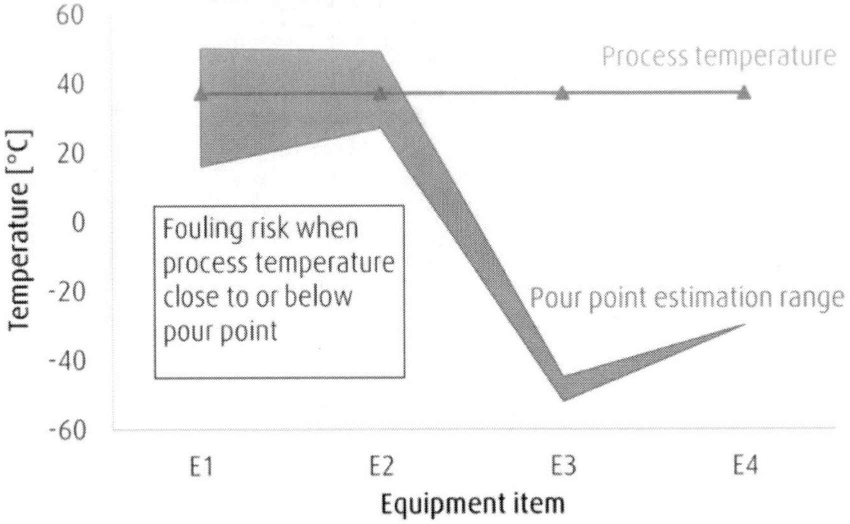

Figure 2: Pour point profile for condensate streams in charge gas compressor intercoolers (base process)

3.4 Evaluation of fouling risk linked to polymerization

As mentioned earlier, polymerization reactions are strongly enhanced by elevated temperatures. This leads to the basic assumption that the polymers found in the foulant material samples were originally formed inside the compressor casing, where temperatures are highest.

The first step in the investigation linked to polymer formation was therefore to look for indications regarding polymerization inside the compressor. Looking at the average outlet temperatures did not support this hypothesis, since they were constantly ranging well below 85°C. In addition, compressor performance data was analyzed in terms of scaled capacity curves. For the given case, no clear indication about polymerization occurring inside the compressor could be found.

As a consequence, other elements of the compressor train were checked more closely. Special interest was brought to the suction drum D3 upstream of the 3rd compressor stage, because this equipment was showing the largest increase in pressure drop. In the investigated plant, the drum D3 receives the cracked gas condensates from the two downstream collection drums D4 and D5, as shown in the flow sketch of Figure 1. We suspected that the continuous recycling and the consequently long residence times of monomer components in the three interconnected drums might promote polymerization.

In order to verify this hypothesis, an average, component-specific residence time was calculated for the three interconnected suction drums D3 to D5. This was done by combining equipment design information and process simulation data in a post-processing step.

The corresponding result is shown in Figure 3. It can be clearly seen that some monomer components are strongly accumulated in the interconnected drums. The residence times, in particular for styrene, are significantly longer than the liquid retention time of a single suction drum due to the constant internal recycling of these species.

These strongly increased residence times lead to high risk of polymer formation in the suction drums, in particular in equipment D3, which is the last element in this recycling cascade.

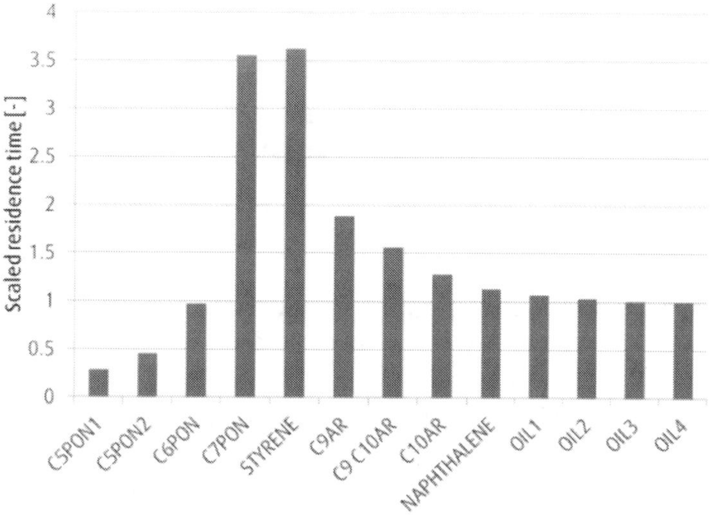

Figure 3: Combined component-specific residence time distributions in suction drums D3, D4 and D5 (base process). Scaled residence time values >1 indicate excessive accumulation of a given component due to plant internal recycles (PON: paraffins, olefins, naphthenes; AR: aromatics).

3.5 Summary of root cause analysis

The investigation led to the following conclusions on fouling root causes for the specific plant:

1. The fouling observed in the second interstage cooler is believed to be originally caused by heavy hydrocarbon components depositing on the exchanger, since the process temperatures are close to or below the estimated pour point temperature range. This is supported by the identification of heavy pyrolysis products in the foulant material samples. The observed polymer formation is supposed to be a consequence of the heavy deposits. The latter will accumulate at the front plate and the tube walls, where they are likely to retain monomers from the gas flow, thereby increasing their residence time and thus facilitating polymerization reactions.

2. In the suction drum upstream of the 3rd compressor stage, polymerization of accumulated monomers is believed to primarily cause the reported fouling. This accumulation is directly linked to the installed routing of cracked gas condensates, which are routed back from other collection drums. Residence times and quantities are significantly increased,

therefore strongly enhancing the risk of polymerization. As for the intercooler, the additional presence of heavy cracking products further increases the probability of solid formation.

Since both root causes are linked to the presence of heavier cracking by-products, it is supposed that the fouling risk was enhanced by the increased propane load and cracking severity conditions at which the plant was operated in recent years, and which were not covered by the original plant design. The impact of furnace feed scenario on condensate quantities and compositions is illustrated exemplarily for intercooler E2 in Figure 4.

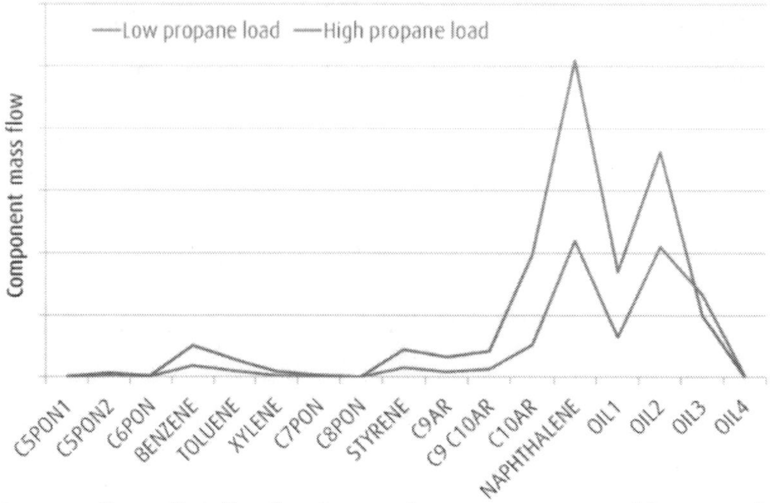

Figure 4: Component mass flow distribution in condensate stream of intercooler E2 for different furnace feed scenarios (PON: paraffins, olefins, naphthenes; AR: aromatics).

4. Developed fouling mitigation measures

Based on the root cause analysis, we defined mitigation measures to reduce the fouling risk linked to both mechanisms. As for the root cause analysis, process simulation and post processing tools were used to evaluate the effect of the proposed process modifications. The following mitigation measures were considered:

1. Recycle pyrolysis gasoline into the water quench system, in order to increase solvent quantities.
2. Optimize previous operation guidelines regarding wash oil quantity and quality for continuous injection into the compressor casing.
3. Modify the process topology by rerouting the condensates from suction drums D5 and D4.

Each measure was evaluated individually regarding its impact on both fouling mechanisms. However, none of these single measures was found to be capable of solving all fouling problems. Instead, a combination of all these measures was tailored which offers an effective fouling mitigation. This follows from Figure 5 and Figure 6, which show pour point profiles and residence time plots, respectively, for this new process setup.

The proposed measures therefore significantly reduce the residence times of potentially polymerizing components in the main process path and provide substantial washing effect at all stages of the cracked gas compressor train.

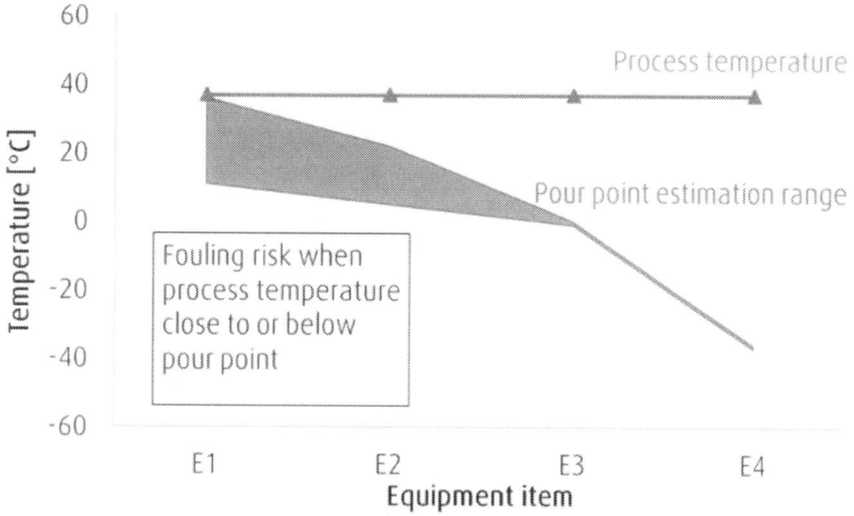

Figure 5: Pour point profile for condensate streams in charge gas compressor intercoolers (optimized process)

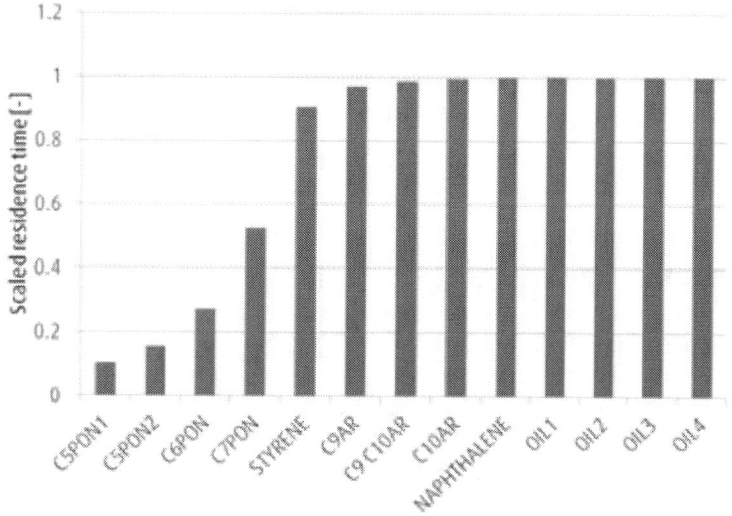

Figure 6: Combined component-specific residence time distributions in suction drums D3, D4 and D5 (optimized process). Scaled residence time values >1 indicate excessive accumulation of a given component due to plant internal recycles (PON: paraffins, olefins, naphthenes; AR: aromatics)

5. Conclusion and general applicability

The paper gives an illustrative example on how process simulation tools can be employed to assist plant operation services in understanding and mitigating fouling problems in water quench and charge gas compressor sections. The numerical tools are building upon standard process simulation methods, combined with an in-house developed post-processing method in order to get quantitative indications on fouling risk. This approach proved to be very effective and useful in a case of an actual plant. The findings described above have been confirmed by an experimental measurement campaign on-site.

The methods developed can readily be applied to other existing plants, independently of the feedstock. They can be used for the troubleshooting of existing fouling issues as well as for the assessment and prevention of future issues. In this manner, before a change in feed or furnace operation, potential fouling problems can be anticipated and taken into account as part of risk evaluation and mitigation strategy. The approach is similarly used for the design of new plants, in particular when increased feed flexibility is requested.

6. References

1. Schmigalle, Holger and Tolergard, Christer (2010), "Fouling Mitigation of Quench Tower System", in *Ethylene Producers Conference, AIChE Spring National Meeting*, San Antonio, Texas.
2. Kolmetz, Karl, Sloley, Andrew W., Zygula, Timothy M., Ng, Wai Kiong and Faessler, Peter W. (2004), "Design Guidelines for Distillation Columns in Fouling Service", in *Ethylene Producers Conference, AIChE Spring National Meeting*, New Orleans, Louisiana.
3. Collin, Gerd, Höke, Hartmut and Greim, Helmut (2012), "Naphthalene and Hydronaphthalenes", in *Ullmann's Encyclopedia of Industrial Chemistry*, Vol. 23, pp. 661-670, Wiley-WCH Verlag GmbH & Co. KGaA, Weinheim
4. Kimball, Russell (2011), "Fouling Mechanisms and Solutions in the Cracked Gas Compressor Train", in *EEPC Ethylene Seminar*, Lisbon, Portugal.
5. Hancock, Jessica M., Rossana, Debby and Arnst, Ted (2013), "Review of Organic Fouling Throughout an Ethylene Train", in *Linde Ethylene Licensor Conference*, Pullach, Germany.
6. Vachon, Jérôme, Hancock, Jessica M. and Robson, Ian (2015), "A chemist's perspective on organic fouling in ethylene operations: update", *Hydrocarbon Processing*, October, pp. 49-53.
7. Emmert, Thomas, Antonas, Sam and Kracker, Gunther (2016), "Full Range Feedstock Flexibility from Naphtha to Ethane", *Ethylene Producers Conference, AIChE Spring National Meeting*, Houston, Texas.

AIChE Paper Number 126a

Carbon Disulfide (CS_2): A New Challenge for Steam Cracker Complexes?

Pascal Coenen

Licensing Technical Manager

Shell Global Solutions International B.V.

Ernesto Angueira

Technical Service Engineer

CRI Catalyst Company LP

Prepared for Presentation at the 2016 Spring National Meeting

Houston, Texas, April 12, 2016

Carbon Disulfide (CS_2):

A New Challenge for Steam Cracker Complexes?

Pascal Coenen

Licensing Technical Manager

Shell Global Solutions International B.V.

Ernesto Angueira

Technical Service Engineer

CRI Catalyst Company LP

Abstract: Chemical companies are continuously exploring new opportunities in order to increase competitiveness. In today's market this can be attained through integration of upstream and downstream, energy recovery, among others. Another opportunity is via feedstock flexibility and its accessibility.

It is this feedstock flexibility that incentivizes companies to look for advantageous opportunities and cheaper imported feeds. However, cheaper raw material introduces new challenges to crackers and downstream operations, especially raw Pygas processing units.

The newest challenge for steam crackers is CS_2 contamination. Recently several companies have reported that they see a substantial increase in CS_2 contamination particularly impacting the Pygas units. Sulfur molecule coordination with Palladium and Nickel catalysts exhibit activity inhibition, such inhibition decreases as sulfur molecules size increase, H_2S>CS_2>Mercaptans>sulfides. Currently, there are multiple technologies that allow the removal of H_2S before reaching a Pygas unit however the same does not apply to CS_2.

Catalyst technology is gearing to develop catalysts with higher tolerance to such a contaminant while maintaining optimum performance.

In this paper, CRI, based upon a real example, will discuss the impact of CS_2 on Palladium catalyst. In addition, this presentation shows a plant example on the performance and the importance of partnering with the operating plant. The plant example will show that a more proactive operating approach instead of a reactive one is required, in order to achieve the desired competitive edge.

Carbon Disulfide (CS_2): A New Challenge for Steam Cracker Complexes?

Pascal Coenen – Shell Global Solutions International B.V. / Ernesto Angueira – CRI Catalyst Company LP

The olefin industry is a very competitive industry where the producers are continuously exploring so-called advantaged feedstocks. Often these advantaged feedstocks offer cost advantages, but simultaneously, often contain some disadvantages. One of these potential disadvantages is the higher level of impurities. In the US – Ethylene Producers Committee (EPC) and European Ethylene Producers Committee (EEPC) many experiences on the impacts have been shared on removal and treating technologies within the various olefin flow schemes.

This paper will discuss CS_2 and its impact on the steam cracker complex with the recent "new trend" in CS_2 levels, impacts observed by olefin producers and what can be done on awareness, monitoring and mitigating its consequences.

What is CS_2?

Carbon disulfide is a colorless volatile liquid with the formula CS_2. Pure CS_2 has a somewhat pleasant "ether-like" odor. Commercial CS_2 is typically contaminated with foul-smelling impurities, such as carbonyl sulfide. Small amounts of carbon disulfide are released by volcanic eruptions and swamps. In organic chemistry it is used to manufacture regenerated cellulose, the main ingredient of viscose, rayon and cellophane. It is also used as an industrial and chemical non-polar solvent.

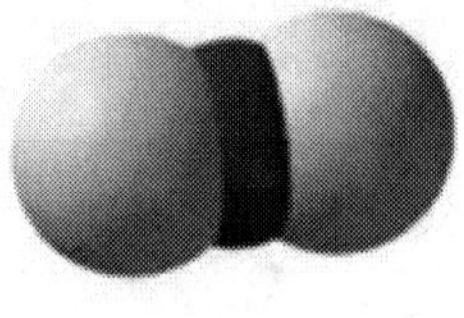

Chlorination of CS_2 is the principal route to produce carbon-tetrachloride.

From a safety point of view, CS_2 is a toxic and highly flammable component. Its toxicity is close to the toxicity of H_2S (see table 1) and as such elevated levels of CS2 are life threatening as it affects the nervous system.

	CS_2	H_2S
LC_{LO} (human, 30 min.) [ppm]	4000	600
IDLH [ppm]	500	100
PEL [ppm]	30 (100 peak)	20 (50 peak)

The flash point of CS_2 is -30°C and auto ignition can occur below 100°C

How is it formed?

At high temperatures CS_2 can be manufactured by combining coke or carbon with sulfur. At lower temperatures, approximately 600 °C, in the presence of silica gel or alumina catalysts natural gas can react with sulfur to produce CS_2.

$$2\ CH_4 + S_8 \rightarrow 2\ CS_2 + 4\ H_2S$$

The reaction of CO_2 and H_2S via COS can form CS_2.

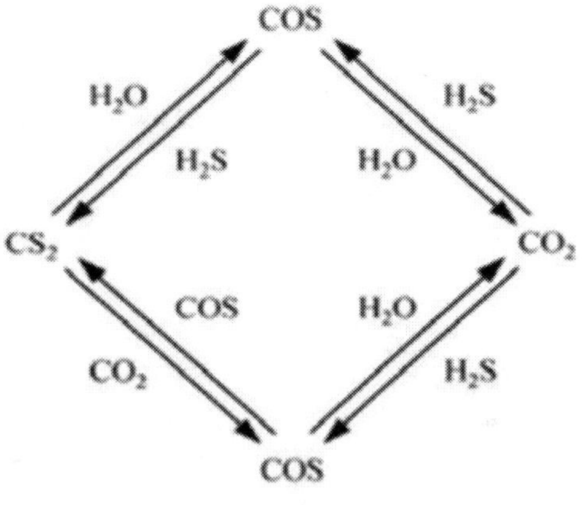

Why and why now?

In the last 2 years, there has been a sudden increase in issues which are related to increase of CS_2 content in the naphtha. The question arises why and why now? To answer, we need to know where the CS_2 could originate from. Some aspects or information are solid however a clear direction on why now is not found. It is well known that CS_2 has tendency to concentrate in C5 stream so light naphtha's are prone to contain higher CS_2 contents. Some off gases can contain CS_2, however these are not likely linked to the issues reported in the steam cracker complexes.

Most plants that have reported issues with increased CS_2 used naphtha as a main feedstock. The source of the naphtha is not clear. Some links are hinting to an increase of naphtha from fracking while others report issues when importing naphtha from Russia.

In Asia back in 2014 the naphtha industry agreed on reporting CS_2 levels in the naphtha. In particular, this was linked to producers in Japan who produce C5 derivatives (e.g. isoprene).

Independent of the source/origin of the naphtha, there are actions an operator can take. First it is some importance to be aware of the change in feed. But how does the operator know? Most operators only measure and specify total sulfur in there feed specification and have no specific CS_2 specification.

Where does it end-up in the steam cracker?

The most likely source of CS_2 to enter the steam cracker is the naphtha feed. The formation of CS_2 from CO_2 and H_2S in the cracking furnaces is negligible as the concentration of CO_2 and H_2S are too low. There have been some studies that conclude the "cracking" of CS_2 in the furnaces is also negligible, which means that almost all CS_2 entering the cracking furnaces will leave the furnaces as CS_2. As mentioned in the previous paragraph, CS_2 has a tendency to concentrate in the C5 stream. So it concentrates in the C5+ stream (pyrolysis gasoline) from the quench water settler and debutanizer bottoms in the steam cracker. For further processing the operator has several options: a) extract the C5's from the pygas in order to process them in a C5 extraction unit (often a so-called isoprene extraction unit) or b) hydrogenate the C5's for recycle cracking or c) send the C5+ stream to the Pygas hydrotreater.

The extraction and hydrogenation units are the units within the steam cracker process that have reported issues with their performance.

The next section discuss the impact of CS_2 on the catalysts used in the Pygas hydrotreaters and what one can do to mitigate the problems associated with increased CS_2 content.

Impact on Pygas catalysts

The catalyst used in the 1st stage of a Pygas hydrotreater unit is typically nickel (Ni) or palladium (Pd). The active Ni or Pd catalysts consist of surface-sulfided metallic Ni or Pd nano-crystallites. The formation of metal sulfides has an impact on the hydrogenation reactions as it retards all hydrogenation rates. It improves the selectivity of converting di-olefins above the selectivity of converting olefins. Unsulfided Ni or Pd catalyst could hydrogenate aromatics and the presence of sulfur strongly suppresses the hydrogenation of aromatics.

Laboratory work on catalyst impact

The impact of sulfur on Ni catalyst has been extensively tested in the laboratory. The impact of spiking the pygas with some CS_2 (20 ppm S) results in a drop of activity, but as can be seen from the graph below a new equilibrium is established.

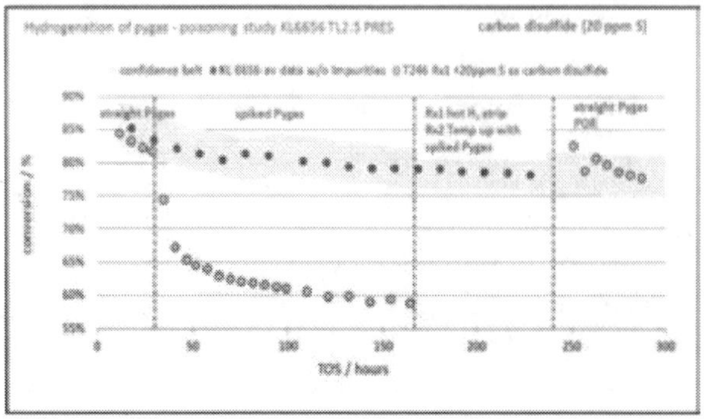

From the experiments it becomes clear that the activity is recovered after a hot hydrogen strip and the CS_2 is no longer present in the feed.

Depending on the amount of CS_2 in the feed the loss of activity can be compensated for by increasing the inlet temperature.

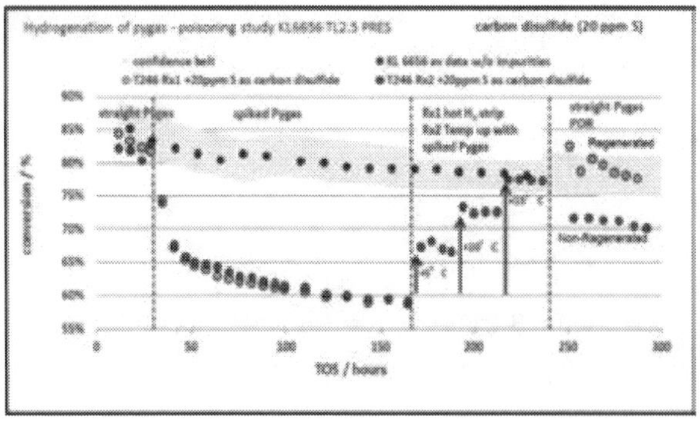

CS_2 acts as a strong inhibitor on the Ni catalyst and has a somewhat irreversible impact without regeneration, but the base activity can be restored to normal level after a regeneration using hot hydrogen stripping.

The impact on the catalysts is dependent on the type of sulfur species. The impact of the various sulfur species is as follows:

H_2S > CS_2 > COS > RSSR > RSH > sulphides, thiophenes

The impact of mercaptans on the activity Pd catalyst is quite limited and even activity recovery doesn't require an increased temperature. For CS_2 the impact on Pd catalyst is such that increasing the temperature partially recovers the activity. For H_2S increasing the temperature (under normal process conditions) cannot recover the drop in activity. Of course a full regeneration using hot hydrogen stripping would be effective.

Catalyst properties such as the Pd content, the catalyst shape/surface and the dispersion of the Pd on the surface play an important role on the catalyst activity and equally on the robustness against poisons (e.g. CS_2).

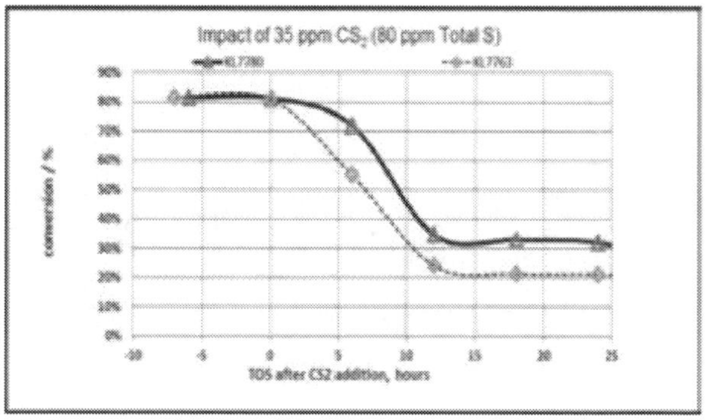

The graph comparing CRI's KL-7780 against KL-7763 shows that the impact of a substantial increase of CS2 has a big impact on both catalysts. However, the KL-7780 has about 150% more remaining activity compared to KL-7763.

Plant example on Palladium Pygas unit

An olefin producer in South America, was experiencing problems maintaining their styrene specification in the 1st stage hydrogenation reactor. The root cause was initially not so obvious as the feed specification was achieved. After some thorough investigation it became clear that although the total sulfur specification was met, there was a substantial increase in the CS_2 content. The naphtha feed specification had no specific CS_2 specification.

CS_2 spikes of up to 9 times the normal levels (10 ppm-wt) were measured.

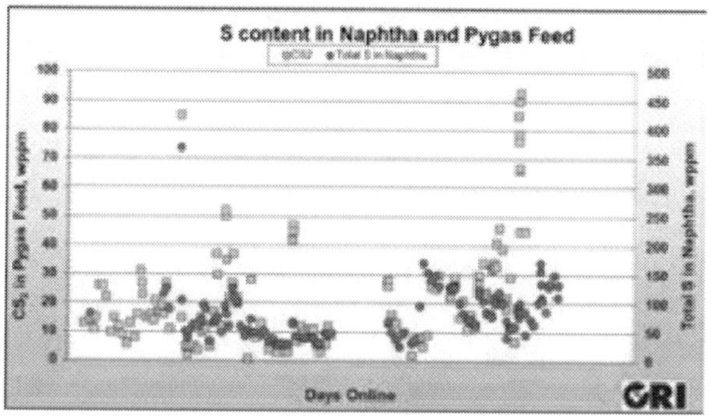

After some discussions with CRI's technical service specialists and some troubleshooting using CRI's proprietary pygas model, the effects of the increased CS_2 could be modelled and some pro-active temperature adjustments could be made by operations. Since the initial incidents the operator is monitoring the CS_2 content in the purchased naphtha. The CS_2 content can fluctuate and the right operating regime can be determined to meet the specification and selectivity objectives of the 1st stage reactor.

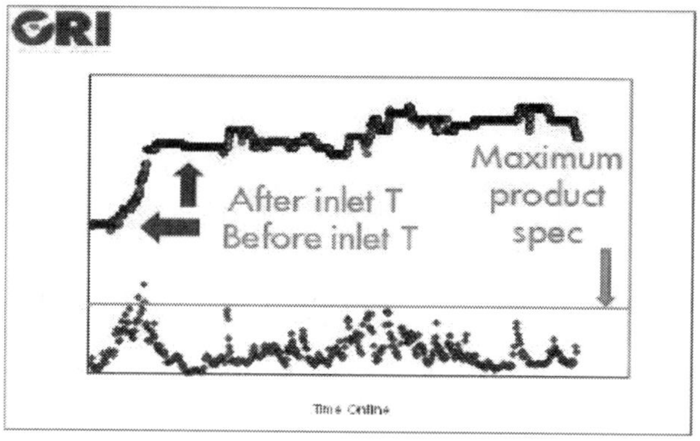

The temperature adjustments were sufficient (see graph above) to stay within the maximum product specification when the naphtha contained higher CS_2 concentrations.

Conclusion

Over the recent years there is an increased frequency of operators reporting issues linked to increased CS_2 in the feed to their steam cracker complex. Although the source is not exactly known operators can take proactive steps. Knowing the sulfur speciation is crucial but not easy. As an extra precaution one could consider to add CS_2 to their feed specification and analysis. A typical number is around <10-15 ppm CS_2.

This article further described the impact of CS_2 on the Pygas hydrogenation unit and how it acts like a "semi-permanent" poison on both nickel and palladium catalyst applied in the 1st stage. One should not neglect the impact on the 2nd stage pygas reactor as underperformance of the 1st stage reactor cascades to the 2nd stage reactor and a substantial shift in exotherm could occur. The customer example shown in this article shows that catalyst developments and service support (modeling and technical support) can help improve performance by shifting the operation to a more pro-active response.

References:

1) Contamination Removal/Treating Technologies: An Overview – Sanjeev Kapur, AIChE, April 2001
2) Carbon Disulfide (CS_2) Mechanisms in Formation of Atmospheric Carbon Dioxide (CO_2) Formation from Unconventional Shale Gas Extraction and Processing Operations and Global Climate Change - Environ Health Insights. 2015; 9(Suppl 1): 35–3 http://www.ncbi.nlm.nih.gov/pmc/articles/PMC4428389/#b5-ehi-suppl.1-2015-035
3) World Health Organization . WHO Air Quality Guide for Europe. 2nd ed. Denmark: 2000. Carbon disulfide; pp. 1–8.
4) 2. U.S. Environmental Protection Agency . Integrated Risk Information System (IRIS) on Carbon Disulfide. Washington, DC: National Center for Environmental Assessment, Office of Research and Development; 1999.
5) 3. U.S. Environmental Protection Agency . Health and Environmental Effects Profile for Carbon Disulfide. Cincinnati, OH: Environmental Criteria and Assessment Office, Office of Health and Environmental Assessment, Office of Research and Development; 1986. EPA/600/x-86/155.
6) Introduction to Sulfur Compounds and Sulfur Removal as Applied to Olefin Unit Design and Operation – JA Reid - AIChE, April 1998.
7) Article “Asia naphtha industry agrees to state carbon disulfide levels in Sikka, Western cargoes” - Platts Singapore – April 2014

AIChE Paper Number: 126b

OIL FIELD ADDITIVES
AND
ETHYLENE FEEDSTOCK CONTAMINANTS

Samuel A. Lordo

Marketing Director, Process

Nalco Champion, an Ecolab Company

M. David Smith

Industry Technical Consultant, Petrochemicals

Nalco Champion, an Ecolab Company

Prepared for Presentation at the 2016 Sprint National Meeting

Houston, TX, April 10-14, 2016

OIL FIELD ADDITIVES

AND

ETHYLENE FEEDSTOCK CONTAMINANTS

Samuel A. Lordo

Marketing Director, Process

Nalco Champion, an Ecolab Company

M. David Smith

Industry Technical Consultant, Petrochemicals

Nalco Champion, an Ecolab Company

Abstract:

The shale gas revolution has fueled dramatic growth in the North American petrochemicals marketplace, sparking expansions and new construction on a scale not seen in decades, particularly in the ethylene. The impacts of the move towards lighter ethylene feedstocks has been well documented in the literature and at Ethylene Producers Conferences, but the issue of oilfield and fracking additives in the shale feeds and their potential as feedstock contaminants for steam crackers is a matter of ongoing interest and study.

This paper discusses some of the common additives used in oil and gas production, particularly shale gas production that is providing an increasing share of ethylene feedstocks. The impacts of these additives on refinery operation have been well documented, so some of these effects are discussed with an eye towards how they might translate to potential issues in ethylene production. Some observations from the ethylene industry are presented.

In addition, this paper re-introduces the EPC Feedstock Contaminants Database as a tool for ethylene producers.

Introduction

With the advent of hydraulic fracturing ("fracking") and horizontal drilling techniques opening up shale plays exploration and development of oil and gas that were previously not economically available, the production of crude oil, natural gas, and natural gas liquids (NGLs) in North America has increased dramatically over the past decade[1], as shown in **Figure 1**.

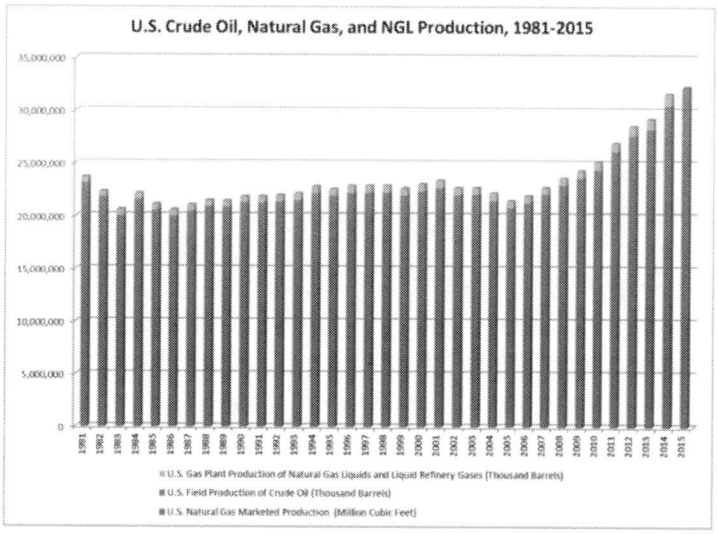

Figure 1: US crude oil, natural gas, and NGL production

Shale oil and gas plays are prevalent throughout the Lower 48 states, from the Marcellus and Utica in the Northeast, to the Bakken in the Northern Midwest, to the Eagle Ford and Barnet in the Southwest, as shown in **Figure 2**.

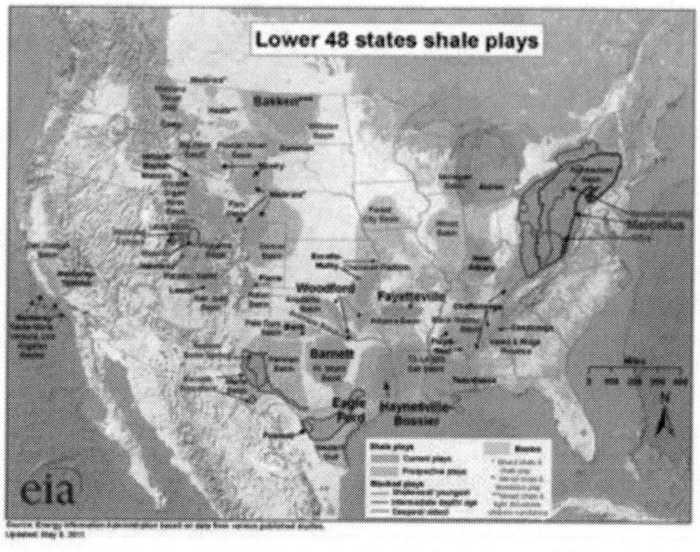

Figure 2: Continental U.S. shale plays

This shale gas "revolution" has fueled dramatic growth in the North American petrochemicals marketplace, sparking expansions and new construction on a scale not seen in decades, particularly in the ethylene market, by providing low-cost feedstocks from NGLs. The impacts of the subsequent move towards lighter ethylene feedstocks has been well documented in the literature, in the media, and at Ethylene Producers Conferences, but the issue of oilfield, fracking, and processing additives in the shale feeds and their potential as feedstock contaminants for steam crackers is a matter of ongoing interest and study. Fortunately, there is some experience with refineries and the processing of shale oil that can has some utility in determining what impacts processing shale gas could have from a contaminant standpoint.

What Is Fracking?

Fracking involves the injection of fluid into a horizontally drilled well of a typical depth of anywhere from 5,000 to 20,000 feet, as shown in **Figure 3**. The fluid injected into the rock through fractures formed using either charges or hydraulically and generally consists of a slurry of water, proppants, and a mixture of chemicals chosen based on the geology of the formation. These will be discussed in more detail below. After the fracking, the fluid mixture, known as "flowback", is removed.

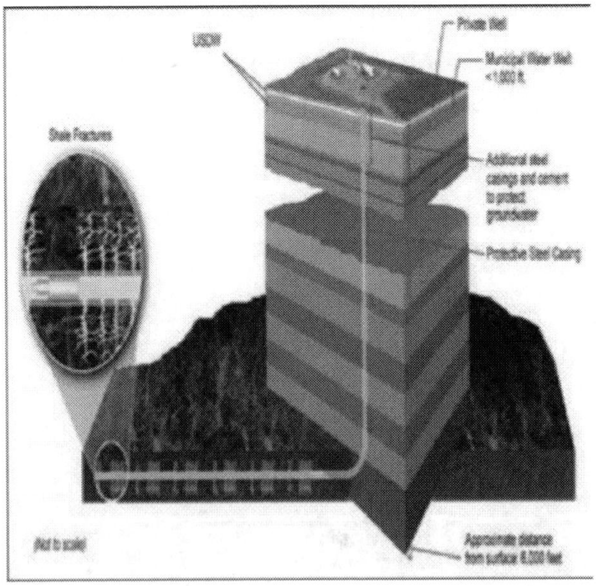

Figure 3: Schematic of a typical fracking setup

The oil or condensate, water, and gas are separated at a Primary Production Facility. The condensate is pumped to pipeline, tanks, or processing tankers (FPSO) for storage, treatment, and sales. The gas is flared, reinjected, or compressed and sent to pipeline for further treatment and sale. The water must be treated further for either disposal or reinjection as per environmental standards, typically an oil & grease of less than 29 ppm, with no visible sheen, is acceptable for injection, while less than 10 ppm oil-in-water is required for surface disposal. Natural gas typically is required to have a dew point of less than 3-7 lbs. H_2O per MMCF of gas; oil is typically required to have <1% BS&W for US processors (<0.5% for overseas) and crude salt spec of 10-50 PTB.

Flow assurance is an important consideration: assurance is expected to provide production throughout a designated flow line 100% of the time. Chemical issues providing impediments to flow assurance include hydrates, asphaltenes, corrosion, paraffin, and foaming.

Production Additives

Fracking Additives

Several chemicals can be used in the process of fracking. Gelling Agents are typically Guar-based materials. Guar-based additives are non-polymeric and water-soluble. Activating Agents are required for viscosity development and to maintain proppant carrying ability; these are usually aluminum or iron compounds. Breakers reduce fluid viscosity on completion and are compounds designed to upset complex / pH equilibrium. Fluid Loss Additives are inorganic particulates that improve fluid efficiency. Proppants are sand or ceramic materials that literally "prop" the fracture open for improved productivity.

Other chemicals used in fracking include Crosslinkers, which are either borates (boric acid and borate sales) or complexes of titanium and zirconium; these are used to achieve higher fluid viscosities. In water-sensitive formations, oil-based fluids such as aluminum or iron phosphate ester gels may be used.

Acidizing has been used extensively to improve oil and gas production in a process known as Matrix Stimulation. The chemistry is dependent on the type of formation. Both carbonate and sandstone formations may use hydrochloric or organic acid systems. Carbonate formations, however, can require special acid systems: alcoholic HCl acids for gas well acidizing or emulsified HCL to retard the acid reaction rate. Carbonate formations also utilize chelant systems such as EDTA (ethylenediaminetetraacetic acid) or NTA (nitriloacetic acid).

Sandstone formations may use mud acid systems that are used for drilling mud damage removal. These could be a combination of HCl, HF, and HB, or fluroboric acid. In place of chelants, overflush fluids are employed. These displace the acid flush away from the wellbore area and are surfactant/mutual solvents that leave the formation water-wet and facilitate flowback.

Primary Production Additives

Oil/water separation is an important part of primary treatment for shale oil Emulsions can be water-in-oil or oil-in-water (sometimes called "reverse emulsions"). Various types of chemistries are used as demulsifiers, emulsion breakers, and water clarifiers. These include polyol esters, cross-linked polyols, polyamine oxylalkylates, sulfonates, oxyalkylated phenols, and either low-oxide or high-oxide resins.

Figure 4: Emulsion breaker testing

Wax growth is also a problem in shale oil. Wax control agents can be solvents, dispersants, surfactants (cationic, anionic, or nonionic), crystal modifiers, or pour-point depressants.

Dissolved gases are major contributors to oilfield, midstream, and downstream corrosion. The three gases of main concern are O_2, CO_2, and H_2S that are naturally present. If there are no soluble gases present, the pH of brine is approximately neutral, and corrosion is greatly reduced. A variety of amine-based chemistries and their derivatives are utilized – primary, secondary, tertiary, diamines, polyamines, etc. – as well as imidazolines, quaternary ammonium compounds (e.g., trimethylalkyl), and phosphate esters.

Natural gas hydrates are ice-like crystals that form when water and natural gas mixtures above 500 psig cool below 60-75°F as shown in **Figure 5**. In a hydrate, there are around 8 "guest" molecules per 46 H_2O molecules. The "first line" of hydrate control is removal of water. This can be achieved via several methods:

- Absorption
    - Glycol dehydration
    - Calcium chloride
- Condensation
    - Chilling
    - Joule-Thomson cooling
- Adsorption
    - Alumina or silica
    - Molecular sieves

Other methods of water removal include membranes, gas stripping, and hydrocarbon displacement. There are also chemical additives that can be used for hydrate control. These include traditionally-used methanol and glycol, as well as a newer family of proprietary chemistries called Low Dosage Hydrate Inhibitors (LDHI) that can be used at dosages nearly 100 times less than traditional "anti-freeze" chemistries[2].

Figure 5: Natural gas hydrates

Scale formation is another flow assurance issue. Scale inhibitors include phosphonates, phosphate esters, polyacrylates, specialty polymers, inorganic phosphates, chelants, and combinations of these chemicals.

Microbiological-Influenced Corrosion (MIC) is corrosion of metals resulting from metabolic activity of bacteria. These bacteria fall into three classifications:

- <u>Aerobic</u> bacteria require oxygen for metabolism and growth.

- Anaerobic bacteria do not require oxygen for metabolism and growth (O_2 is toxic).
- Facultative bacteria are "switch hitters" that can thrive in either environment.

The biocides used to combat MIC fall into two main groups: Oxidizing and Non-Oxidizing. Oxidizing biocides include chlorine, bromine, and chlorine dioxide. Non-oxidizing biocides include glutaraldehyde, THPS, amines, diamines, quaternary amines, isothiazolines, and acrolein.

A summary of common oil production applications is shown in **Figure 6**, and a summary of common condensate applications is shown in **Figure 7**.

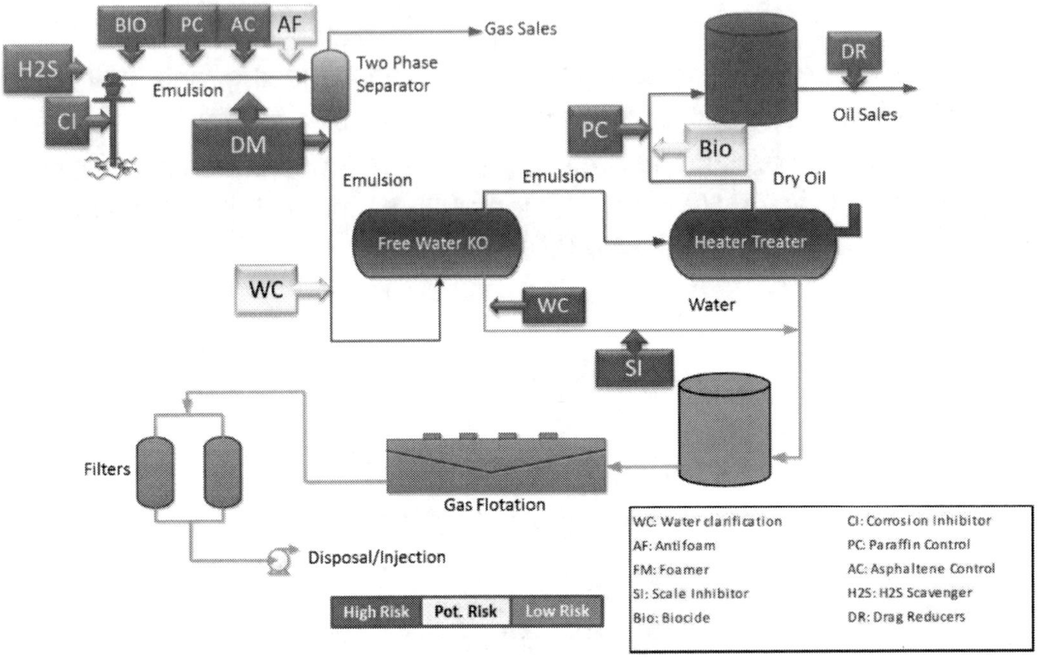

Figure 6: Common oil production applications

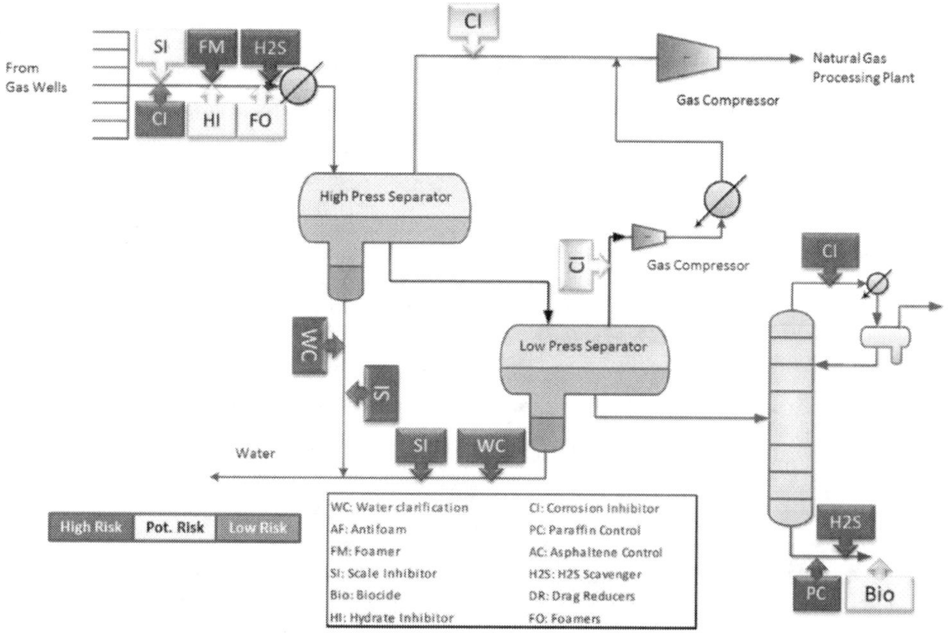

Figure 7: Common condensate production applications

Impacts to Process Units

Impacts to Refineries

The impacts of various upstream additives on refineries are fairly well understood. The key impacts are amine salting (from "tramp amines"), exchanger or furnace fouling, microbiological-induced corrosion, and feedstock compatibility. A summary of various additives and their potential impacts on refinery operations is shown in **Table 1**.

Table 1: Production additives and potential impacts

Applications	Chemistries	Impact On Process Units
Surfactants	Acids (citric, lactic, etc.); Isopropanol	Potentially
Cross Linkers	Boric acid(s), Al, Zr, KCl, NH4Cl, CaCl2	Yes
Breakers	Ammonium Persulfate	Potentially
Scale Control	Phosphate esters, Phosphonates, glycols	Potentially (Fouling)
Biocides	Gluteraldehyde, etc.	Potentially (WWT)
Gels	Guar Gum, hydroxyethyl cellulose, phosphate esters	Phosphate ester (yes); others no impact
Corrosion Control	Formic Acid, KI, Sb, Imidazolines, Quat Amines, Amides	Potentially
Pipe Cleaners	HCl, HF, HBr, DDBSA	Yes
Paraffin Control	Polyalphaolefin polymers	Some (chlorinated versions)
H_2S Scavengers	Triazines	Yes (depends on type)
Corrosion Control, Biocide	Quantanary Amine Chloride salts	Potentially
Emulsion Breaking	Water treating polymers	Low probability (water- based)
Emulsion Breaking	Metal salts ($ZnCl_2$, $AlCl_2$, poly(aluminum) chloride)	Yes - corrosion
Emulsion Breaking	Oxyalkylate resins	No
Paraffin control (Dispersants & Inhibitors)	Solvents, Polyalphaolefin polymers	Some (chlorinated solvents)
Scale Control, Corrosion Control	Phosphate esters, Phosphonates	Yes – corrosion, fouling
Scale Control	Acrylic acids	Yes - corrosion
Antifoams	Polyglycols, silicones, fluro-silicones	Potentially
Biocides	Gluteraldehyde and others	Potentially (WWT)

High Risk	Pot. Risk	Low Risk

Impacts to Ethylene Producers

The impacts of the various additives to ethylene producers are less well documented and still largely rely on anecdotal evidence rather than broad data and thorough understanding. There are, however, some inferences that can be drawn from the refinery impacts.

Tramp amines that are causing amine salting in crude unit overhead systems would crack to form ammonia in ethylene cracking furnaces[3]. This could cause a buildup of NH_3 in the quench water loop, causing an increase pH and stabilizing emulsions. Ammonia that goes with the cracked gas could collect on the dryer beds, shortening time between regenerations, or exit with the ethylene product as a contaminant. In plants where the dryer beds are regenerated to the fuel gas system, an increase in NO_x emissions would result during regen. In such a situation, the excess ammonia can be difficult to purge from the system.

Fouling in the refinery exchangers and, in particular, the crude furnaces, could relate to episodes of fouling that have been reported in the convection sections of ethylene furnaces over the past couple of years. This is an area not typically prone to fouling. Solids have also been noted in the storage tanks for liquid feeds whose origin was a fracked shale play. These could be similarly related and perhaps could be traced to proppants, corrosion byproducts, or other solids. Some of the materials identified in deposits from furnaces are consistent with some of the metallic elements of fracking and production chemicals.

Oxygenates in the feed could result in furnace effluent containing more organic acids (such as acetic acid). This could lead to an increase in corrosion in such locations as the dilution steam system, primary fractionator overhead, or the process gas compressor train. While most compressor and interstage coolers are currently not treated for corrosion, anecdotally there has been an increase in such programs in the years following the implementation of shale gas feeds. Another impact of oxygenates in the feed is an increased carbonyl concentration. One such carbonyl, acetaldehyde, can lead to aldol polymer formation and fouling in the caustic tower system. Carbonyls can also concentrate in the crude C_4 product (debutanizer overhead)[4].

Other impacts from the extraction and processing of shale oil and gas are still unknown and under investigation. It is the hope of the authors that the description in this paper of the chemicals used and their refinery impacts will lead to a better understanding and further discovery of the impacts on the ethylene producers.

The Feedstocks Contaminants Database

In 2008, the EPC Feedstocks Contaminant Subcommittee presented a new tool for collecting information on various contaminants from papers presented at the EPC over the years[5]. The database was used by the authors of this paper to attempt to relate some of the known experiences of ethylene producers in shale gas era to additives used in the fracking and production processes. The tool is available online at the official website of the Proceedings of the Ethylene Producers' Conferences; an EPC login is required. The Feedstocks Subcommittee is planning to update the database in 2016.

Summary

Having become available via the process of fracking, shale gas has transformed the North American petrochemicals industry. Fracking and subsequent processing of shale oil and gas involves the use of a variety of chemicals to provide flow assurance. While the impacts of many of these additives on refinery operations are well known and well understood, the petrochemical industry is still discovering and investigating the impacts on ethylene production.

References:

1. United States Energy Information Association (2016), "U.S. Field Production of Crude Oil), www.EIA.gov.
2. Kelland, Malcolm A., Thor M. Svartaas, Jorunn Ovsthus, and Takashi Namba (2000), "A New Class of Kinetic Hydrate Inhibitor," *Annals of the New York Academy of Sciences*, Volume 912, pp. 281-293.
3. Hines, Dwight, Dr. Ed Mohundro, Sam Valleru, Scott Miller, and Sarah Miller (2006), "Ammonia in the BASF FINA Petrochemicals Limited Partnership Steam Cracking Plant at Port Arthur Texas," AIChE Spring National Meeting, EPC Conference, Session T2003.
4. Hood, David and Steve Coleman (2007), "Minor Components, Major Issues: An Overview of Contaminants in Ethylene Plants," AIChE Spring National Meeting, EPC Conference, Paper 51c.
5. Bernard, Andre and Sheri Snider (2008), "Ethylene Plant Contaminants: A Database and Interactive Tool," presented at the AIChE Spring National Meeting, EPC Conference, Session 220.

AICHE ABSTRACT 126c[1]

CAUSTIC TOWER PERFORMANCE CHANGES FROM LIGHTER FEEDSTOCKS

Scott Cheney
Operations Engineer
BASF

Alex Saleeby
Account Manager
GE Betz

Prepared for Presentation at the 2016 Spring National Meeting
Houston, TX. April 10 – 14, 2016

CAUSTIC TOWER PERFORMANCE CHANGES FROM LIGHTER FEEDSTOCKS

Scott Cheney
Operations Engineer
BASF
Alex Saleeby
Account Manager
GE Betz

Abstract: Caustic tower performance is critical in the production of high quality ethylene for effective removal of acid gases from the cracked gas. However, an undesired polymerization reaction in caustic towers can result in accumulation of polymer, "red oil", resulting in fouling, plugging and overall decreased asset availability. At the BASF TOTAL Petrochemicals LLC, ("BASF TOTAL Petrochemicals") cracker in Port Arthur, a joint venture between BASF Corporation and Total Petrochemicals & Refining USA, Inc., a shift toward lighter feedstocks due to market conditions resulted in higher red oil levels and increased inhibitor demand. This report investigates caustic tower optimization for the reduction of red oil formation while decreasing inhibitor demand.

Introduction

Caustic scrubbing towers are installed in many olefin production facilities in order to remove acid gases (H_2S, CO_2 and mercaptans) that are present in the cracked gas as a result of the thermal cracking processes. These acid gases must be removed for facilities to meet product quality specifications.

These caustic towers generally operate on the basis of counter-flow of the charge gas against a sodium hydroxide (or similar) of specified concentration. The resulting aqueous reaction produces carbonates and sulfides that are removed with spent caustic to further processing units. The tower is usually situated between stages of the cracked gas compressor (CGC) and is comprised of several circulated sections of varying caustic strength each. Spent caustic is processed through various types of equipment, depending on the facility, but includes coalescers, strippers, neutralization, oxidation and biological treatment units.

Proper operation of the caustic tower is critical to maintaining production rates and product quality which can be impeded by the build-up of polymeric material within the column or downstream units. This paper will describe the operational changes that the BASF Total Petrochemicals facility in Port Arthur, TX made to improve red oil control, and reduce chemical additive usage, particularly as the plant shifted to lighter feedstocks.

Tower Fouling

Polymeric material is generated in caustic towers through two major mechanisms: aldol condensation and free radical polymerization.

Free radical polymerization occurs due to hydrocarbon decomposition (initiated by an oxygen source, metal catalyst or heat) of olefins, which become free radicals and propagate with other olefin monomers until the polymer is terminated by another free radical [1].

The aldol condensation reaction occurs between molecules having carbonyl functionality, most commonly acetaldehyde and its oligomers. This reaction is catalyzed by the hydroxyl ions in the caustic water to form polymers. The aldol polymers generally have a reddish color, from which the industry has adopted the term "red oil" as the descriptor for the most common type of caustic tower fouling.

Both mechanisms are a function of heat, residence time and concentration of the reactive species. Additionally, the quality of the fresh caustic (low chlorides and low oxygen) and of the dilution water (demineralized and deaerated) play a significant role in severity of red oil formation reactions [2].

Polymer accumulation in caustic scrubbing towers will result in increased pressure drop as tower internals are plugged by the red oil and ultimately leads to decreased acid gas treatment efficiency – potentially leading to reductions in production rate or product quality consistency. Red oil also has major negative effects on downstream spent caustic processing equipment, resulting in fouling and plugging that can lead to maintenance downtime and expenses.

Control Strategies

As a result, tower control strategies usually involve temperature reduction, frequent blowdowns or skims, and quality control of the caustic and dilution water fed to the tower.

However, many times these control strategies are either not sufficient or not available, thus many facilities use chemical additives to mitigate red oil fouling in caustic services.

One such "additive" is simply to add gasoline to the tower (in either continuous or campaign fashion) to dissolve the hydrocarbon red oil, making it easier to remove from the tower via skims or similar. However, this has potential to create emulsions and foaming in certain towers and can place extra load on downstream process equipment.

The other two types of chemical additives are antipolymerants and antifoulant dispersants. These additives are both injected into the caustic tower at controlled dosing rates, but have different mechanisms, depending on the configuration and goals of the site [3].

Antipolymerants work by terminating the polymeric reaction before the hydrocarbons grow long enough to create viscous, fouling material. For aldol condensation, these have been called "carbonyl killers" since they react with the carbonyl polymers to form inert oximes that can be removed in the aqueous phase. Other additives bind with free radical molecules to prevent that type of polymerization. These additives do not contribute to tower foaming, but are ineffective at removing existing red oil build-up.

Antifoulant dispersants are chemicals with a polar and non-polar end, designed to keep polymer particles (formed by either mechanism) in the aqueous solution so they cannot drop out to foul the tower. The specific design of the chemicals prevents additional polymerization. These are removed in solution with the spent caustic and processed through the downstream equipment. This type of additive can remove existing build-up but also can create foaming issues, so it often dosed into towers along with anti-foam additives.

BASF Total Petrochemicals Port Arthur Cracker

At the Port Arthur site, the caustic towers are fed in series by cracked gas from the third stage of the compressor after it travels through a compressor aftercooler, separation drum and preheater (Figure 1).

The Port Arthur facility is equipped with a nitrogen-blanketed fresh caustic tank. The caustic used meets the low chloride specifications and the dilution water sourced from plant LP condensate. The towers are equipped with standard sieve-type trays. The site uses two coalescers to separate the spent caustic from hydrocarbons before processing through a fuel gas stripper to remove benzene prior to a storage tank. A high pressure, high temperature wet air oxidation (WAO) unit is used to manage sulfur prior to neutralization and sending to the wastewater treatment unit.

Initially the plant utilized only one caustic tower, consisting of two caustic circulations sections and a wash water section. However, due to early issues with carbon dioxide excursions, a second tower, consisting of just one caustic circulation section, was added upstream of the existing tower. Therefore the current arrangement has strong, intermediate and weak caustic sections; the site has not experienced carbon dioxide breakthrough issues since this addition.

Red oil polymer and gasoline comprise the hydrocarbon phase but are continuously removed via skims at the bottom of both towers; these skims are processed through coalescers with the rest of spent caustic, and is ultimately sent to the Pyrolysis Gasoline unit.

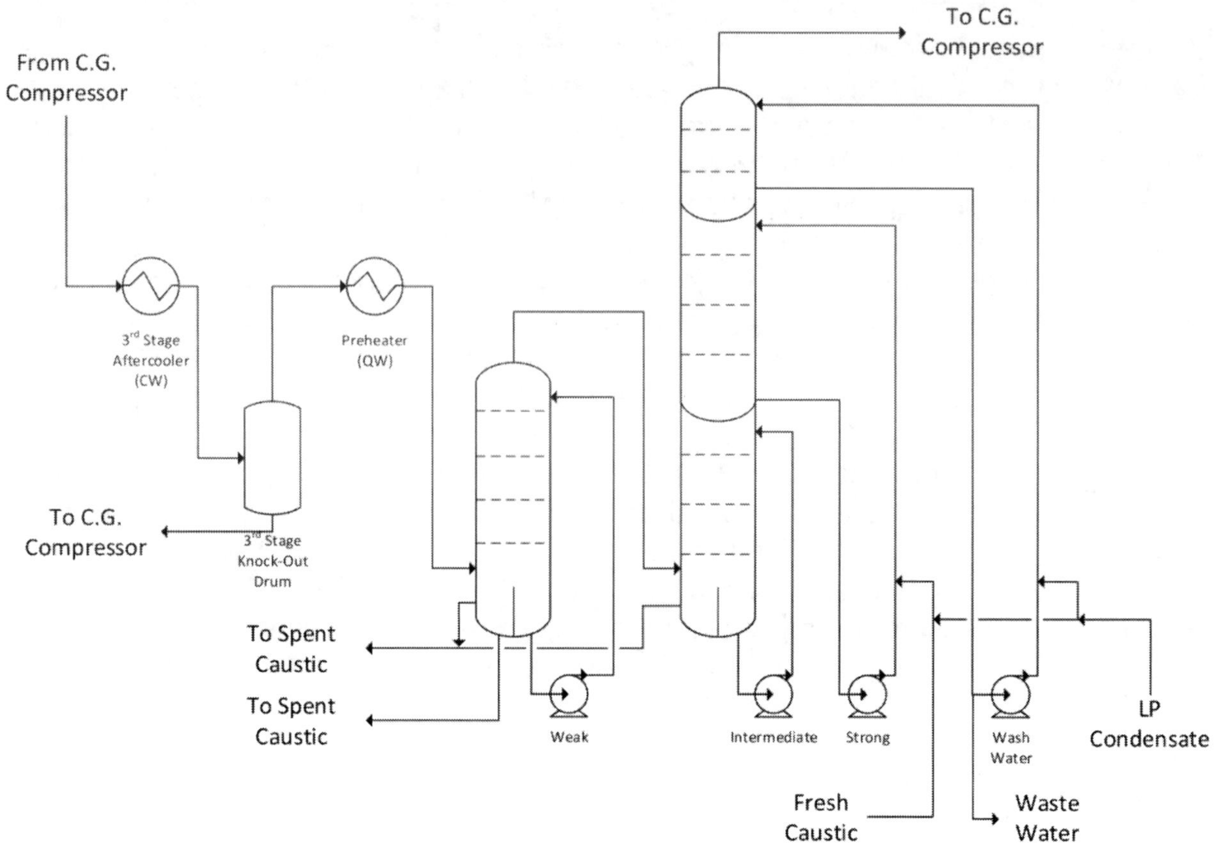

Figure 1: Process Flow Diagram of the Caustic System

BASF Total Petrochemicals has used an antipolymerant in caustic tower operations at the Port Arthur cracker since start-up in 2001. An initial test was performed by GE on the caustic to determine which type of polymerization was predominant in the column and a maintenance dose established. The antipolymerant in use is GE Betz's Petroflo* 20Y15 (*Trademark of General Electric Company; may be registered in one or more countries), which acts as a carbonyl scavenger, terminating the polymerization reaction before the red oil polymer chains grow long enough to cause fouling.

GE adjusts additive dosing as needed to maintain red oil levels near KPI limits. Samples from caustic recirculation lines, tower skim lines, and the coalescers are collected and checked for visual appearance. Caustic section samples are also tested for red oil content. Water wash section samples are tested for residual caustic. These are monitored regularly, along with other operating conditions throughout the caustic system and have provided excellent control of red oil since start-up.

Changes in Red Oil

The Port Arthur facility started up as a naphtha cracker; however, due to changing market conditions in the early 2010's the facility began to incorporate increasing amounts of butane as feedstock. Market conditions continued to change as a result of hydraulic fracturing opening untapped domestic gas resources, and in 2013 the Port Arthur facility began incorporating ethane as one of the feeds to the furnaces. As this transition to lighter feedstocks occurred, along with increases in plant capacity, red oil production increased, requiring an increase of caustic tower additive rates. This defies general industry findings that increased ethane cracking should result in reduced acetaldehyde and thus less red oil formation [4].

A sampling program was created in 2014 to test the cracked gas feed for acetaldehyde concentration. No tight correlation could be established between feedstock and acetaldehyde concentration during this time due to a fairly consistent feed slate. Acetaldehyde concentration feeding the caustic tower generally stayed in the 50-150 ppm range (Figure 2). Outlet acetaldehydes have consistently been 0 ppm.

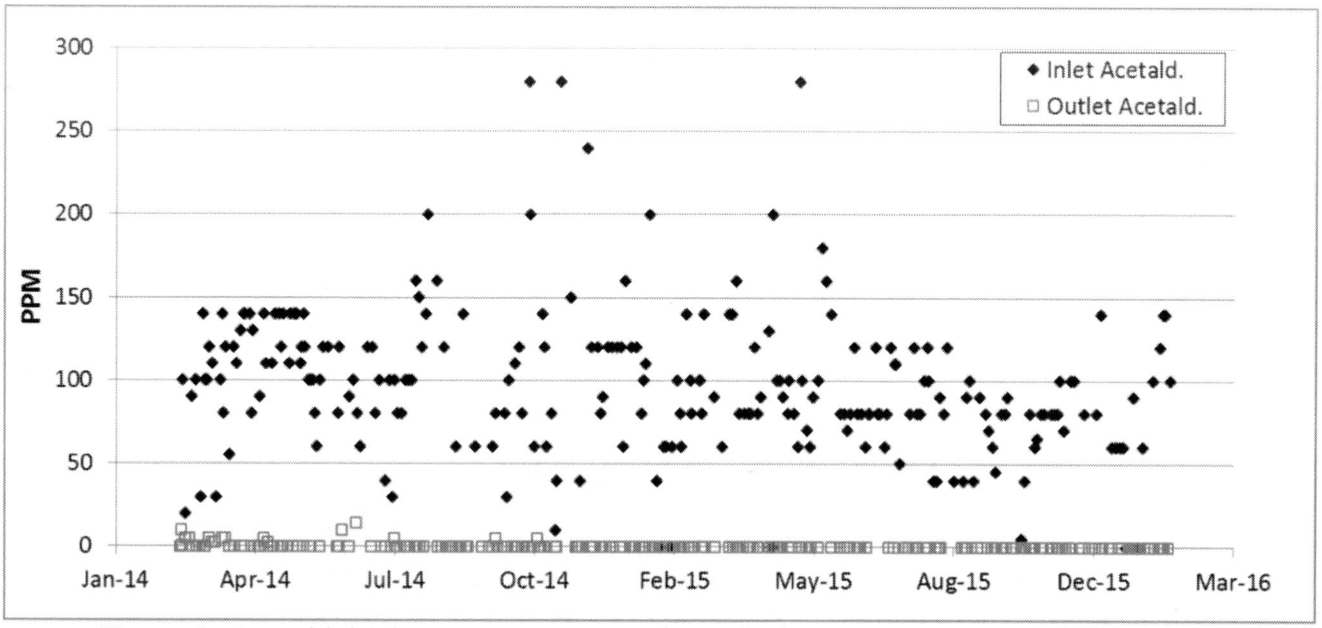

Figure 2: Acetaldehyde Concentrations of Cracked Gas Feed to the Caustic Tower

In early 2015, the site began investigating the increased red oil concentration and the potential correlation this might have with the transition to lighter feedstocks. Specifically of interest was the impact that gasoline-range molecules have on the tower, which are much less abundant in the cracked gas under lighter feed slates. To test this, the site began reducing the preheat of the cracked gas to the towers.

By a happy coincidence, the 3rd stage aftercooler became partially plugged, concurrent to this gradual reduction of superheat. This led to elevated temperatures at the 3rd stage knock-out drum, and contributed to the desired reduction in preheat, and increased gasoline-range molecules present in the caustic tower. As the superheat fell below 10°F, GE noticed that skim samples were improving in quality, showing a larger gasoline layer, as well as lighter caustic color.

Additionally, red oil concentration started to drop and GE started reducing inhibitor dosage. This observation was contrary to typical expectations that 2-phase operation results in foaming and increased polymerization [5]. BASF Total Petrochemicals worked with GE Betz to develop this improvement and understand this phenomenon in more detail.

Case Study

The first test involved allowing continued sub-optimal operation of the CGC aftercooler, and reduction of the caustic tower superheat to 5°F. The cracked gas experiences roughly 5°F of temperature loss through the towers, resulting in a tower outlet temperature slightly lower than the feed to the system (Figure 3).

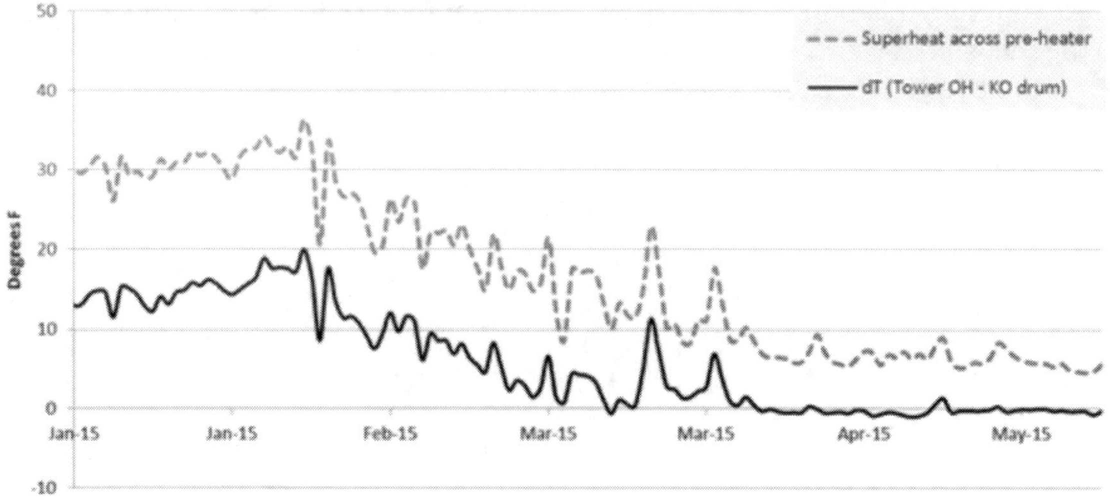

Figure 3: Cracked Gas Superheating

As expected, condensation of hydrocarbons in the caustic towers increased notably. Due to the elevated temperature at the upstream separation drum, many gasoline-range molecules in the cracked gas did not condense until reaching the caustic tower. This was observed via the gasoline coaslescer flowrates, and the appearance of the skim samples and the samples taken of each circulating section. Early samples were cloudier and more yellow in color. Later samples, after preheat was reduced, were clearer and showed better phase separation.

Red oil concentration during this same time span decreased significantly (Figure 4). Previously the site had seen year-over-year increases in red oil concentration in the caustic, leading to increased inhibitor demand each year. The impact of gasoline condensing in the towers brought red oil levels to historic lows.

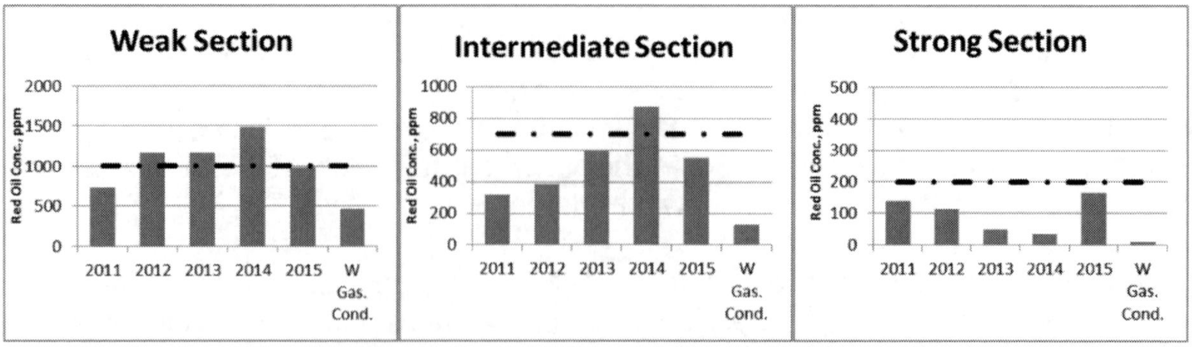

Figure 4: Red Oil Concentration in Each Section, With and Without Gasoline Condensation

Based on red oil concentration measurements during this test, the site was able to reduce inhibitor dosing by roughly 30%. This trend was still on a downward slope when the test had to be finalized (due to required cleaning of the compressor aftercoolers) (Figure 5).

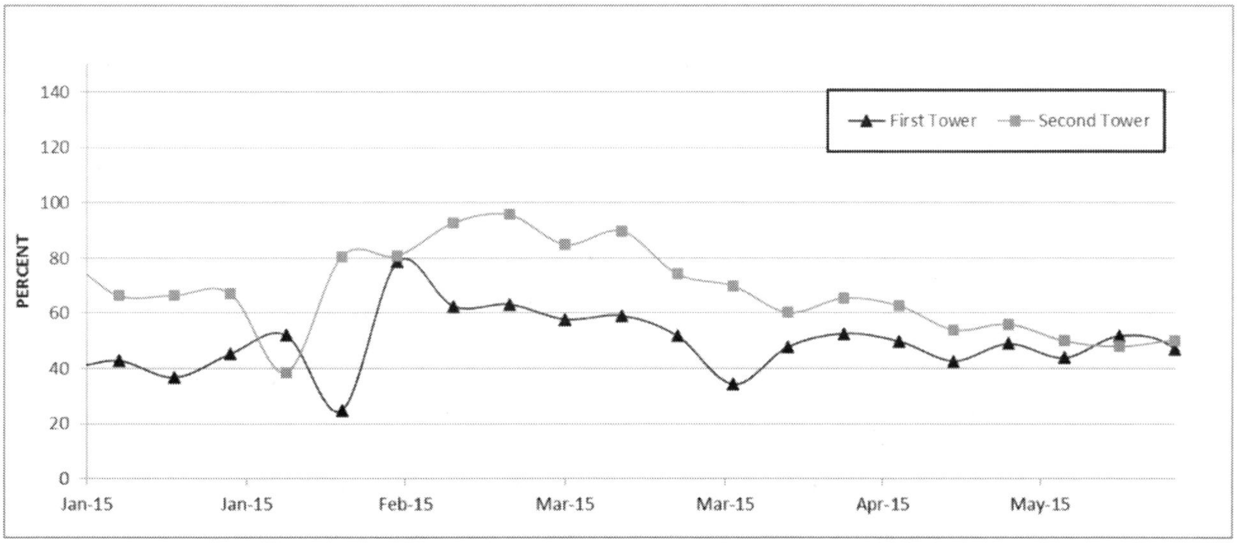

Figure 5: Inhibitor Demand Reduction over Time

After cleaning of the compressor aftercooler, the temperature of the 3rd stage knock-out drum was drastically reduced, leading to more gasoline-range molecules dropping out in that drum, and resulting in the a loss of gasoline condensation in the caustic towers. Inhibitor dosages had to be raised to previous levels to provide the same level of red oil control (Figure 6), despite keeping tower temperature at historically low levels (105°F).

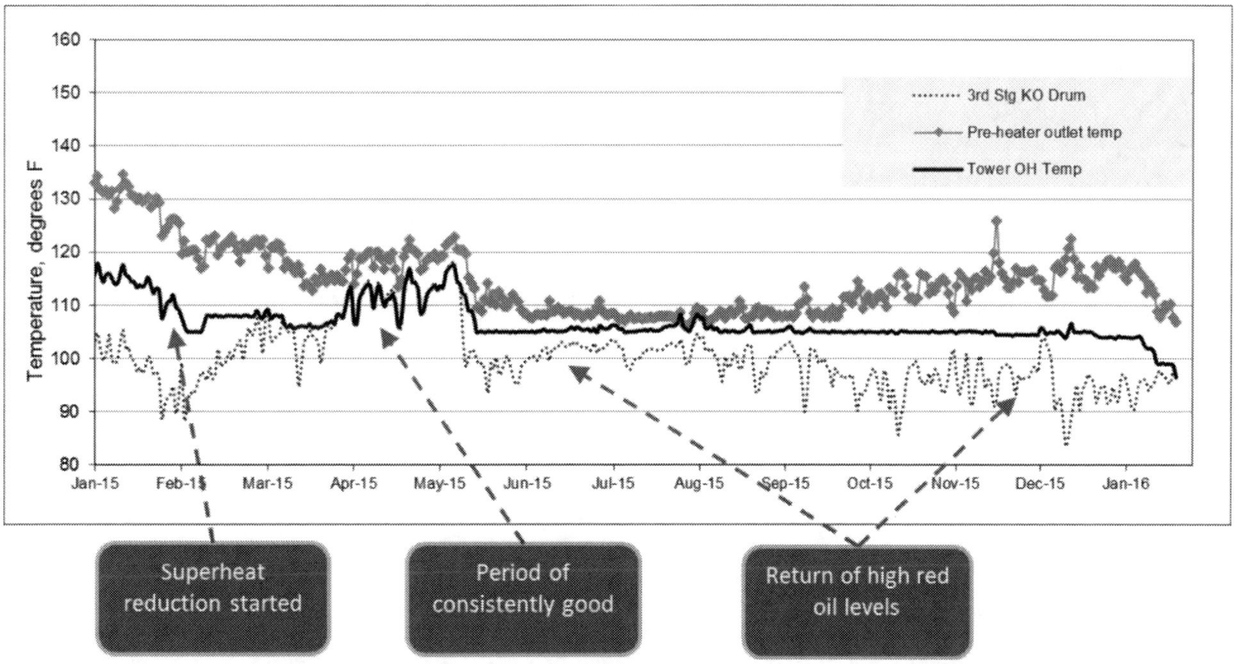

Figure 6: System Temperatures Before and After Test

This is significant because the effect of lower temperature on carbonyl polymer kinetics has been discussed before [6]. This showed that the reduction in red oil was not resulting just from the temperature reduction; instead, the result appears to be directly related to the superheat applied and the condensing of gasoline-range hydrocarbons in the column.

There is nothing especially novel about this concept either, since some sites use an online gasoline wash to minimize fouling in the tower. The polymeric reaction happens in the aqueous phase, and the additional gasoline helps with separation, bringing the short-chain polymers out of the reactive phase and removing them via skim, thus reducing the need to be inhibited using additive.

The Port Arthur facility was able to adjust process parameters to use latent gasoline-range molecules in the cracked gas for this same purpose. Previously, when the facility used primarily naphtha feedstocks, there may have been sufficient inherent gasoline wash at those conditions. However, after the switch to higher amounts of butane and ethane, this benefit mostly disappeared, and red oil concentrations started to rise as a result. It is likely that a switch back to naphtha feeds and the new, lower temperature conditions would result in overwhelming the caustic towers with gasoline-range hydrocarbons.

A second test was then performed to confirm the conclusions and also test the limits of the tower operations. With the aftercooler still running normally, the caustic tower outlet temperature was gradually reduced below the site's historical lows. This was done by again reducing cracked gas feed superheat at the preheater. Similar hydrocarbon condensing was observed at these new conditions and the site has begun inhibitor reductions again, and are ongoing at the time of this report.

Secondary Impacts

Running at these historically low temperatures (95°F) presented a number of additional concerns and benefits. The main concerns were carbon dioxide breakthrough, foaming and overwhelming the downstream coalescers. Viscosity of the caustic increases as temperature decreases, potentially resulting in mass transfer limitations for carbon dioxide capture thus limiting tower performance; no CO_2 breakthrough has been observed so far under these conditions (Figure 7). Normal CO_2 and H_2S concentration to the tower is 25 ppm and 50 ppm, respectively.

Foaming and coalescer capacity issues both could result from excessive gasoline in the tower, but neither was observed at the temperatures during this test. Instead, coalescer performance appeared to improve; due to the additional gasoline condensation, the specific gravity of the hydrocarbon phase decreased, leading to better phase separation.

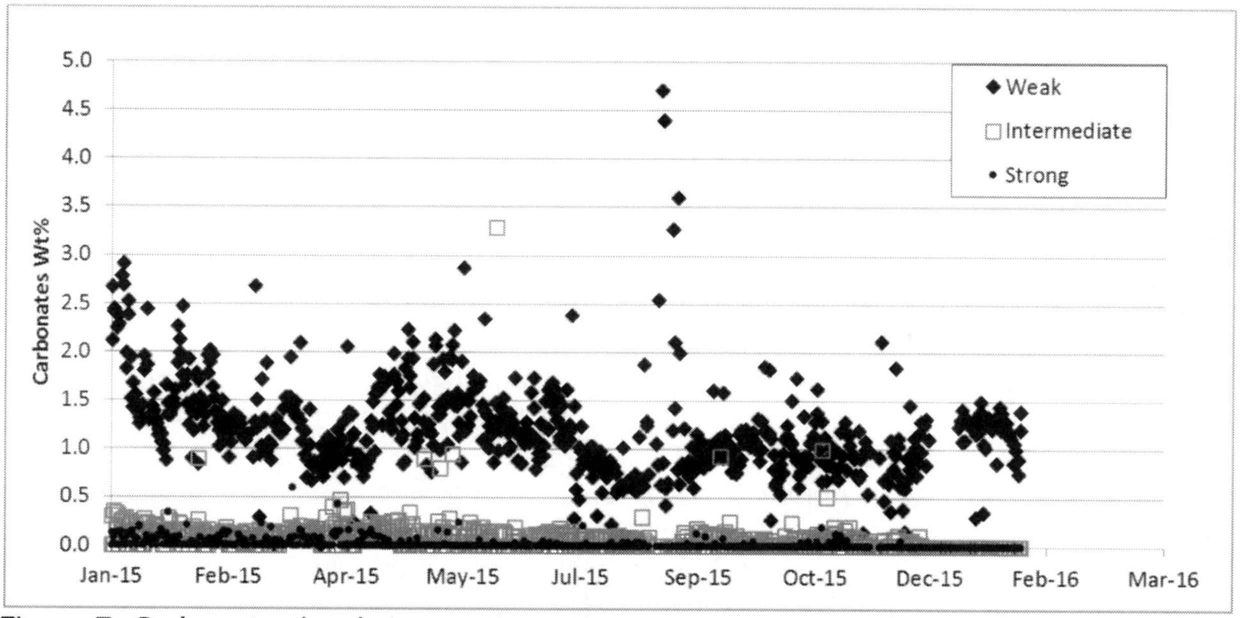

Figure 7: Carbonates levels in caustic sections - no negative impact on CO_2 removal from reduction in operating temperatures

Additionally, lower temperatures lead to better efficiencies in the 4th stage of the cracked gas compressor, as well as lower discharge temperatures in that stage (where fouling occurs at >200°F). The removal of gasoline-range molecules in the caustic tower reduces feed flow to fractionation towers, potentially providing debottlenecking at those towers, but also reducing aromatics content in those towers. These aromatics can help to prevent fouling in the down-stream reboilers, thus the Port Arthur site is doing additional monitoring of reboiler health and fractionation additive rates as caustic tower parameters are changed.

Conclusion

Lighter feedstocks result in less gasoline-range molecules that are fed to the caustic tower and then condense there before being skimmed to the caustic/gasoline coalescers. With less gasoline-range molecules available in the cracked gas, temperatures must be adjusted to prevent condensing upstream of the caustic tower and encourage condensing within the tower. Gasoline condensation allows for red oil to dissolve in the liquid gasoline, which is rapidly skimmed from the tower, thus removing it from the reactants where additional polymerization would occur. This can be done without emulsions or foaming, and results in less polymerization inhibitor demand.

Remaining Questions

Several questions remain at the conclusion of the site's current efforts.

- What is the minimum inhibitor dosing that can be applied? Based on the initial test in 2015, it appeared that dosing could be reduced as much as 50%, but better temperature control of the CGC aftercooler is needed to provide the long-term stability required for optimizing dosage rates.

- How far can superheat be lowered? At what point does CO_2 absorption or circulation rates suffer due to viscosity increase? In this test, the superheat of the cracked gas feed to the caustic tower was always kept at least 5°F due to a leaking control valve. During the next outage this valve will be repaired and allow for additional testing of tower limits.
- What is the true correlation at the Port Arthur site between feed slate and acetaldehyde levels? How does the facility compare to industry norms? Can acetaldehyde levels be reduced through changes in furnace operation? The site will continue sampling acetaldehyde data and build a correlation to the selected feed slates. The mechanisms and factors related to acetaldehyde production in furnaces are complex and will be a topic of consideration for future studies.
- Can skim frequency be reduced without impacting inhibitor demand? The Port Arthur facility operates with continuous skims, which results in sub-optimal utilization of caustic, but provides immediate removal of contaminants. Additional campaigns are needed to determine the best way of controlling skims.
- How does this method of red oil control compare against injection of aromatic gasoline into the tower recirculation lines? The quality of the added hydrocarbons may be superior to the latent hydrocarbons but may not provide additional benefit. However, added hydrocarbons may be necessary for even lighter feed slates to provide quantity.

References

1. Blaschke, Marilyn W. (2003), "Caustic Tower Fouling: Identifying the Causes," *15th AIChE Ethylene Producers Conference*, pp 209-221

2. Dionne, D, S. Kurukchi and J. Gondolfe (1999) "Caustic Tower Revamp Using High Capacity Second Generation Ripple Tray Technology," *11th AIChE Ethylene Producers Conference*, pp 275-290

3. Mullenix, David, et al. (1996), "Control of Carbonyl Polymer Fouling in Caustic Tower," *8th AIChE Ethylene Producers Conference*, pp 90-100

4. Burke, John M. and Jerry Branson (1996), "Session 19 Questions and Answers," *8th AIChE Ethylene Producers Conference*, pp 190-192

5. Maugans, Clayton, Mike Howedeshell and Steve De Hann (2009), "The Effects of Caustic Tower Operations and Spent Caustic Handling on the Zimpro Wet Air Oxidation (WAO) of Ethylene Spent Caustic," *21st AIChE Ethylene Producers Conference*

6. Hammond, Charles and Vance Ham (2009), "Caustic Tower Operation Considerations for Effective Performance," *21st AIChE Ethylene Producers Conference*

7. Payne, David (2009), "Caustic Tower Operation at Westlake Chemical," *21st AIChE Ethylene Producers Conference*

8. Brown, Michael G., Michael J. Kennedy and Casey L. Mueller (1995), "Environment Subcommittee Questionnaire/Survey Data: Spent Caustic Treatment Survey," *7th AIChE Ethylene Producers Conference*, pp 290-306

9. Grover, Rajiv and Hassan M. Gomaa (1993), "Commercially Proven Technologies for Spent Caustic Management," *5th AIChE Ethylene Producers Conference*, pp 366-381

10. Mo, Hua and David Dixon (2014), "Improved Caustic Tower Operation by Complex System Evaluation," *26th AIChE Ethylene Producers Conference*

AIChE Paper Number: 126d

HYDRATES IN ETHYLENE PLANT COLD SECTION

Peter Feld

Senior Global Technology Manager

Dow Benelux B.V.

John Holderness

Manufacturing and Engineering Technology Fellow

The Dow Chemical Co.

Prepared for Presentation at the 2016 Spring National Meeting
Houston, TX, April 10-14, 2016

HYDRATES IN ETHYLENE PLANT COLD SECTION

Peter Feld

Senior Global Technology Manager

Dow Benelux B.V.

John Holderness

Manufacturing and Engineering Technology Fellow

The Dow Chemical Co.

Abstract

Hydrates have been an operational issue for ethylene plants for years, causing plugging at various locations. This has caused excessive pressure drop, flaring, production losses, and plant down time.

Hydrates are ice-like crystals, composed of water and gas molecules that are not chemically bonded. Water molecules can form a structure around smaller gas molecules. Hydrates typically form at higher pressures and at relatively high temperatures (above freezing point). Hydrates can be broken by heating up, depressurization, and/or adding inhibitors (like alcohols).

Typical locations for hydrate formation in ethylene plants are liquefied gas feed systems, cracked gas dryer inlet and cold section (cold box, demethanizer, C2 splitter). This paper provides background of some known hydrate locations in the cold section of ethylene plants including potential causes, detection, and removal.

With increased awareness and operational experience the hydrate issues have decreased over time, however still happen from time to time. Some examples and operational experiences of hydrate formation and mitigation in ethylene plant cold section are described.

Introduction

Hydrates have been an operational issue for ethylene plants for years, causing plugging at various locations. This has caused excessive pressure drop, flaring, production losses, and plant down time.

Hydrates are ice-like crystals, composed of water and gas molecules that are not chemically bonded. Water molecules can form a structure around smaller gas molecules. Hydrates typically form at higher pressures and at relatively high temperatures (above freezing point). Hydrates can be broken by heating up, depressurization, and/or adding inhibitors (like alcohols).

Typical locations for hydrate formation in ethylene plants are liquefied gas feed systems, cracked gas dryer inlet and cold section (cold box, demethanizer, C2 splitter).

Hydrate

History[1,2]

Hydrates have been first detected and described over 200 years ago. Clathrate hydrates were first documented in 1811 by Sir Humphry Davy. He studied and experimented with oxymuriatic acid gas and declared it a chemical element (chlorine). During his studies he also found that crystals that were considered to be solidified chlorine were not and actually water was a primary component of this substance. "It is generally stated in chemical books that oxymuriatric acid is capable of being condensed and crystallized at low temperature; I have found several experiments that this is not the case. The solution of oxymuriatric gas in water freezes more readily than pure water, but pure gas dried by muriate of lime undergoes no change whatsoever at a temperature of 40 below 0° of Fahrenheit". His student Michael Faraday conducted further studies and reported twelve years later an analysis that corresponds to the formula Cl2·10H20 (actual formula is closer to Cl2·8H20).

Hydrate was purely of academic interest (other gas hydrates were discovered and described) until the 1930s when the problem of gas pipeline blocking was found to be an industrial nuisance. Natural gas pipelines plugged with ice-like crystals, especially in winter. It was first assumed that water vapor in the natural gas did condense and then freeze. However, the ice-like substance also formed at temperatures well above the freezing point of water. When the ice-like substance was heated, it was found to release natural gas. Therefore, it was realized that it was a stable crystalline substance composed of water and gas that formed under high pressure. It was called a gas hydrate given the similarity to chlorine hydrate.

For the natural gas industry limitations on water content in gas were implemented to mitigate the issues. Other findings were that alcohols and glycols are effective inhibitors for hydrate formation and these were widely used in the gas industry.

Further study continued to develop phase diagrams water (liquid, ice), hydrocarbons (vapor, liquid) and hydrates[3,4]. Also the structure of the hydrate as gas molecules inside a structure of water molecules was determined and three different structures have been found. Various calculation methods to predict hydrate formation for gas and gas mixtures have been proposed. Recently simulation programs based on the different calculation methods have been developed and are available to predict hydrate formation.

Structure

As indicated the gas hydrates (or clathrate hydrates) are ice-like crystals made up from water and gas molecules. Water molecules can form structures (through hydrogen bonding) with several cavities. When a minimum number of these cavities are filled with gas molecules, the crystalline structure will becomes stable and solid gas hydrates will form even at temperatures above the freezing point. When the gas hydrate melts the gas is released.

Three forms of gas hydrate structures are known: Structure-I, structure-II and structure H.

Each structure has a unit cell made from a number of water molecules. Differently sized cavities are present and some can host gas molecules. Structures I and II host different gas molecules, structure H requires a mixture of small and large molecules. More background is given below[5].

As shown in the figure below, the two common forms of gas hydrates are known as structures-I and II, which have been investigated with X-ray diffraction methods by von Stackelberg and Müller (1954). They found that the unit cell of structure-I (sI) is a 12 Å cube consisting of 46 water molecules which has two types of cavities. The two small cavities are pentagonal dodecahedra (5^{12}), whereas the six large cavities are tetradecanhedra ($5^{12}6^2$) having two opposite hexagonal faces and twelve pentagonal faces. The smaller cavities are almost spherical, whereas the larger cavities of structure-I are slightly oblate. The unit cell of structure-II (sII), which is a 17.3 Å cube with 136 water molecules, also contains two types of cavities. The 16 smaller cavities are distorted pentagonal dodecahedra and the 8 larger cavities are hexadecahedra ($5^{12}6^4$) having 4 hexagonal faces and twelve pentagonal faces. The latter cavities are almost spherical in shape.

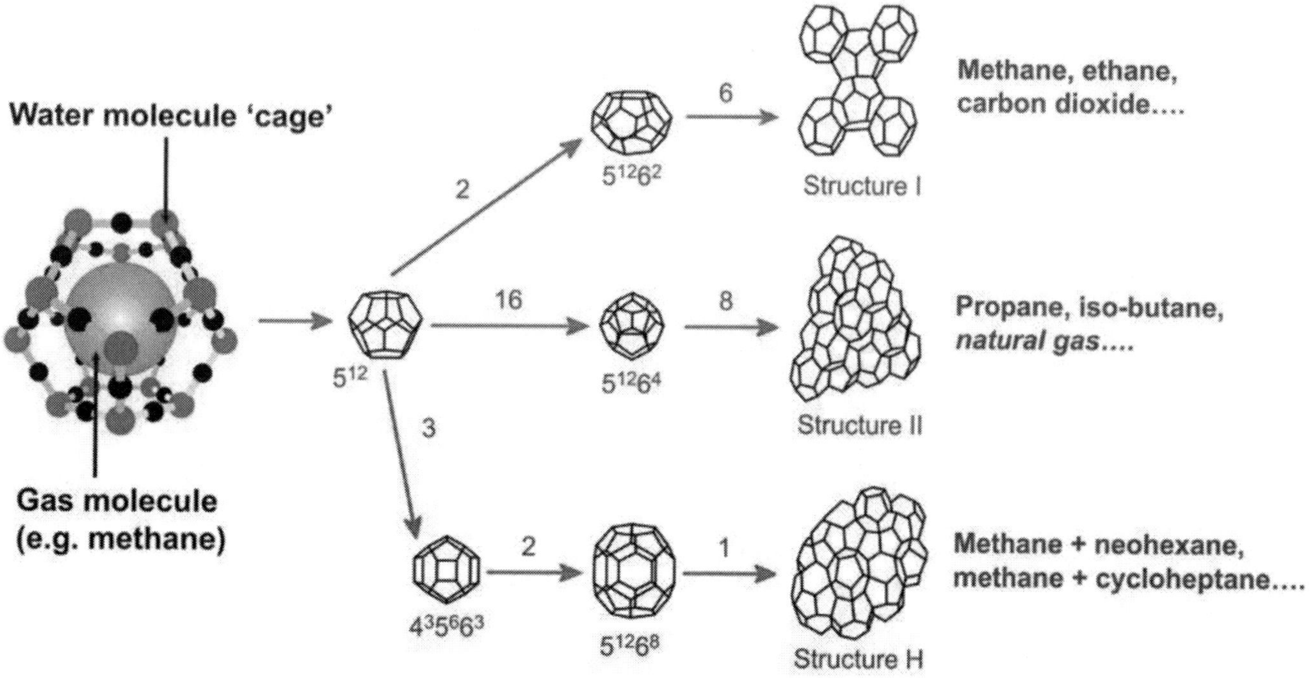

The gas hydrate structures

Ripmeester et al. (1987) reported a new hexagonal hydrate structure which requires both large and small molecules to stabilise the structure. They proposed that the new structure be known as structure-H (sH). According to the above authors, the unit cell has 34 water molecules forming a hexagonal lattice with parameters a = 12.26 Å and c = 10.17 Å. The structure-H has three different types of cavities, three 5^{12} cavities which is common to all known hydrate structures, two new 12 faces $4^3 5^6 6^3$ cavities and one new large $5^{12} 6^8$ cavity. The $4^3 5^6 6^3$ cavity has three square faces, six pentagonal faces, and three hexagonal faces, whereas the $5^{12} 6^8$ cavity has 12 pentagonal faces and eight hexagonal faces. The first two cavities accommodate the small molecules which act as help gases. The large cavity in this structure can accommodate even larger molecules, so molecules in the size range of 7.5 to 8.6 Å can potentially form gas hydrates.

Phase diagram

Phase diagram for water and hydrocarbons will indicate the pressure temperature conditions where solid, liquid and vapor water, and hydrocarbons can exist and additionally the hydrate phase.

A typical phase diagram for water and light hydrocarbons is presented below[6]. The critical point C indicates the area where there will only be a single supercritical phase. The point Q2 indicates where hydrocarbon liquid, hydrocarbon gas, hydrate, and water can exist. The point Q1 shows where hydrocarbon liquid, hydrate, water and ice can exist. For Q1 temperature will be around 0°C, for Q2 it will be somewhat higher. Then line between Q1 and Q2 indicates equilibrium between gas, water, and hydrate.

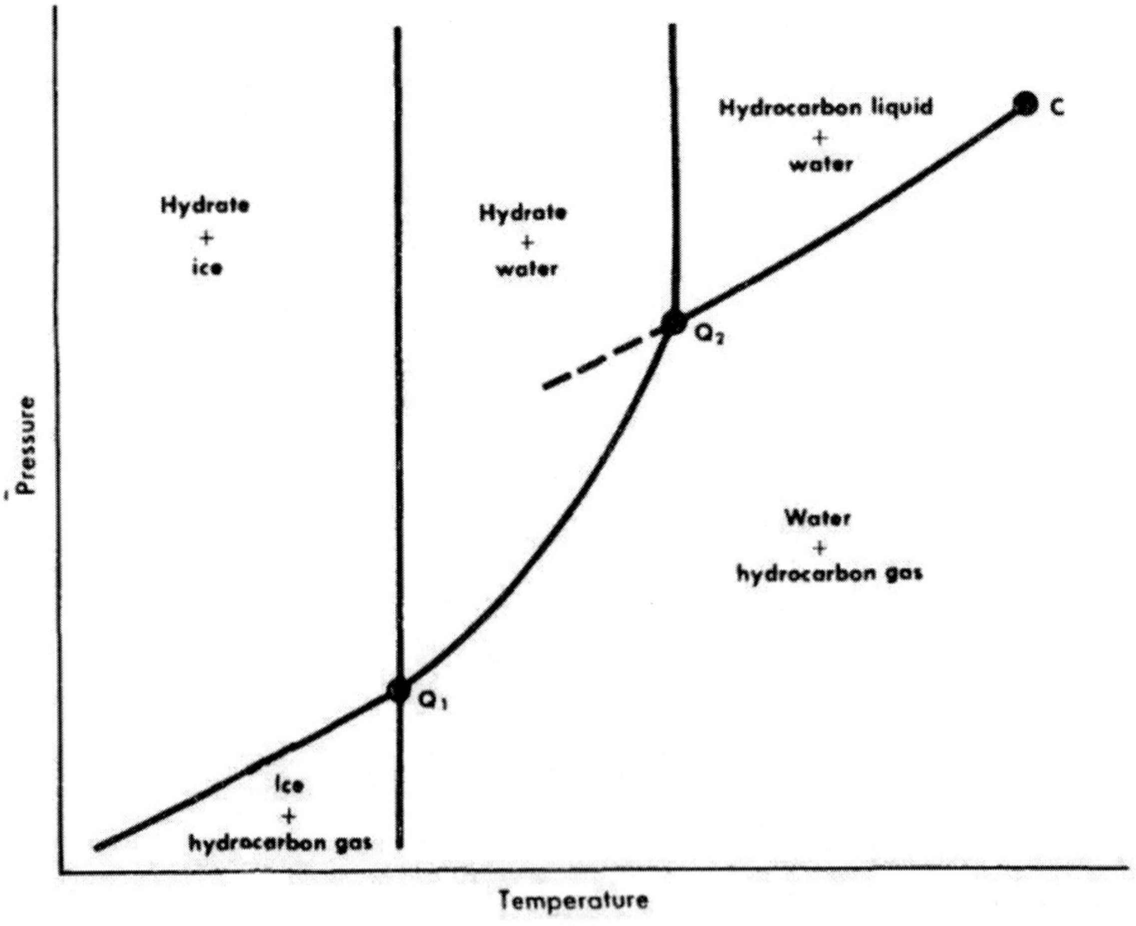

Currently several simulation programs are available to indicate the region of hydrate formation.

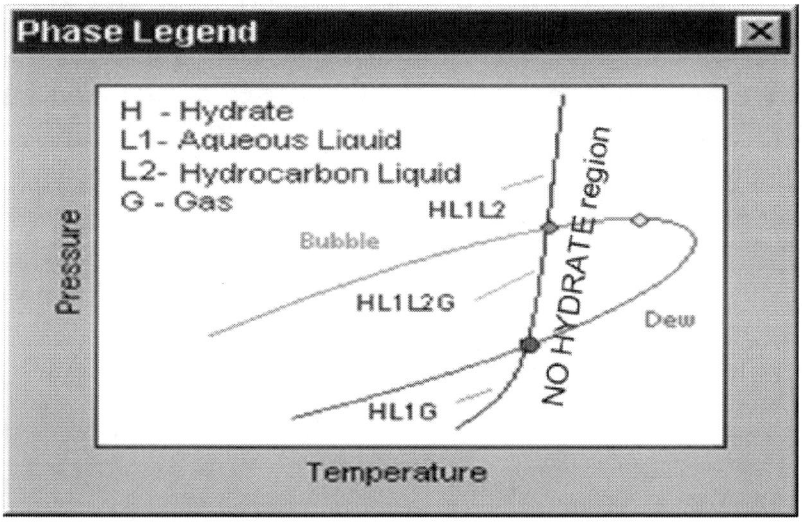

Quite often these programs also allow to estimate for a given pressure, temperature, and composition the maximum allowable water content before sold phase hydrates (and/or ice) may form. An example of some results for a C2 splitter is presented in table 1 below.

Table 1: Maximum allowable water content for C2 splitter system.

Description	Ethylene/ Ethane (mole %)	Temp (C)	Pressure (bara)	Max. allowable water content (mol ppm)
C2 splitter Feed	80/20	-45	11	19
C2 Splitter Feed tray	80/20	-62	4.5	8
C2 Splitter Bottoms	0/100	-54	4.7	8
C2 Splitter Overhead	100/0	-77	4.1	2

Potential causes for hydrates in ethylene plant cold section

The separation is done at higher pressure and low temperatures, so there is potential for hydrate formation if water is present, even in low quantities. However also ice may be formed and is sometimes difficult to determine if the issue is ice or hydrate formation. The cracked gas dryers are installed to keep moisture out (below 0.5 ppm) during operation.

In the cold section the more sensitive location for hydrate formation is often the C2 splitter. Small amounts of water can pass though the cold section and may accumulate or directly cause issue in the C2 splitter. Other locations are the demethanizer feed chilling and tower.

Cracked gas dryer

The most obvious cause of hydrate (or ice) formation in the separation section is operational issues of the dryers leading to water breakthrough. Some more specific causes are:

- Aging giving a run time lower than the regeneration time
- Analyzer malfunction
- Regeneration with inadequate flow or temperature
- Regeneration heater steam leakage
- Bed support issues

There are also less obvious causes for misoperation. Examples are:

- Bypass of the cracked gas dryer.
  This may happen if vent lines from inlet and outlet are connected to a common header that can be blocked.
- Liquid carry over due to inadequate gas/liquid separator upstream of the dryer
  Liquid carried over onto the bed leads to early bed aging and possibly liquid running down the dryer vessel wall.
- Poor gas distribution or channeling
  This may be caused by operating a low rates or poor inlet distributor design.

Water left after maintenance

When equipment or lines are inspected, repaired or constructed, hydrotesting is used. Alternatively water may also leak in (rain, condensation) when equipment is not adequately protected or covered. In all these cases it is important to remove all free water before drying. Low point drains need to be available and used.

Even when all free water is removed drying should not be taken lightly. For adequate drying low pressure is essential. Higher flow may cause higher pressure and as such less effective drying. On the other hand too low flow is also ineffective. This can be the case for large equipment like heat exchangers.

Warm temperatures (hot air or nitrogen) can significantly improve the timing required for drying.

To determine whether the drying is complete dewpoint readings should be taken at various spots. Two methods are possible.

- Determine absolute dewpoint. This is done after reducing purge flow to a minimum and after some time to allow for equilibrium to establish.
- Determine dewpoint difference between dry gas and outlet gas. This is done during purge and is a measure of water pick-up.

These two methods should not be mixed. When measuring dewpoint during purge the value is much lower than the equilibrium absolute dewpoint.

Water formation in hydrogenation reactors

Another potential source of water is the hydrogenation reactors. Here water may be formed from CO and hydrogen. The amount of water formed is low, however water may accumulate downstream and cause hydrate formation over time.

Hydrate mitigation

Cracked gas dryer

Perfect operation and monitoring of the cracked gas dryers and associated equipment are essential to avoid water breakthrough. Mitigations to avoid water breakthrough are:

- Performance monitoring, timely bed replacement
- Run both on analyzer and time
- Automation of regeneration
- Optimize bed loading
- Drain inlet header low point(s) before swapping dryers
- High efficiency gas liquid separator before dryer

Water left after maintenance

The best prevention for this is to avoid hydrotesting and use of alternative testing methods.

When hydrotesting is performed all water should be removed immediately from all low points and witnessed by plant personnel.

Also adequately covering equipment is important.

Drying procedure should cover all low point for dew point, the dewpoint measurement method and targets and tracking. In case targets are not reached alternatives methods can be considered like vacuum drying, not drying, or methanol filling.

Water formation in hydrogenation reactors

If water is formed in hydrogenation reactors, mitigation is not possible, only removal by installing a secondary dyer after the reactors.

Hydrate removal

Deriming: Temperature/Pressure

To understand deriming process for hydrate removal the temperature/pressure conditions where hydrate cannot form must be known. As indicated previously there is a (low pressure and higher temperature) region at which hydrate cannot exist. Changing operating conditions (increasing temperature/decreasing pressure) to these conditions can break the hydrate. However this is not always possible without interrupting production.

In case production is interrupted and hydrate is removed by increasing temperature with a hot gas, it is important not only to consider temperature but also pressure. The lower the pressure, the "easier" (at lower temperature) to break the hydrate.

Inhibitors

Alternative for deriming with a hot gas is the use of a hydrate inhibitor. There are two types of inhibitors: Thermodynamic and kinetic (anti-agglomerants).

Thermodynamic inhibitors remove the hydrate by lowering the melting point. It may take significant amount and direct contact is required. Alcohols and glycols are typical thermodynamic inhibitors. In ethylene plant methanol is quite often used. Advantages are that it is easily available and very effective in hydrate removal. However there are a few important disadvantages to consider.

Dry methanol can be quite corrosive, especially for alumina alloys (used for brazed aluminum heat exchangers in cold boxes). Adding small amount of water (if not already present) will reduce corrosiveness, without affecting the hydrate inhibitor effectiveness significantly.

Due to the (non ideal system) low relative volatility of methanol in hydrocarbons systems, methanol can end up and contaminate product streams (ethylene, propylene, Crude C4) which all have a low (ppm level spec) for methanol. Using specific injection location (C2 splitter feed instead of reflux) and removal of methanol/hydrate after the injection location can mitigate these issues. Also C2 and C3 recycle stream can get contaminated with methanol leading to CO formation in furnaces, with potential impact on hydrogenation reactors and methanator.

It should be noted that application of methanol is limited to -98 °C, the freezing point of methanol. Using methanol below these temperatures will even increase the plugging. At lower temperature heavier alcohol like n-propanol is used. However n-propanol is less effective, as it is quite viscous at these conditions. Use of n-propanol is limited to -127 °C, the freezing point.

In ethylene plants the lowest temperature can be below -127 °C and for these temperatures a mixture of alcohols can be used. When mixing alcohols the binary system has an eutectic point at which the freezing point is lower than the individual freezing points. Below is some info for the methanol/ethanol systems. Risk of (partial) de-mixing needs to be considered for the application at very low temperatures.

Kinetic inhibitors will suppress the hydrate formation. In some pipeline system some low dosage hydrate inhibitors (LDHI) have been used. In ethylene plants this has not been applied.

Hydrate examples

Ethylene gas desuperheater

After a turnaround some performance loss of an ethylene desuperheater (in ethylene refrigeration system) was noted. One year later tube leakage developed causing a plant outage to repair the exchanger. It was discovered that some tubes were plugged with hydrates/ice. This caused a failure of some tubes. Some trapped gas was observed while removing the plugging.

Further investigation showed that tubes were sagging in the middle and water could easily collect there. The equipment was opened during the previous turnaround and even though no hydrotest was performed water could have gotten in due to condensation and/or rain. Drying of the exchanger was done as part of the larger piping system, giving very low flows through the exchanger. Also dewpoint measurement was not accurately taken.

Learning is to inspect all the tubes before closing these exchangers and if needed blow out water from individual tubes.

Demethanizer

Due to dryer misoperation water ended up in the demethanizer tower. After start-up at first nothing was noted. Only at rates above 75% an issue with the demethanizer (dp increase and separation loss) was noted. A hydrate was suspected and several attempts were made to break the hydrate by injection of methanol in the feed lines and/or reflux. When these were not successful a tower scan was performed. This scan showed a plugging issue on tray just below the lowest feed location (at higher rates).

As a last attempt the tower was heated up (limited reflux) and to increase methanol concentration at this location the lowest of two feed lines was blocked, filled with methanol and then opened again to push methanol into the tower. This did not break the hydrate.

As this was a dual downcomer tray it was suspected that one of the downcomers was fully plugged, so methanol was not able to reach the hydrate as it went down the other downcomer. This was also supported by the fact that the issue appeared at 75% rates.

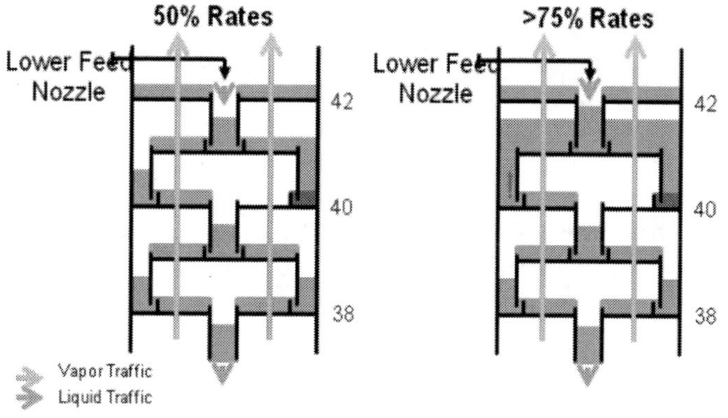

The tower was taken out of operation, all liquid removed, and the tower was heated with nitrogen. Based on tower weight, available nitrogen flowrate, and temperature, the time to heat up was estimated to be 8 hours. During the heat-up it was observed that temperature did gradually increase in the tower, except at the tray where hydrate was suspected. Here the heat-up did stop for 20 minutes around 32°F, before increasing. This could be melting of the hydrate (and/or ice).

After restart of the tower the operation the operation was back to normal which proved that there has been a hydrate issue rather than a mechanical issue.

C2 splitter

For years plant suffered from hydrate issue. As C2 splitter was also plant limit and also potential for green-oil fouling, it was not fully understood if and when hydrate issues occurred.

As the symptoms are the same, it is difficult to make a distinction between overloading a column, a hydrate issue, and green-oil fouling.

For high pressure C2 splitters it is known that water can accumulate over time on trays in the bottom section and then start forming a hydrate. This was supported by the fact that methanol injections also work when injecting at the feed tray.

To detect hydrate plugging various parameters were investigated and modeled. It was found that an easy, simple (only one instrument) and accurate method is to correlate the top (clean) tower section pressure drop with the total/bottom (pressure drop). There is a linear correlation between the two, except when a hydrate is developing and total pressure drop is much higher. In a graph the region where the hydrate problems occur are clearly visible. Alarm based on the correlation (red line) was implemented for early (also at low rates) detection.

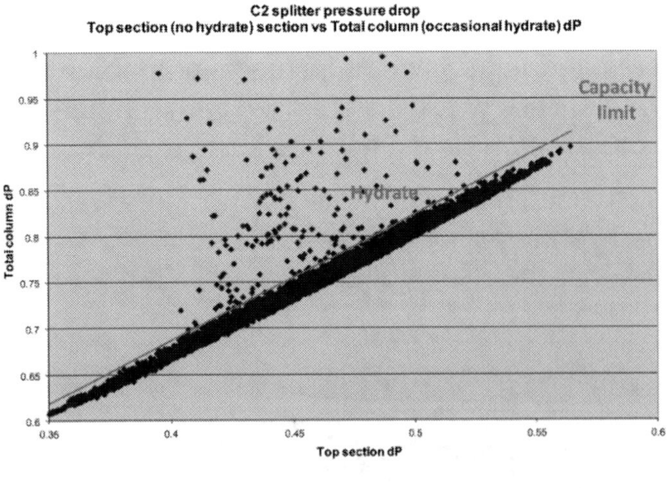

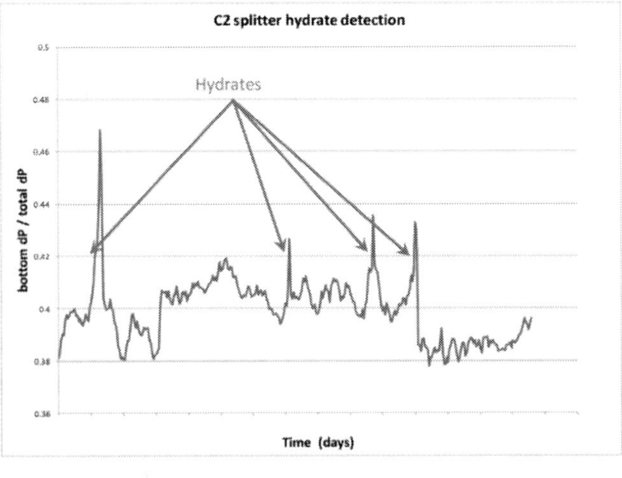

The cause of hydrate was determined to be liquid (water) carry over to the dryer and water formation in the hydrogenation reactors.

Summary

The theory and equilibrium of hydrate formation is well understood. Good engineering tools exist to predict the formation of hydrates and the operating conditions necessary to avoid their formation. Plants are designed to properly remove water and avoid the undesirable operating conditions. Despite this, hydrate formation impacting cracker operation is still, unfortunately, not an uncommon event.

When a hydrate event does occur, proven methods are available to remove the hydrate. However, this often presents a risk to product quality or that the ethylene plant has to be taken offline.

With most hydrate events, when investigated, the root cause and initiating events can be determined. However, the ethylene plant, by its very nature (wet cracking section followed by cryogenic separation), presents a high risk of hydrate formation. To avoid hydrates the ethylene plant must be operated flawless every minute of every day. Therein lays the challenge.

References

1. L. Pauling, R.E. Marsh, "The structure of chlorine hydrate", The Gates and Crellin laboratories of chemistry, California Institute of Technology, Pasadena, California, Communicated December 31, 1951.
2. M. Faraday, "Experimental Researches in Chemistry and Physics", Taylor and Frances, 1859.
3. E. Dendy Sloan Jr., Carolyn Koh, "Clathrate Hydrates of Natural Gases", 3rd Edition, CRC Press, 2007.
4. T. Letcher, "Chemical Thermodynamics for Industry", chapter 6, Royal Society of Chemistry, 2004.
5. Centre for Gas Hydrate Research, "What are Gas Hydrates?", Institute of Petroleum Engineering.
6. W.D McCain Jr., "The properties of Petrolem Fluids", 2nd Edition, PennWell Books, 1999.

444087 Cooling Tower Inspection and Repairs

Abstract:

Cooling water systems are a vital part of the operation of an Olefins plant. However, cooling towers often do not receive the same inspection and maintenance attention as other equipment in a unit, yet an unplanned failure can have just as significant an impact on plant operation.

This paper presents failure mechanisms within cooling towers, the inspection techniques and requirements for detection, and the repair and replacement strategies to insure long term reliability.

AIChE PAPER NUMBER 127b

UNIQUE PRE-CLEANING AND PASSIVATION CHALLENGES IN COOLING WATER SYSTEMS

Richard Fruit, Ph.D.
Technical Manager: Water
Athlon Solutions, LLC

Michael Kudijaroff
Maintenance Engineer
Williams

Anthony Maciejewski.
Area Manager
Athlon Solutions, LLC

Prepared for Presentation at the 2016 Spring National Meeting
Houston, Texas April 10 – 14, 2016

UNIQUE PRE-CLEANING AND PASSIVATION CHALLENGES IN COOLING WATER SYSTEMS

Richard Fruit, Ph.D.
Technical Manager: Water-
Athlon Solutions, LLC

Michael Kudijaroff
Maintenance Engineer
Williams

Anthony Maciejewski.
Area Manager
Athlon Solutions, LLC

Abstract: Pretreatment of cooling water systems is essential to maximize the service life and reliability of heat exchange equipment. Usually, pretreatment consists of two phases for newly fabricated or extended lay-up of piping and heat exchangers pre-cleaning and passivation.

Many times economics, discharge limitations, logistics and time requirements combine to reduce the effective application of pre-cleaning and passivation to prepare piping and exchangers for longer usable life. In the case study presented, the pre-cleaning and passivation was even more challenging for a number of reasons:

- Emergency shutdown: No ability to protect equipment against localized corrosion and microbiological growth for 11-month shutdown.
- Expansion of the plant was undertaken during this shutdown so new piping and exchangers were included as part of pre-cleaning and passivation along with piping and exchangers exposed to shut down without protective actions.
- Time and excessive congestion prevented pre-cleaning and passivation of individual exchangers so procedures could not be optimized for each of the two situations; shutdown/unprotected old equipment and new equipment with fabrication fouling.

Consequently, pre-cleaning and passivation of the entire system was required that could satisfy requirements for both types of equipment. This paper depicts the operational, chemical and monitoring methods used to pre-clean and passivate both new and existing equipment. The result was an effective pre-clean/passivation under difficult, complex circumstances leading to successful startup and operation after removal of manufacturing debris from the new equipment, "chip" scale from the old equipment and passivation of old and new equipment.

Pretreatment of Cooling Water Equipment

Pretreatment of cooling water systems is necessary to maximize the service life of heat exchange equipment. Usually, pretreatment consists of two stages

1. Pre-cleaning, to remove the accumulation of foreign matter
2. Passivation or pre-filming, to promote the rapid formation of an inhibiting film

Pre-cleaning is important because it prepares the surface for the passivating stage. After the surface has been cleaned, passivation minimizes initial startup corrosion by preparing for effective application of a corrosion inhibitor program. Economics, discharge limitations, and time requirements dictate whether pretreatment can or should be applied to the entire system or to individual exchangers.

Pre-Cleaning

Pre-cleaning of new cooling systems removes grease, oil, corrosion products and mill scale to prepare for rapid, uniform passivation. Failure to effectively pre-clean can cause increased corrosion and fouling that can result in heat transfer limitations in critical heat exchangers, premature failures, and high maintenance costs.

Normally, a detergent-like treatment consisting of tri-sodium phosphate, dispersant, surfactant and azole (if copper alloys present) is circulated through the cooling system with anti-foam as needed. The high pH of the tri-sodium phosphate combined with surfactants and dispersants in hot water remove light rust and hydrophobic materials left from fabrication and construction. Temperature is maintained at 120 - 160°F (50 - 70°C) for 8 – 24 hours depending on water temperature as higher temperatures contribute to faster removal of foulants. When the program is completed, heavy blowdown is required for rapid removal of the pre-cleaning chemicals.

Although optimal results are achieved when pre-cleaning is done at 120 - 160°, it is difficult to achieve these temperatures unless individual exchangers are being treated with relatively small solution volumes. Lower temperatures can be compensated for by circulation for longer times. Even at lower temperatures, pre-cleaning significantly improves results.

Passivation

Since most corrosion control methods rely on barrier film formation, effectiveness depends on the rate of film formation, which is related to the concentration of the inhibitor. Passivation is the rapid formation of a uniform film that immediately stifles corrosion. Once established, the film can be sustained through continuous maintenance inhibitor levels.

Passivation of cooling water equipment is recommended following initial commissioning/pre-cleaning, chemical/mechanical cleaning, and start-up after turnaround or inspection. Passivation is also recommended to respond to low pH excursions or process leaks depending on the severity and system exposure interval.

Use of polyphosphates, such as pyrophosphate, is important as they effectively remove corrosion products, while promoting a protective film. Tetra-Potassium Pyrophosphate (TKPP), along with dispersants, surfactants, azoles and molybdate are circulated at concentrations of 100 - 600 ppm phosphate as PO_4. This method can be effective in rapidly providing an inhibitor film in 4-6 hr at a temperature of 120°F followed by restoration of the control limits of the continuous inhibitor/dispersant/biocidal treatment program.

Pretreatment is vital for the carbon steel cooling water equipment, which is more susceptible to corrosion. Other exchanger tube materials such as copper alloys also benefit from pretreatment. In general, pretreatment, followed by well-controlled inhibitor/dispersant chemistry with good microbiological control will minimize corrosion for improved heat transfer, longer service life, and reduced plant maintenance. Getting a good start with effective pretreatment is just the beginning of extending heat exchanger service.

Individual Exchanger Pre-Treatment

Pretreatment of isolated individual exchanger or banks of exchangers using dedicated circulation/filtration equipment with ability to achieve and control optimum temperatures and chemistries has the potential to provide optimal results. Utilization of chemical cleaning companies with these capabilities in collaboration with the water treatment company provides the following benefits;

1. Reduces equipment cleaning time and minimizes generated waste
2. Cleaning company/water treatment professionals ensure job efficiency
3. Customization for common problems in collaboration with water management company
4. Full flow filtration for continuous flushing and removal of suspended solids

It is not uncommon that economics, discharge limitations, and time/logistic requirements require deviation from this ideal as to whether pretreatment should be applied to the entire system or to individual exchangers.

Pre-Treatment Challenge

The challenges in the design and implementation of online pretreatment for the entire system included the following circumstances or limitations compared to the preferred individual exchanger pre-treatment process;

1. Existing equipment shut down for 11 months with no treatment or circulation to protect against threat of localized corrosion and microbiological growth
2. New lines/exchangers with fabrication foulants
3. Pre-treatment must be done online for entire system, old and new.
4. Pretreatment temperatures will be 90 - 100 ^{o}F vs. recommended 120 – 160 ^{o}F
5. Existing system corrosion products and cooling basin debris/solids
6. Limited recirculation rate until Day 24 of the process

The objective is to pre-clean all new or re-tubed equipment and remove as much of the corrosion products and/or "chip" scale from the old equipment to prepare for effective passivation phase.

Phase Zero: Preparation for Pre-Cleaning

In order to make up for these limitations, an additional step was added and emphasized. The following actions were taken as Phase Zero;

1. Thoroughly cleaned cooling tower basin
2. Filled cooling basin with clarified water and circulated
3. Fabricated and installed an alternate cooling water return line to bypass tower fill
4. Maximum blowdown rate to flush solids from system was limited by wastewater treatment plant.
5. Insure functional exchanger backflush and blowdown valving present.

These actions reduced suspended solids loading on the system before initiating the Pre-cleaning and passivation phases. This is important to improve access of cleaning chemicals to surfaces in the existing equipment, especially biocides to eliminate or prevent iron reducing and/or sulfate reducing bacteria that may be encountered after 11 months in potentially anaerobic, stagnant conditions.

Monitoring: Critical for Length of Process

Because of the necessary length of the process at reduced temperatures, monitoring and control of the chemical parameters shown in Table 1 was important to the success of the program. The graphs below show progression through the process with compliance to the control limits listed in the tables, which was a result of the dedicated and persistent collaboration between client and water management company.

Table 1: Monitoring of Chemistry

Parameters	Control Range
pH	8.5 – 9.5
Temperature, ^{o}F	< 130 ^{o}F
Δ Unfiltered - Filtered Phosphate	< 1.0 ppm
Calcium as $CaCO_3$	< 200 ppm
Iron as Fe	< 5.0 ppm
Ortho-Phosphate as PO_4	150 – 250 ppm – Pre-Cleaning Phase
Poly-Phosphate as PO_4	150 – 200 ppm – Passivation Stage 1
Tolyltriazole (TTA)	> 2.0 ppm, 10 – 15 ppm Passivation
Deposit Control Polymers	15 ppm
Molybdate as MoO_4	50 – 100 ppm – Passivation Stage 2
Anti-Foam	As needed to control foaming
Free Chlorine	0.2 – 0.5 ppm
Total Chlorine	0.8 – 1.5 ppm
Microbiological Testing	< 1,000 Colonies/ml., SRB & IRB Testing

Phase 1 – Pre-Cleaning

After the cooling system had been circulated with heavy blowdown for four (4) days to allow both filtered and unfiltered iron levels to drop to ≤ 5.0 ppm as Fe from a maximum of 35 ppm (estimated at over a ton of Iron Oxides and/or Phosphates), chemical addition was initiated to achieve the following control limits shown in Table 2:

Table 2: Pre-Cleaning Control Limits

Parameters	Control Range
pH	8.5 – 9.5
Temperature, ^oF	< 130 ^oF
Calcium as $CaCO_3$	< 200 ppm
Iron as Fe	< 5.0 ppm
Ortho-Phosphate as PO_4	150 – 250 ppm
Tolyltriazole (TTA)	10 – 15 ppm
Deposit Control Polymers	50 ppm
Surfactant (Anti-foam as needed)	As needed to maintain slight foam
Isothiazolin Biocide	100 ppm
Free Chlorine	0.2 – 0.5 ppm
Total Chlorine	0.8 – 1.5 ppm
Microbiological Testing	< 1,000 Colonies/ml., SRB & IRB Testing
Visual Inspection	Applied Grease to Coupons

Generic descriptions of the chemical cleaning ingredients are provided below;

1. Phosphate: Tri-Sodium Phosphate, Na_3PO_4
2. Surfactant: C_{14} – C_{16} Sodium Alkyl Sulfonates (Anti-foam as needed)
3. Deposit Control Polymers: High charge density Ca-, Fe- Phosphate co-polymer and ter-polymers with strong general suspended solids dispersion capability.
    a. Acrylic Acid/Acrylamido-Methyl-Propyl Sulfonate: AA/AMPS (35 ppm)
    b. AA/AMPS/Sulfonated Styrene: AA/AMPS/SS (15 ppm)
4. Copper Alloy Inhibitor: Tolyl-Tri-Azole (TTA)
5. Microbiocide: Isothiazolin

Chemical residuals were maintained for seven (7) days until grease-coated coupons indicated that pre-cleaning was complete. Backflushing of tube-side water exchangers was done along with blowdown from the bottom of shell-side water exchangers to assist in removing suspended solids. Figure 1 shows the filtered (soluble) and unfiltered (total iron) levels as the process progressed (0.22 micron filter pads).

During the Phase Zero, circulation was stopped for >12 hours to repair a leaking valve on a 42-inch line. Maximum makeup and blowdown was continued through the basin and corresponds to the precipitous drop in iron levels just prior to initiation of Pre-Cleaning.

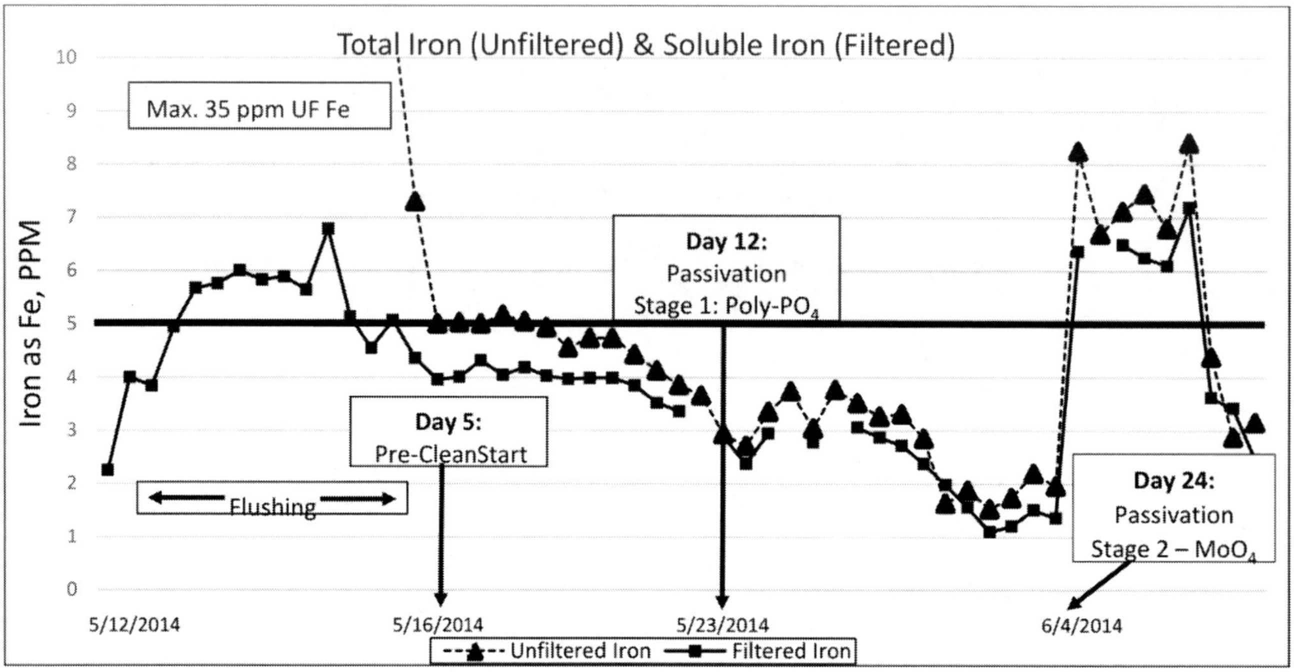

Figure 1: Filtered and Unfiltered Iron as Fe

As shown, much of the suspended iron oxides/phosphates were removed during the 4-day Phase Zero Preparation for Pre-cleaning. Removal of solids/"chip" scale was key to improve penetration of the cleaning/passivating chemistry, especially microbiocides to eradicate iron reducing and sulfate reducing bacteria under deposits in the existing equipment that was stagnant for 11 months.

Aggressive backflushing of tube-side water exchangers and bottom blowdown of shell-side water exchangers was also stressed to improve pre-cleaning. Backflushing and blowdown was done twice/week. Where backflush valving/piping were deficient, repairs were made to enhance solids removal. Figures 2 and 3 show the piping/valving arrangement necessary for backflushing tube-side water exchangers.

The increase in iron oxide levels on Day 24 was a result of getting all recirculating pumps in operation, which suspended additional suspended solids for removal by blowdown.

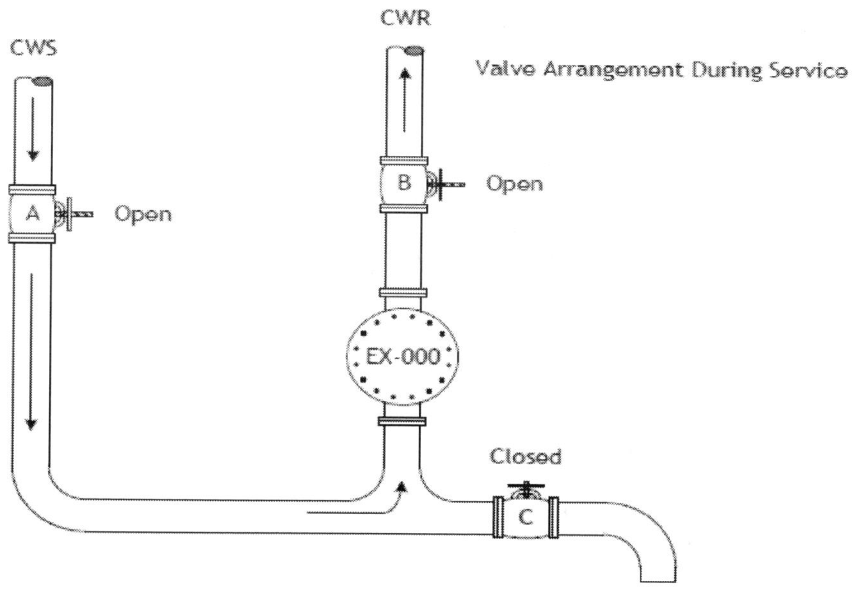

Figure 2: Backflush Valve/Piping during Service

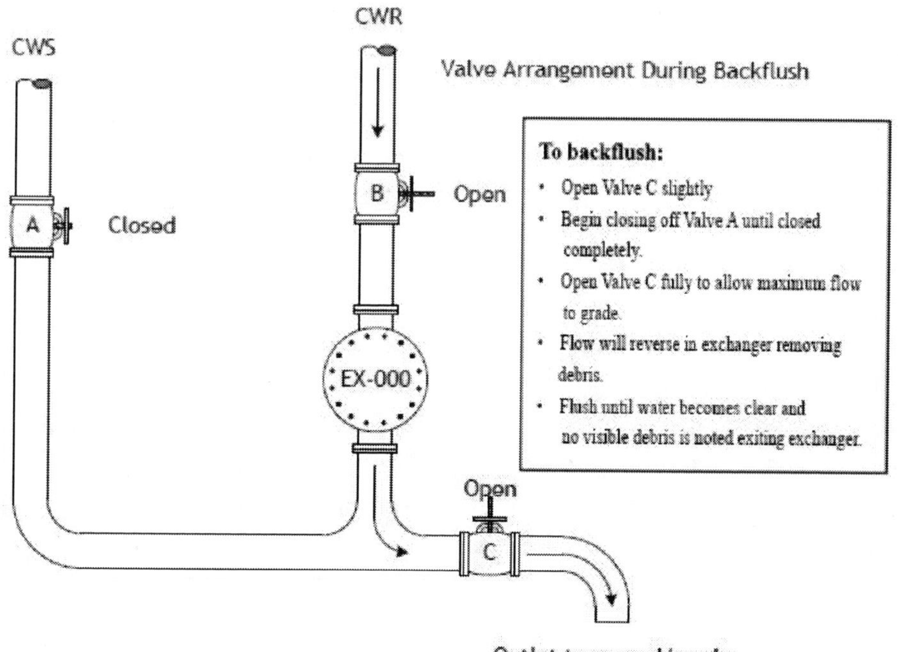

Figure 3: Backflush Valve/Piping during Backflushing

Phase 2: Passivation – Stage 1

On Day 12, when grease-coated coupons were observed to be clean and ortho-phosphate levels had been reduced, the passivation phase was initiated with a reduction in pH range and addition of poly-phosphate to the remaining ortho-phosphate for removal of "chip" scale, corrosion products and passivation.

Table 3: Passivation Control Limits

Parameters	Control Range
pH	7.6 – 8.2 (Max 8.5)
Temperature, ^{o}F	85 - 100 ^{o}F
Calcium as $CaCO_3$	< 200 ppm
Iron as Fe	< 5.0 ppm
Ortho-Phosphate as PO_4	50 – 100 ppm
Poly-Phosphate as PO_4	150 – 200 ppm
Tolyltriazole (TTA)	10 – 15 ppm
Molybdate as MoO_4	50 – 100 ppm (Stage 2)
Deposit Control Polymers	50 ppm
Surfactant	As needed to maintain slight foam
Anti-Foam	As needed to control foaming
Isothiazolin Biocide	100 ppm
Free Chlorine	0.2 – 0.5 ppm
Total Chlorine	0.8 – 1.5 ppm
Microbiological Testing	< 1,000 Colonies/ml., SRB & IRB Testing

Generic descriptions of the chemical cleaning ingredients are provided below;

1. Poly-Phosphate: Tetra-Potassium Pyro-Phosphate, $K_4P_2O_7$
2. Molybdate: Sodium Molybdate: Na_2MoO_4 (Stage 2)
2. Surfactant: $C_{14} - C_{16}$ Sodium Alkyl Sulfonates
3. Deposit Control Polymers: High charge density Ca-, Fe-Phosphate co-polymers and ter-polymers with strong general suspended solids dispersion capability.
   a. Acrylic Acid/Acrylamido-Methyl-Propyl Sulfonate: AA/AMPS (35 ppm)
   b. AA/AMPS/Sulfonated Styrene: AA/AMPS/SS (15 ppm)
4. Copper Alloy Inhibitor: Tolyl-Tri-Azole (TTA)
5. Water-based Anti-Foam
6. Microbiocide: Isothiazolin

Figure 4 shows the ortho- and poly-phosphate levels were in control during Stage 1 of the Passivation phase. The difference in filtered and unfiltered phosphate in the first few days of the circulation/flushing stage is indicative of iron phosphate suspended solids until iron levels had been reduced. After that, the nearly identical values for filtered/unfiltered phosphate indicates that iron phosphate solids had been removed or were being dispersed effectively.

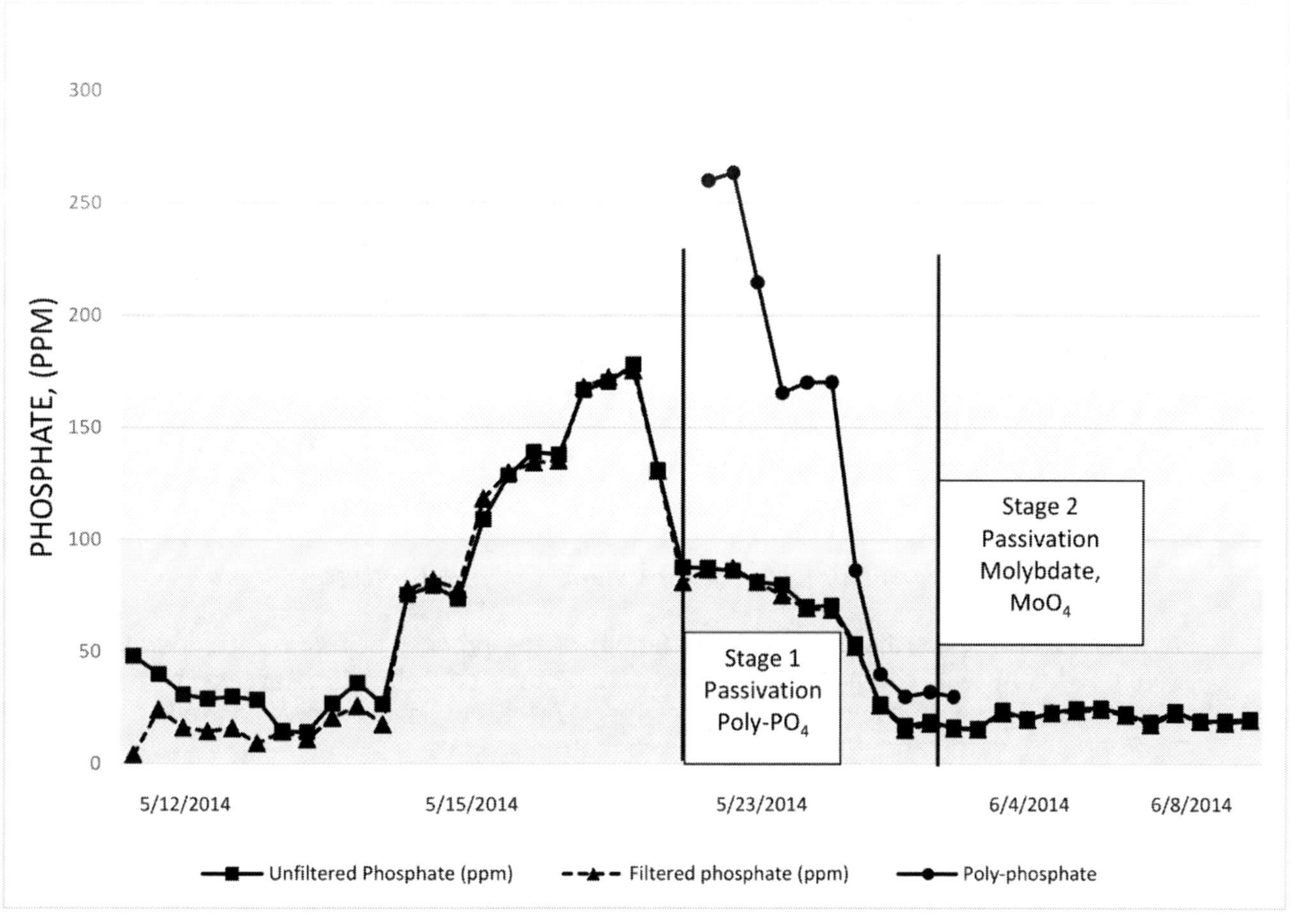

Figure 4: Phosphate Levels during Pretreatment Process

Phase 2: Passivation – Stage 2

After 12-days at these conditions with continuing backflushing of exchangers, phosphate-based passivation was considered complete based on time, temperature and phosphate levels used. With mildly passivating ortho- and poly-phosphate levels in the 20 – 50 ppm range, molybdate was added to further passivate the system. Molybdate is effective in inhibiting concentration cells, localized corrosion likely to be prevalent under deposits in the old equipment. This additional passivation stage was intended to assist in controlling localized corrosion under deposits by repairing the passivating film under these deposits.

Figure 5 shows that Molybdate levels were in the control range of 50 – 100 ppm MoO_4 during Stage 2 Passivation.

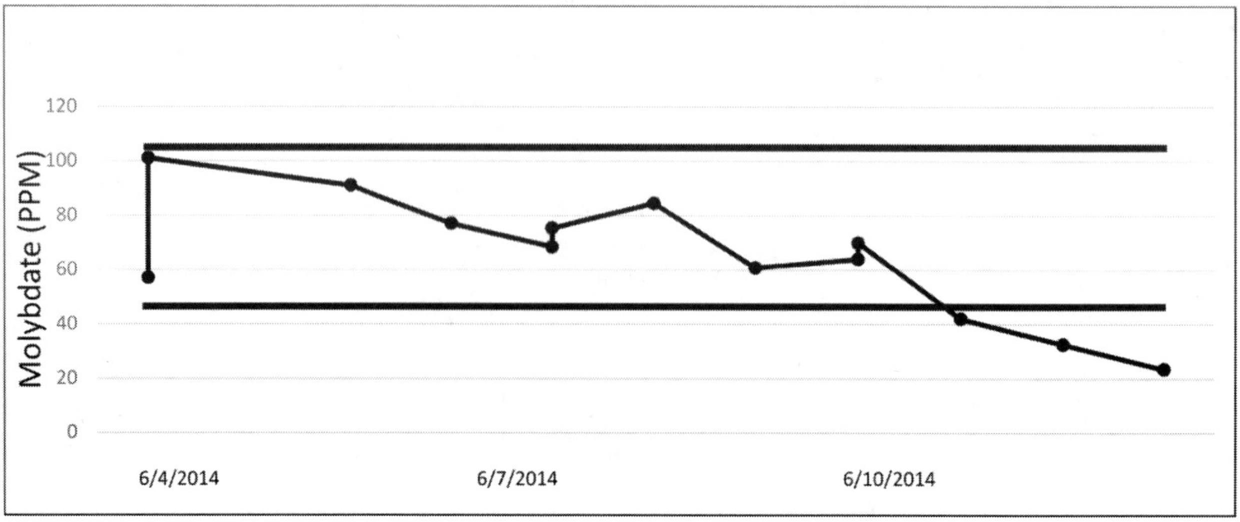

Figure 5: Molybdate Passivation Control

Control of pH, Deposit Control Polymers and Copper Inhibitor (Azole)

As mentioned previously, the necessary length of the process at reduced temperatures made continued control of chemical parameters within limits critical to the success of the program. Figures 6, 7 and 8 show that pH, polymer and azole levels were controlled within the appropriate control ranges for the entire pretreatment process

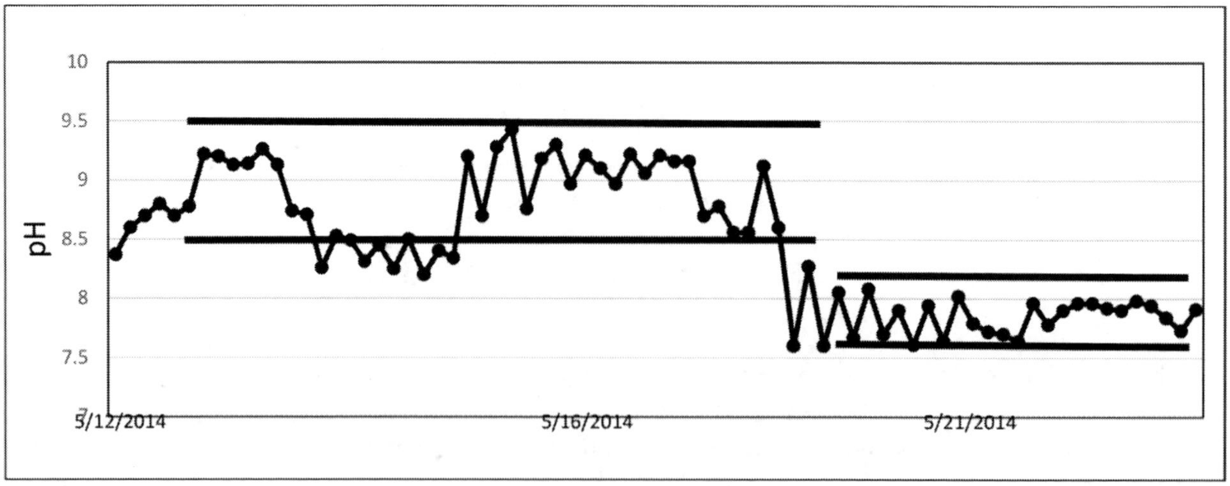

Figure 6: pH Pre-treatment Control

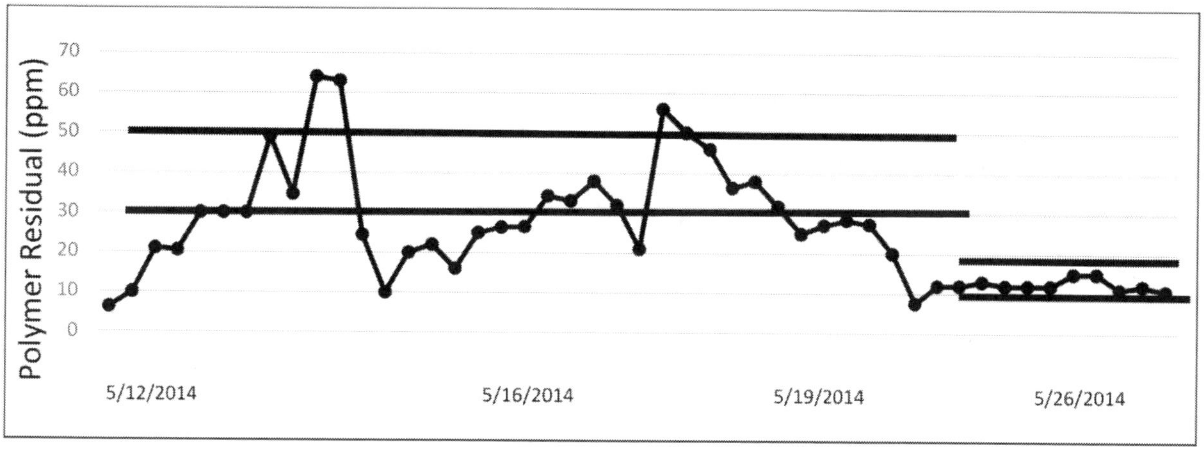

Figure 7: Polymer Residual Pre-treatment Control

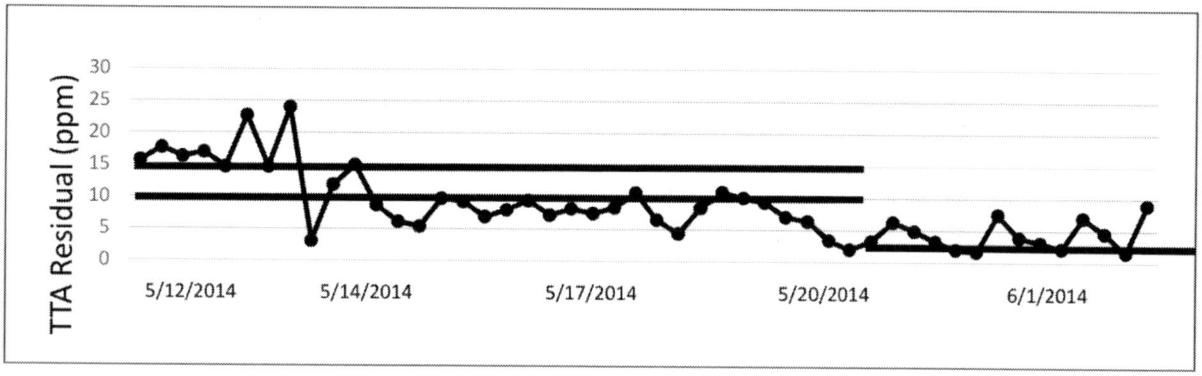

Figure 8: Copper Alloy Inhibitor Pretreatment Control

Results

Bottom line, the plant operated at record production levels with the added capacity from the expansion through the hot, Gulf Coast summer months.

As might be expected, all exchangers were not in pristine condition;

1. Significant "chip" scale was present in outlet channel heads, which was removed from a number of exchangers that were opened prior to initiation of a production heat load.
2. First pass exchangers that had been re-tubed had significant plugging with chip scale and were hydro-blasted prior to startup.

3. While a number of exchangers show some flow restriction, they were sufficiently clean to permit operation at record production levels until cleaning during cooler times.

After plant startup in June, 2015, a flow/temperature study was conducted by the water management company to quantify the degree of flow restriction and its effect on skin temperature and velocity compared to design. The results of this study will allow;

1. Valving adjustments to balance flows to achieve better approach to design.
2. Prioritization of cleaning schedule.

The client and water management company continue collaborative efforts to maximize heat exchanger efficiency, reliability and service life.

Summary

In summary, the dedicated, persistent collaboration of a water management company with the operations and maintenance staff of the client met the challenge of pretreatment of old equipment and fabrication-fouled new equipment under less than ideal conditions. The combination of operational, chemical and monitoring/control over the 30-day period provided results that allowed operation at record levels with continued improvement ongoing.

AIChE PAPER NUMBER 127C

INNOVATIVE SOLUTION FOR AT HEIGHT INSPECTION – MAINTENANCE – REPAIR WORK SCOPES

Howard Wall
General Manager
MHT Access Services Inc.

Steve Mc Elwee
Business Development
MHT Access Services Inc.

Clint Linebarger
Operations Manager
MHT Access Services Inc

Prepared for Presentation at the 2016 Spring National Meeting
Houston, Texas April 10 – 14, 2016

INNOVATIVE SOLUTION FOR AT HEIGHT INSPECTION – MAINTENANCE – REPAIR WORK SCOPES

Howard Wall
General Manager
MHT Access Services Inc.

Steve Mc Elwee
Business Development
MHT Access Services Inc.

Clint Linebarger
Operations Manager
MHT Access Services Inc

Abstract: With every inspection, maintenance, and turnaround program there are a multitude of work scopes that have to be planned and executed within critical scheduling restraints. Sometimes not all the work scopes are completed due to scheduling and budget overruns.

This presentation will introduce innovative approaches for at height work to help inspection, maintenance, and turnaround programs reduce cost, shorten work scope completion time, and provide a positive impact on the turnaround start-up schedule. Pre-Turnaround planning and turnaround work scopes will also be discussed.

Introduction

As Petro-Chemical facilities age the cost of maintaining pressure equipment, associated piping and support structures to safe industry standards increase. Some examples of age include damaged insulation, coating damage, mechanical damage, metal loss, fatigue, etc. While inspection and maintenance departments work to help keep the facility safe, management is tasked with completing the increasing work load with the same or reduced budget. As a result, managers are forced to prioritize inspection and maintenance items, starting with high priority items and working down the list. Some of the lower priority items are often not completed and pushed to the next turnaround budget. Either way, until these items are addressed, a potentially hazardous condition could arise for personnel, equipment and the environment.

The number one reason facilities end up with huge backlogs of inspection and maintenance items, directly relates to the overwhelming costs associated with elevated access. As an example, a chemical facility along the Houston Ship Channel reported scaffolding costs running between 30% and 50% on each job involving elevated access.

Each time a crew member works at elevations above grade level, the potential risk of accidents occurring, multiply greatly. In the year 2014 OSHA documented 349 deaths associated with falls with the majority being from ladders and scaffolding.

Inspection and Maintenance work items not being addressed in a timely manner can lead to unnecessary risks for equipment and personnel. A few examples are;

1. Damaged insulation leading to Corrosion Under Insulation (CUI)
2. Damaged pipe supports near pressure equipment creating excessive external stress on nozzles and shell
3. Coating failure and corrosion issues
4. Leaking flanges

Equipment and associated piping in areas away from ladders, landings, and platforms are the most at risk. These areas are usually overlooked due to the added time and expense needed to erect scaffolds to perform inspection and maintenance work.

Often, this determines the way facilities inspect piping systems. For as long as Thickness Monitor Locations (TMLs) have been used to assess piping circuits, managers have been tasked with keeping costs to a minimum. The thought process of assigning TMLs on piping circuits with all parts being equal, (i.e. temperature, pressure, product, metallurgy etc.), the metal loss, or corrosion rate throughout the circuit should be consistent. This is the reason why TMLs are assigned in areas where they will not need scaffolding or the use of a man lift to inspect. The problem with this approach is that it does not take account for areas of accelerated flow or erosion corrosion. The rule of thumb with any piping monitoring program is to place TMLs on the first directional change fitting discharging from equipment, and then place the balance of TMLs on fittings and pipe throughout the circuit. The equipment being; exchangers, drums, towers, headers, etc. Unfortunately, some of these areas are never inspected and the true analysis of the piping circuit cannot be determined.

Pressure Relief Valves are often not tested during a Turnaround due to costs associated with scaffolding access. Pressure Relief Valves (PSV) protect equipment and piping systems from exceeding Maximum Allowable Working Pressure (MAWP). PSV valves not serviced on a regular interval can lead to malfunction from sticking caused by corrosion, fouling, etc. This potentially leads to over pressuring equipment to dangerous limits.

Towers are equipment are very seldom externally inspected 100%. API 510 inspectors perform external inspections on towers from the ladders and the landings. Since these ladders and landings are usually located on one side of the tower, 40% to 50% of the tower where the inspector cannot see is never inspected. In order to externally inspect towers 100%, scaffolds or some other means of access would need to be utilized.

Inspection of pipe supporting spring cans is another issue that is often overlooked due to costs associated with access. These support devices were engineered and strategically placed to support the piping system and to relieve stress in an effort to extend the life cycle of the system. These pipe support devices should be inspected periodically for coil fatigue, functionality and proper load settings. Piping support failure can cause catastrophic damage to equipment, the environment and also personal injury or even loss of life.

Piping resting on steel beams without pipe shoes can be susceptible to Touch Point Corrosion (TPC), which can hole through creating a potential hazardous failure. Corrosion under Insulation (CUI) is another related issue. These inspections are often limited or overlooked due to the associated cost of scaffolding.

Externally inspecting aboveground storage tanks can prove to be challenging and is seldom performed correctly. Often times these tanks are externally inspected from the ground looking up, and from stairs which wind up one side of the tank to the roof. Unless the inspector finds low thickness readings or an area of concern from his or her vantage point, issues can be missed. Not to mention, the safety aspect of walking on a tank roof without handrails installed along the perimeter.

Time Line

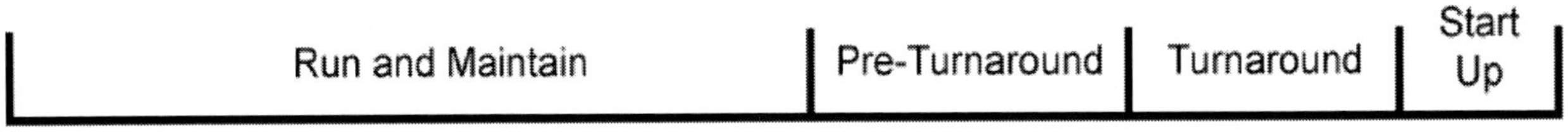

Most of the items listed above can be addressed during regular run and maintain and pre-turnaround operations. Other items will need to be addressed during a turnaround. Furthermore, all of these work scopes can be completed without the high cost of scaffolding.

Alternative Access Methods for Performing Inspection, Maintenance and Repair at Heights

Rope Access teams are generally much smaller than most maintenance and inspection crews due to the lack of need for conventional methods of access, (scaffolding, man lifts, etc.) and their "full turn-key" capabilities. Technicians accessing the work face also perform the work scopes. Utilizing Rope Access reduces man hour's resulting in less exposure to risk which has in return given the rope access industry as a whole, an impeccable safety record. Since the inception in the US market around the year 2000, there has not been a single fatality recorded in the Rope Access industry.

The incorporation of rope access technologies provides industry facilities with 20% to 40% or more in costs savings vs. scaffolding costs.

With the utilization of rope access technologies, obscure areas that may have been over looked for years due to access issues and costs, can now be accessed to provide a complete inspection, maintenance & reliability program. Rope Access provides multi-disciplined personnel that can complete a wide variety of work scopes which include but are not limited to; non-destructive testing, general maintenance, pipe supports, field welding, insulation and coating services. Rope Access teams can be tailor made to fit a wide range of work scopes which provides a synergistic approach to inspection, maintenance & repair work at heights.

Utilization of Rope Access saves time and money. Rope Access teams do not have to wait for scaffolds to be erected or have to wait on the scaffold contractor to inspect the scaffold and sign the green tag before work can begin. The time and budget savings can then be applied to more inspection and maintenance items which will reduce the overall inspection and maintenance backlog.

A rope access technician is performing insulation and lagging repair.

Utilizing rope access techniques and procedures, industrial facilities are now able to install TML's, and perform complete inspections on towers, vessels, exchangers, drums, etc. and associated piping. Previously these inspections would have been incomplete with TML's installed only where accessible off of ladders, landings and stairwells. At the completion of these inspections, Rope Access Technicians are also capable of performing preventive maintenance and repairs, all without scaffolding.

Inspection and maintenance can now be performed in a timely manner. Rope Access teams can easily access at height locations while repairing damaged coating, damaged insulation, leaking flanges without the expense of scaffolding.

Pressure Relief Valves (PSV) need to be serviced and inspected during each turnaround. Rope Access teams can remove and/or help with the removal of PSVs. They can assist crane operators and riggers by attaching slings or chokers to the valve, remove bolts, signal the crane operator to lower the valve to ground, and installing blind flanges. After the valves have been serviced, Rope Access teams can re-install the PSVs, all without scaffolding.

Touch Point Corrosion (TPC) is a non-uniform type of corrosion that occurs in piping systems where resting on supports. It is thought the causes for this type of corrosion originates from galvanic, pitting, and crevice corrosion. Because this type of corrosion is non-uniform, localized areas can cause severe damage to pipe with unexpected leaks. These TPC areas are often overlooked due to the costs associated with providing scaffolding at each support. Rope Access teams can quickly and safely provide a cost effective means to inspect piping systems for TPC, even lifting the pipe to perform coating repair while installing Teflon spacers between the pipe and support to stop the affects from TPC.

Corrosion Under Insulation (CUI) can happen on equipment and piping systems where moisture is trapped between the metal surface and the insulation. Studies show this type of corrosion is usually localized and more aggressive than most other types of corrosion. Temperature plays an important role with CUI. Per API-510 the susceptible temperature range for carbon steel is between 10° F and 350° F while austenitic stainless steel temperature range is between 140° F and 400° F.

Without inspecting for CUI, areas of aggressive corrosion can hole through causing product escaping from system. Utilizing Rope Access can help soften the expense of visually inspecting equipment and piping systems for damaged insulation where moisture intrusion could occur. The areas where damaged insulation is located can then be further evaluated for extent and severity of CUI damage, also without the use of scaffolds.

Rope Access can provide for thorough external inspection of above ground storage tanks by being able to inspect the roof including the edges without fear of falling and inspect the shell top to bottom 360° around. This type of inspection provides close visual inspection with UT thickness readings.

Rope Access Technician installing weather proof Inspection ports to thickness monitoring locations (TMLs) 360° around the vessel.

Within each facility, pipe support hangers are often located in obscure areas difficult to see from the ground level or landings. By combining rope access services with pipe support inspection specialists, facilities are now able to incorporate a full 4 phase auditing program which ranges from the "hands on" inspection of each assembly where all inspection and design data is recorded, as well as repair/adjust/replace criteria is established. Then all critical and non-critical adjustments are completed, followed by completing any required repair/replacement items. The entire inspection and maintenance program is documented in a formal reporting program complete with photographs of each assembly, inspection, maintenance & repair summary and line drawings. With the incorporation of this 4 phase pipe support program, industry facilities are now able to cost effectively inspect and monitor the life cycle of their pipe supports and perform all required maintenance and repair activities.

Rope Access Technician performing an initial "hands-on" audit of piping supports that had gone years without being inspected.

Rope Access Technicians performing maintenance and repair on high energy piping supports.

Case Study

Inspection and Pipe Repair Project

Performing work in pipe racks can prove to be challenging due to the amount of scaffolding costs involved to gain access. A facility in Orange, TX found a solution to this problem by incorporating the use of Rope Access to complete an inspection and repair project on a 5,000 foot, 4" ammonia line connecting the tank farm to the process area. It was pre-determined that there was already touch point corrosion issues on each of the I-beam supports in the pipe rack, as well as CUI from damaged insulation. The scope of work for this project was completed in two phases.

Pre-Turnaround Work Scope

- Remove superfluous insulation and perform CUI and coatings inspections on the entire length of the pipe (5,000 ft.).
- Formal report generation and develop repair/replace criteria for the turnaround.

Note: the pre turn-around insulation removal and inspection was completed in 20 working days with a three man rope access inspection team.

Turnaround Work Scope

- Cut out approximately 800 feet of rejected piping and re-weld in sections of new pipe (74 TIG welds).
- Perform Liquid Penetrant Testing on all weld root passes, and X-Ray 10% of welds.
- Perform lead based paint abatement on approximately 1,000 feet of piping at areas of coating damage with corrosion issues.
- Recoat 1,000 feet of pipe with a 3 part epoxy coating system.
- Install approximately 125 each, I-Rod pipe shoe kits with Teflon base in order to eliminate touch point corrosion issues.
- Perform hydro test of the line before putting into service.

The turnaround work scope above was completed in 24 working days with no safety incidents. By utilizing rope access procedures and working methodology the cost of the project was completed for $445,344.00 compared to the competing quote of $895,000.00 utilizing scaffolding.

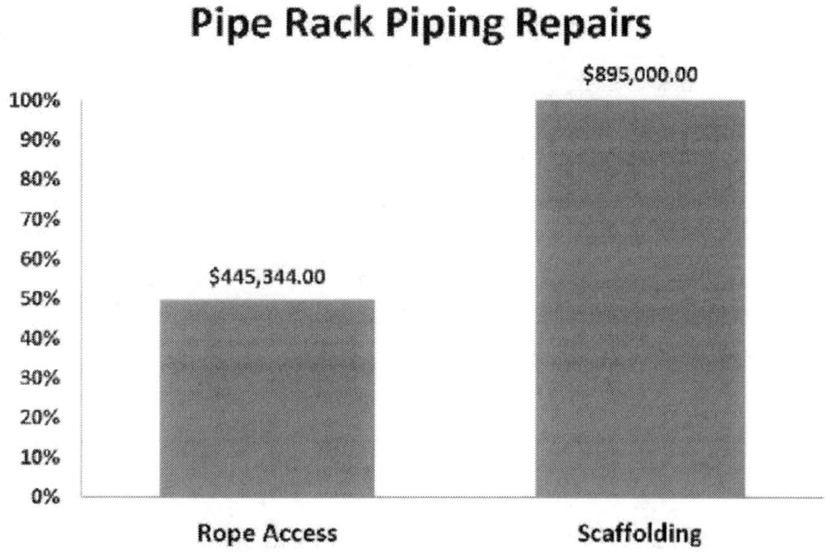

A Rope Access Technician installing I-Rod pipe shoe kits in order to eliminate touch point corrosion issues.

Rope Access Technicians hauling section of piping into the pipe rack for welding.

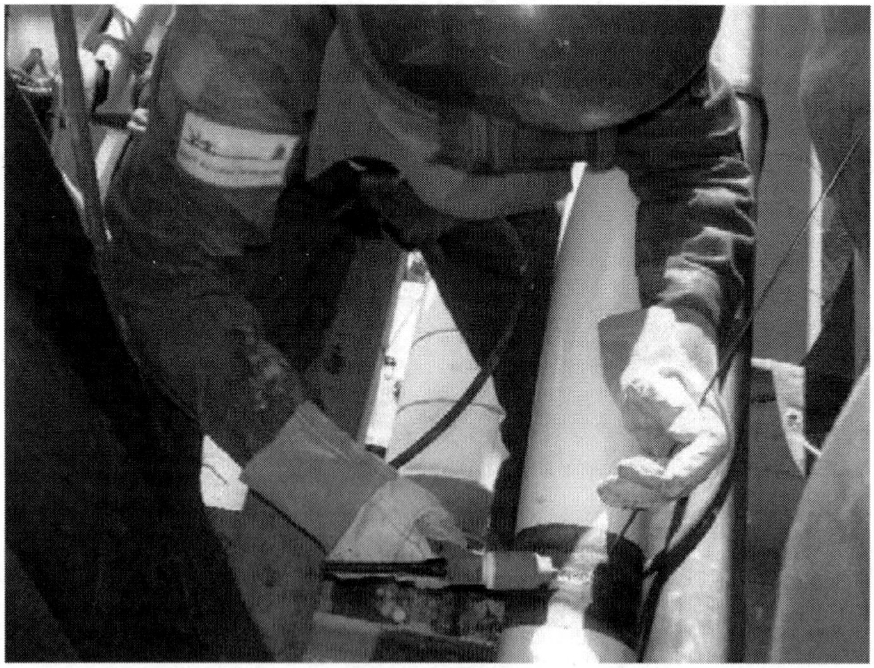

A Rope Access Welder performs TIG welding on piping in pipe rack.

Flare Stack Inspection

Scaffolding flare or emission stacks for inspection and maintenance can be costly. In some cases the scaffolding can cost more than completing the work scope itself. One Geismar, LA olefin facility during the pre-turnaround planning decided to try an alternative access method for the inspection of their 260′ free standing flare stack and perform any maintenance repair work discovered during the inspection.

There was a limited time frame to perform the work scope since access to the flare stack would not be available during the initial shutdown of the facility and during the startup which drastically reduced the 35 day window.

The facility needed a safe and efficient means to perform this inspection and remedial actions as an alternative to the time and cost associated with erecting scaffolding and dismantling scaffolding within the reduced time frame.

Turnaround Work Scope

- Inspect the standoffs centralizers in the annulus between the flare pipe and the outer structure for anomalies.
- Take ultrasonic thickness readings on the flare pipe.
- Magnetic particle inspection on any weld anomalies.
- Photo / video inspection documentation.
- Inspect attachments for integrity (i.e. risers, ladders and landings)
- Perform any welding repairs (if any)
- Deliver formal inspection report.

To accomplish this task, the facility decided to utilize rope access procedure and work methods. A four man rope access team completed the inspection in 6 working days which included one day of safety training. The inspection revealed no anomalies requiring remedial action. There were no safety incidents and the inspection cost was $27,864.00. During the pre-turnaround planning the facility obtained a quote for scaffolding in the amount of $75,600.00. The utilization of Rope Access provided for a cost saving in excess of $47,736.00.

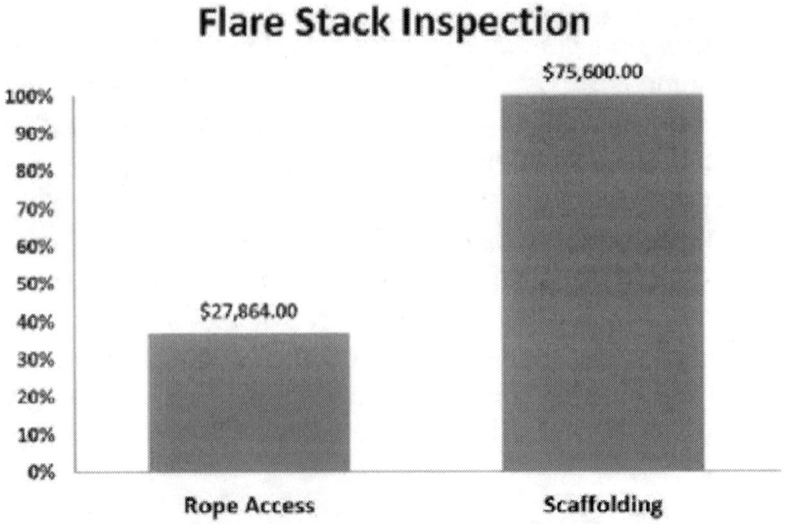

A Rope Access Technician inspects ladder, landing and riser attachments for integrity.

A Rope Access Technician performs internal visual inspection while obtaining ultrasonic thickness reading on the flare pipe.

Conclusion

Industrial Rope Access entered the US market around 2000 and has over the years proven to be a safe, cost effective alternative to completing work scopes at above grade levels. There has not been one reported fatality in the United States. Recognizing this impeccable safety record US industries are rapidly beginning to recognize the values that rope access can offer to the everyday operations of their facilities.

Rope Access Teams in most cases require less personnel due to the multi-disciplined aspect, where the same technicians that access the work face also perform the work tasks. This also provides for minimal disruption to other operations and a smaller foot print, while minimizing man hours equaling less exposure to risks.

Rope access provides a synergistic approach between inspection and maintenance / repair at heights with 20% to 40% or more in cost savings. This value added approach, eases the budgeting pains and reduces the production and safety concerns of facility managers and engineers. More of the prioritized inspection and maintenance items are able to be addressed within budgets, in a shorter time frame, which can increase equipment reliability and safety, reduce turnaround time and increase plant production.

AIChE PAPER NUMBER 127d

UTILIZING ULTRASONIC INLINE INSPECTION TECHNOLOGY TO MANAGE ETHYLENE FURNACE CONVECTION SECTION RELIABILITY

Darrell W. Meinen
Technical Advisor
Quest Integrity Group, LLC

Alex Grigsby
Ceda Group

Prepared for Presentation at the 2016 Spring National Meeting
Houston, TX, April 10-14, 2016

Utilizing Ultrasonic Inline Inspection Technology to Manage Ethylene Furnace Convection Section Reliability

Darrell W. Meinen

Technical Advisor

Quest Integrity Group, LLC

Alex Grigsby

Ceda Group

Abstract: Ultrasonic technology has been available for decades, but has not been frequently utilized in the inspection of ethylene furnaces, primarily due to furnace inaccessibility and lack of familiarity with advancements in the technology. Ultrasonic technology can now be applied to conduct full inspection of ethylene furnace convection sections with minimal modification to the furnace. Today's technology allows for customizable tool access to each pipe, further increasing the value and accuracy of ethylene furnace inspections.

Conducting ultrasonic in-line inspections on ethylene furnaces can be utilized to evaluate cleaning effectiveness, determine tube condition and ultimately calculate remaining life and fitness-for-service, thereby minimizing unplanned outages. This presentation will outline the furnace cleaning process and ultrasonic method of ethylene furnace inspection, using case studies of completed ethylene furnace inspections.

History

Introduction

Ultrasonic inspection technology has been utilized to non-destructively inspect pipes for decades, and has proven itself to be an accurate and reliable method of inspection. With advancements in electronics processing, battery life, data storage and miniaturization, the technology has been adapted to inspect the interior of a pipe segment. Further enhancements have allowed free swimming, untethered tools to be developed to inspect pipes down to 3" diameter. This technology has been widely utilized in process furnaces for the refining industry for many years.

Figure 1. Multi-Segment Tool Navigating a Return Bend

Ethylene Furnace Challenges

There are a number of reasons this technology has not been widely adapted to ethylene furnaces. Factors including cleaning the convection section pipes, access to individual passes and a lack of exposure to the technology have limited the use of ultrasonic inspection technology within ethylene furnaces. In this paper, we will discuss the first two major obstacles and attempt to address the last challenge.

Access to Individual Process Passes

Most ethylene convection section designs consist of a multi-pass coil, which is welded to a common manifold. This is done to reduce the number of flanges within facilities as these present potential sources of containment loss. Recently, tools have been developed which can be temporarily inserted into the manifold by either removing the end cap or an inlet pipe spool. The same method of access to the convection section outlet manifold is also utilized, thus allowing for dedicated access to each pass.

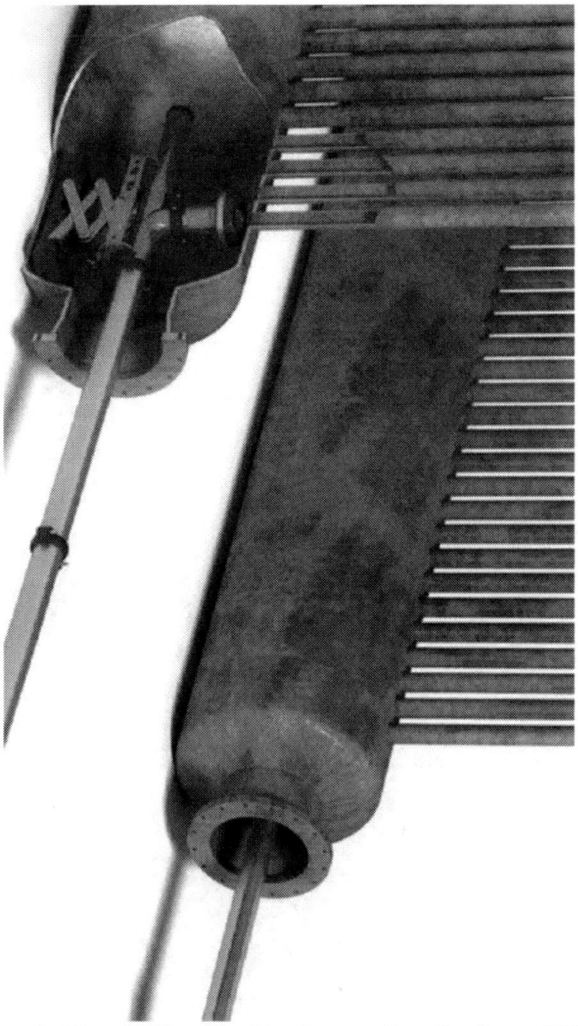

Figure 2. Steady Flux Tool in Convection Section of Piping

Because there are many different furnace designs, custom development is required for each individual tool. However, many plants have multiple furnaces with essentially the same mechanical design. Therefore, a single set of tools can often be utilized on many furnaces within a facility.

This tool is designed with both inspection and cleaning constraints in mind. If material contamination is a concern, the components of the tool can be manufactured from a number of different materials. This tool is also designed with both size and weight in mind to allow for easy manipulation in the field. By implementing this approach, access is granted for each individual pass without the need to cut or re-weld each of the branch connections.

Pipe Cleaning

Due to the sensitivity of the ultrasonic technology, it is important to clean the interior of the pipes prior to inspection. This is generally accomplished with either a mechanical or chemical process for the convection section, as the tubes are not visible and steam/air decoking techniques cannot be monitored. The pigging process was initially developed for pipeline cleaning and utilized large permanent structures for launch and retrieval along with pumps, storage tanks, etc. In the 1990's, this process was tailored to the small diameter pipes (3"-6"), which are common in furnaces. These advancements include self-contained trucks which include all the pumps, controls, tanks and filters. These trucks carry the necessary temporary piping/hoses and launch retrieval equipment. Additionally, special custom designed cleaning pigs were developed to address the unique challenges of varying density deposits on the internal pipe wall.

Figure 3. Self-Contained Mobile Pigging and Decoking Truck

Figure 4. Cleaning Pigs

Typically, water is used as the cleaning medium to push the pigs through the pipes. This water can be treated as necessary to avoid contamination issues. The dirty water can then be disposed of using the facility waste water treatment because nothing new has been added to the water that would not typically be present in the waste water system. Attention may need to be given to volume and treatment rate. However, this is typically not necessary with the volumes needed and because the water is recirculated during the process.

Inspection Process

Once the cleaning is completed, the same equipment is utilized for the inspection process. This minimizes down time because no additional mechanical equipment is necessary. The "smart pig" is introduced to the pipe the same way the cleaning pigs are introduced. When running the inspection tool, additional care is required with respect to pump operations to ensure the tool is pushed at the proper speed for the inspection process. This is generally slower than what is used for the cleaning tools. The inspection tool contains multiple ultrasonic transducers, batteries, on board signal processing and data storage.

Once the inspection tool completes a run through the piping circuit, the data is then immediately downloadable to a PC for additional processing by a certified ultrasonic technician. Looking from the inside of the pipe outward allows for measurement of both the radius and the wall thickness. The external heat transfer surfaces (fins or studs) do not interfere with the reading, as they are external to the tube itself. It is possible to occasionally see these appendages. However, this noise can be eliminated from the data to reduce confusion.

The data from the inspection tool provides the owner with both wall thickness and radius data for the pipe, while simultaneously providing information regarding the cleaning efficiency. If the pipes are not sufficiently clean, the inspection data quality suffers significantly. However, with the inspection tool data, the exact location of internal fouling can be reported to the cleaning company for additional cleaning. Reinspection can then be accomplished in a clean coil.

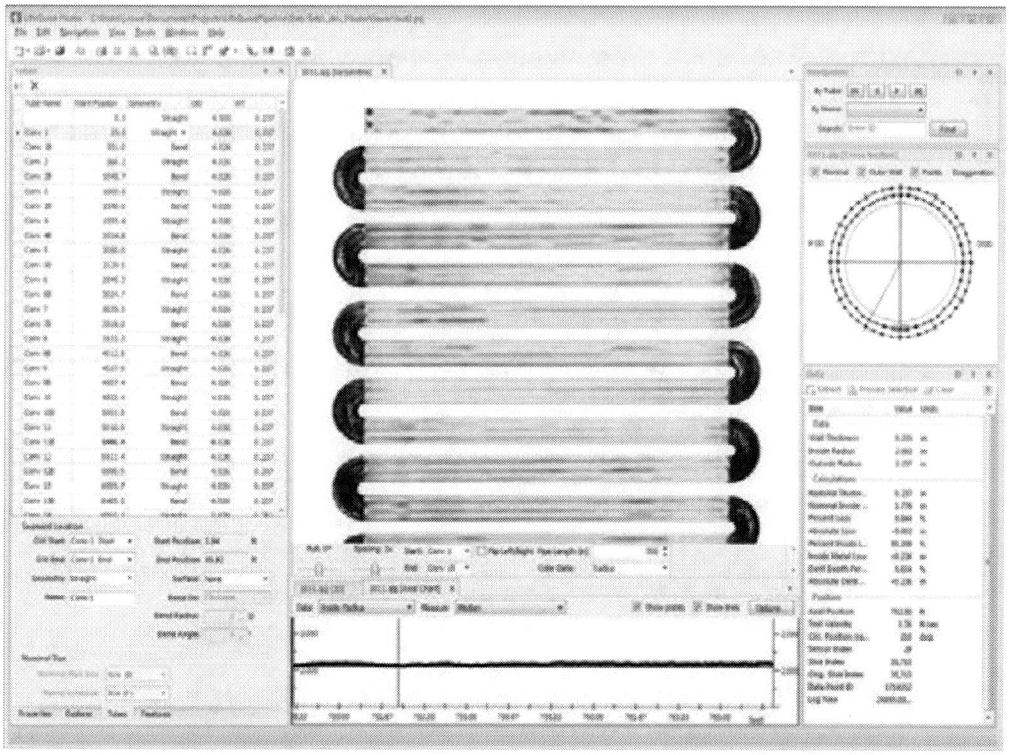

Quest Integrity Group
Proprietary Information
Patent Protected

Figure 5. Inspection Results

The typical inspection process will allow for complete 100% data collection of all convection pipes in a single furnace within a single shift. Once the inspection is completed and it is confirmed that the data is acceptable, the cleaning/pumping equipment can be de-mobilized from site allowing for other mechanical work to be completed. Inspection reports are generated using the data recovered from the tool, and access to the furnace is no longer required.

Inspection Results / Assessments

The inspection results are developed by a certified technician. The technician, with the aid of software, will analyze the data and report on the pipe wall thickness (internal or external corrosion or erosion) as well as the radius (swelling, bulging, denting, ovalizing). These results can then be compared to the facility's allowable thickness criteria, which allows focused repair or replacement of individual pipes. Additionally, enough data is often collected to allow for full API 579 Fitness-for-Service (FFS) engineering assessments in accordance with Part 4, 5 and 10 (General Metal Loss, Local Metal Loss and Creep assessment). From these evaluations, future inspection intervals can be set, repair/replacement plans can be made and the asset risk of failure can be better understood and managed.

Case Study

The following is a good example of a convection section we were asked to inspect at the last minute. This particular heater was scheduled for routine cleaning of the convection section. Upon cleaning, it was discovered that it was "raining" inside the heater. The source of the water was the cleaning water, but the exact location and extent of damage was unknown.

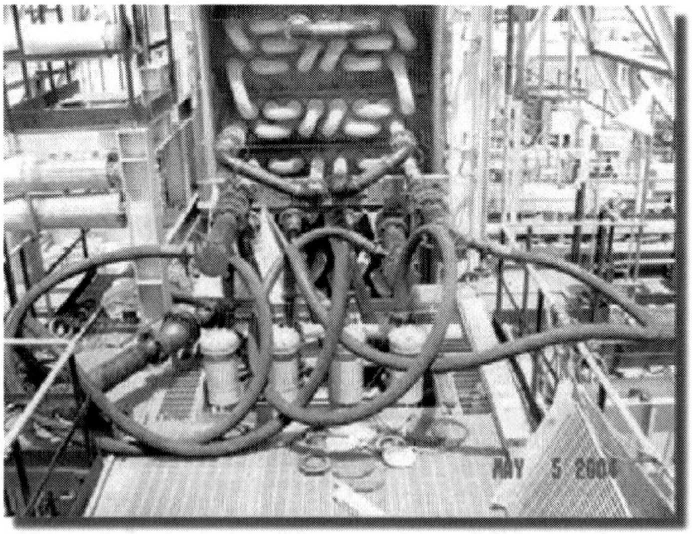

Figure 6. Convection Section Set to Pig

An inspection of the convection section was quickly arranged with no additional piping or modification required. The inspection resulted in the following wall thickness data.

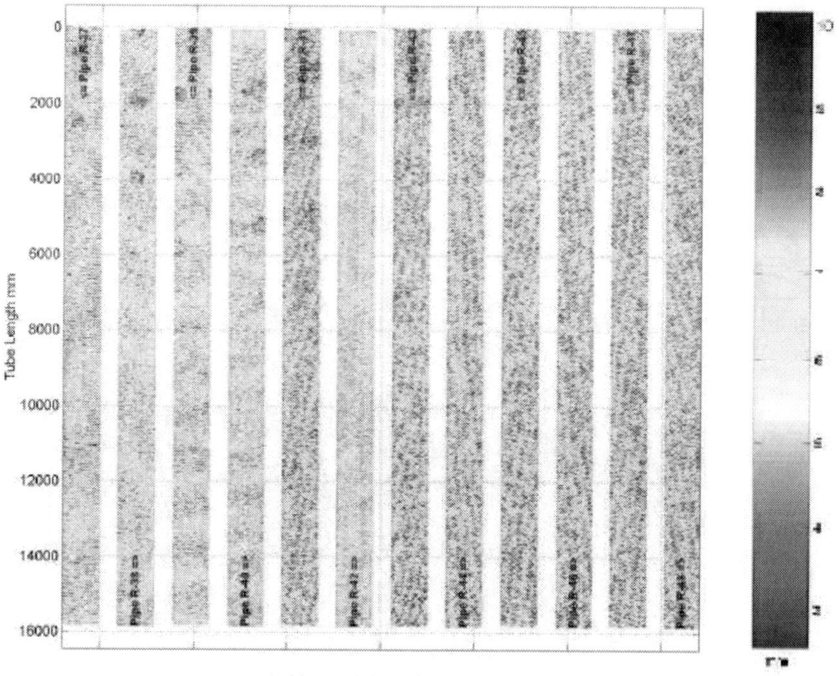

Figure 7. Inspection Results

With some careful study a pattern can be seen with respect to the damage location, red areas.

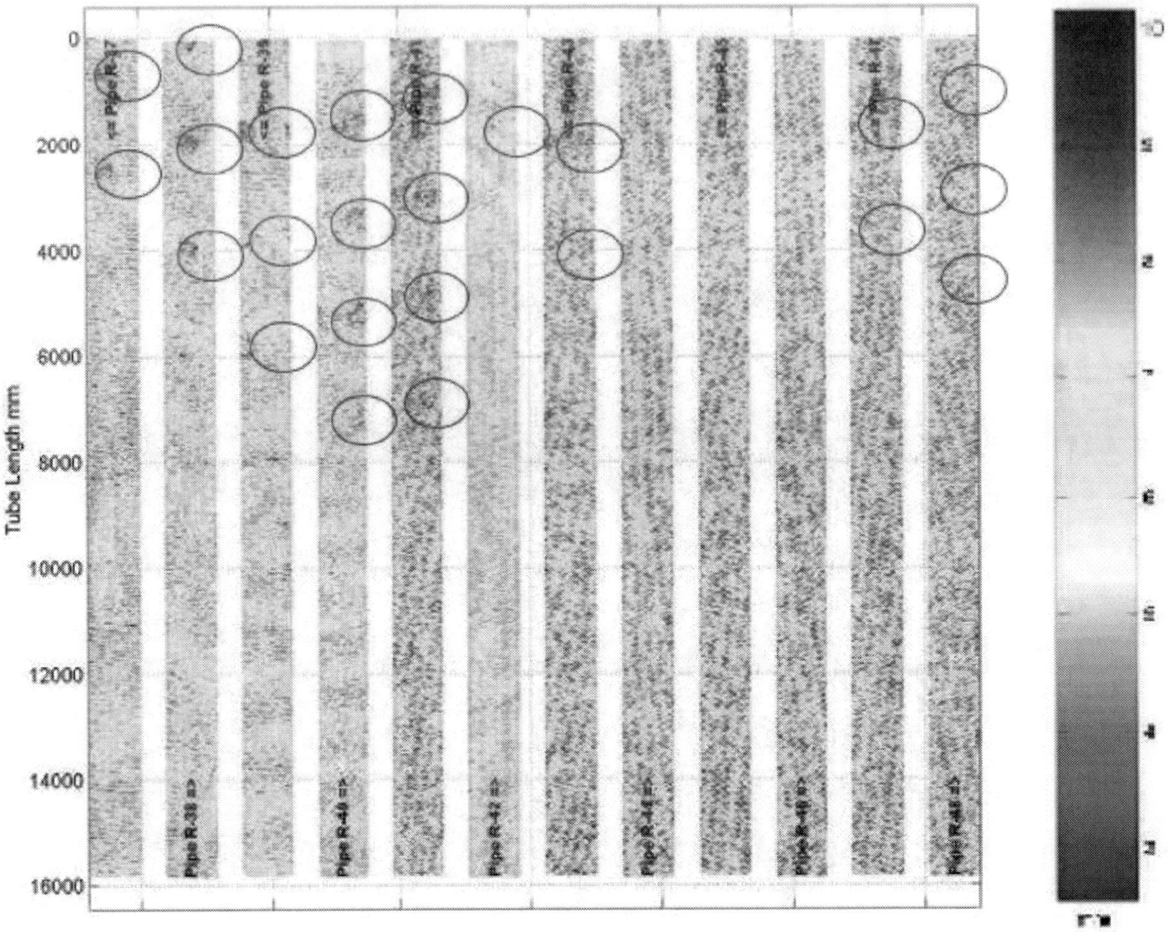

Figure 8. Inspection Results

The patters were then overplayed with the fabrication drawings and it was determined that the location corresponded to a soot blower installation. The subsequent root cause was determined to be a faulty valve on a soot blower allowing for continuous water leakage. This water combined with the sulfur produced form the combustion process to create sulfuric acid, which then caused a hole to develop.

Figure 9. Field Findings

Identifying this damage allowed for rapid repair and return of the unit to normal operation.

AIChE PAPER NUMBER 127E

CASE STUDY ON INSPECTION OF ETHYLENE FURNACE RADIANT TUBES

Kevin Stevens
Senior Consultant
Quest Integrity NZL Ltd

Charles Thomas and Andy Saunders-Tack
Quest Integrity NZL Ltd

Amy McLeod, Catherine Bishop and Milo Kral
University of Canterbury, Christchurch, New Zealand

Bob de Lorenzo
Quest Integrity LLC

Prepared for Presentation at the 2016 Spring National Meeting
Houston, Texas April 10 – 14, 2016

CASE STUDY ON INSPECTION OF ETHYLENE FURNACE RADIANT TUBES

Kevin Stevens
Senior Consultant
Quest Integrity NZL Ltd

Charles Thomas and Andy Saunders-Tack
Quest Integrity NZL Ltd

Amy McLeod, Catherine Bishop and Milo Kral
University of Canterbury, Christchurch, New Zealand

Bob de Lorenzo
Quest Integrity LLC

Abstract: A tube crawler has been developed to quantify the carburization in 25%Cr-35%Ni alloy tubes used in furnaces for ethylene production. Eddy current probes are used to detect the increase in magnetic permeability associated with chromium depletion in the alloy matrix. Diameter measurements detect the swelling associated with carburization, which means the crawler can also be used on higher alloy tubes. The eddy current probes can also detect cracking along the brittle chromium carbide phases. The design, physical set-up and operation with National Instrument's Labview of the crawler will be described, along with the data analysis procedure in Mathworks Matlab and Quest Integrity's LifeQuest reformer software, typical results and reports.

Remaining life assessments are made using Abaqus Finite Element Analysis (FEA) of the coil and hanger geometry, and changes in temperature during startup, typical production runs, and decoke cycle. The FEA incorporates plastic and creep material properties. Mechanical testing of ex-service HP 50 alloy has been undertaken to improve a material property data-base for carburized material. Work is underway on mechanical testing of ex-service 35Cr-45Ni alloy. The mechanical test data correlates the crawler magnetic permeability measurements with tube hardness, ultimate tensile strength, 0.2% yield stress, and elongation to failure under tensile testing and creep testing regimes. Optical and electron microscopy has been used to characterize the ex-service material, with Energy Dispersive Analysis of X-rays (EDAX), image analysis and Electron Back Scattered Diffraction used to quantify the carbide phases through the tube wall, which can be correlated with the magnetic response and tube swelling. Finite Element modelling of the carbon diffusion and carbide precipitation using phase equilibria predicted by Thermocalc for HP Nb alloy, is being used to incorporate damage accumulation rates caused by the carburization into the global FEA models.

Introduction

Ethylene is a feed stock used for producing polyethylene plastics, anti-freeze used in automotive engines, and surfactants and detergents. Global ethylene production is in excess of 140 million tonnes [1] per year and it is produced by the hydro-cracking of naphtha, hydro wax, mixed gas-oil or mixed gas within radiant furnace tubes with temperatures typically between 1000 and 1120°C. The tube alloys are typically HP alloy with a nominal composition of 25%Cr-35%Ni-1%Nb-balance Fe, or a higher alloy of 35%Cr-45%Ni-balance Fe. The alloys are selected for carburization and oxidation resistance, along with mechanical strength and creep resistance at elevated temperatures. The diffusion of carbon from the inside of the tubes causes the precipitation of brittle chromium rich carbides and a loss of ductility. Coke formation inside a tube causes a reduction of flow rate that makes it necessary to undertake a temperature cycle down to 950 to 1000°C every 20 to 50 days (depending on the feedstock and age of the tubes), with the flow of a steam/air mixture used to burn the coke off. During the decoke cycle, the higher thermal expansion coefficient tube alloy will contract around an intact coke plug with a lower expansion coefficient, ideally causing the coke to spall off the tube, but it also creates higher stresses and creep and potentially plastic strains in the tube [2, 3]. Attempts to weld repair tubes can be impossible, if the tubes develop micro-cracks during the thermal cycles up to in excess of 1800°C due to the loss of ductility due to carburization [4].

The radiant tubes typically last for 4 to 6 years in service, before being replaced at planned outages. However some operators run furnace tubes until failure occurs, and then replacement needs to be mobilized at short notice. A set of tubes for a furnace can cost approximately US$1.6M, and have to be ordered 6 to 12 months in advance. With ethylene costing approximately US$900 to 1,300 per tonne, and a margin of US$300 per tonne [5], with a typical furnace producing 60,000 to 80,000 tonnes of ethylene per year, the production from each furnace represents revenue margin of US$10M to 20M/year. The economics necessitate decisions to be made about the length of time tubes are kept in service before replacement, and the risk profile of the plant. Plants often have scheduled maintenance outages at which it is possible to inspect furnace tubes for carburization levels using eddy current techniques [6-11] or diameter measuring systems mounted onto tube crawler robots, without the use of scaffolding. Conversion of the carburization measurements into a remaining life estimate requires an in-depth understanding of furnace geometry and hangers, operating temperatures, decoking cycles, mechanical strength of the tube alloy and coke, and creep properties of the tube and mechanical property changes with carburization. The only way to meaningfully bring these factors together in a model is by the use of FEA techniques. This paper summarises the inspection and assessment methodology developed at Quest Integrity.

Initial data collection and review

The first stage of the process is to assess the feasibility of an inspection by confirming details about the furnace, including:

- furnace geometry drawing and manufacturer. Tube lengths, diameters and tube to tube spacings are checked to confirm that a crawler can navigate the tubes. Current crawlers cover the range of 98.3 mm (3.87") to 182.1 mm (7.17") outside diameter, but a smaller tube crawler to cover down to 2.5" has been designed.
- material alloys on the various tube passes. Depending on the alloy type we use Eddy Current Testing (ECT) to measure carburization (or just for crack detection) or just diameter measurements.
- sizes of trenches, clearances at bottom or top of furnace that affects crawler automatic collision avoidance/stop mechanism.
- tentative shutdown dates and duration.
- Health & Safety requirements.

If an inspection is feasible, then a scope is suggested that defines the number of furnaces and tubes for inspection, alloy type, deployment considerations, and options for a more detailed life assessment using Finite Element Analysis (FEA) and testing of ex-service material if available. This approach allows plant operators to assess the economics of inspection and replacement based on tube condition vs. replacement on a time or plant history based schedule, or at a failure.

More detailed data collection and review

If an inspection and FEA work is to go ahead then more detailed data is required for the remnant life assessment. Some of this data can be collected during a site visit for the inspection. The data reviewed includes:

- Hanger settings and types, guide tube locations and clearances.
- Temperature data e.g. thermocouple or infrared pyrometry surveys. Ideally temperatures provided on all tubes and passes at different stages of the decoke cycle, showing variations with the height in furnaces and feedstock.
- Burner locations.
- Gas pressure variations across tube passes.
- Length of service of furnace tubes, number of cold and warm starts, and shutdowns.
- Inspection history including any baseline data for tube diameters, otherwise rely on nominal manufacturer tube diameters or diameters at least carburized locations.
- any metallurgical reports on tube failures.
- History of tube replacements, blockages, tubes touching furnace floors and/or failures.
- Ideally a survey of cold and hot hanger settings/displacements and guide tube displacement helps to validate the FEA model. Surveys several years apart would show the axial creep. Details on any hanger adjustments during service.
- Collection of drawings of elbows and reducers not shown in sufficient detail in global furnace geometry.
- Axial length of the tubes to validate mean creep strain rates used in FEA.
- Weld geometry and procedure if required.

- Current plant tube replacement strategy, e.g. based on diametric strain at bottom of furnace, tube axial elongation, past plant experience with specific alloys etc.

Inspection process

The successful completion of an on-site inspection needs to consider and include:

- H&S, checking and completion of training certificates (confined spaces etc), job safety analyses, site specific induction, PPE requirements and visas etc.
- The shutdown length, furnace availability and shift length with advance notice of the shutdown required to confirm field technician availability with site deployment of 2 or 3 person crew. Typically an inspection can be completed in 1 to 2 days working a normal shift, but a larger crew can be deployed to work 24 hours if required.
- Power (cable length to sockets, plug types, and wattage, safety transformers) and lighting etc. Protection of electrical equipment from rain etc.
- Access for a laptop which is used to control the tube crawler using Labview. Data is viewed in real time on strip-charts on the laptop screen.
- Data analysis and plotting of raw diameter and ECT data in LifeQuest Reformer software and Matlab. Production of 3d contour plots of tube diameters up tube height. Calculation of maximum diametric and elliptical strains on each tube, and listing in tables.
- Fitting of an electromagnetic FEA model at multiple frequencies to ECT data for 25%Cr-35%Ni alloy tubes to determine the carburization thickness. Examination of plots for any evidence for cracking in ECT data.
- Typically need to deliver formal report within 1 week of inspection. Hold site close out meeting if possible to discuss data plots before leave site.
- Optional inspection and condition assessment of convection section tubes using ultrasonics.

Figure 1 shows an animation of the inspection process, with the 2 halves/shells of the crawler being clipped together around the tube in Figure 1(a), then driven to the top of the furnace tubes as shown in Figure 1(b), using Labview on a laptop communicating with the crawler via a long usb cable.

Figure 1(a)

Figure 1(b)

Figure 1: (a) Crawler being placed on the tubes, (b) crawler near the top of the furnace, see full animation at http://www.questintegrity.com/services/inspection-services/steam-reformer-inspection/.

Figure 2 shows typical three dimensional contour plot results for the tube diameters in a in a 5 pass Technip design furnace after 13 years, with the diameter increased from a nominal 110.0 mm for new tubes. The tubes are typically 6m long. Two tubes have a much lower diameter of 113 mm, because they have been cut and replaced.

The resulting inspection report typically contains figures for every tube showing the tube diameter and the ECT data. Typical data for a 35%Cr-45%Ni alloy tube is shown in Figures 3 and 4. The change in diameter caused by the welds is clearly visible at 2.2m in

Figure 3. The most bulging/ellipticity is aligned with the eddy current probes on the hottest side of the tubes facing the gas burners. The ECT data in Figure 4 shows no evidence for cracking along the grain boundary carbides. The increase in normalized probe inductance (L/L0) shown in Figure 4 indicates an increase in magnetic permeability associated with carburization. The calibration is different for 25%Cr-35%Ni alloys compared to 35%Cr-45%Ni alloy tubes [6].

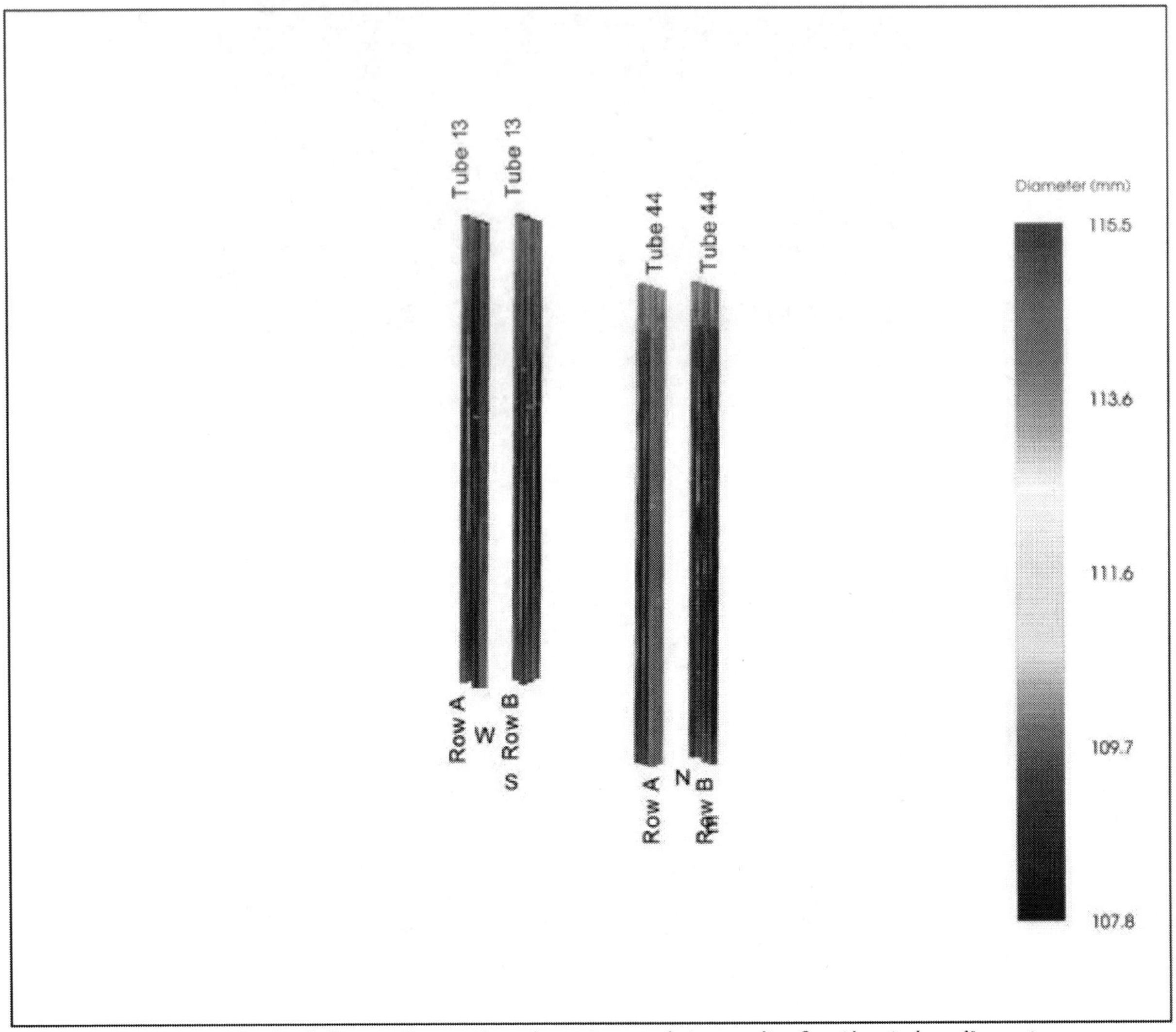

Figure 2: Typical three dimensional contour plot results for the tube diameters of the 5th pass tubes in a 5 pass furnace after 13 years of tube service.

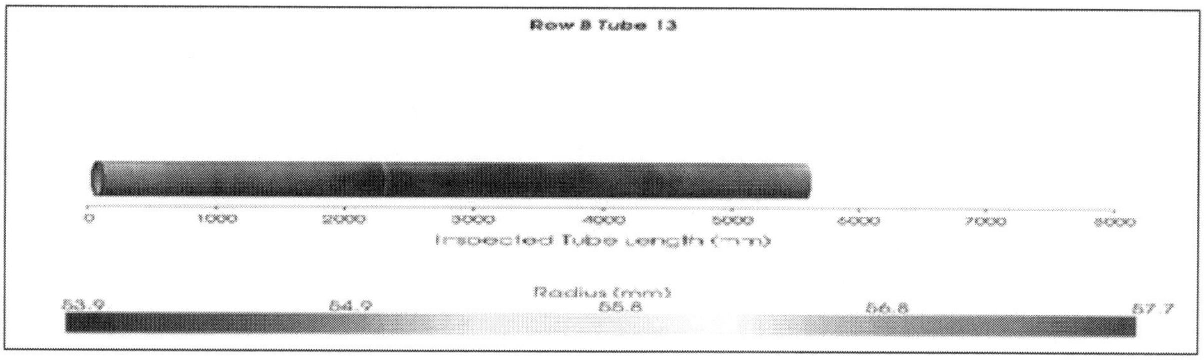

Figure 3(a)

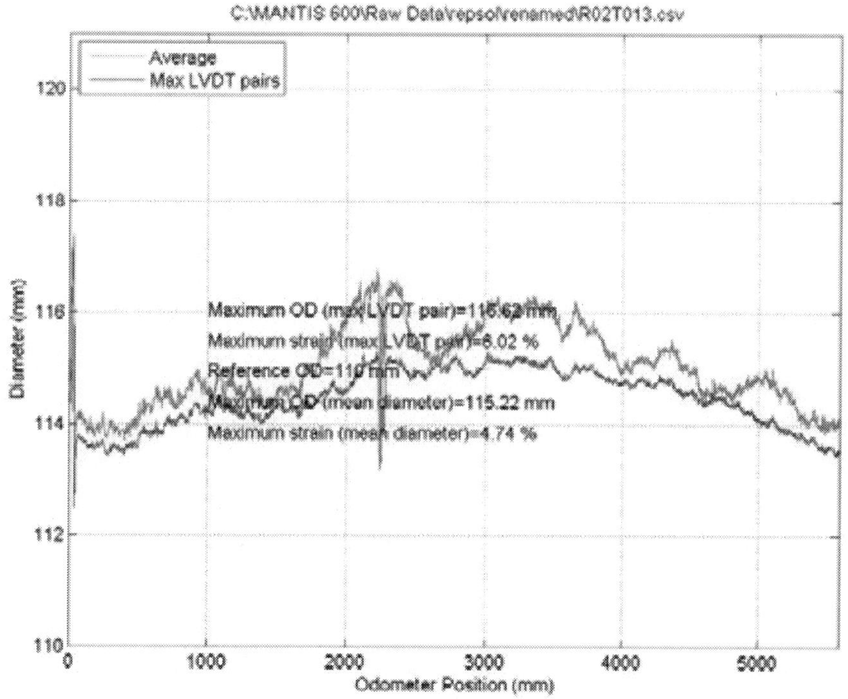

Figure 3(b)

Figure 3: (a) 3D view of radius, (b) XY diameter plot for row B, pass 5, tube 13, showing the average and maximum diameters and strains.

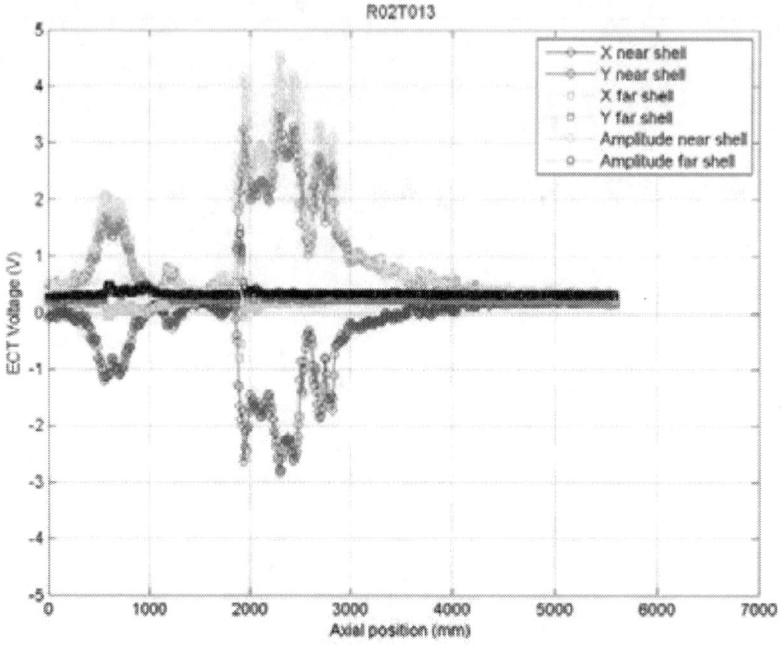

Figure 4(a)

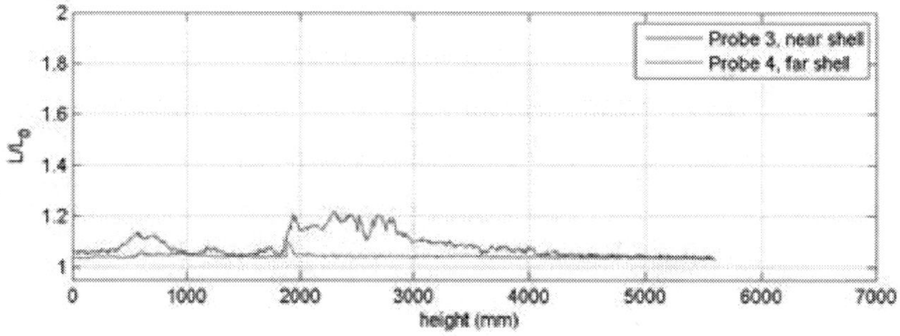

Figure 4(b)

Figure 4: Row 2, tube 13, (a) ECT voltage scans at 100 Hz; (b) normalized probe inductance (L/L0) changes at 100 Hz. The near shell probe is on the hotter side of the tubes facing gas burners. The far shell probe is between the tube rows.

Figure 5 shows variations in mean and elliptical strain for the pass 3, 4 and 5 tubes. Elliptical strains have been measured on the 5th pass tubes as high as 8.7%. The data shown provides an obvious grading of the respective conditions of the tubes, and prioritization for replacement.

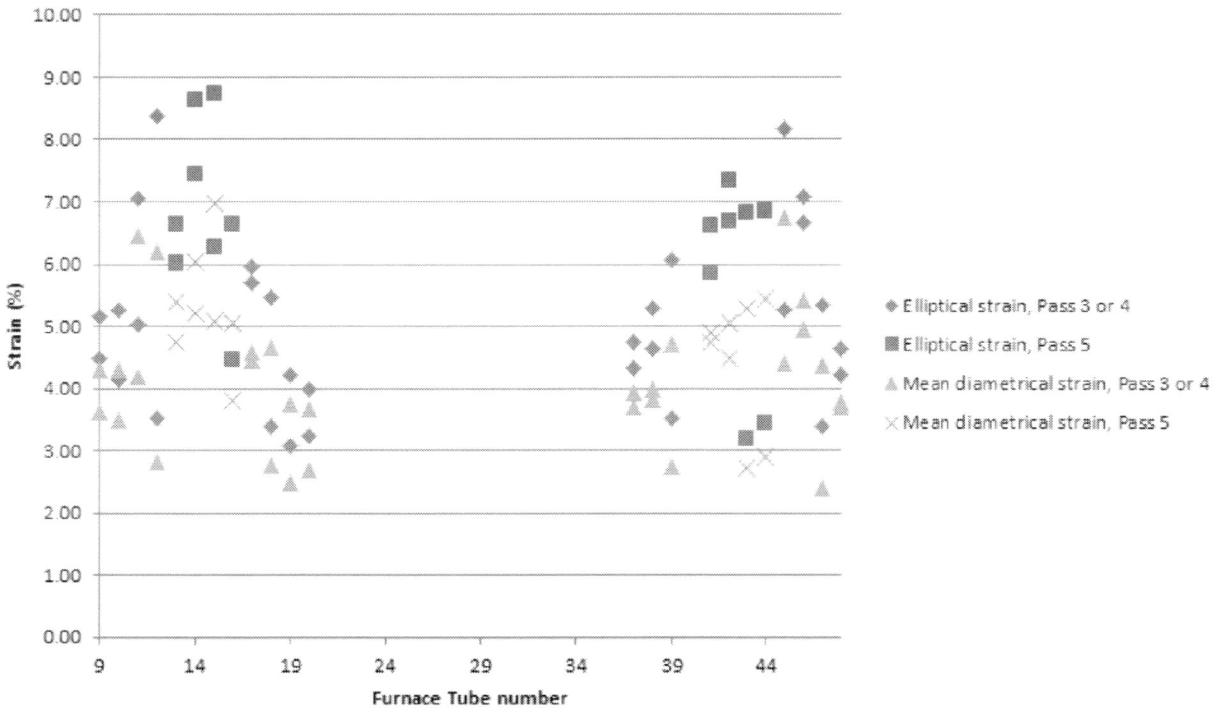

Figure 5: Strain calculated from the LVDT's using the maximum LVDT pair ("elliptical profile") or the mean diameter. The horizontal axis is the furnace tube number, e.g. tubes 1 to 8, 21 to 36, and 49 to 56 are the inlet pass 1and 2 tubes of nominal OD 71.0 mm which were not inspected; tubes 9 to 12, 17 to 20, 37 to 40 and 45 to 48, are the pass 3 and 4 tubes of nominal OD of 97.0 mm; tubes 13 to 16 and 41 to 44 are the pass 5 tubes of nominal OD 110.0 mm. There are 2 data points for each tube number as there are 2 banks of tubes.

Detailed remnant life assessment

There have been only a limited number of publications on the life assessment of ethylene pyrolysis tubes. Several authors [12-15] have published FEA studies of the stress and strain state in ethylene pyrolysis tubes. Several methods have been used to incorporate the effects of carburization damage accumulation upon creep strain rate or to employ Kachanov-Rabotnov type equations in steam reformer tubes to model the creep damage and combine it with the carburization front propagation in ethylene pyrolysis tubes. The Quest Integrity approach is to model the diffusion and carburization process [16] and adjust the creep strain rates. The FEA model in Abaqus CAE can be commenced before an inspection, but completion of the FEA can typically take 1 to 2 months, and includes:

- Construction of the geometry and meshing.
- Rainflow counting of temperature data, to calculate the maximum temperatures, decoke cycle temperature, duration of heating and cooling, and ramp rates.
- Literature search for new materials properties for the temperature dependence of the coefficient of thermal expansion, Young's modulus of elasticity, ultimate tensile strength, 0.2% yield strength, elongation to failure, parametric creep stress rupture strength, and parametric minimum creep rate.

- Mechanical testing of ex-service material, if available, to measure hardness, remaining ductility, yield stress, ultimate tensile strength and creep strain rates at various temperatures across the tube wall.
- Extraction of the rate of creep strain accumulation and diameter swelling and ellipticity due to carburization for the remnant life assessment. A global model is used to predict the axial creep strains and bending stresses, and a sub-model is used to predict the creep strains created by a decoke cycle.
- Sensitivity of the results to decoke cycle and coke mechanical properties.
- Microstructural characterization of ex-service material using optical and electron microscopy to show carbide types and distribution, and chromium depletion in the matrix which is correlated with an increase in magnetic permeability.
- FEA modelling of welding process (if requested).

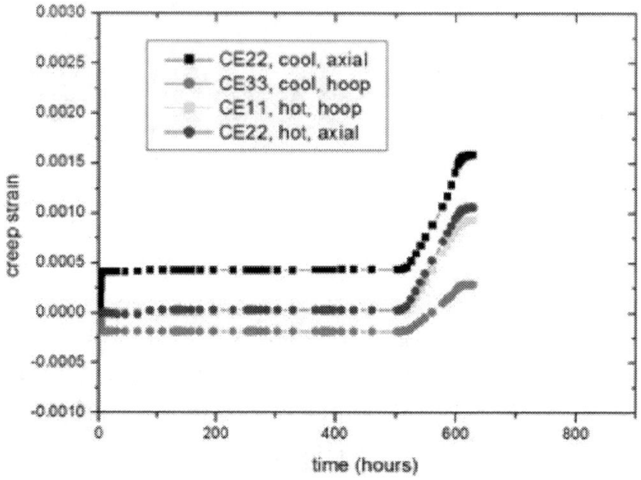

Figure 6: Equivalent creep strains for coke with a Young's modulus of 20GPa, with 1000°C at start of run, 1100°C end of run, 1000°C decoke, coke tensile strength of 1 MPa, compressive strength of 45 MPa. 25°C temperature gradient around the tube. 500 hour production run followed by a 100 hour decoke cool.

Figure 6 shows the creep strains predicted by the FEA in the hottest section of an outlet tube accumulated by a cold start followed by a decoke cycle, for a Lummus SRT VI design furnace which has a 2 pass geometry with 4 inlet tubes per outlet tube. Strains are summed to a volume swelling caused by carburization, and accumulated over multiple decokes and compared to strain limits to estimate a remaining life.

Conclusions

Typical results for a carburization inspection have been presented, and the Quest Integrity remnant life estimation methodology has been outlined. The reports produced can be used by plant operators to improve the plant tube replacement strategy.

Acknowledgements

The authors would like to thank the New Zealand Ministry of Science and Innovation and Callaghan Innovation for a PhD scholarship for A. McLeod.

References

1. Special report "Global ethylene capacity continues advance in 2011", Oil and Gas Journal, Vol. 110, Issue 7, 2011.
2. D. Jakobi and R. Gommans, "Typical failures in pyrolysis coils for ethylene cracking", Materials and Corrosion, 54(11), pp881-886, 2003.
3. R. Gommans, S&C 3rd European Ethylene Symposium, 21-22 May 2001, Numbrecht.
4. B. Li and G. Lai, "Automatic Orbital GTAW for Repair Welding of Ethylene Furnace Tube Coils", NACE Paper C2012-001235, NACE Conference 2012.
5. Special report "US Olefins first half 2013", Oil and Gas Journal, Vol. 111, Issue 9, 2013.
6. K.J. Stevens, A.J. Tack, C.W. Thomas and D.M. Stewart, "Through-Wall Carburization Detection in Ethylene Pyrolysis Tubes", Journal of Physics D: Applied Physics 34, 2001, p814-822.
7. K.J. Stevens and W.J. Trompetter, "Calibration of Carburization Measurements in Ethlyene Production Tubes using Ion Beam Analysis", J. Phys. D:Applied Physics 37 (2004), p501-509.
8. McLeod, A.C.; Bishop, C.M.; Stevens, K.J. and Kral M.V. Microstructure and Carburization Detection in HP Alloy Ethylene Pyrolysis Tubes, Metallography, Microstructure and Analysis, 4 (4): 273-285, 2015.
9. K.J. Stevens, A. McLeod, C.W. Thomas, A. Saunders-Tack, B. de Lorenzo and M.V. Kral, "Inspection and Remnant Life Assessment of Ethylene Cracker Tubes", Ethylene Middle East Technology Conference Oct 2014, Bahrain.
10. S.H. Khodamorad and D.F. Haghshenas, "Inspection of carburization and ovalness in ethylene cracking tubes by using a semi-robot", Engineering Failure Analysis 25, pp81-88, 2012.
11. I.C. da Silva, R.S. da Silva, J.M.A Rebello, A.C. Bruno and T.F. Silveira, "Characterization of carburization of HP steels by non destructive magnetic testing", NDT&E International 39, pp569-577, 2006.
12. C. Zhou and T. Zhao, "Creep stress calculation model for ethylene cracking furnace tubes under consideration of carburization", J. Pressure Equipment and Systems 3, pp131-135, 2005.
13. L.M. Shen, J.M. Gong and Y. Jiang, "Numerical simulation of damage for Cr35Ni45Nb steel ethylene cracking furnace tube at action of coupled multiple factors", Journal of Materials Engineering, Vol. 12, pp77-82+88, 2012 (in Chinese).
14. M.L. Zhu, Y. Yin, T. Wang, S.T. Liu and J.S. Zhang, "Numerical simulation on stress and creep damage fraction in ethylene cracking tube", Journal of Dalian University of Technology, 41(2), pp192-196 (2001) (in Chinese).
15. J.M. Gong, S.T. Tu and K.B. Yoon, "Damage assessment and maintenance strategy of hydrogen reformer furnace tubes", Engineering Failure Analysis 6 (1999), p143-153.

16. A. Chauhan, M. Anwar, K. Montero, H. White and W. Si, "Internal Carburization and Carbide Precipitation in Fe-Ni-Cr Alloy Tubing retired from Ethylene Pyrolysis Service", J. of Phase Equilibria and Diffusion, Vol. 27, No. 6, p684-690, (2006).

AIChE PAPER NUMBER 127F

IMPROVING COMBUSTION EFFICIENCY, RELIABILITY, AND SAFETY IN LARGE FIRED HEATERS UTILIZING MULTIPLEX LASER BASED MONITORING AND DAIGNOSTIC SYSTEMS

Dr. Andrew Sappey

Chief Technology Officer

Zolo Technoloiges Inc.

Greg Heilbrunn

Vice President Industrial Markets

Zolo Technologies, Inc.

Prepared for Presentation at the 2016 Spring National Meeting
Houston, Texas, April 10 – 14, 2016

IMPROVING COMBUSTION EFFICIENCY, RELIABILITY, AND SAFETY IN LARGE FIRED HEATERS UTILIZING MULTIPLEX LASER BASED MONITORING AND DAIGNOSTIC SYSTEMS

Dr. Andrew Sappey
Chief Technology Officer
Zolo Technoloiges Inc.

Greg Heilbrunn
Vice President Industrial Markets
Zolo Technologies, Inc.

Abstract: Operators are often challenged to maintain well-balanced, uniform combustion across a large fired heater in order to optimize efficiency, reliability, production yields, and safety and to reduce emissions. While most large furnaces and heaters have several individual sensors available to monitor temperature, O2, and CO, typically at the furnace exit, the number of such sensors available for use directly inside the firebox is very limited. This paper describes a novel but well-proven wavelength-multiplexed, laser-based monitoring and diagnostic technology that measures temperature, oxygen, and carbon monoxide directly in the high temperature zone of combustion furnaces and heaters. Because of the fiber-coupled nature of the system, multiple locations (up to 24 lines of sight) can be measured with a single system.

Utilization of wavelength-multiplexed tunable diode laser absorption spectroscopy (TDLAS) laser systems for continuous reporting of multiple combustion gases across the radiant and convective zones of a combustion process represents a first-of-its-kind application for TDLAS technology. It allows operators to tune burner controls to achieve fuel savings, increase production yields, lower excess O2, monitor CO, reduce NOx and CO2 emissions, and extend furnace tube/coil life through well-balanced temperature and oxygen profiles across the entire envelope of the heater.

This paper describes the design, use, and value of wavelength-multiplexed TDLAS laser technology as applied to large furnaces.

Introduction:

Ethylene cracking furnaces use a significant amount of energy to convert a feedstock of ethane, LPG or naphtha into ethylene and propylene. The cracking process yields are very sensitive to a number of critical factors including propylene/ethylene ratios; conversion rates, Coil Outlet Temperature (COT), residence time, and coking rates. Balanced combustion in the furnace can have a large impact on many of these factors and therefore the profitability of the plants. A typical ethylene furnace has sensors installed in different areas for the purpose of monitoring the combustion process. However, the number of sensors available for use directly inside the fired heater is very limited. Historically, the lack of measurement data directly in the combustion zone makes it very difficult to optimize temperature, O2, and CO profiles for combustion and process improvements in the combustion zone.

Imbalances in temperature, O2, and CO across the furnace can adversely affect ethylene production as well as combustion and process efficiencies. Production run times can be limited, variability in process and coil outlet temperatures can reduce process yields, and fuel efficiency and emissions can be adversely affected.

Until recently, there was no means to simultaneously measure key combustion constituents directly inside the furnace. This paper addresses the benefits of *in-situ,* real-time combustion monitoring to maximize production yields and runtimes, reduce energy usage, improve asset reliability and safety, and reduce emissions.

Overview and Architecture Of Multiplexed TDLAS Laser Systems

Utilization of multiplexed Tunable Diode Laser Absorption Spectroscopy (TDLAS) has provided *in-situ* combustion measurements in various industrial applications including coal-fired power plants, steel making, and chemical and refinery applications since 2006 with over 70 installations worldwide. Multiplexed TDLAS is well proven, accurate and reliable and is designated as a best practice by API for large fired heater applications.

Multiplexed TDLAS combustion diagnostic systems simultaneously measure the average temperature, O_2, and CO in real-time, directly in an ethylene cracking furnace along multiple laser paths. There are no probes to insert, no sensitive electronics near the furnace and no in-field calibration required. These real time measurements provide a more accurate and reliable representation of the furnace conditions than traditional single-point thermocouples, tube temperature measurements, ZrO_2 sensors and traditional TDLAS sensors as illustrated in Figure 1.

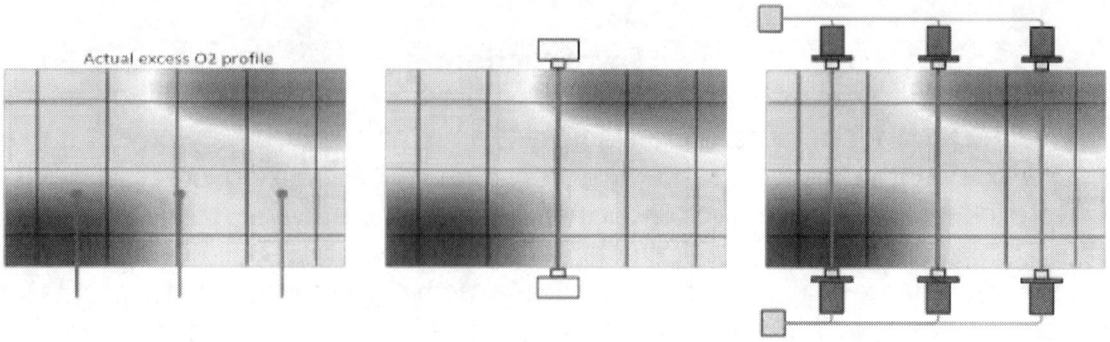

Figure 1: Multiple paths provide a better representation of the actual combustion zone profile

Multiplexed TDLAS Systems combine multiple lasers onto a single optical fiber. In the case of the most recently advanced multiplexed TDLAS system, pairs of laser heads incorporating automatic alignment are installed in strategic locations along the furnace walls to send (pitch) and receive (catch) the laser signals along a line of sight. The combined light is sequentially transmitted to each of the paths where a pitch head sends the laser light across the furnace. Light is collected by a catch head and routed back to the control rack via optical fiber. In the control rack the combined laser light is separated so that the light absorbed by each constituent is measured to determine the average concentration of O_2, CO and temperature along each laser path. Multiple paths are arranged to provide combustion information corresponding to the combustion control parameters for each zone in the furnace.

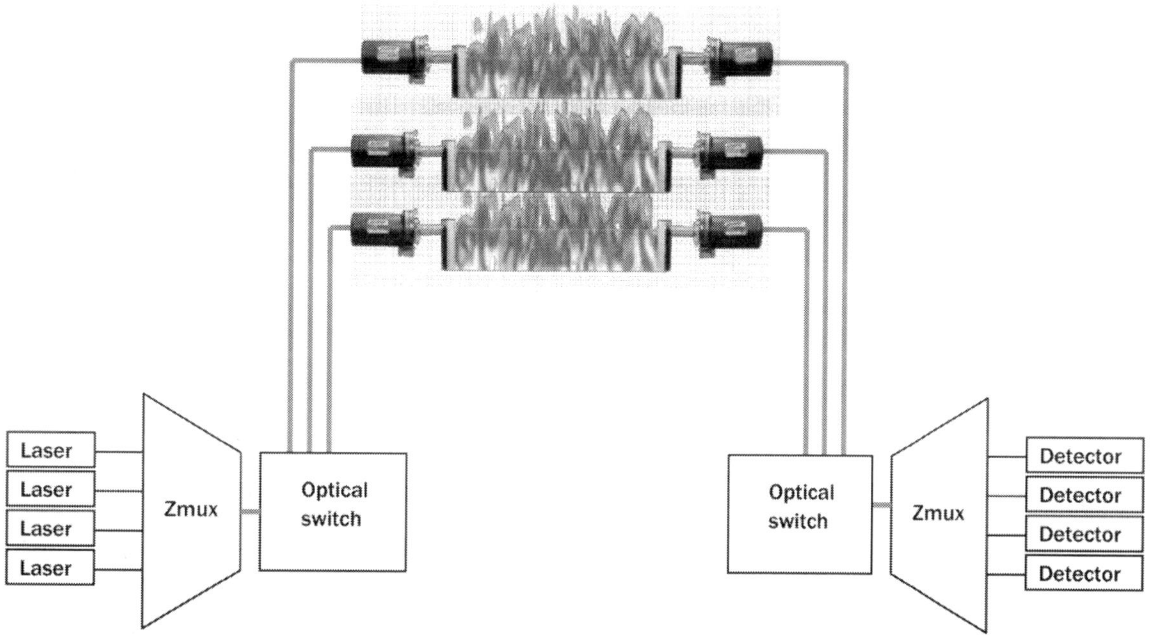

Figure 2a: Schematic of a multiplexed TDLAS system

The control rack for this system is typically NEC Class 1, Div 2 and ATEX compliant and can be located away from the furnace. The control rack houses all of the sensitive electronics such as the lasers, detectors and computer as shown in Figure 2a above. The automatic alignment heads mounted to the side of furnace do not require any cooling nor purge air and automatically maintain alignment of the laser beam across the furnace for high reliability and accuracy.

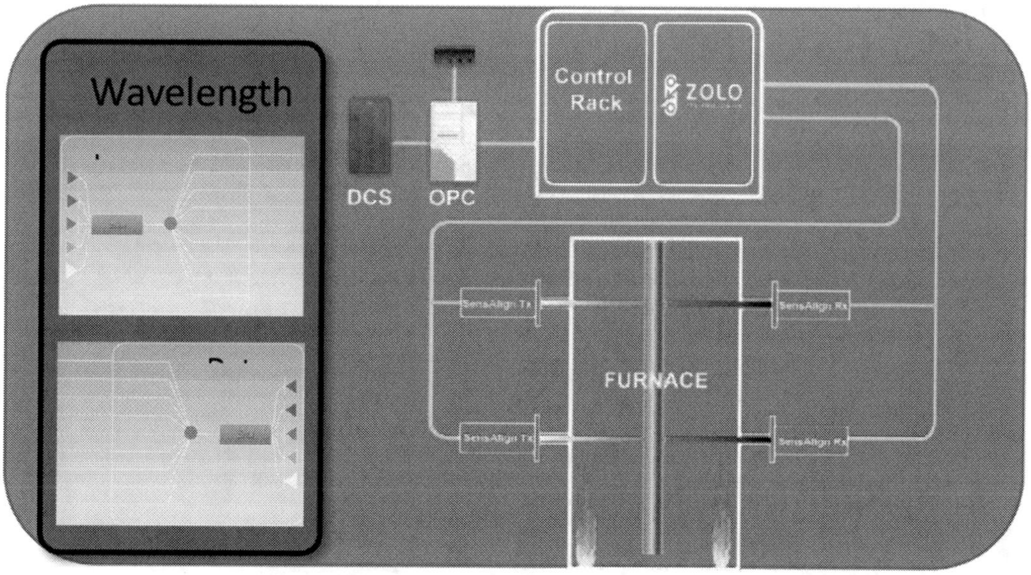

Figure 2b: Multiplexed TDLAS System Architecture

The optimal layout for measurement paths depends upon the specific burner configuration and controls, process tube/pass configuration and spacing, and the process and combustion monitoring objectives of the specific cracking furnace. A typical layout is shown in Figure 3 with path locations generally at an elevation in the upper furnace; paths over each row of burners and several cross paths to provide balance information. In addition, a single path may be located at the exit of the radiant section into the convective section to provide a more representative measurement of overall excess O_2 than a single point O_2 sensor. An example is shown below in Figure 3.

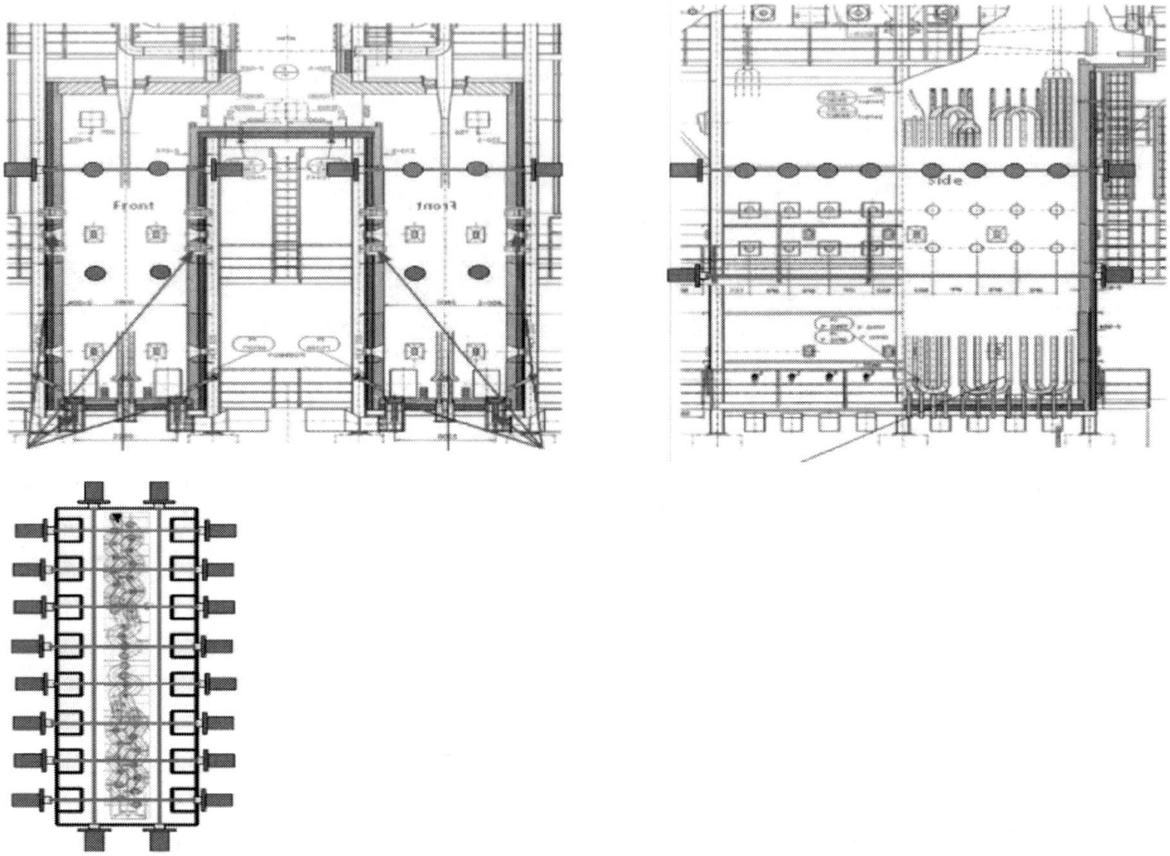

Figure 3: Typical multiplexed TDLAS path layout for an ethylene cracking furnace

Each system can accommodate up to 24 measurement paths from a single control rack. This provides a cost effective means to provide multiple path measurements from multiple adjacent furnaces (chambers/cells) with a single multiplexed TDLAS system.

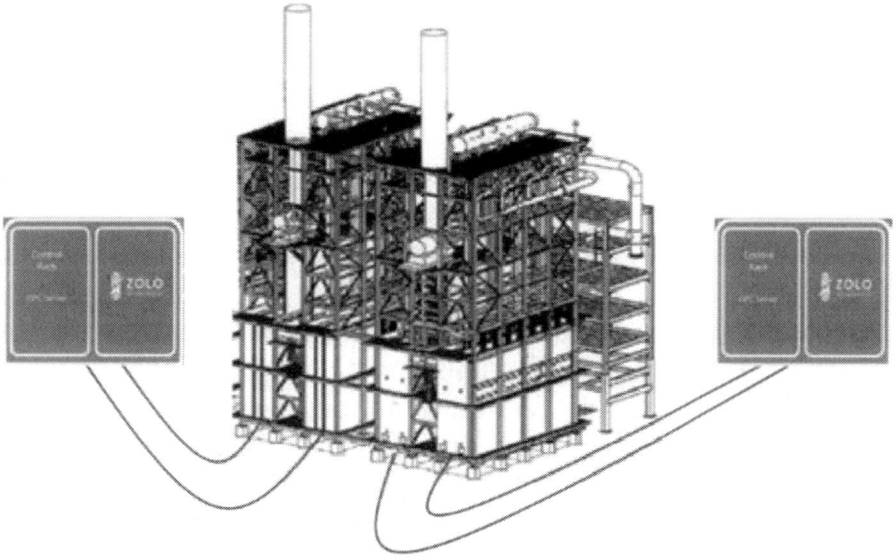

Figure 4: A single system can service multiple adjacent furnaces.

The system interface provides real-time balance information for each measurement path across the cracking furnace. Measurement rates range from 3-10 seconds/path, depending on automatic path alignment requirements, with a complete profile of the furnace updated typically within a few minutes, depending on the number of paths. Operators can use information from the system interface to make adjustments to the air/fuel ratio of the burners to improve the temperature and oxygen balance in the furnace. The real-time measurements can also be integrated directly into the combustion control system in order to achieve the greatest benefits available with automated control. An example of the interface is shown as Figure 5 showing the balance of each constituent across the furnace.

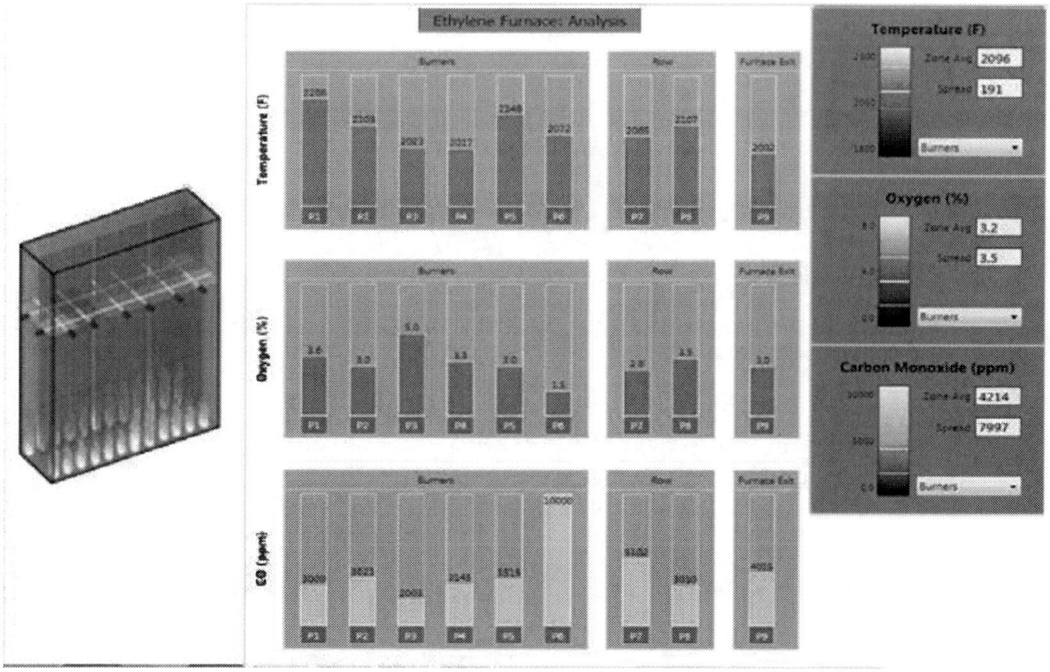

Figure 5: Multiplexed TDLAS system interface showing balance information for each constituent

Production Run Times:

Production cycles are typically limited by localized "hot spots" in the furnace that create increased coking rates inside the process tubes. These "hot spots" are often created by non-uniform combustion profiles which are difficult to identify with limited traditional sensors downstream of the furnace or even periodic tube wall temperature (TWT) measurement. Over time, coke forms inside the process tubes and when a single pass or tube reaches the maximum TWT (or a pressure drop limit) as shown in the Figure 6, the plant will begin a de-coking procedure. During the production run, the TWT increases over time as well as the length of the coil until the maximum TWT is achieved. Localized "hot spots" can accelerate coke formation (which is exponentially related to temperature) and force a premature de-coking procedure, thereby shortening the production cycle. A single coil with a high TWT can trigger a premature de-coking cycle even though the other coils are within acceptance temperature ranges.

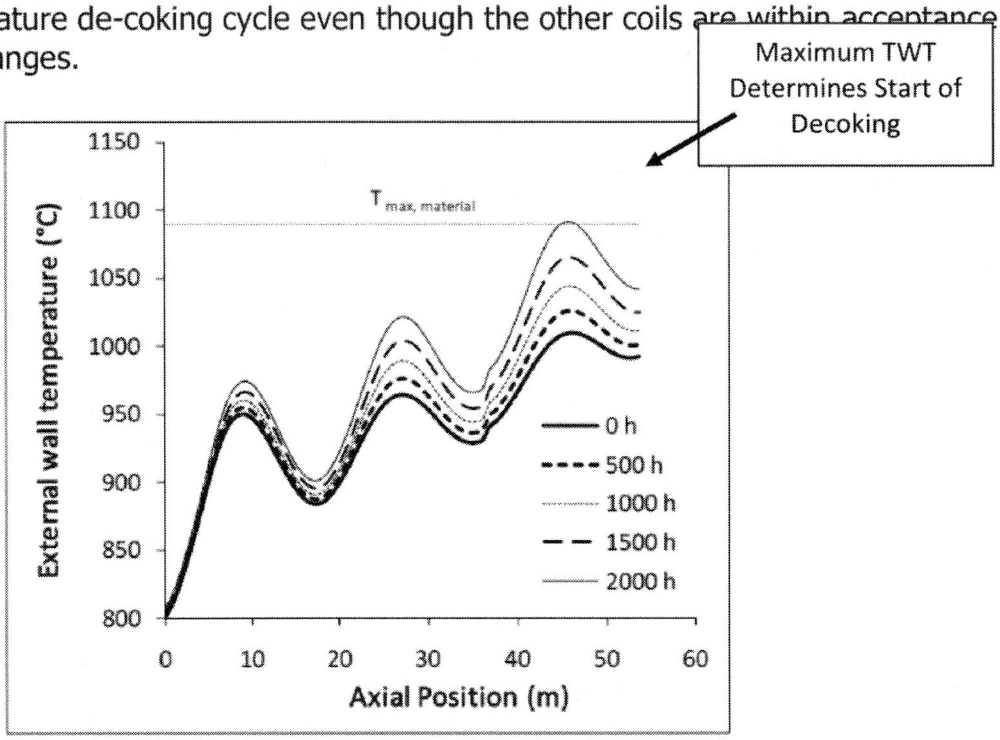

Figure 6: External TWT profile as function of time

To prevent premature decoking, a multiplexed TDLAS system can be installed to provide real-time, in-situ temperature, O_2 and CO measurements to manage the flue gas profile and thereby the tube wall temperatures (TWT). By reducing the localized "hot spots", a more uniform combustion gas temperature profile is created, which results in a more uniform coking rate for all passes and permits longer operation between de-coking cycles. The coking rates are very sensitive to the COT (which is influenced by the flue gas temperatures) as shown in Figure 7. For example, a relatively small reduction of 15°F (7°C) in the COT (i.e. removing a TWT hot spot) can reduce the coking rates by more than 20%. Consequently, a small reduction in the coking rate can have a significant impact on the production run length. Figure 8 shows that a reduction of only 20% in coking rates can increase the run length by 30%. For a cracking furnace with a 40 day production length this would result in 12 more days of production each cycle (about 4 more days of production overall per year). The extra 4 days of production would easily pay the capital costs of a multiplexed TDLAS system. Additionally, in cases where plants have limited product demand, they could operate fewer days to produce the same product and therefore save energy costs.

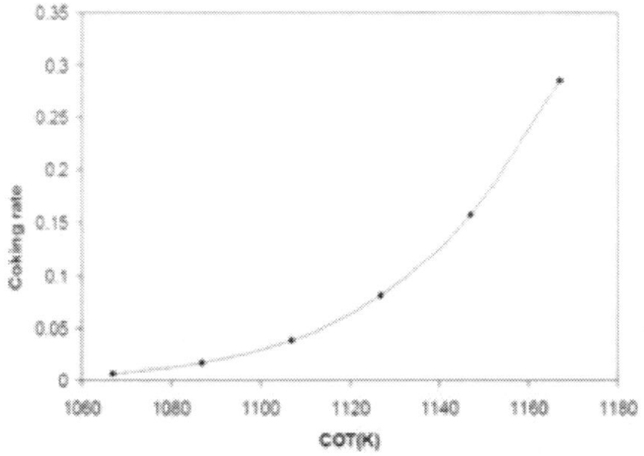

Figure 7: Impact of COT on Coking Rate (2)

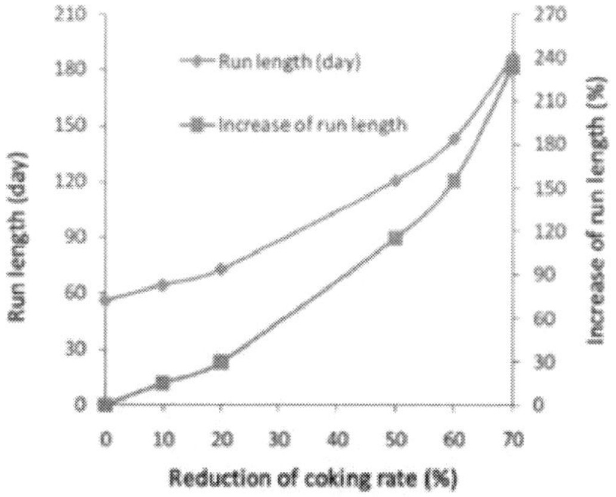

Figure 8: Impact of Coking Rate on Run Length (1)

Full scale commercial multiplexed TDLAS systems have been installed on operating Steam Methane Reformers (SMR's) in Hydrogen and Ammonia production plant applications to balance the flue gas profiles by changing (biasing) the burner settings and therefore influencing the TWT profile. Reformer furnaces are very similar to cracking furnaces with a similar fuel and process tube arrangement. While reformers typically have multiple rows of process tubes, ethylene furnaces have only a single row. Figure 9 below shows actual plant data demonstrating a ~50 F decrease in the spread of TWT across the reformer using the information by the multiplexed TDLAS system to balance flue gas temperature. The aim was to reduce "hot spots" and therefore reduce "at risk" tubes to improve process efficiency, and extend furnace tube life.

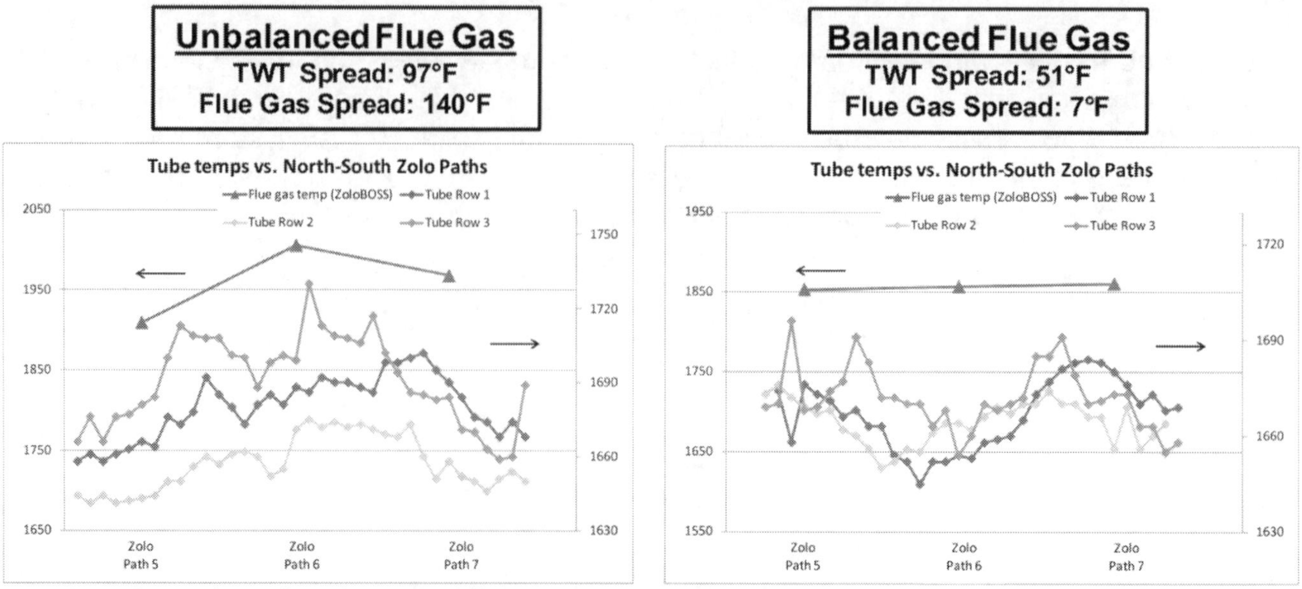

Figure 9: Balanced combustion using flue gas temperature measurements can reduce "hot spots" and therefore reduce coking rates.

Process Yields:

The ethylene process yield depends on a number of factors including: propylene/ethylene ratios; conversion rates; dilution steam ratio, residence time and feedstock characteristics. However, in all cases, a more uniform temperature profile across the furnace and a more uniform heat flux profile along the tube length will result in better process temperature control thereby increasing ethylene yields. The coil outlet temperatures (COT) across the furnace can vary substantially with a wide spread in values for each coil (minimum to maximum). Typically, operators will manage the COT with changes to firing rates and/or process flow rates for each coil and monitoring of TWT. However, it is difficult to obtain consistent measurements for TWT using traditional pyrometer technology. Very little information is available on real-time heat flux profiles.

However, a multiplexed TDLAS system provides highly accurate, reliable and repeatable combustion flue gas measurements. Additionally, combustion gas profiles can be correlated to burners (or groups of burners) in order to balance the temperature, CO and O_2 profiles directly in the furnace. Groups of burners or individual burners can be adjusted manually (or biased) to better balance the combustion flue gas profile. Figure 10 illustrates an unbalanced flue gas temperature profile across the furnace and the same furnace balanced using a multiplexed TDLAS system in a path configuration similar to Figure 3.

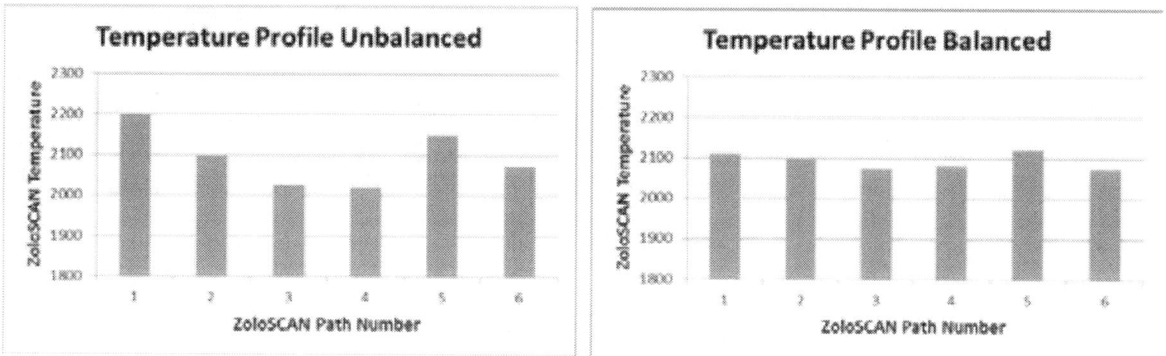

Figure 10: Balanced combustion can reduce "hot spots" and therefore reduce coking rates.

A more uniform flue gas profile would create more uniform cracking temperatures from coil to coil across the furnace. Even with the same "average" COT for each coil, a more uniform COT profile across all coils could increase the yield by 0.25-0.5%. While the temperature profile is important, also having the O_2 and CO profiles (which cannot be obtained from traditional optical pyrometers) allows the user to dictate the changes needed to be made to specific burners or groups of burners in order to maintain a uniform temperature profile.

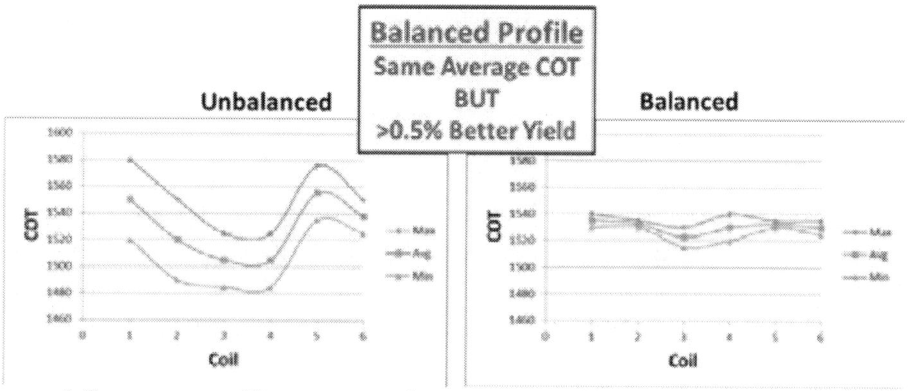

Figure 11: Balanced flue gas profiles can produce more uniform COT profiles and higher yields.

The feedstock characteristics or quality (ethane, propane or naptha) can also impact the yields. Cracking conditions may be different for each feedstock and therefore combustion requirements may need to change. Variable fuel sources or qualities (refinery gas and/or natural gas) can make it a challenge to maintain uniform and optimal combustion in the cracking furnace. Real-time combustion information from a multiplexed TDLAS system can provide valuable information to adapt or compensate for these changing feedstock and/or fuel situations.

Combustion Efficiency and Emissions:

A more uniform combustion profile (temperature, O_2 and CO) in the furnace permits operation at lower overall excess O_2 levels while avoiding high CO levels. Since there is less overall air, less energy is used to heat-up the N_2 in the extra combustion air. At plants with high fuel costs (i.e. EU), a small improvement in fuel efficiency can result in significant fuel savings. Furthermore, firing less fuel at lower excess O_2 levels also reduces NOx and CO_2

emissions. A reduction in overall excess O_2 levels of only 1% can result in savings of 0.4% in fuel costs.

Multiplexed TDLAS technology can be used to identify localized areas of high (or low) O_2. The adjustments to the burners or air/fuel ratio can be made to create a more balanced O_2 profile. With a more balanced profile, the overall excess O_2 levels can be safely lowered without the risk of high CO as shown in Figure 12 below.

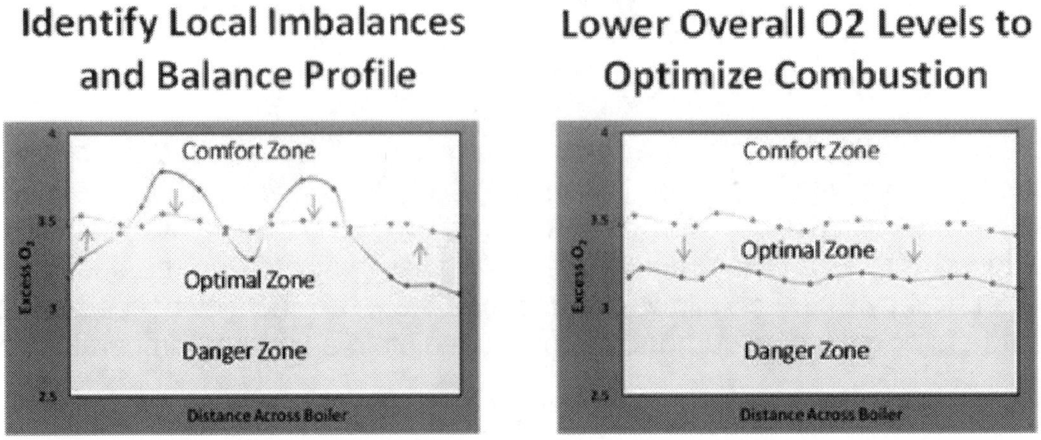

Figure 12: A more balanced O_2 profile can permit operation at lower overall excess O_2 levels

With a better representation of the O_2 levels in the furnace from a multiplexed TDLAS system, the plant can operate at a Set Point (SP) much closer to the optimum level without the fear of high CO levels. Lower O_2 levels can not only improve the combustion efficiency but also reduce NOx as shown in Figure 13. Depending upon the burners and the level of excess O2, NOx levels can be reduced by 5-20% as shown in Figure 14.

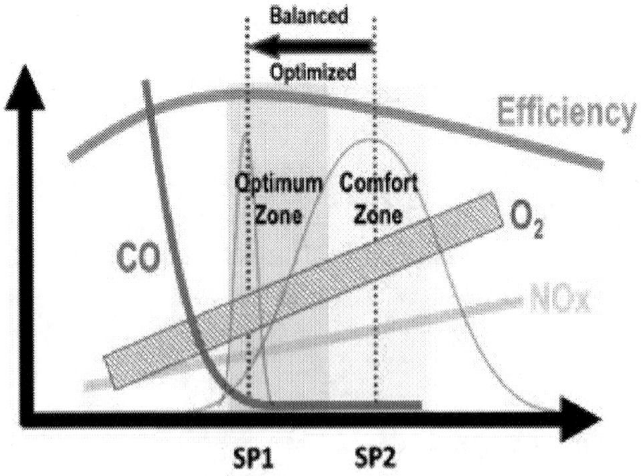

Figure 13: A more balanced O_2 profile can permit operation at lower overall excess O_2

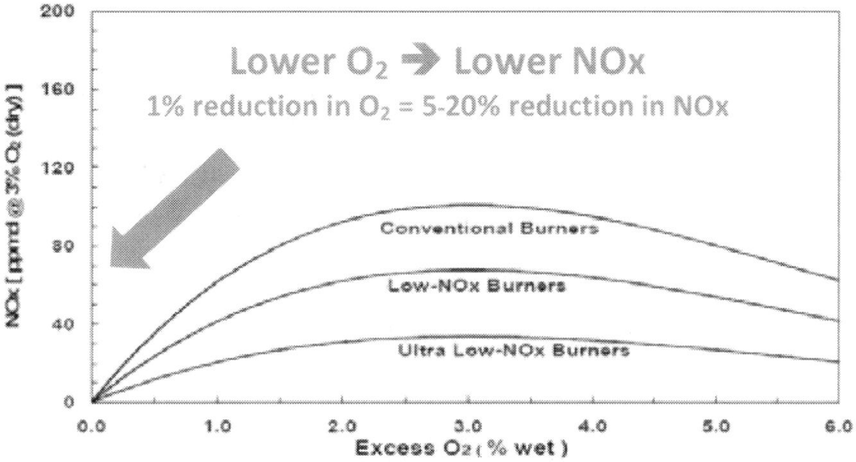

Figure 14: Predicted NOx for Ethylene furnace with 100% methane fuel. (3)

Coil Tube Life:

The inner surface metal of the tubes is slowly removed by both the coking and decoking process, leading to pitting and erosion, and a shortened run life of the tubes. Carburization from the coking also can change the properties of the tube metal and lead to deterioration and/or damage of tube metallurgy. Since the carbon diffusion is thermally activated, carbon from the coke, especially at high temperatures, diffuses into the metal of the tubes thereby changing its properties. Thermal stresses are greatest during start-up, shut-down and during the de-coking process. As a result, the mechanical properties, mostly the creep and thermal shock resistance properties, are altered to the point that the tube material becomes very brittle. The tube can fail due to thermal shock as shown in Figure 15. More uniform temperatures, elimination of "hot spots", and fewer de-coking cycles will increase the life of the tubes.

Figure 15: Example of damaged tube due to thermal shock

Safety/Sentry:

Ethylene furnaces typically have a wide array of sensors located on the fuel and air pipes both before and after the combustion zone. However, there are very few sensors available to measure directly inside the furnace. Over 70% of furnace explosions occur during start-up or shut-down, when operator involvement is significant. The *in-situ* measurement capability of a multiplexed TDLAS system provides real-time status of the furnace so that operators can identify poor combustion conditions or dangerous situations directly from the safety of the control room. For example, excessive CO levels measured *in-situ* by a multiplexed TDLAS system can be a signal of a potentially dangerous combustion condition. A sudden increase in the temperature as measured by a multiplexed TDLAS system can provide an early warning of a process tube leak which could prevent further damage to adjacent tubes. Quantitative measurements can be configured to trigger alarms if certain limits are exceeded.

"Excessive coke deposits inside the furnace coils can negatively impact plant safety and productivity. Coke is known to have a tendency to deposit heavily where the diameter of the coil changes and in the coil's bends. Along the reactor coil, the wall temperature profile varies strongly because the firing nozzles are placed at the bottom and side walls of the furnace, creating a non-uniform flux profile. The non-uniform distribution of temperature and pressure makes coke deposit heavily on certain sections of the coil. Those sections are subjected to particularly high temperatures by the insulation effect of the coke and so-called "hot spots" can arise. Because hot spots decrease the durability of the coils and increase the coke growth rate, they can jeopardize safe and continuous operation of the furnace." (2)

Ethylene Furnace Combustion Monitoring and Diagnostic Technology Comparisons

Table 1: Combustion Monitoring

	Traditional Sensors	Traditional TDLAS	Multiplexed TDLAS
Proven, Reliable & Accurate	Proven, but prone to operator or calibration errors	Yes	Yes
Maintenance	High	Low	Low
In-situ measurements directly in furnace	Limited by high temperatures. Extractive sampling can provide erroneous O_2 & CO measurements.	Yes	Yes
Auto-alignment of laser beam	No	No Relies on divergent beam to maintain	Yes Actively aligned collimated beam for

		alignment which lowers sensitivity.	maximum power, sensitivity and measurement reliability.
Simultaneously measures multiple constituents from a single sensor	No	No Single constituent (O_2 or CO) for each sensor.	Yes Temp, O_2, and CO measured simultaneously from a single measurement
Multiple measurement paths from a single sensor	No	No Only 1 constituent per sensor.	Yes Up to 24 paths from a single sensor to measure in multiple areas.
Multiple furnace measurements from a single sensor	No	No	Yes A single multiplexed TDLAS system can service multiple adjacent furnaces with up to 24 total paths
Maximum measurement temperature	1500°F (815°C)	2750°F (1500°C)	3500°F (1925°C)
Periodic, in-field calibration required	Yes	No	No
Representative measurement of combustion area	Poor Single point only.	Fair Average concentration across a single path.	Best Average concentration across multiple paths.
Combustion balance information	No	No	Yes Multiple paths provide balancing information across furnace

Table 1: Comparison of traditional sensors, Traditional TDLAS, and Multiplexed TDLAS

Ethylene Plant Savings with Multiplexed TDLAS Technology

The benefits of real-time combustion monitoring and control in ethylene furnaces can be significant. Table 2 shows the potential benefits for a single furnace producing 100,000 Mtons/yr ethylene for a plant on the Gulf Coast USA. The benefits of better combustion using a multiplexed TDLAS system has simple payback of less than 1 year.

Table 2: Summary of Multiplexed TDLAS Value

Summary of Multiplexed TDLAS Value		
Ethylene Production	100,000	Mtons/Yr
	25,368	lb/hr
Base Run Time	40	Days
Fuel Usage	1,800,000	mmbtu/yr
Fuel Cost	4.50	$/mmbtu
Run Time Improvement		
Reduced coking rate	20%	
Extra Production Days y	3.9	Days
Less Fuel Usage n	21,164	mmbtu
Value of Increased Run Time	$705,467	per Year
Process Yield Improvement		
Process Improvement	0.25%	
Extra Production	250	tons/year
Value of Extra Yield	$150,000	per year
Combustion Efficiency Improvement		
Fuel Savings	0.40%	Percent
Fuel Savings	7,200	mmbtu/yr
Value of Combustion Efficiency	$32,400	per Year
Total Value	**$887,867**	**per Year**

Conclusion

Multiplexed TDLAS combustion monitoring systems can assist ethylene cracking furnaces and large fired heater operators to improve run times, process yields, process efficiency, and equipment life through real-time combustion monitoring and diagnostics. These systems provide temperature, O_2, and CO measurements directly in the furnace which can be used to balance the combustion flue gas profiles. The ability to measure and control furnace temperature, O2, and CO gas profiles in a narrow and precise range across the full area of the furnace can achieve improved equipment productivity, reliability, emissions reduction, and safety.

References

1. G.-Y. Gaoa, M. Wangb, C. Ramshawb, X.-G. Lia, H. Yeung, "Optimal operation of tubular reactors for naphtha cracking by numerical simulation"
2. Dr. ir. Kevin M. Van Geem, et al, "Effect of Operating Conditions and Feedstock Composition on Run lengths of Steam Cracking Coils", Laboratory for Chemical Technology, University of Ghent
3. Kunz, Robert G. "Burner NOx from Ethylene Cracking Furnaces", RGK Environmental Consulting

444036 CCPS Process Safety Metrics A Global Approach

Abstract:

The use of global metrics is essential for companies and industries who strive to understand their process safety incidents and to improve performance. These metrics are an important component in the comparison and evaluation of site to site, or unit to unit, performance within a company, and to help determine where focus is needed and resources should be allocated. In addition to internal comparisons, global process safety metrics can assist companies (and industries) in benchmarking their performance with others or in comparison to an industry average or goal.

In 2007, CCPS published **"Process Safety Leading and Lagging Metrics…You Don't Know What You Don't Measure"** (available for free download at http://www.aiche.org/sites/default/files/docs/embedded-pfd/CCPS_ProcessSafety2011_2-24-web.pdf). The definitions established in the document were a key input to the creation of the existing ANSI / API standard (ANSI/API RP 754) which was finalized and released in April 2010, with CCPS support. In 2011, CCPS updated their metrics publication with the intent of aligning CCPS and API documents. CCPS went further in 2013 to develop the use of process safety leading and lagging indicators by publishing the results of a member company survey "Process Safety Leading Indicators Industry Survey" (free download at http://www.aiche.org/sites/default/files/docs/pages/8404_leading.web_v2.pdf). This publication provides an update on the chemical, petroleum and other process industries in the use, direction and effectiveness of leading indicators to help drive performance improvement and the reduction in the number and severity of process safety incidents.

API is in the final reviews of the revisions to RP 754. CCPS was part of the review panel and worked closely to provide input to the revisions. Once the final document is issued, CCPS plans to review and update the **"Process Safety Leading and Lagging Metrics…You Don't Know What You Don't Measure"** and align information so that these documents can be used consistently on a global basis. In addition, CCPS is in the process of updating the Reactive Chemistry Worksheet tool to version 4.0 and updating the Process Safety Incident Evaluation Tool with an app for each to be developed following their releases.

This presentation will discuss the CCPS metrics tools, and ongoing work towards global metrics.

445220 Control Valve Packing Follower Failure

Abstract:

The authors will detail the incident of a control valve packing follower failure in cold service. While attempting a repair, the follower failed on a process service @ 465 psig / -150 degrees Fahrenheit which was eminent based on an operations / maintenance action. The failure potential of failing without any interaction was highly possible due to the broken follower.

Some of the topics that will be covered by the authors are:

1. The root cause analysis findings from the incident.

2. The immediate corrective actions taken as a result of the findings.

3. The long term corrective action items and the path forward to implement them.

AIChE Paper Number: 164c

Evaluation and Prevention of Potential Cold Embrittlement Hazards in Petrochemical Facilities

Erick Comeaux
Manager, Engineering & Technical Services
Williams Olefins, LLC

Prepared for Presentation at the 2016 Spring National Meeting
Houston, TX, April 10 – 15, 2016

AIChE Paper Number: 164c

Evaluation and Prevention of Potential Cold Embrittlement Hazards in Petrochemical Facilities

Erick Comeaux
Manager, Engineering & Technical Services
Williams Olefins, LLC

ABSTRACT

A summary of contemporary practices and approaches for evaluating and preventing potential cold embrittlement hazards within ethylene production facilities will be presented. The practices and approaches will be discussed in the context of API 579 Parts 3 and 9, ASME Section VIII Division 1, and ASME B31.1. These three bodies of knowledge have been recognized by OSHA as RAGAGEP in the U.S. on fitness for service requirements that govern the design, fabrication, inspection, testing and certification of pressure vessels exceeding internal or external operating pressures of 15 psig and piping.

The primary materials that will be discussed include carbon and low alloy steels, 400-series stainless steels, duplex stainless steels and 300-series stainless steels. The discussion will include considerations of grain size, steel chemistry, heat treatment, environmental conditions, process conditions, and their relative effects on material toughness and impact strength. A brief summary of fracture mechanics will also be included, primarily to highlight the differences between ductile and brittle fractures and to draw a clear connection between the key factors that drive material toughness and how they can also influence the mode of fracture.

Potential cold embrittlement vulnerability varies according to a chemical plant's vintage and design, and maintenance considerations can have an enormous impact on the suitability of these materials for cold temperature service. This paper will walk readers through a practical means of assessing the suitability for service of existing equipment, properly specifying materials of construction for new designs, and apposite maintenance practices for preserving pertinent toughness qualities for existing equipment in cold temperature service.

Keywords: Cold Embrittlement, Brittle Fracture, ASME SA-212, Fitness for Service

1. Introduction

Perhaps, the most infamous case of cold embrittlement hazards was the sinking of the R.M.S. Titanic luxury ocean liner, in 1912. Unfortunately, this tragedy was only one of many devastating catastrophes in history, resulting from unidentified and unmitigated cold embrittlement hazards in the manufacturing and transportation industries, as well as in the military. The primary objective of this article is to help preclude future incidents and near misses associated with cold embrittlement, by highlighting present-day practices and approaches for evaluating and preventing potential cold embrittlement hazards within ethylene production facilities, though these principles can be applied across all industries.

In a nutshell, the cold embrittlement concern for any manufacturing or transportation industry is this: carbon steels and other low allow steels exhibit a dramatic change in ductility at "low" temperatures, and the definition of "low" depends on a number of critical factors. In other words, all carbon steel is not "good to -20°F," as many have often heard. On the other hand, some carbon steels can be good to well below -20°F. Every material displays a unique transition range (*on a toughness versus temperature plot*) that illustrates a prodigious decrease in notch-toughness properties below this relatively narrow transition range of temperatures. Above this transition temperature range, materials are capable of absorbing large amounts of energy and fail in a ductile manner. Below this transition temperature range, however, the same materials fail catastrophically in a brittle, or cleavage, manner because their ability to absorb energy from an impending stress is remarkably diminished. Fortunately, low temperature brittle fracture is not common, but the risk does pose a convincing case to carefully and deliberately coordinate materials of construction with the full range of potential operating conditions.

While catastrophic low temperature brittle fractures are documented back several decades, it wasn't until the seventies that industry began to grasp and recognize this phenomenon. This enlightenment then spawned several reexaminations of past disasters that seemed to potentially fit the failure mechanism, and the findings of these investigations, coupled with developing research into low temperature brittle fraction mechanics, motivated organizations to revisit long-established codes and standards. The first reference to low temperature brittle fracture hazards in industrial codes and standards was the American Society of Mechanical Engineers (ASME) "Boiler and Pressure Vessel Code, Section VIII, Division 1," 1987 revision. Even then, there was no consistently accepted methodology to evaluate existing equipment for suitability in low temperature service until January of 2000, when the American Petroleum Institute (API) issued "API 579 Recommended Practice for Fitness-for-Service." Since that time, Recognized and Generally Accepted Good Engineering Practices (RAGAGEP) have been updated to continually enhance our ability to analyze equipment and evaluate its fitness for low temperature service. For instance, in 2007, a joint ASME/API committee published API 579-1/ASME FFS-1, which set forth many of the tools and techniques referenced in this paper. [1]

Process Hazard Analyses (PHAs) are used throughout industry, as the primary means of identifying and managing potential process hazards. While it's critical to include considerations for materials of construction—*namely, potential cold embrittlement in this case*—in those PHA reviews, the scope of a typical PHA may not be sufficient to thoroughly scrutinize a facility's vulnerability to this potential hazard. Similar to Layers of Protection Analysis (LOPA) reviews and Alarm Management rationalization sessions, a compelling case exists to dedicate a focused evaluation to the low temperature properties for the materials of construction utilized in processing plants. API 579 Parts 3 and 9, ASME Section VIII Division 1, and ASME B31.1 have been recognized by OSHA as RAGAGEP in the U.S. on fitness for service requirements that govern the design, fabrication, inspection, testing and certification of pressure vessels exceeding internal or external operating pressures of 15 psig and piping. These three bodies of knowledge contain the parameters that should be employed when evaluating any given pipe or vessel's fitness for service in low temperature applications. [2, 8]

2. Fundamentals of Cold Embrittlement

Brittle failure is the single most undesirable failure mode because it happens with little or no warning—*often referred to as a "break before leak mode"*—and typically materializes in a catastrophic manner, with the vessel or pipe either splitting rapidly or shattering in pieces. In the process industry, this type of disaster most always results in a sudden and complete release of the entire toxic and/or flammable contents of the vessel or pipe and is frequently followed by a resulting fire or explosion. In the military, this phenomenon was responsible for battleships latterly breaking in half and sinking during World War II. Brittle fracture can, but does not necessarily have to, be related to low temperatures; however, all low temperature induced failures are brittle fractures. [3]

Minimal Accepted Temperatures (MAT), or Minimal Design Metal Temperatures (MDMT), are the temperatures above which materials can be in service without risk for low temperature induced brittle fracture. The API 579, Sections 3 and 9, fitness for service standard walks the user through computing MATs, or MDMTs, and determining whether or not the material in question is fit for low temperature service. The methodologies employed by the standard are derived directly from empirical impact energy transition curve data for various materials, illustrating the relationship between impact toughness and temperature. During a ductile failure, the material is shown to absorb a greater amount of energy from impact, which slows the progress of the failure through a gradual plastic deformation. Ultimately, a greater amount of strain energy is required to force the material into failure, and the relatively slow yield response of the material provides both visible indications of a pending failure and additional time to attempt a response. Brittle fractures, on the other hand, manifest a completely different set of impact data. Since the failure mechanism consists of a rapid and sudden cleavage through slip planes, grain boundaries and regions of flaws and cracks, the failure response is quite rapid and provides no visible indication prior to the failure. Furthermore, with brittle fracture, the materials fails well below its known yield strength. [4, 5]

For the purposes of incident investigations, brittle fractures manifest impeccably distinct characteristics that leave little room for uncertainty. Using both macroscopic and microscopic observations, the brittle fracture displays little to no evidence of yielding deformation or a shear-lip. In addition, directional Chevron markings (indicating the direction of propagation) and stress concentrators (cracks or flaws) provide key forensic indicators of a brittle fracture. Upon the engagement of more advanced analytical techniques, investigators will also find that brittle tension failures exhibit a 90° fracture, and torsion failures exhibit a 45° fracture; whereas, ductile tension failures exhibit a 90° shear, and ductile torsion failures exhibit a 90° fracture. Ultimately, Scanning Electron Microscopy (SEM) is often employed to provide the final, definitive verdict on brittle fracture, and these micrographs unmistakably distinguish between brittle and ductile fractures, with brittle fractures revealing a non-dimpled fracture surface, with sharp, jagged topography. [3]

Much like the combustion pyramid, low temperature induced brittle fractures require three conditions to be present to occur: a susceptible material with a stress riser (crack or flaw), an applied stress above the minimum stress level, and temperatures below the Nil-Ductility Transition Temperature. Alarmingly, an applied stress equivalent to only forty percent of the Maximum Allowable Working Pressure (MAWP) on vessels built to 1998 or earlier codes, or thirty-five percent on vessels built to 1999 through present day codes, is sufficient to initiate a low temperature induced brittle fracture in susceptible materials with a stress riser present. Materials most susceptible to cold embrittlement include steels with a high carbon loading by weight, steels with high loading of sulfur or phosphorous contaminates, non-stress relieved steels, grades of steel with coarse grains and thick steel plate. On the other hand, austenitic stainless steels and nickel-, aluminum- or copper-based alloys tend to be resistant to cold embrittlement. [6]

ASME SA-212 is another material that is highly susceptible to cold embrittlement that deserves a discussion completely dedicated to itself. ASME SA-212 is a specification of carbon steel plate that was withdrawn by ASME, in 1967, and completely phased out of production. Because the speciation was only for carbon steel plate, all of the material went into boiler and pressure vessels, and was not fabricated into piping. After a number of low temperature induced brittle fractures, the specification was replaced by ASME with A-516 for fine grain and A-517 for coarse grain material. The finer grain A-516 was deemed suitable for low temperature service, while the coarse A-517 material was not. Studies over the years have shown that is it not uncommon for vessels fabricated of SA-212 to have MDTs well above ambient conditions, indicating that they are not suitable for use. Therefore, the concern with continued use of vessels fabricated of SA-212 is that they must be properly evaluated under the current fitness for service methodologies of API 579, Sections 3 and 9, to ensure that they are suitable for their present intended service. [7]

3. Cold Embrittlement Hazard Analysis Approach

Preparation for Study

As with any form of process hazard evaluation, the quality of the assessment is directly proportional to the diligence applied to the preparation. Prior to commencing any facility-wide cold embrittlement evaluation, it is imperative that the following process safety information be assembled and verified, at a minimum:

Process Description

- Plot Plan and Equipment List
- PFDs with Material & Energy Balance
- P&IDs for Areas of Interest
- Process Mediums
- Low Temperature Alerts, Alarms & Interlocks
- Potential Auto-Refrigeration Scenarios
- Critical Exposure Temperature (CET)
- Low Temperature Near Miss Reports
- Low Temperature Incident Reports
- Low Temperature PHA Scenarios
- Pressure & Temp Trends from Historian

Equipment Files

- Specifications (U-1) & Repairs (R-1)
- Year of Fabrication & Any Repairs
- Design & Typical Operating Pressure
- Design & Typical Operating Temp
- Maximum Allowable Working Pressure
- Weld Procedures & Repairs
- Historical Post Weld Heat Treat Data
- Nominal Wall Thickness
- Corrosion Allowances
- Charpy Impact Testing
- Equipment Arrangement & Condition

For the initial screening exercise, fired and heated services (~200°F or greater) can be potentially excluded. It is not unheard of, however, to observe Minimum Accepted Temperatures (MAT) between 80°F and 100°F—especially for vessels fabricated of ASME SA-212—so caution should be exercised when excluding services under 200°F. For good measure, the rationale for all excluded services should be documented to ensure that reasonable vulnerabilities are not inadvertently overlooked. For ethylene plants, the areas of highest concern are the refrigeration systems, from the charge gas compressor through fractionation, and storage bullets and spheres.

Phase I of Study: Evaluation of Normal & Non-Steady State Process Conditions

Once the initial process safety information has been assembled and verified, the evaluation team should commence a process hazard analysis review session to identify and discuss:

1. Services to be excluded from the cold embrittlement hazard evaluation
2. Normal temperature ranges for each identified cold service
3. Startup & shutdown low temperature conditions (remember to include the initial introduction of hydrocarbons into empty vessels after a turnaround)
4. Process upsets that could result in low temperature excursions (low temperature alerts, alarms and interlocks, low temperature near miss and incident reports, and low temperature PHA scenarios make for an excellent start on this effort but should not be considered an exhaustive list of all potential vulnerabilities. Both

likelihood and potential severity should be quantified and documented, so that the team can ensure that the ultimate mitigation is appropriate for the risk.

Since auto-refrigeration excursions are experienced more commonly during startups, shutdowns and process upsets than during steady state operations, special emphasis should be placed on fully exploring non-steady state conditions that could result in sharp decreases in temperature. A process simulator and experienced user are absolutely essential at this point, as this initial phase of the evaluation establishes the fundamental basis for the entire study. Regardless of how rigorous the mechanical evaluation is executed in subsequent phases of the study, an overlooked non-steady state condition at this point could render the entire effort critically flawed and leave serious vulnerabilities unaddressed.

The deliverables from this initial phase of the facility-wide cold embrittlement study should include:

1. Plot plan and equipment list, indicating each piece of equipment included or excluded from the study, along with documented reasons for all exclusions
2. A chart summarizing:
    a. potential temperature ranges for normal, startup/shutdown and process upset conditions for each piece of equipment to be evaluated
    b. narrative describing the initiating events for any upset conditions of concern
    c. existing low temperature safeguards in place (alerts, alarms and interlocks)
3. Prioritized plan of the order in which services should be evaluated

Phase II of Study: Level 2 Mechanical Evaluation

The next phase of the cold embrittlement process hazard evaluation is a Level 2 mechanical evaluation and is based on the methodology described in API 579 Recommended Practice for Fitness-for-Service, Section 3, Assessment of Existing Equipment for Brittle Fracture.

The Level 2 mechanical evaluation involves the preparation of temperature/pressure equilibrium curves for the process fluid in the intended service, as well as minimum allowable temperature (MAT)/pressure curves for the metallurgy under consideration. The MAT/pressure curves are determined according to Type A, B, C or D materials, which are all defined in the API 579, Section 3, fitness for service standard.

If the Critical Exposure Temperature (CET), or lowest expected process temperature, curve falls at or below the Minimum Accepted Temperature (MAT) curve, the vessel fails the Level 2 screening and requires a further Level 3 mechanical evaluation to determine whether or not the vessel or piping is at risk for brittle fracture.

Phase II of Study: Level 3 Mechanical Evaluation

The Level 3 mechanical evaluation involves a much more intense level of scrutiny of the actual piping or equipment, to determine whether or not the vessel or piping is at risk for brittle fracture. Tools and techniques employed during this phase of the study include Finite Element

Analyses (FEA), Phase-Array Ultrasonic Testing (PAUT), Wet Magnetic Particle Inspection, and low temperature hydrostatic testing.

In the Level 3 mechanical evaluation, a critical crack size is determined for each piece of equipment that failed the Level 2 mechanical evaluation. The physical inspection techniques referenced in the previous paragraph are then used to determine if any flaws equal to or larger than the critical crack size are present in the equipment. If such a flaw is identified, the equipment fails the Level 3 screening and requires replacement or appropriate mitigation to remain in service. Examples of appropriate mitigation would include heat treating or grinding to remove the flaws and then, repairing the voids. In any case, if repairs are made the equipment, the original physical inspection technique must be reapplied, to ensure that the repair was effective.

Ideally, only a small portion of the equipment should progress to the stage of a Level 3 mechanical evaluation. However, because of the requirement for the equipment to be out of service, and the considerable resources that will be required to expedite the inspection process, it is not recommended to begin a facility-wide cold embrittlement study in the midst of a turnaround, if the timing can be avoided. Rather, it is strongly recommended that such a study commence well in advance of a scheduled turnaround, so that the scope of inspections can be well defined before the unit ceases operations, and so that the necessary resources can be properly coordinated in advance. In addition, it is imperative that contingencies be developed, in the unlikely and unfortunate event that a major piece of long-lead-time equipment fails the Level 3 mechanical evaluation and cannot be returned to its intended service. This would be of particular concern with any equipment fabricated of ASME SA-212.

4. Cold Embrittlement Preventive Measures

Once the facility-wide cold embrittlement hazard evaluation is complete, and the list of residual cold embrittlement vulnerabilities has been compiled, an optimal mitigation strategy will need to be selected from a number of viable alternatives. Of course, replacement of the piping or vessel with a more suitable material of construction is always the most preferable. However, a number of complicating factors can render this option impractical, especially for large equipment or long piping runs, with prolonged lead times. Fortunately, replacement of individual problematic components can often be utilized to provide an equivalent level of protection. In addition, equipment can sometimes be Post Weld Heat Treated (PWHT) to achieve an acceptable level of protection; in some cases, this can even be accomplished in situ, without having to remove the equipment for shipment to an external shop. In cases where flaws are detected in excess of the critical crack size, an attempt can be made to grind the flaw out of the vessel wall and repair the void, but these repairs must always be inspected to ensure that the remedy was effective.

In other cases, full or partial replacement of the equipment, PWHT, or vessel repairs may all be deemed impractical—at least in the short term. In these situations, solid engineering and administrative controls must be implemented to prevent the CET from ever dipping below the MAT of the vessel or piping. Emergency Shutdown Devices (ESDs) and other forms of automated interlocks can be implemented to isolate vulnerable systems from unacceptably low temperatures. Depending on the risk ranking assigned to the process upset scenario that initiates the low temperature excursion, these automated controls can reside in either the Basic Process Control System (BPCS) or a Safety Instrumented System (SIS), though configuration in an independent SIS is preferable. Control system alarms and alerts can also be used to form an integral part of the protective strategy, but unless the risk ranking is extremely low, these administrative controls should always function in support of—*and not in place of*—automated interlocks. In any case, it is imperative that standard response procedures be implemented to guide operations personnel through appropriate means of recovering from low temperature excursions, and these procedures need to account for the fact that the wall temperature will lag behind the process equilibrium temperature during re-pressurization.

Finally, engineering records and maintenance files need to be updated to clearly document the final disposition of each piece of equipment and to ensure that newly implemented protective measures are both understood and preserved.

5. Conclusion

Cold embrittlement related catastrophes are relatively uncommon, but the hazard still poses a serious potential vulnerability that needs to be properly assessed and evaluated in all process plants. While standard Process Hazard Analyses (PHAs) and Layers of Protection Analyses (LOPAs) should consider potential cold embrittlement scenarios, they are insufficient means in and of themselves to comprehensively measure a site's level of vulnerability. At some point in the life of every process facility, a focused cold embrittlement study is needed to fully scrutinize the materials of construction used throughout cold service applications and to determine that the materials are suitable for their intended applications. While such a comprehensive study can be a costly and resource-intensive effort, proper planning and a solid project management plan will greatly enhance both the efficiency and effectiveness of the effort. Due to the complex nature of the applicable RAGAGEP, engaging an experienced and licensed professional resource, with a healthy list of references from similar studies and multidisciplinary capabilities, is highly recommended. Furthermore, their expertise should also be utilized in developing a robust management program to ensure that the risk does not increase over time as equipment ages and the process changes.

References

[1] API 579 Recommended Practice for Fitness-for-Service, Section 3, Assessment of Existing Equipment for Brittle Fracture, 2007.

[2] W.S. Jacobs, ASME Code Material Requirements for Low Temperature Operation, Section VIII, Division 1 and Division 2, 8th ed., 1999 Addenda.

[3] Benac, Daniel J., Cherolis, Nicholas, and David Wood, Managing Cold Temperature and Brittle Fracture Hazards in Pressure Vessels, Journal of Failure Analysis and Prevention, Vol 15, Iss 6 (2015).

[4] Bansch, Rowe, Considerations for Operating Pressure Vessels Below -20°F, Industrial Refrigeration Consortium: Research & Technology Forum, May 2-3, 2012.

[5] J.E. McLaughlin, J.R. Sims, Assessment of Older Equipment for Risk of Brittle Fracture, Service Experience and Life Management: Nuclear, Fossil, and Petrochemical Plants, PVP-vol. 261, ASME, 1993.

[6] King, Ralph E., Auto-Refrigeration/Brittle Fracture Analysis of Existing Olefins Plants—Translation of Lessons Learned to Other Processes, Journal of Hazardous Materials 142 (2007) 608–617.

[7] Vega, Alejandro, "ASTM A-212 Pressure Vessel Steel—A Case Against Continued Use," PV Newsletter, Vol 4, Iss 1 (2011).

[8] J.R. Messinger, Cold Service Risk Assessment of Existing Bimetallic De- Methanizers, in: AICHE 2001 Spring National Meeting of the 13th Ethylene Producers Conference, Houston, TX, April 22–26, 2001.

AIChE Paper Number 164d

Voting architectures to meet Safety Integrity Levels (SIL) Challenges and considerations

Menno van der Bij
Deputy Manager Instrument Department
Technip Stone & Webster Process Technology

Jelle-Gerard Wijnja
Technology Manager Ethylene Furnaces
Technip Stone & Webster Process Technology

Paul de Roode
Sr. Process Control Engineer
Technip Stone & Webster Process Technology

Prepared for Presentation at the 2016 Spring National Meeting
Houston, 10-14 April 2016

Voting architectures to meet Safety Integrity Levels (SIL) Challenges and considerations

Menno van der Bij

Deputy Manager Instrument Department

Technip Stone & Webster Process Technology

Jelle-Gerard Wijnja

Technology Manager Ethylene Furnaces

Technip Stone & Webster Process Technology

Paul de Roode

Sr. Process Control Engineer

Technip Stone & Webster Process Technology

Abstract

This paper addresses some of the challenges and considerations for implementation of safety instrumented systems (SIS) for ethylene cracking furnaces.

To ensure stable operation, many process variables are continuously controlled from a distributed control system (DCS). To ensure that safe operation is maintained, a SIS is implemented to prevent operation outside the design constraints of the cracking furnace.

When the reliability of the control functions can be improved the demands on the safety system will be reduced. Under very specific conditions as defined in the standards, instruments for both control and safety functions can be shared while still achieving the required safety integrity level (SIL).

The selected measurement principle might be perfect for control, but not suitable for a safety function when voting architectures (i.e. more than one instrument) are required. By careful evaluation of the applications a choice can be made between diversity and / or redundancy.

From safety point of view a high reliability of the SIS is required, but at the same time high availability shall be maintained to prevent spurious trips. These may be conflicting objectives when selecting voting architectures e.g. for ESD valves. Reducing the number of valves by employing partial stroke testing sounds attractive, but may introduce a false sense of safety.

Introduction

Safety Integrity Level (SIL) is generally about safety and instrumentation but can also include mechanical protection devices. Many papers start with explaining the meaning of SIL. There are many good papers published and numerous books written on this topic. When you are interested in a comprehensive discussion on SIL, please consult the bibliography at the end of this paper. Within this paper we will only give a "short" introduction to the background of safety instrumented systems (SIS) and SIL.

Safety has always been considered when designing a cracking furnace. Lessons are learned from mistakes made and industry has developed good engineering practices by publishing standards.

Incidents and near misses suggest that a 100% safe design may be very difficult to accomplish. Nevertheless, we are challenged and should stay motivated to achieve the 100% safe target as close as possible. This is reflected in the as low as reasonably practical (ALARP) model, which will be discussed later in this paper.

Sometimes this means that we have to invest in hardware and operational procedures. But we also see that with a good understanding of the process and related hazards, safety can be increased, and even cost reduced, by removing unnecessary safeguards and/or safeguards that may introduce new hazards.

But always keep in mind as someone said:

"If you think safety is expensive, try an accident next time"

Abbreviations and acronyms

ALARP	As low as reasonably practical (IEC 61508-5 Annex C)
BFW	Boiler feed water
CCPS	Centre for chemical process safety
DC	Diagnostic coverage
DCS	Distributed control system
ΔP	Differential pressure
ESD	Emergency shutdown
FG	Fuel gas
GWR	Guide wave radar
HAZOP	Hazard and operability (study)
HPSSH	High Pressure Steam SuperHeater
HFT	Hardware fault tolerance
IEC	International electrotechnical commission
ISA	The instrumentation, systems and automation society
LOPA	Layer of protection analysis
MP	Medium pressure
PE	Programmable Electronic
PFD_{avg}	Average probability of failure on demand
PLC	Programmable logic controller
PST	Partial stroke testing
RRF	Risk Reduction Factor
SFF	Safe failure fraction
SIL	Safety integrity level (IEC 61508 / 61511)
SIF	Safety instrumented function (safety loop)
SIS	Safety instrumented system (ESD system that contains all SIF's)
SRS	Safety requirement specification
TLE	TransferLine Exchanger

Definitions

Consequence	The direct, undesirable result of an accident, usually measured in health/safety effects, environmental damage, loss of property or business costs
Frequency	Number of occurrences of an event per unit of time; used with LOPA
Hazard	An inherent physical or chemical characteristic that has the potential for causing harm to people, environment or property (source CCPS)
Hazard scenario	Specific, unplanned event or sequence of events that has an undesirable consequence (source Primatech)
Likelihood	Number of occurrences of an event expressed either as frequency or probability without using numbers. Used during HAZOP studies to provide an order of magnitude
Probability	The expression for the likelihood of occurrence of an event or an event sequence during an interval of time or the likelihood of the success or failure of an event on test or on demand. By definition, probability must be expressed as a number ranging from 0 to 1
Risk	Likelihood x Consequence Risk is a measure of human injury, environmental damage, or economic loss in terms of both the incident likelihood and the magnitude of the loss or injury (source CCPS)

What are the Process design objectives?

From an instrument engineer's perspective, a cracking furnace is a "simple" piece of equipment. Hydrocarbons in, hydrocarbons out.

The furnace mixes hydrocarbons with steam, heats the feed up, and then cools it down and finally sends it to somewhere else. Waste of energy! Does not make much sense if you do not understand the process.

If the instrument engineers have no knowledge of or do not understand the process, how can they be expected to select the correct instrumentation, control and safeguarding system?

It is therefore important that those responsible for implementing a SIS are aware of the process design objectives. For this purpose, at least the following documents shall be published and available before beginning to design a SIS:

- Piping and instrument diagrams
- Process description
- Process control description
- Process safeguarding description and cause & effect diagram
- Alarm and trip setting list
- Safety assessment reports
- Safety requirement specification

The following controlled variables are safeguarded to prevent furnace operation outside the safe window. When any of these thresholds are exceeded, the SIS will bring the furnace in a total or partial shutdown (hot standby).

- LL Feed flow to the furnace
- LL dilution steam flow to the furnace
- HH and LL fuel gas pressure at burners
- HH temperature of high pressure export steam
- HH and LL combustion chamber pressure
- HH and LL steam drum level

Within this paper voting architectures for fuel gas emergency shutdown (ESD) valves and two process systems will be discussed:

- Dilution steam flow
- Steam drum level

Hereafter follows a short description of the process objectives and possible hazards as a result of deviating from the process design intent.

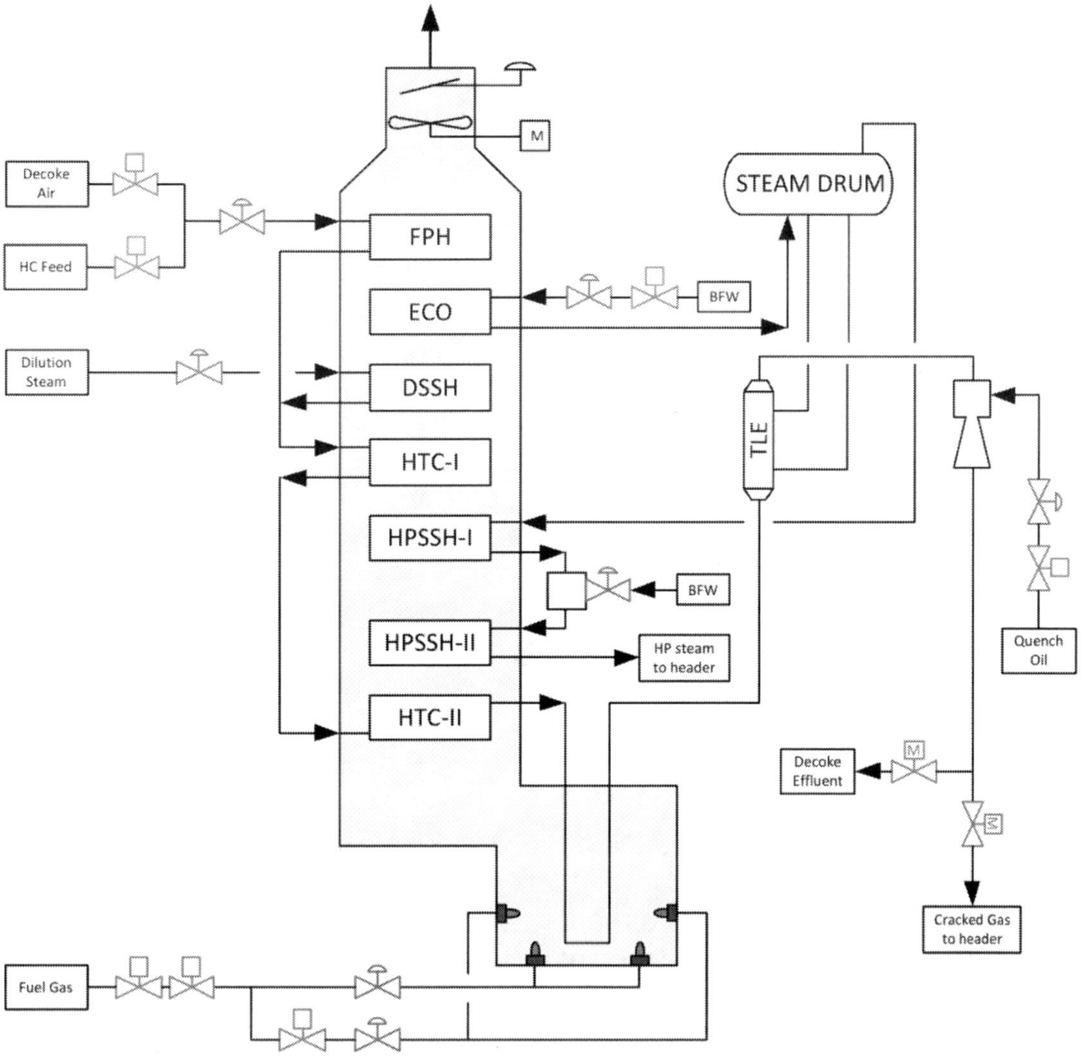

Figure 1

Dilution steam

Via multiple parallel control valves dilution steam is routed to the cracking furnace. In the furnace convection section, before entering the radiant coil, the hydrocarbon feed and dilution steam are preheated up to temperatures where cracking reactions are about to start.

Depending on the mode of operation the function of dilution steam varies and the effect of loss of dilution steam can differ.

The ethylene furnace has three modes of operation:

1. Cracking mode - normal mode of operation.
   In ethylene cracking furnaces dilution steam is mixed with hydrocarbons before entering the radiant coil, the location where the cracking reactions take place. The function of dilution steam is to lower the hydrocarbon partial pressure, minimizing the occurrence of secondary condensation reactions meaning,
   - maximizing the yield towards valuable lower olefins and
   - minimizing coke formation, resulting in a longer run length

   Additionally, at elevated temperatures, the presence of steam is removing carbon deposits (coke) from the coil to form carbon monoxide, hydrogen and carbon dioxide by way of the water-shift reaction. Less coke is resulting in a longer on stream time.

   The amount of steam added during normal operation is depending on the type of hydrocarbon feed and cracking severity, typically varying from 0.3 kg steam per kg ethane up to 0.8 kg steam per kg vacuum gasoil.

   With increasing coke thickness over the run both the maximum tube metal temperature and the venturi pressure ratio increases. When reaching the end of run criterion the coke layer needs to be removed by decoking.

2. Decoking mode
   With the furnace being hot and by the addition of decoke air the coke layer is removed via burning. Besides being a cooling medium dilution steam is also acting as a diluting agent (reducing O_2 partial pressure) in order to moderate initial burning of coke and prevent hot spots which would lead to damage of the radiant coil. Instead of dilution steam also MP steam can be used during decoke.

3. Standby mode
   In between different modes of operation the furnace is kept in hot condition by keeping the burners in operation. Dilution steam is acting as a cooling medium and keeps the furnace hot in a safe way.

During initial furnace start-up, before routing steam to the furnaces, the dilution steam system is warmed up and drained in order to minimize water being send to the convection section. As more burners are lit and the temperature in the furnace is rising, gradually the dilution steam flowrate is increased up to the furnace standby conditions.

Flow Control

The dilution steam flowrate to be controlled varies from the minimum case as set by the decoke case (or normal operating case for cracking light feedstock) up to the maximum case normally set by the hot steam standby case (or normal operating case for cracking heavy feedstock). The typical dilution steam mass flow turndown requirement is around 1:2 (excluding start-up and shut-down). For ease of start-up it is recommended to increase the turndown of the steam flow measurement.

Dilution steam generation

The available dilution steam pressure at the furnace inlet depends on the plant design. In liquid crackers it is determined by the approach with quench oil whereas in gas crackers the approach with the MP steam condensation temperature is determining. In both types of crackers MP steam is used to superheat the dilution steam by at least +10-15°C. Typically the dilution steam condition at the furnace inlet is around 7.5 bara, 185°C. In general dilution steam contains organics including phenols. The organic compound have a tendency for polymerization.
B**ecause** of the importance of dilution steam to the safe operation of the steam cracker MP steam is used as a back-up. In case of loss of dilution steam pressure the MP steam pressure controller will act.

Possible hazards

In general a higher than required dilution steam flow, besides being an operational upset, is not regarded as a hazard. Depending on the furnace design (e.g. convection section material selection) less or no flow of dilution steam can lead to:

- Fouling of the convection section as a consequence of vaporization of liquid hydrocarbons at too high wall temperatures
- Mechanical damage of part of the convection section due to exceeding of design temperatures (e.g. Dilution Steam SuperHeater convection section bank). Tube rupture will lead to release of hydrocarbons to the atmosphere
- Rapid coke formation inside the radiant coil and possible overheating of the tubes. This may result in blockage and consequently instantaneous tube failure (loss of containment, fire in radiant box)

Steam drum level

Depending on furnace design, mode of operation, type of feedstock and cracking severity, the gas temperature leaving the radiant coil and entering the TransferLine Exchanger (TLE) can vary from 750°C to 890°C. The cracked gas is rapidly cooled in the TLE to stop the pyrolysis reactions, minimizing the residence time thus maximizing the yield towards valuable olefins. The recovered heat is used for high-pressure steam generation.

In the ethylene cracking furnace the high pressure steam generation system consists of multiple TLE's, one steam drum and interconnecting risers and downcomers.

High pressure boiler feed water (BFW) is routed via the convection section economizer to the steam drum. In the TLE cracked gas (tube side) is cooled by evaporating high-pressure water (shell side) from the steam drum under natural circulation. The steam/water mixture is separated into water and steam phase in the steam drum. Typical steam drum operating conditions are 110 bara + 318°C. In the convection section the saturated steam coming from the steam drum is superheated against flue gas to the required outlet temperature before being routed to the superheated high pressure steam export header (~510°C). The export steam outlet temperature is controlled via the injection of BFW using a desuperheater.

The level in the steam drum is controlled by using level control by balancing the BFW supply with the export steam flow. At initial furnace start-up the steam drum needs to be filled with BFW and pressurized up to the pressure required for lining up the superheated high pressure steam export line to the header.

Possible hazards

Having no level in the steam drum will result in the TLE shell side running dry leading to mechanical damage and total failure of the TLE's. A too high level in the steam drum will result in water carry over to the convection section High Pressure Steam SuperHeater bank (HPSSH) leading to thermal shock and fouling.

What do the standards tell us?

Standards reflect good engineering practice. They have been developed over many years. Some are existing for more than 50 years, others recently published for the first time. There are national and international standards. And although they all have the same objective they differ in content and sometimes even contain contradicting requirements. This suggests that the best solution for one application (location) may not be the best for another. This may sound strange, but local circumstances can be diverse, safety perceived differently in other industries and the characteristics of the fluid to be heated creates different hazards in case of loss of containment.

This is a first indication that compliance with a standard may not be adequate to fulfill the safety target. Standards are a good starting point and can be used as check lists to achieve a safe design. But only with additional safety assessments the main target can be achieved:

Design a cracking furnace with a risk As Low As Reasonably Practicable. **ALARP!**

ALARP model

Because ALARP may be interpreted according to the local culture, the concept and targets have to be placed in time and perspective. A plant designed 30 years ago may still meet all the safety targets set at that time, but circumstances and perception of safety may have changed. This is one of the reasons that a safety assessment e.g. a hazard and operability (HAZOP) study, has to be repeated several times in the process plants life. (See lifecycle concept from IEC 61511-1)

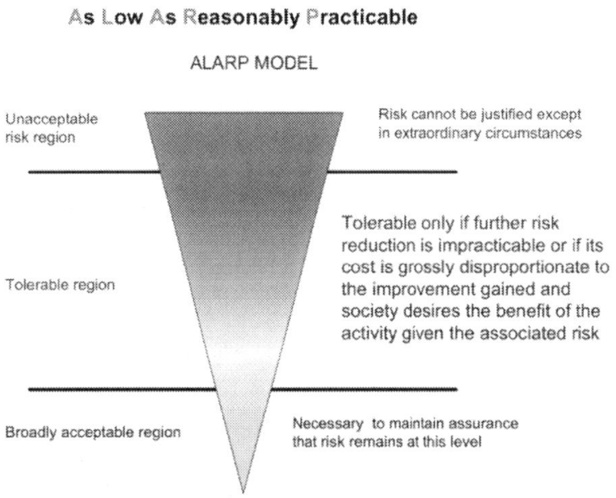

Origin: The UK Health and Safety at Work act 1974

For new plants, the objectives are more clear; ALARP as per today's expectation's.

These expectations may come from:

- The Government and other local authorities
- Local community
- People working on site
- Operating company

What are the Safety Objectives?

The safety objectives can be expressed as risk reduction targets (see ALARP model). Risk reduction targets have to be assigned for:

Personnel safety: This includes people on site but can extend to the public, outside the fence. On-site includes operating and maintenance crew. Own staff, but also contractors and visitors. In the context of process safety it addresses hazards from process upsets, not as a result of for example falling.

Environmental protection: Environmental consequences are typically caused by loss of containment and can affect flora and fauna. With safety assessments, possible effects within the fence and over the fence are identified. It is important to know if the effects are permanent or temporary, what is required to clean up and how much time this will take.

Asset loss: Asset losses take into account damaged equipment; cost for replacement but also the labor cost for demolishing and replacement. In addition direct production losses should be taken into account but also knock on effects on other units could be considered.

Risk reduction targets are the acceptable (ALARP) frequency of an undesired event.

Who will decide these acceptable risk targets?

Asset loss is clearly something for the management of the operating company.

But what about personnel safety or the environment?

This is also the responsibility of the operating company, but the targets will (should) be set by the community i.e. the local authorities.

Some countries have set a target risk reduction for loss of life. For example the frequency of an incident resulting in a single fatality shall not occur more than once in 100,000 years.

Safety assessment; HAZOP Study

Many tools are available for performing safety assessments. The most common in the process industry is the HAZOP study.

The purpose of a HAZOP study is to identify undesired events (hazards), as a result of operating the plant outside its design constraints.

It is important that the HAZOP study report clearly describes the initiating event(s), undesired resulting event, severity of the undesired outcome and the expected frequency of occurrence. The ultimate consequence of the hazard shall be described assuming no safeguards are in place.

The combination of severity and frequency will give us a risk ranking. Generally 3 regions are identified: Acceptable, ALARP and unacceptable. The next step is to identify independent protection layers. The HAZOP study team may agree that with these protection layers the design is acceptable or ALARP. In case the design is unacceptable, the team shall give the recommendation that additional protection layers are required.

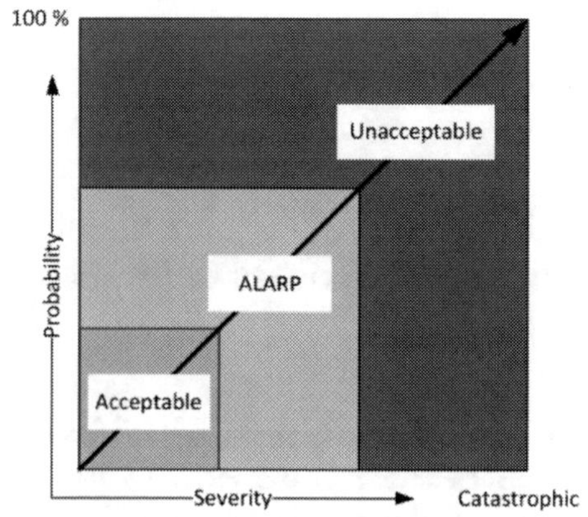

Note: Although this is not to be expected with cracking furnaces, the team may also decide to reject the design as unacceptable even with additional safeguards. In that case the process design has to be changed all together.

When the design is acceptable or ALARP, it is common practice to proceed with a detailed risk analysis to determine the required risk reduction (strength) of the protection layers. One of such methods is a layer of protection analysis (LOPA).

Layer of Protection Analysis

IEC 61511 part 3 suggests several methods for determining the required strength of the SIS. One of the methods is LOPA. With a LOPA it can be determined if the risk is acceptable or ALARP. With LOPA it can also be determined if a SIS as an additional safety layer is needed and the required SIL.

The LOPA method is extensively discussed in the CCPS publication "LOPA, Simplified process risk assessment".

Control system as safety layer

Failure of a control loop can be the cause of a process upset. For example, if the fuel gas (FG) pressure control loop fails and fully opens the control valve, the pressure can increase beyond the safe limits. In this example, failure of the control loop is the initiating event.

If the pressure controls of a common fuel gas knock-out drum fails, the pressure at the battery limit of the furnace may reach the design pressure. In this case, the FG pressure control loop can be considered as a protection layer.

It is important that this difference is identified during the HAZOP study and clearly stated in the HAZOP study report.

Safety requirement specifications.

At the beginning of this paper it is already mentioned that without knowledge of the process, it will be difficult for the instrument engineer to select the required safety reduction by implementing and selecting instrumentation.

Even more important are the requirements for the SIS.

The standard IEC 61511 gives guidance for preparing a safety requirement specification (SRS). This document shall contain all information for the instrument engineer to design the safety system meeting the targets defined in the safety assessments (e.g. HAZOP study and LOPA).

Some of the aspects that shall be addressed in the SRS are:

- Response time e.g. closing time of ESD valves
- Maintenance override switch; required or maybe not allowed
- Target risk reduction i.e. SIL

Safety Integrity Level?

Risk reduction can be expressed as a SIL. Each higher SIL requires or provides an additional risk reduction of a factor 10.

The IEC 61508 and 61511 standards include 4 levels. When a risk reduction of more than 100,000 (SIL 4) is required, the design is generally considered as unacceptable and the conceptual design should be reviewed and/or additional independent layers of protection should be provided such that the demand on the safety instrumented function (SIF) under consideration will become lower.

For the low demand mode of operation, SIL is defined as follows:

Safety integrity levels: Demand mode of operation		
SIL	PFDavg	RRF
4	10^{-4} to 10^{-5}	10,000 to 100,000
3	10^{-3} to 10^{-4}	1,000 to 10,000
2	10^{-2} to 10^{-3}	100 to 1,000
1	10^{-1} to 10^{-2}	10 to 100

PFD_{avg} = Average probability of failure on demand
RRF = Risk Reduction Factor

This means that when a safety instrumented function (SIF) shall have a SIL 1, it should be able to provide a risk reduction of a factor 10 or higher. On the other hand, when the installed SIF has a SIL 1, it will provide a risk reduction between 10 and 100.

How do we arrive at required safety integrity levels?

The determination of the required SIL, starts with a HAZOP study followed by a so called SIL assessment. Several methods to perform a SIL assessment are described in IEC 61511-2004 part 3. One method is a LOPA as explained earlier in this paper.

Once the required SIL is known, the required hardware fault tolerance (HFT) of the SIF can be determined.

Other requirements to meet the required SIL are:

- Average probability of failure on demand (PFD_{avg}) of the SIF
- Systematic capability of the components of the SIF

Note: These requirements are further explained in IEC 61508 and IEC 61511.

Hardware fault tolerance

The standard IEC 61511-2004, contains a table with requirements for the HFT.

An HFT of 1 means that the failure of a single device:

- Will not result in an unsafe situation
- Would not render the safety system ineffective i.e. it can still bring the process in a safe state when a demand is placed on the safety system

Different voting architectures are available to achieve this e.g. 1oo2 voting or 2oo3 voting.

The hardware constraints within IEC 61511-1 chapter 11.4.4 are defined as follows:

Table 6 – Minimum hardware fault tolerance of sensors and final elements and non-PE logic solvers		
SIL	*Minimum hardware fault tolerance (see 11.4.3 and 11.4.4)*	Typical architecture
1	*0*	1oo1; 2oo2
2	*1*	1oo2; 2oo3
3	*2*	1oo3; 2oo4
4	*Special requirements apply (see IEC 61508)*	

Note: For additional explanation and comments, refer to IEC 61511-2004.

An alternative method is described in IEC 61508-2010.

The hardware constraints within IEC 61508-2 chapter 7.4.3.1 are defined as follows:

A subsystem can be regarded as type 'A' if, for the components required to achieve the safety function:

1. *The failure modes of all constituent components are well defined* ***AND***
2. *The behavior the subsystem under fault conditions can be completely determined* ***AND***
3. *There is sufficient dependable failure data from field experience to show that the claimed rates of failure for detected and undetected dangerous failures are met (see 7.4.7.3 and 7.4.7.4)*

A subsystem shall be regarded as type 'B' if, for the components required to achieve the safety function:

1. *The failure mode of at least one constituent component is not well defined* ***OR***
2. *The behavior the subsystem under fault conditions cannot be completely determined* ***OR***
3. *There is insufficient dependable failure data from field experience to support claims for rates of failure for detected and undetected dangerous failures are met (see 7.4.7.3 and 7.4.7.4)*

The architectural constraints of either table 2 or table 3 shall apply to each subsystem carrying out a safety function.

Table 2 – Hardware safety integrity: architectural constraints on type 'A' safety related subsystems:

Safe failure fraction	*Hardware fault tolerance* [2)]		
	0	*1*	*2*
< 60%	*SIL 1*	*SIL 2*	*SIL 3*
60% - <90%	*SIL 2*	*SIL 3*	*SIL 4*
90% - 99%	*SIL 3*	*SIL 4*	*SIL 4*
≥ 99%	*SIL 3*	*SIL 4*	*SIL 4*

Notes:

1. *See 7.4.3.1.4 for details interpreting this table*
2. *A hardware fault tolerance of N means N+1 faults could cause a loss of the safety function.*

Table 3 – Hardware safety integrity: architectural constraints on type 'B' safety related subsystems:

Safe failure fraction	*Hardware fault tolerance* [2)]		
	0	*1*	*2*
< 60%	*Not allowed*	*SIL 1*	*SIL 2*
60% - <90%	*SIL 1*	*SIL 2*	*SIL 3*
90% - 99%	*SIL 2*	*SIL 3*	*SIL 4*
≥ 99%	*SIL 3*	*SIL 4*	*SIL 4*

Notes:

3. *See 7.4.3.1.4 for details interpreting this table*
4. *A hardware fault tolerance of N means N+1 faults could cause a loss of the safety function.*

Safe failure fraction (SFF) is defined as:

$$\mathrm{SFF} = \frac{\lambda^{DD} + \lambda^{SD} + \lambda^{SU}}{\lambda^{DD} + \lambda^{DU} + \lambda^{SD} + \lambda^{SU}}$$

λ^{SD}	Safe failure rate, detected (revealed safe)
λ^{SU}	Safe failure rates, undetected (unrevealed safe)
λ^{DD}	Dangerous failure rates, detected (revealed dangerous)
λ^{DU}	Dangerous failure rates undetected (unrevealed dangerous)

Voting architectures

The most common voting architecture is 1oo1. For initiators (transmitters) this means that there is only one (SIF) transmitter. If it fails it can cause a nuisance trip or may not respond when there is a safety demand. For a valve it means that there is only one valve e.g. to isolate the process. The valve may close due to a failure (loss of signal) causing a nuisance trip or may not respond to a safety demand (e.g. ball does not move) causing an unsafe situation.

The objective of alternative voting architectures may be:

- To increase availability e.g. 2oo2 voting
- To increase safety e.g. 1oo2 voting
- To increase both safety and availability e.g. 2oo3 voting

With 2oo2 voting on transmitters, both transmitters need to detect an unsafe situation before a safety action by the SIF will take place. If either transmitter is unhealthy (i.e. not able to detect an unsafe situation), the SIF will take no action when there is a demand.

With 1oo2 voting on transmitters, only one of the 2 transmitters need to detect an unsafe situation for the SIF to take action. As long as at least one of the two transmitters is operational, safety is maintained.

With 2oo3 voting on transmitters, two transmitters need to detect an unsafe situation before a safety action by the SIF will take place. If one transmitter is unhealthy, the SIF will still take action when there is a demand, however when two have failed no action may take place when there is a demand.

The most common architectures found with cracking furnaces are, 1oo1, 2oo3 and 1oo2 (mainly with ESD valves).

Median or average

When 2oo3 voting is selected, it can be considered to use the signals from the 3 safety transmitters also for control. Note that this is permitted under strict requirements as prescribed in IEC 61511-1.

When these 3 signals are connected to the distributed control system (DCS) either the average or median value of the 3 can be selected as process variable input to the controller. If for example the measured value of all 3 transmitters is at 50% of scale, than both the average and median value is also 50%. If one transmitter fails or drifts away from the other two, e.g. the signals become 50%, 50% and 0%, then the average will be 33% whereas the median remains 50%. With median selection the reliability of the control loop will be improved while with the average value the controller may drive a control valve to an undesired position resulting in a demand on the SIS.

Dilution steam low-low flow trip

Figure 2 shows a simple sketch of the dilution steam flow control loop. In addition, depending on the location of the flow meter and battery limit variations, a pressure and temperature measurement may be required for density compensation of the flow measurement.

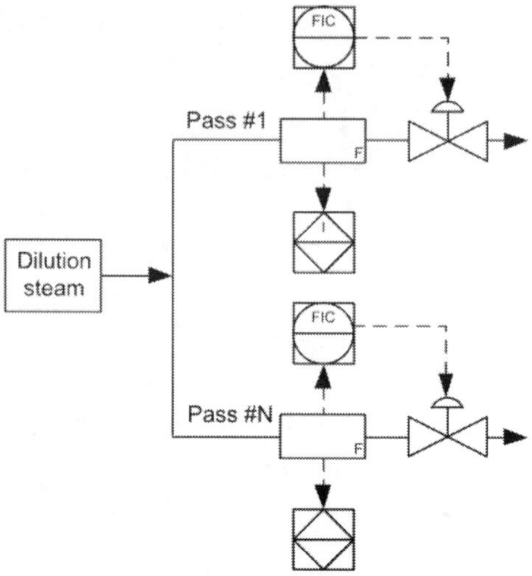

Figure 2

In the cracking mode, loss of dilution steam has as consequence blocked coils and tube rupture may be the result. To prevent this incident to occur, a low-low dilution steam flow trip is implemented. Although it may be required for control purposes, pressure and temperature measurements are not always needed for density compensation of the SIS flow measurement. If required, however, they shall be independent from the instruments used for density compensation of the flow indication in the DCS.

Note: Failure of a pressure or temperature measurement may cause erratic flow indications. Within the DCS measures can be taken to limit the impact or exclude faulty inputs. Within the SIS it may cause a nuisance trip or renders the safety function ineffective. The impact of faulty density corrections shall be taken into account when determining the SIL that can be achieved.

The safety assessments shall address the need for a trip based on low low dilution steam flow in the common header to the furnace or for each individual pass.

To meet the required SIL, the voting architecture has to be determined based on the required HFT. For SIL 1 and 2 (based on IEC 61508), a HFT of 0 is sufficient i.e. a voting architecture of 1oo1 can be chosen.

For a 1oo1 voting architecture two, independent, flow measurements are required: one for control (DCS) and one for safety (SIF).

The most common flow measurement principle is the **flow orifice** with 2 differential pressure (ΔP) transmitters. Although the orifice plate is a common device for both flow measurements it is not considered as a source for a common cause failure.

Separate flange tappings are required with individual process isolation (root) valves. According good engineering practice, the transmitters are installed below the process connections. The impulse lines to the transmitters will be filled with water.

Sometimes 3 or even 4 transmitters are required. In principle many transmitters can be installed but the number of independent tappings is limited and space to install the transmitters (in a protection box) may be restricted.

At locations with ambient temperatures below 0 °C, the water in the impulse lines shall be protected against freezing (possible blockage) and winterizing (heat) tracing is required. The tracer for the control loop transmitter impulse lines shall be independent from the tracer for the SIF transmitter impulse lines. When multiple transmitters are required for the SIF, also the tracers of the respective impulse lines shall be independent from each other. Both electrical and steam tracing is possible, but with electrical tracing it is easier to identify a trace heating failure and generate an alarm in DCS.

Another cause for blockage of the impulse lines is the quality of the dilution steam. If the dilution steam has been generated from process condensate it may contain traces of hydrocarbons, polymerization is likely to occur. Blockage of the impulse lines can be slow but also fast (in weeks) and the rate may be different for each leg of the impulse lines. This can result in erratic readings, nuisance trips or no trip while there is a demand.

The best, obvious solutions are to improve either the quality of the dilution steam or use another source e.g. medium pressure (MP) steam.

Diaphragm seals with capillaries may be another solution, but the relative high steam temperature could become a constraint in selecting the appropriate capillary fluid. Even with seals, it may still be possible that root valves become blocked. With 2oo3 voting systems, space may also become a limitation.

Another option is to select a **vortex flow meter**.

Some companies consider vortex flow meters as first choice for any flow measurement application. Higher turndown while maintaining high accuracy are the main reasons. Compared to ΔP flow meters, a vortex flow meter does not require impulse lines hence no winterizing may be required.

On the other hand, compared to ΔP head flow meters, the number of measurements per primary element is limited. Initially 2 vortex flowmeters in series where required for two independent measurements. Today most vendors can supply vortex meters with dual sensors. These vortex meters generally share a common shedder bar (bluff body). Sensor designs are different between vendors and an assessment is required to determine that sufficient independency is available between the two measurements for control and safety.

Some companies demand 2oo3 voting for any trip causing a furnace total shutdown. If sharing transmitters is not permitted, in total 4 sensors are required. This requires as a minimum 2 dual sensor vortex meter in series. The manufacturer has to be consulted for the straight length required between the vortex meters to maintain an acceptable accuracy.

Compared to the ΔP flow meters, which only requires one obstruction (the orifice plate), two obstructions (2 shedder bars) have to be installed. This will create 2 times a pressure loss, in total higher than the permanent pressure loss of a single orifice plate. Since dilution steam pressure at battery limit is typically just sufficient to arrive at an acceptable pressure profile it is important to check the hydraulics for the maximum dilution steam flow rate before proceeding with vortex meters.

At this point it is key to look at the most important criteria for selecting a sensor for control and for safety functions. For control, repeatability may be important in addition to accuracy and response time. For a safety function, accuracy may be less important where response time is.

For a 2oo3 application where the 3 sensors can be shared for control and safety, two vortex meters are required: one with a dual sensor and one with a single sensor. A shorter straight length between the vortex meters can be accepted compared to the straight length required for the upstream vortex meter. Because of the shorter straight length, the uncertainty will be higher. When median selection of the 3 values is used in both the DCS and SIS, the measurement of the downstream vortex meter may be cancelled out and will only become functional when either sensor of the upstream vortex meter "fails" (demands a trip).

Note: Even with shorter straight lengths is it possible to decrease the additional uncertainty by adjusting the meter factor "k". This adjustment is based on actual installed straight length. The manufacture's installation instructions shall be consulted for the corrections to be applied.

When 4 measurements are required, one of the sensors of the upstream vortex meter can be used for control, and for the SIF, the second together with the two sensors of the downstream vortex meter can be used in a 2oo3 voting architecture. It should be noted that the failure of the control sensor (e.g. fail/drift high) will, although not immediately, place a demand on the safety system when the control valve is (slowly) driven to its closed position.

Note: Some plants use FOUNDATION Fieldbus™ (FF) as standard for control applications. At present, FF is not available for safety functions. In case of 2oo3 voting a choice has to be made. Sharing transmitters for control and safety requires all 3 transmitters to have analogue output signals; otherwise 4 transmitters are required.

Steam drum low-low level trip

Figure 3 shows a simple sketch of the high pressure export steam control loop. The so called 3 element level control loop consists of a measurement for the boiler feed water supply to the steam drum and desuperheater, the export steam flow and the level in the steam drum. The controls are addressed in many papers and books. For all measured parameters, flow and level, repeatability is more important than the absolute accuracy.

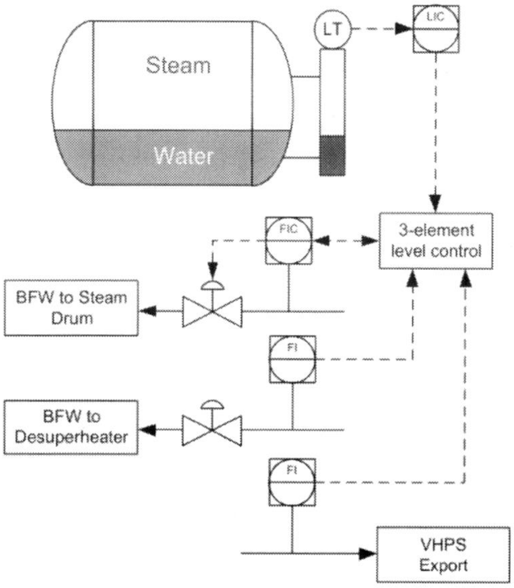

Figure 3

From a safety point of view, overflow of the steam drum and an empty steam drum are the hazards to be addressed. Overflow, high level, can be prevented by either closing the boiler feed water supply or increasing the blowdown.

An empty steam drum requires removal of the heat input to the "boiler" by shutting off the fuel gas supply to the furnace.

Traditional system design includes:

- Transmitter(s) for level control
- Level switch(es) for HH-level trip
- Level switch(es) for LL-level trip

The disadvantage of (level) switches is that their availability is difficult to verify other than by regular testing. When the level switch (e.g. float switch) is replaced with a transmitter, the measurements for safety and control can be continuously compared with each other.

For this reason, majority of new built plants have one transmitter for control and at least one transmitter for safety.

It is common practice to measure the level of the water in the steam drum not directly, but by measuring the level in a so called reference chamber. This reference chamber (Figure 4) has an open connection to the steam drum. Based on the principal of communicating vessels, the level in the reference chamber will be the same as in the steam drum. This however is only true when the density of the steam and water in the reference chamber is the same as in the steam drum.

During initial filling of the steam drum, the level in the reference chamber is close to the level in the steam drum.

If the steam drum is operating at e.g. 323°C the density of the water in the steam drum is 663 kg/m^3. The steam in the reference chamber is condensing. The temperature in the reference chamber is less than in the steam drum. Therefore, the water is heavier than the water in the steam drum and circulates back into the steam drum. If that temperature is 292°C the density of the water is 739 kg/m^3. Consequently, if the level in the steam drum is 1 meter, the level in the reference chamber is 0.89 m, an error of 10%. (See Figure 5)

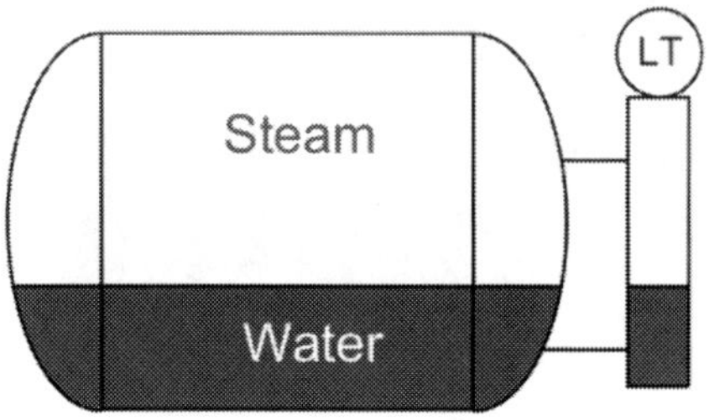

Figure 4

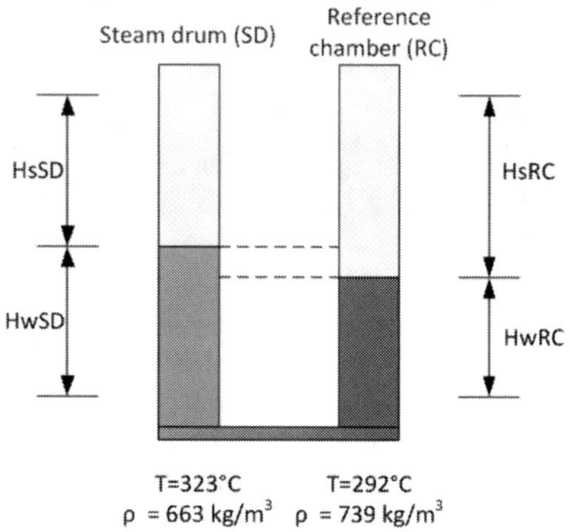

Figure 5

Note: The change in steam density from cold to operating temperature is also a cause of errors when not considered in the selection / calibration of the level instrument or otherwise compensated for.

For steam drum level measurement, the following principles may be considered:

- Hydrostatic head with ΔP transmitter
- Buoyancy with displacer type level transmitter
- Reflection with guided wave radar (GWR)
- Reflection with magnetostrictive type level transmitter
- Interval (point) measurement with electronic probes

Each of the measurement principles will give a different output. For example, a displacer will indicate the correct level in the steam drum if calibrated for the steam drum water normal operating density; at start-up however it will have an offset. On the other hand, the GWR will measure the level in the reference chamber, independent from the density of the water. During start-up it will indicate the correct level in the steam drum; however during normal operation it will indicate a lower level by 5% to 10%.

To be able to compare the control transmitter output with the safety transmitter output, they have to be of the same measurement principle.

When 3 transmitters are required for 2oo3 voting, the consequences of selecting different measurement principles shall be carefully evaluated; see above comparison between displacer type measurement and GWR measurement principle.

Note: Before selecting the architecture and type of level instruments, requirements from local boiler codes shall be observed. Many boiler codes demand 1 or 2 glass type level gauges. Others permit, at least one, glass type level gauges to be replaced with a magnetic level gauge. Several manufactures offer magnetic level gauges in combination with a GWR or magnetostrictive measurement which make installation more cost effective.

Because none of the level measurement principles will provide correct level measurement from start-up (low temperature) to normal operating, three options are to be considered:

- Calibrate the level instrument for operating density and accept deviations until boiler is at operating temperature.
- Calibrate the level instrument for operating density and correct deviations until boiler is at operating temperature.
- Accept offset (e.g. with GWR) and take this into account with determining the trip settings and controller set point.

For density correction of a displacer level transmitter the pressure or temperature in the boiler are needed. Correction in the control system is generally sufficient. If the steam drum is always operating at the same temperature, the correction can be switched off during normal operation.

If density correction is also included within the safety system, than the failure of the pressure or temperature measurement has to be take into account when making the SIL implementation verification for the complete SIF.

In principle as many level instruments as required can be installed on the steam drum. Each level instrument shall have its dedicated nozzles and isolation valves. In cold climates, trace heating may not be required when instruments are properly insulated and drained when the cracking furnace is out of operation. However, if tracing is required it shall be applied to each level instrument independent from each other.

The SIF may encompass the following components:

- Initiator e.g. level transmitter
- Intrinsic safe galvanic isolator, analogue input
- ESD PLC analogue input card
- ESD PLC logic solver
- ESD PLC binary output cards
- Intrinsic safe galvanic isolator, binary output
- ESD valves (valve, actuator, solenoid valve)

The required SIL applies to the complete SIF.

For each of the components of the SIF, the PFD_{avg} has to be calculated and added together. The sum shall be less than required SIL. For example, if SIL 2 is required, the sum shall be less than $1E^{-2}$. (Risk reduction > 100). Equations to calculate the PFD_{avg} can be found in IEC 61508 and ISA-TR84.00.02-2002 – Part 2.

The architecture for the initiator and ESD valves do not have to be the same. The following are typical architectures that can be found in the process industry.

Initiator	ESD valve	Typical Max. SIL
1oo1	1oo1	1
1oo1	1oo2	2
1oo2	1oo2	3
2oo3	1oo2	3

Note: The actual SIL that can be achieved depends among others on the test interval of the initiators and ESD valves.

A low-low steam drum level should result in removing the heat input to the furnace by closing fuel gas ESD valve(s). Most furnace safety standards require 2 ESD valves in series for isolation of the fuel gas to the burners. (See Figure 6)

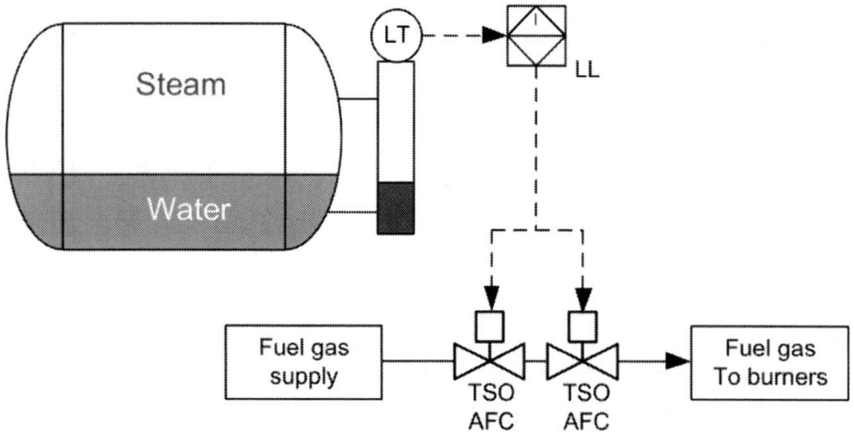

Figure 6

With the same architecture e.g. 1oo1 voting, the SIL that can be achieved with initiators is generally one level higher than with the ESD valves(s). This is because the failure rate of valves is higher than that of a transmitter and, without interrupting the process, transmitters can be tested more frequently than valves.

Consequently, changing the voting architecture for the initiators e.g. from 1oo1 to 2oo3 will lower the PFD_{avg} of the initiators (increase the SIL) and at the same time reduces the nuisance trips caused by transmitter failure. However, this may not substantially improve the

SIL of the SIF or reduce nuisance trip rates of the SIF because of the higher PFD_{avg} of the ESD valves.

ESD valves and partial stroke testing

In most applications, the ESD valves are the weakest link. Changing the voting architecture from 1oo1 to 1oo2 will generally result is a substantial improvement. However adding more valves will not improve the SIL much further due to the common mode failure.

Note: For more information on common mode failures (ß-factor), see IEC 61508 part 6 annex D and SINTEF report A17956.

Partial stroke testing (PST) is promoted to improve the SIL of ESD valves. ANSI /ISA standard ANSI/ISA TR96.05.01-2008 provides guidelines and limitations of PST such as "*For fail-closed applications, the test does not validate that the block valve will seat or that it is leak tight*".

With the fuel gas ESD valves, closure of the valve and seat tightness is very important to prevent any combustible gas mixtures to build-up in the fire box when burners are not ignited.

Where possible, a full stroke test at sufficient test intervals is strongly recommended over improving the SIL with additional PST. Some furnace standards require an annual full stroke test of both ESD valves. If this is not feasible a full bypass with the same ESD valves has to be installed. Within these standards, PST is not discussed or accepted as a means to reduce the full stroke test frequency.

If PST is applied, the following should be taken in consideration:

- The diagnostic coverage (DC) claimed shall be justified; i.e. which failures can be detected. For a fuel gas ESD valve a DC of more than 30% may not be realistic.
- PST is not addressed in the IEC standards. Any claims may have to be approved by local authorities.
- According IEC 61598, DC can only be claimed if the test interval is within the time frame required to repair, which is typically 8 to 24 hours.
- PST can be used to increase the SIL, but not to reduce the hardware fault tolerance required.
- Criteria have to be defined to judge when based on PST valve repair is required.

Summary and conclusions

Before considering which voting architectures may be required, a good understanding of the required performance of the instruments is needed. An instrument or device used in a safety critical application shall be effective, available and reliable.

Compliance with many different standards may be required. These requirements may be conflicting with each other. Only a thorough safety assessment can determine the minimum requirements to achieve ALARP and justify deviations from the standards.

Diverse instrumentation may not necessarily improve the reliability of a safety instrumented function. Measured values may differ from each other, and comparison of the signals to find faulty instruments may become difficult.

When selecting a voting architecture, the complete safety instrumented function shall be considered. A transmitter with SIL 2 has no practical meaning if not all other components in the loop meet the requirements for a SIL2 application and the total PFD_{avg} is lower than the target.

Disclaimer

This paper is based on present knowledge and dated issues of referenced standards. Those responsible for the operation and design of a cracking furnace shall ensure that the safety of the cracking furnace meets the local and globally acceptable "as low as reasonably practical" (ALARP) level. Compliance to an international standard may not be sufficient and it may be even required to deviate from a standard to achieve the target safety level. Safety assessments for all applications, considering the local circumstances, operational procedures, environmental conditions and the public shall be applied, justified and documented.

The material appearing in this presentation is for general information purposes only. Technip S.A. and its affiliated companies ("Technip") assume no responsibility for any errors or omissions in these materials. Technip makes no, and expressly disclaims any, representations or warranties, express or implied, regarding the materials contained in the presentation, including any implied warranties of merchantability or fitness for a particular purpose. Under no circumstances shall Technip, the other sponsors, presenters and any of their respective partners, officers, directors, employees, agents or representatives be liable for any damages, whether direct, indirect, special or consequential, arising from or in connection with the use of materials and information contained in the presentation

Referenced standards

IEC 61508 ed. 2010	Functional safety of electrical/electronic/ programmable electronic safety related systems
IEC 61511 ed. 2004	Functional safety — Safety instrumented systems for the process industry sector
ANSI/ISA-TR84.00.02-2002 – Part 2	SIL Evaluation Techniques Part 2: Determining the SIL of a SIF via Simplified Equations
ANSI/ISA TR96.05.01-2008	Partial Stroke Testing of Automated Block Valves

Bibliography

Reliability prediction method for safety instrumented systems – PDS example collection	Report A17956 2010, SINTEF
LOPA Simplified process risk assessment	CCPS, ISBN0-8169-0811-7
Practical industrial safety, risk assessment and shutdown systems for industry	D.Macdonald, Elsevier, ISBN 07506 58045
Hazops, trips and alarms	D. Macdonald, Elsevier, ISBN 07506 62743
Safety integrity level selection, including layer of protection analysis	E.M.Marszal, E.W. Scharpf, ISA, ISBN 1 55617 7771
Safety instrumented systems verification	W.Goble, H.Cheddie, ISA, ISBN 1 55617 909X
Safety shutdown systems: Design, analysis and justification	P.Gruhn, H.Cheddie, ISA, ISBN 1 55617 6651
Control systems safety evaluation and reliability	W.Goble, ISA, ISBN 155617 6368

RSI
create, evaluate, operate

Developing Abnormal Situation Management Skills in Ethylene Plant Console Operations

2016 AIChE Ethylene Producers Conference

Ethylene Unit Safety Session

April 11-14, 2016

NOVA Chemicals

Agenda

- State of Console Operator training in the US (lack of competency standards)
- World class training example at NOVA Chemicals Joffre
- Comprehensive console operator training on an ethylene simulator
- OTS benefits are tangible ($$$) as they lead to safe, reliable and environmentally friendly operations

NOVA Chemicals

RSI
create, evaluate, operate

Positive shift in safety and training

Ethylene producers increasingly rely on high-fidelity simulators to improve the performance of their console operators, giving them real-life ethylene plant operating experience on a simulator, built to accurately reflect the operating conditions of the actual plant.

This represents a significant improvement from the current operator training culture.

NOVA Chemicals

RSI create, evaluate, operate

Who trains whom?

Of all the professional, trade and administrative disciplines required to operate an ethylene plant, managers, accountants, engineers, electricians, pipefitters, machinists, administrative assistants,

.. console operators, who unarguably occupy the most **safety-critical** position in the plant, are the only ones trained and brought all the way to "qualification" by their coworkers.

NOVA Chemicals

RSI create, evaluate, operate

Who trains whom?

For the billions of dollars that are invested in ethylene plants and the rigorous standards inherent in engineering, design, instrumentation, automation and safety systems, the day to day, round-the-clock control of these plants rests in the hands of console operators for whom neither the state nor the industry impose uniform competency standards.

NOVA Chemicals

RSI create, evaluate, operate

Cause of error in our industry

From; Guidelines for Preventing Human Error in Process Safety, Center for Chemical Process Safety, AIChE

"... a systems perspective is taken, which views error as a natural consequence of a mismatch between human capabilities and demands, and an inappropriate organizational culture."

NOVA Chemicals

RSI create, evaluate, operate

Texas Department of Licensing and Regulation

TDRL

- Hair stylists must successfully complete 1500 hours of training in a state licensed institution
- Only after successfully completing a state sanctioned, written and demonstration test, does the TSDLR issue the Cosmetology Certificate
    - Cosmetology: 1500 hours (37 weeks – 10 months)
    - Esthetics: 750 hours
    - Nail Technology: 600 hours
    - Hair Weaving: 300 hours
    - Hair Braiding: 35 hours

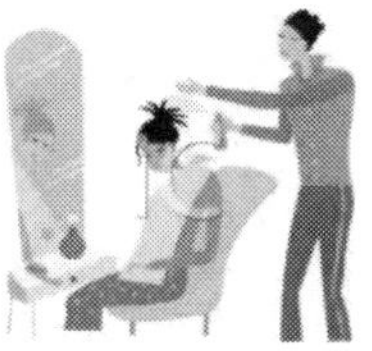

NOVA Chemicals

RSI create, evaluate, operate

In stark contrast

Console operators on the other hand are often in complete, sole control of processes...

...more costly and with more instrumentation than a Boeing 787 Dreamliner

- Under hundreds or thousands of pounds of pressure
- Fast dynamics
- Dangerous chemicals and volatile hydrocarbons

... and yet it is not uncommon for console operators to have achieved this level of responsibility with:

- No more than a high school diploma
- Brought to qualification by a coworker

NOVA Chemicals

RSI create, evaluate, operate

A Pocket of Excellence in our Industry NOVA Chemicals Joffre

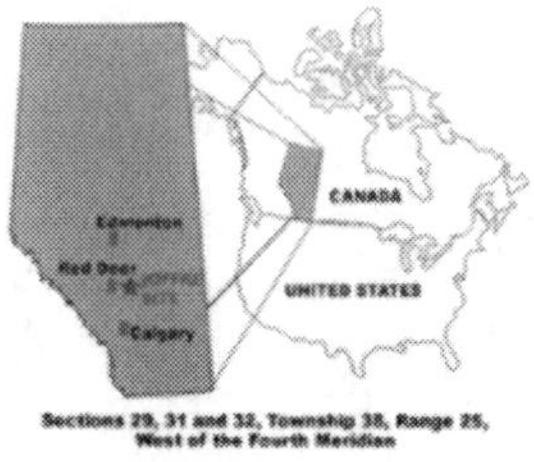

NOVA Chemicals

RSI
create, evaluate, operate

NOVA Joffre; RSI simulators for training and certification

1. Ethylene 1 (E1) – Finishing (1998) – Cracking (2011)
2. Ethylene 2 (E2) – Finishing (1999) – Cracking (2012)
3. Ethylene 3 (E3) – Finishing and Cracking (2000)
4. Polyethylene 2 (PE2) - Process (2001)
5. Pipeline (2012)
6. Polyethylene 1 (PE1) – (2016)

NOVA Chemicals

RSI
create, evaluate, operate

NOVA Joffre's panel operator training process

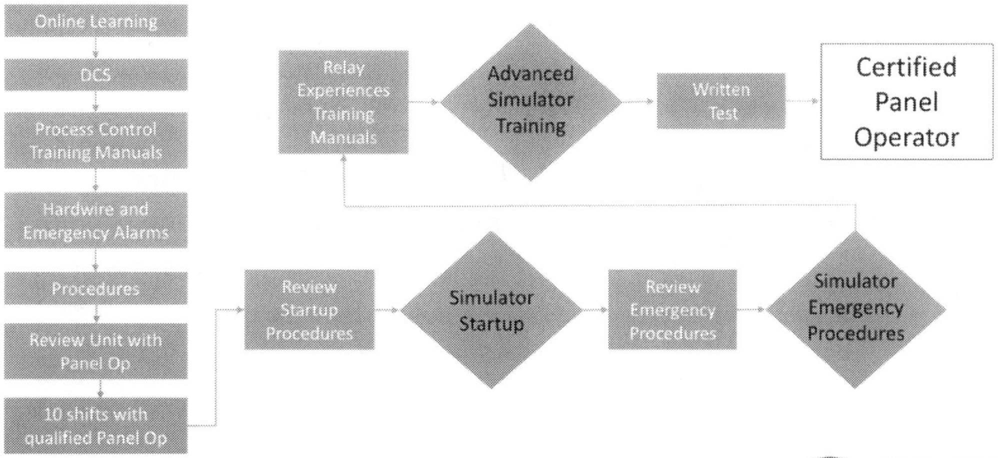

NOVA Chemicals

RSI create, evaluate, operate

E2 Finishing simulator training curriculum

Instructor's Station

- Startup
  - Day 1 Startup Refrigeration systems and WHB
  - Day 2 Cracked compressor start and feed in
- Emergency procedures and trip applications
  - A quick review of key points of procedures and Apps before practicing on simulator (3 days)
- Advanced malfunctions
  - Decision making exercises
  - Failures of valves, transmitters, pumps, compressors, equipment fouling (2 days)
- Refresher training (3 days per year)

NOVA Chemicals

RSI create, evaluate, operate

Additional simulator benefits

1. Deployment of ASM multi-window HMI system in E3 (2000)
2. Adoption of modern HMI graphic standards
3. Startup design problem identified
4. Emergency procedure test and validation
5. Controller tuning and verification of transmitter ranges
6. Safety logic validation

NOVA Chemicals

RSI create, evaluate, operate

Additional simulator benefits

Environmental Stewardship; Significant flare reductions

1. E3 Startup
2. E1 and E2 startups and shutdowns
3. Propylene refrigeration startup

NOVA Chemicals

RSI create, evaluate, operate

Additional simulator benefits

Like a "Time Machine" novices gain experience that typically takes a decade or longer

NOVA Chemicals

RSI create, evaluate, operate

Comprehensive console operator training on ethylene simulator (hot-side)

RSI Companion Training Manual™

1. **Normal Operations** →
   - Ethane dryer switch
   - Increase feed, adjust dilution steam
   - Adjust furnace pressure
   - Adjust feed for high venture pressure
   - Switch furnace and de-coke
2. **Instrument Troubleshooting** →
   - False high level in flash drum
   - False high dilution steam flow
   - False high level in VHP steam drum
   - False low furnace draft pressure
   - Decrease VHP BFW to transfer line exchanger
3. **Advanced Control - SIS** →
   - Maintenance of feed meter
   - Maintenance of VHP steam meter
   - Maintenance of water stripper overhead flow
4. **Non-shutdown Emergencies** →
   - Loss of induction fan
   - Loss of quench water pump
   - Loss of expander
   - Loss of ammonia
   - Tube rupture in primary transfer line
   - Loss of steam drum level
   - Furnace tube rupture
5. **Emergency Shutdowns** →
   - Loss of ethylene feed
   - Loss of dilution steam
   - Loss of cooling water
   - Loss of quench water pumps
6. **Normal Startup and Shutdown**
   - Furnace startup
   - Furnace shutdown
   - Feed system startup
   - Dilution steam startup
   - Quench water startup

NOVA Chemicals

RSI create, evaluate, operate

Cause and effect matrix

Cause and Effect Matrix Reactor Section

Scenario	Resid MBPD	ICGO MBPD	HCGO MBDP	GO MBPH	Rx Temp °F	FC112/113/114/115 – Feed Pass - SP	FC212/213/214/215 – Feed pass - SP	FC112/113/114/115 – Feed pass - OP	FC212/213/214/215 – Feed pass - OP	TC111 – Oil furnace outlet - SP	TC211 – Oil furnace outlet - SP	TC111 – Oil furnace outlet - OP	TC211 – Oil furnace outlet - OP	TC134 - H_2 furnace outlet - SP	TC234 - H_2 furnace outlet - SP	TC134 - H_2 furnace outlet - OP	TC234 - H_2 furnace outlet - OP
Start	44.0	4.0	2.5	0.0	810/820												
1A	40.0	3.0	2.0	0.0	800/810	↓	↓	↓	↓	-	-	↓	↓	-	-	↓	↓
1B	46.0	4.0	2.5	0.0	810/820	↑	↑	↑	↑	↑	↑	↑	↑	↑	↑	↑	↑
1C	30.0	2.0	1.5	0.0	780/780	↓	↓	↓	↓	↓	↓	↓	↓	↓	↓	↓	↓
1D	0.0	0.0	0.0	800	700/700	↓	↓	↓	↓	↓	↓	↓	↓	↓	↓	↓	↓

Scenario	TI 150 – F101 inlet temperature - PV	TI 250 – F102 inlet temperature - PV	WABT11 – F101 Average bed temperature - PV	WABT12 – F-102 Average bed temperature - PV	XOT11 – F101 Exotherm	XOT12 – F102 Exotherm	XOT21 – F201 Exotherm	XOT22 – F202 Exotherm	LC113A/B – F-103 level - OP	LC213A/B – F-203 level - OP	LC145/145A – F-107 level – OP	AR113 H_2 Density
Start												
1A	-	-	↓	↓	↓	↓	↓	↓	↓	↓	↓	↓
1B	-	-	-	-	-	-	-	-	↑	↑	↑	↑
1C	↓	↓	↓	↓	↓	↓	↓	↓	↓	↓	↓	↓
1D	↓	↓	↓	↓	↓	↓	↓	↓	↓	↓	↓	↓

Scenario	PDI132 – Recycle compressor delta pressure - PV	MC701 – Recycle compressor speed	PIC130 – System Pressure Controller - SP	PIC130 – System Pressure Controller - OP	FC156 – System pressure purge stream - SP	FC156 – System pressure purge stream - OP	Catalyst withdrawal	Reactor Bed Density	SI102 – Ebullation pump speed - PV	SI103 – Ebullation pump speed - PV	SI202 – Ebullation pump speed - PV	SI203 – Ebullation pump speed - PV
Start												
1A	-	↑	-	↑	↑	↑	↑	↑	↓	↓	↓	↓
1B	-	↓	-	↓	↓	↓	-	↑	-	-	-	-
1C	-	↑	-	↑	↑	↑	↑	↑	↓	↓	↓	↓
1D	↓	↑	-	↑	↑	↑	↑	↑	↓	↓	↓	↓

NOVA Chemicals

RSI create, evaluate, operate

Conclusion

- Simulators are used in the airline, power and process industries to develop the highest proficiency standards and safety practices
- NOVA Chemicals simulator training program is a model of training excellence and environmental stewardship
- With simulators console operators practice process upsets and learn proper corrective actions
- Develop diagnostic skills to accurately troubleshoot problems, and learn optimization techniques to maximize profit
- Approximate 5:1 payback in a high-fidelity simulator is

NOVA Chemicals

RSI create, evaluate, operate

AIChE PAPER NUMBER: XXx

PROPYLENE PIPELINE PUMP FIRE INCIDENT

Michael Kudijaroff

Maintenance Engineer

Williams

Prepared for Presentation at the 2016 Spring National Meeting

Houston, Texas April 10 – 14, 2016

PROPYLENE PIPELINE PUMP FIRE INCIDENT

Michael Kudijaroff

Maintenance Engineer

Williams

Abstract: An incident occurred which led to a pump fire on the olefins plant Propylene Pipeline Pump. The olefins plant was down at the time of the incident due to a power outage the week prior. The minimum flow system had been lined up to a storage bullet in order to circulate propylene and the pump was started from an idle state 3 hours prior. There were no injuries, or environmental impacts as a result of the incident, however there was significant damage to the pump, the spare pump, both motors, and collateral damage to nearby equipment. This paper describes the investigation findings determined by the basic steps of the Cause Mapping method.

Background

The Williams ethylene unit located in Geismar, Louisiana is a multi-step separation & distillation purification process to produce a high purity ethylene product stream. Constructed in 1966 with the process and design license by Lummus Technology, the original plant was designed to crack ethane/propane and butanes to produce 600 million pounds per year ethylene plus co-products. Several modifications over the years have increased nameplate ethylene capacity to 1.95 billion pounds per year plus co-products. Polymer Grade Propylene is a significant co-product produced at the facility.

A portion of the Polymer Grade Propylene (PGP) is sent to the propylene pipeline by means of one of two high-pressure propylene pipeline pumps. The pumping system is referred to as GA-821/ GA-821S and is responsible and essential for maintaining the operating pressure on the propylene pipeline system.

Incident

On May 24 2015 the olefins plant experienced a pump shaft tandem seal failure that resulted in a fire on one of the two high-pressure propylene pipeline pumps. The emergency brigade team responded well to the incident. The fire resulted in extensive damage to instrumentation and electrical services in the immediate area. There were no reported injuries from the fire, or subsequent emergency response activities. There was no off-site impact as a result of the fire. The olefins plant was down at the time of the incident due to a power outage the week prior.

The preliminary framework for moving forward comprised of:

1) Investigating the incident.
2) Damage assessment.
3) Inspection & testing of piping/equipment/instrumentation/electrical systems within the immediate area.
4) Repairing and returning the high-pressure propylene pipeline pumping system to normal operation.

Incident Investigation

Williams has adopted the Think Reliability Cause Mapping methodology in circumstances where incidents result in a high severity event or have the potential for a high severity (near miss). The investigation assembles a cross functional team that includes the subject matter experts within the organization.

The Root Cause Mapping methodology used for the investigation process consisted of three (3) Steps:

1) identifying the problem and creating the "problem outline". The problem outline is a summary of the basic information about the event. The outline captures what people see as the problem, when it happened, where it happened and the specific impact to each of the organizations overall goals.

2) involved an analysis to understand and identify why the overall goals were negatively impacted. The analysis starts with one of the goals that was negatively impacted. The investigation begins with one Why-question, then adds another and another. By working backwards, a simple beginning can form a detailed analysis. It is during this step that the "Cause Map" (visual dialogue) is developed.

3) identifying solutions. A solution is a specific action that controls a particular cause. A thorough, evidence based cause-effect analysis is the starting point for finding solutions. The evidence is added directly to the bottom of the cause that it supports or disproves. Possible solutions can be proposed for any one of the causes. The best solutions are then selected from the possible solutions. From this an action plan with owners and due dates is defined.

As Seen From the Problem Outline

The Safety goal was negatively impacted as a result of the fire as well as the high potential for injury as a result of event. Property and Equipment goal was negatively impacted due to damage in the immediate area as well as damage to auxiliary electrical equipment. The Labor and Time goal was negatively impacted due to the unintended resources and man hours that were required for repairs.

A Summary of the Analysis:

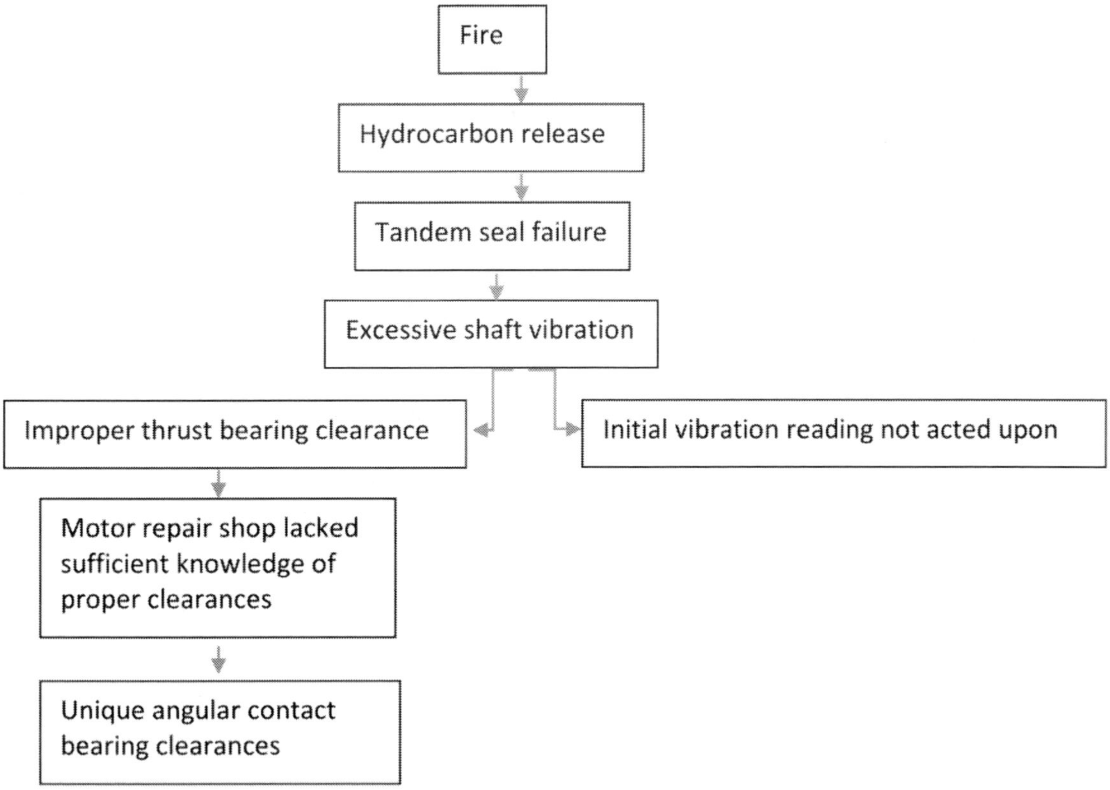

Upon completing the analysis, the following solutions were identified:

1) A training opportunity was identified for our reliability technicians in Level 1 vibration monitoring and associated software.

2) A comprehensive review of the facility's vibration logs was necessary to determine if any other "lurking" high vibration indicators were present in vertical motors.

3) Development of an Enterprise-wide Continuous Improvement advisory to address examination of clearances of spring configured angular contact thrust bearing assemblies.

Lessons Learned

On 24-May-2015 a pump fire occurred on the Propylene Pipeline Pump as a result of a tandem seal failure. The pump had been started up from an idle state 3 hours prior. There were no injuries or environmental impacts. There was significant damage to the pump motors and collateral damage to nearby equipment. The pump motor had been installed in April 2014 but was in a non-running state due to other facility issues. The motor had been pro-actively repaired to return all equipment to a zero time state run-condition. A defect was introduced into the pump motor during the repair at an outside motor repair shop. A vibration reading taken in March 2015 revealed that the pump motor had increased vibration above the alarm threshold. While the data system captured the vibration event the technician did not write a corrective work order. This we believe was due to significant organizational fatigue. There is also an opportunity for certification training of the Vibration Technicians.

Paper 192 A: DMTO and Beyond: DICP's Sustainable Innovations in Technologies for on-Purpose Production of Light Olefins from Methanol

Zhongmin Liu
Dalian Institute of Chemical Physics, Chinese Academy of Sciences
Zhongshan Road 457, Dalian, 116023, China
Email: liuzm@dicp.ac.cn

Mao Ye
Dalian Institute of Chemical Physics, Chinese Academy of Sciences
Zhongshan Road 457, Dalian, 116023, China
Email: maoye@dicp.ac.cn

Prepared for Presentation at
American Institute of Chemical Engineers
2016 Spring Meeting
11th Global Congress on Process Safety
Houston, Texas
April 10-14, 2016

UNPUBLISHED

DMTO and Beyond: DICP's Sustainable Innovations in Technologies for on-Purpose Production of Light Olefins from Methanol

Zhongmin Liu
Dalian Institute of Chemical Physics, Chinese Academy of Sciences
Zhongshan Road 457, Dalian, 116023, China
Email: liuzm@dicp.ac.cn

Mao Ye
Dalian Institute of Chemical Physics, Chinese Academy of Sciences
Zhongshan Road 457, Dalian, 116023, China

Keywords: methanol-to-olefins, catalyst, process development

Abstract

The rapid development in past thirty years has triggered a significant increase of the light olefins (including ethylene, propylene, and butylene) production capacity in China, and it is expected the demand will continue to expand in the coming years. In view of China's tremendous coal reserves, the Dalian Institute of Chemical Physics (DICP) developed a fluidized bed methanol-to-olefins technology (DMTO). The DMTO technology was successfully commercialized in 2010 in Shenhua Baotou's coal-to-olefins (CTO) plant (with an ethylene and propylene production capacity of 600 kt/a). So far seven commercial units using DMTO technology have been put into stream in China. As part of the DICP's ambition of sustainable innovations for on-purpose production of light olefins from methanol, a dual functional catalyst was developed for converting methanol to light olefins and cracking C4+ to ethylene and propylene simultaneously. By use of this catalyst, a second generation MTO technology (DMTO-II) has been developed and successfully implemented into a recent started up CTO plant with an ethylene and propylene production capacity of 670 kt/a. This paper will introduce the research and development (R&D) activities at DICP on DMTO and DMTO-II technologies, which covers the zeolite and catalyst synthesis, process development, reactor scale-up, and commercial operations. Beyond that, DICP's strategy of innovative catalyst and process research for on-purpose production of light olefins from methanol will also be highlighted.

1. Introduction

The rapid development of China's economics in the past three decades has triggered a significant increase of the demand of light olefins (including ethylene, propylene, and butylene) and thus the production capacity in China. From 2005 to 2010, China's production capacity of ethylene was increased from 7.58 to 15.19 Mt/a while the production capacity of propylene was increased from 8.86 to 15.83 Mt/a. It is expected the demand of light olefins will continue to expand in the coming years. It was estimated that at the end of 2015 China's production capacity of ethylene could be further increased to 27.00 Mt/a and that of propylene increased to 24.00 Mt/a. Ethylene and propylene are conventionally produced from petrochemical feedstock via naphtha thermal cracking and fluid catalytic cracking (FCC) processes. It is since the oil crisis in 1970s that there have been continuously growing efforts in finding alternative resources other than oil to produce light olefins (Chang, 1984; Chang and Silvestri, 1977; Liang et al., 1990). Researchers in Mobil discovered that a class of synthetic shape-selective zeolites, namely ZSM-5, could be used to convert methanol to light olefins. Methanol-to-olefins (MTO) opens an important way to produce light olefins from non-oil resources since methanol can be readily produced from coal, natural gas, and biomass via synthesis gas by existing and proven technologies (Chang, 1984; Chang and Silvestri, 1977; Keil, 1999; Tian et al., 2015). Later on, scientists from the Dalian Institute of Chemical Physics (DICP), Chinese Academy of Sciences found that a silicoaluminophosphate catalyst (SAPO-34), which was first synthesized by Union Carbide in 1980s, could greatly improve the selectivity to light olefins in MTO process (Liang et al., 1990). Since then, MTO process has attracted considerable interests and the researches span catalyst synthesis, reaction mechanism, reaction kinetics, process development, and reactor scale-up.

In August 2010, the world's first commercial MTO unit (600 kt/a of ethylene and propylene production) was successfully started up in Shenhua's Baotou coal-to-olefins plant in north China (Liu et al., 2014). This unit is based on the MTO process developed by DICP, and its success represents an important milestone and breakthrough of MTO process development. In this paper, the research and development (R&D) activities at DICP on MTO technologies will be introduced, which covers the zeolite and catalyst synthesis, process development, reactor scale-up, and commercial operations, and focuses on the DMTO and DMTO-II process. Beyond that, DICP's strategy of innovative catalyst and process research for on-purpose production of light olefins from methanol will also be highlighted by two examples.

2. Catalyst development

2.1 Synthesis of Silicoaluminophosphate (SAPO) molecular sieves

SAPO molecular sieves were first synthesized by Union Carbide Corporation in 1984. Among them, SAPO-34 which has the CHA structure shows excellent catalytic performance in MTO reaction. The CHA framework topology is characterized by the cylinder-like cages with eightmembered ring openings. SAPO-34 thus has small pore, medium acidity, and high thermal/hydrothermal stability, and is the ideal breeding ground for hydrocarbon pool intermediates in the MTO reaction.

Scientists at DICP had made great efforts to the synthesis and crystallization mechanism of SAPO-34. In the late 1980s, scientists at DICP first reported the excellent catalytic performance of SAPO-34 in the MTO reaction and its high stability during regeneration. The SAPO-34 synthesized at DICP could keep more than 80% relative crystallinity after steam treatment at 800 °C for 45 h. These efforts confirmed that SAPO-34 is a promising MTO catalyst for industrial application.

In order to further make the SAPO-34 catalyst available for industrial applications, researches on the suitable templates for SAPO-34 synthesis had also been conducted since templates may influence the morphologies, microstructures, and acidic properties of the catalyst. It has been found that triethylamine (TEA), diethylamine (DEA), and their mixtures would be good choice. On one hand, the amines were much cheaper than the traditional tetraethylammonium hydroxide (TEAOH). On the other hand, the template has a lower boiling point (below 100 °C) than water, and the amine could be easily separated from the mother liquid through flash distillation and reused for the synthesis. This could significantly reduce the cost for catalyst synthesis and waste treatment.

In addition, DICP also conducted researches to control the catalytic performance of SAPO-34 via altering the Si coordination environments. They studied the mechanisms governing the nucleation, crystallization, and Si distribution for a better control of composition and Si environmental of the synthetic product, and it has been found that SAPO-34 with low Si content and single Si(4Al) species shows longer lifetime and lower selectivity of propane (thus slow coking rate). The SAPO-34 synthesized at DICP is featured by the rapid crystallization, adjustable acidity, and on-purpose particle size.

In 1993 scientists at DICP synthesized SAPO-34 by employing a 1.0 M^3 autoclave in order to supply catalyst for the pilot-scale experiments. In 2005, DICP built a pilot plant for catalyst synthesis in which two 2 M^3 autoclaves were used.

2.2 Synthesis of fluidized bed MTO catalyst

The fluidized bed MTO catalyst with SAPO-34 as the active component is prepared by spray drying method, which has the following characteristics: spherical, particle size distribution similar to FCC catalyst, sufficient mechanical strength, and good catalytic performance. In 2005, roughly 25 tons of MTO catalyst was produced and supplied for the DMTO demonstration unit (16 kt/a MeOH feed). In 2008, a 2000 t/a MTO catalyst plant (containing molecular sieve synthesis and catalyst manufacturing) was constructed and put into stream. The manufacturing process is composed of mixing, milling, spraying, drying, calcination, packaging, and storage.

Establishment of quality controlling system and standard test procedure is critical to keep the excellent performance of MTO catalyst. Five quality control points were set for SAPO-34 synthesis process: raw materials, purity of the solid in the slurry, content of distilled amine solution, collected waste liquids, and SAPO-34 molecular sieves product. The latter includes crystal phase, composition, morphology, textural properties, solid content, and catalytic performance. Meanwhile, four quality controlling points were set for the catalyst particle manufacturing: raw materials, particle size distribution of the milled slurry, particle size distribution and morphology of the semiproduct, and calcined product. The parameters for final quality controlling of the calcined product includes solid content, composition, morphology, particle size distribution, attrition index, textural properties, bulk density, and catalytic performance.

3. DMTO technology

3.1 MTO reaction in fluidized bed reactor

The MTO reaction over SAPO-34 catalyst is highly exothermal, which will release about 196 kcal heat when 1 kg methanol feed is completely converted at 495 °C. A feature of MTO reaction over SAPO-34 is the rapid catalyst deactivation due to coke deposition. Figure 1 shows the methanol conversion and selectivity to ethylene and propylene in a fluidized bed reactor without external catalyst circulation. As can be seen, a rapid drop of methanol conversion occurs after the coke content on catalyst reaching a certain *critical* value. In order to maintain high activity of catalyst in the reactor, the in-line combustion of coke is required. Another feature of MTO reaction over SAPO-34 is that the selectivity to ethylene is very sensitive to the coke content on catalyst. At the *critical* coke content on catalyst, the selectivity to ethylene becomes the maximum. The selectivity to propylene basically keeps stable with the increasing coke content. This is very important for MTO reactor operation.

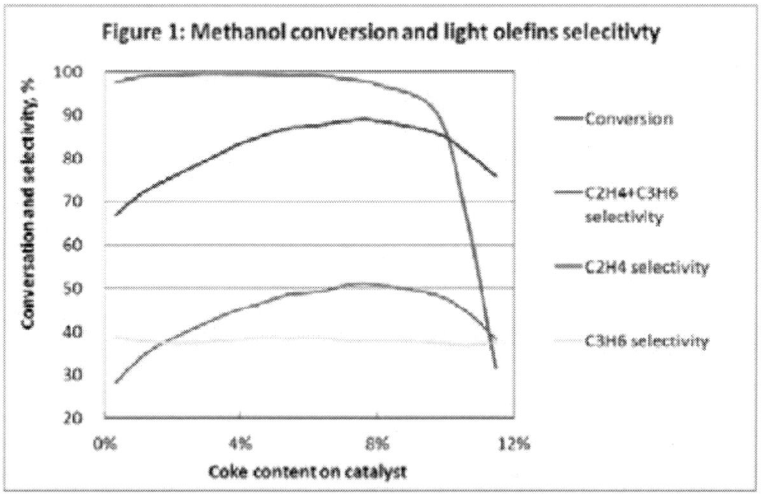

The particle size of DMTO catalyst is similar to that of fluid catalytic cracking (FCC) catalyst, which belongs to Geldart A type of particles. Despite the good fluidization performance, the great quantity of fine catalyst particles can provide large surface area and benefits the heat transfer in the fluidized bed. The temperature difference in the DMTO fluidized bed can be readily reduced to just several degrees via optimal operation.

3.2 MTO fluidized bed reactor scale-up

The scale up of fluidized bed reactor is one of big challenges in fluidized bed process development. The DMTO fluidized bed reactor was scaled up step by step from micro-scale fluidized bed to commercial scale. In the micro-scale reactor (tube diameter is roughly ~1 cm), catalytic performance of DMTO catalyst was evaluated, and the optimal operation conditions were identified. Note that the coke content on catalyst is a function of catalyst residence time (RT), an optimal catalyst RT in pilot-scale reactor (tube diameter is ~10 cm) can be established based on the time-on-stream (TOS) in the micro-scale fluidized bed reactor. In addition, the reaction kinetics of DMTO reaction has also been studied in the micro-scale facility. In the pilot-scale experimental facility, the reactor is connected to the regenerator via standpipes, thus a complete catalyst circulation between reactor and regenerator can be tested. In the pilot-scale reactor, the catalyst stripping was tested. The fluidization performance of DMTO catalyst was also checked in the pilot-scale reactor. The methanol conversion is above 99% for both micro-scale and pilot-scale reactors, while the typical selectivity (in nonaqueous products) to ethylene and propylene is above 80% in both micro-scale reactor and pilot-scale reactor. In 2006, a semi-industrial DMTO demonstration unit was constructed in Huaxian, Shanxi, which consists of fluidized bed reactor, fluidized bed regenerator, quench tower, and sour water stripper. The aims of the demonstration unit are to study the hydrodynamics and reaction performance of DMTO turbulent fluidized bed reactor and provide data for commercial reactor design. The diameter of the reactor in the DMTO demonstration unit is ~1 m, with a methanol feed rate of 2.5 t/h. The unit was running for 1200 hours, and an average selectivity to ethylene and propylene was 79.21%, and the methanol conversion was higher than 99%. Furthermore, the ethylene/propylene ratio could be adjusted in the range of 0.8 to 1.2 by varying operation conditions. The data from this demonstration unit was implemented in the design package of commercial DMTO units.

3.3 DMTO commercial units

Figure 2. The schematic diagram of DMTO process

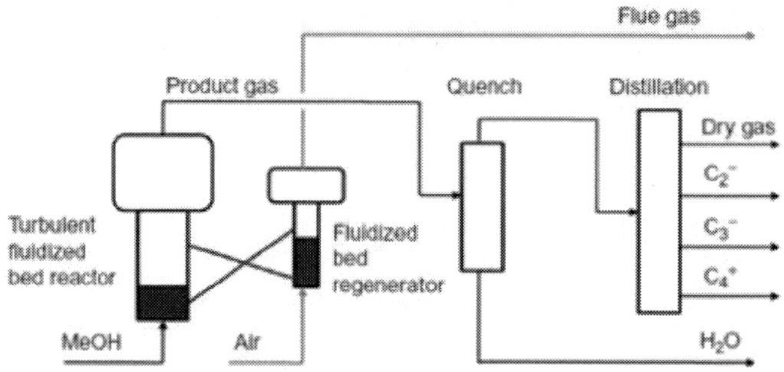

In May 2010, the world's first DMTO commercial unit was constructed in Baotou, North China by Shenhua group. The schematic diagram of DMTO unit is shown in Figure 2. The capacity of light olefins production is 0.6 Mt/a, where the reactor is a shallow fluidized bed with a diameter of 11 m and bed height of 3 m. The unit was successfully started up in August 8, 2010. The heat released by the MTO reaction was used to heat the reactor and regenerator during the start-up, which can on one hand accelerate the start-up process, and on the other hand reduce the energy consumption. Figure 3 illustrates the methanol conversion and selectivity to light olefins in the commercial unit. As can be seen, the methanol conversion reaches ~100%, and the selectivity is higher than ~80%. Typical production distribution in DMTO unit is shown in Table 1. In March 2011, a 72-hour calibration of the Shenhua Baotou DMTO unit was conducted, and the methanol consumption of 2.96 tons per ton product of ethylene and propylene had been obtained.

Table 1. Product distribution for DMTO reactor in commercial units

Product	Selectivity, %
H2	0.20
CO2	0.03
CO	0.12
CH4	1.74
C2H6	1.56
C2H4	41.40

C3H8	3.00
C3H6	39.61
C4	10.92
C5+	1.42
总计	100.00

In total 19 DMTO units have been licensed in China, with a light olefin product capacity of 11.26 Mt/a. And so far 8 DMTO units have been brought into stream. The majority of these DMTO units were constructed together with coal-to-methanol units for methanol supply. Several units built in the East China import methanol from overseas as their feedstock. Figure 4 shows the locations of these DMTO units. The smooth start-up and steady operation of 8 DMTO units indicate that the DMTO technology is becoming a main stream technology for light olefins production in China.

Figure 3. The conversion and selectivity to light olefins in commercial DMTO unit

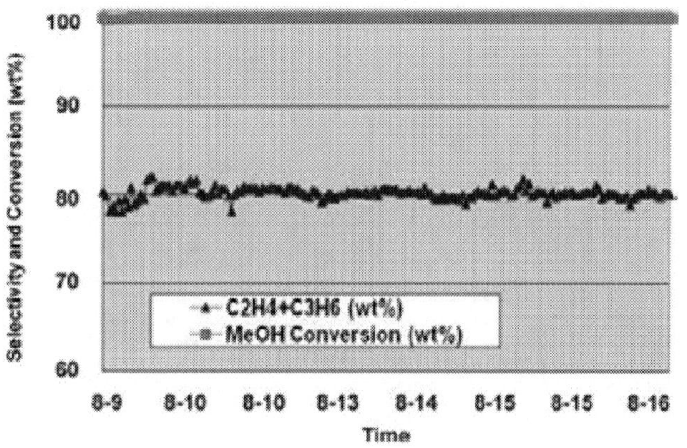

Figure 4. The location distribution of licensed DMTO units in China

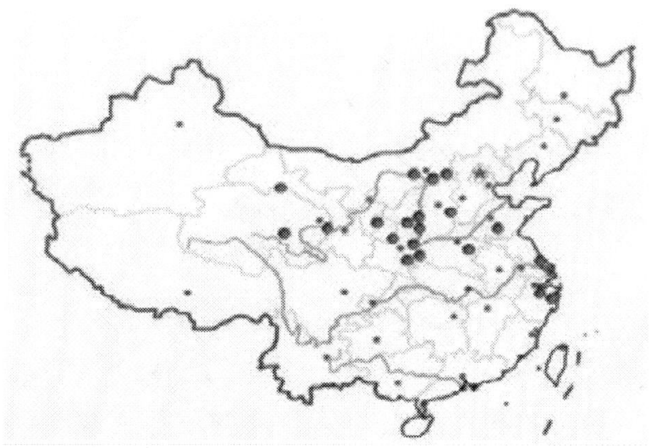

4. DMTO-II technology

4.1 DMTO-II process

DMTO-II is the second generation DMTO process. In this process the byproduct of C4+ was separated from the product and further converted to ethylene and propylene in a second fluidized bed reactor. Figure 5 shows the scheme of DMTO-II process. The advantage of the DMTO-II process is that a double-functional catalyst is used, which can catalyze both the MTO reaction and C4+ conversion reaction. Thus only one regenerator for catalyst regeneration is required.

In 2009, the DMTO demonstration unit was revamped by adding a second bubbling fluidized bed reactor for C4+ cracking. From 2009 to 2010, the DMTO-II process had been verified in the demonstration unit. The selectivity to ethylene and propylene increased from 79.21% in DMTO process to 85.68% in the DMTO-II process with 60 wt % of C4+ recycling, and the methanol consumption for producing one ton ethylene and propylene reduced from 2.96 to 2.67 tons.

Figure 5. The schematic diagram of DMTO-II process

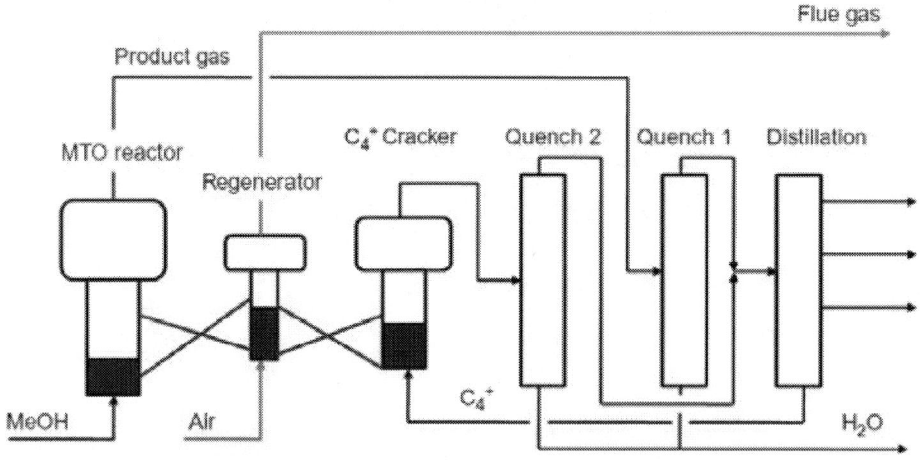

Figure 6. The commercial DMTO-II unit

Table 2. Typical results for C4+ cracker in DMTO-II commercial unit

Product selectivity,%	Condition 1	Condition 2
	Low catalyst circulation rate	High catalyst circulation rate
C2H4	15.74	20.00
C3H6	25.34	32.69
Conversion, %		
C4	66.48	75.63
C5+	65.29	78.76

4.2 DMTO-II commercial unit

In October 2010, DICP licensed the first DMTO-II unit (0.67 Mt/a polyethylene and polypropylene product) to Pucheng Clean Energy Chemical Co. Ltd. In December 2014, the first commercial DMTO-II unit was successfully started up. Table 2 shows the typical results for the commercial DMTO-II unit. Figure 6 shows the photo of the unit.

5. Sustainable development of DMTO technology

DMTO has been widely accepted as the advanced technology in ethylene and propylene production. Beyond that, however, the sustainable development of DMTO technology and its derivatives continues by many other industrial orientated technologies. In the following, several examples are introduced to illustrate DICP's strategy in sustainable development of chemical processes.

5.1 Fundamental research on catalysis

As one of the leading energy research institutes in China, DICP is always committed to develop new catalyst as well as processes for the sustainable, efficient, and clean usage of traditional fossil fuel like oil and coal. In the past years, coal-to-chemicals (CTC) was one of the focuses in DICP, and many breakthroughs have been achieved in our fundamental research. For example, a recent developed bifunctional catalyst that can convert syngas to light olefins with very high selectivity, which exceeds the mechanistic limitation of <60% conversion for Fischer-Tropsch catalysts. As to MTO process, a lot of fundamental research has been done to understand the reaction mechanism, such as the formation of first C-C bond, the induction period, the intermediate product in the reaction, the deactivation kinetics, and diffusion inside cages and channels in zeolites. Based on these research results, the on-purpose improvement of MTO process and optimization of catalyst can be achieved to satisfy different demands in the market.

5.2 DMTP process

As propylene is also highly demanded in market, the on-purpose production of propylene stimulates our research on methanol-to-propylene (DMTO) process development. In this process a catalyst with dual function which can convert methanol to light olefins (ethylene and propylene) and further transform the ethylene is used. And a special designed fluidized bed reactor is developed to maximize the performance of catalyst. The pilot experiments with the scale of 300 t/a methanol feed have been completed. Figure 7 shows the scheme of the DMTP process. Figure 8 shows the concentration of ethylene at the inlet and outlet of the reactor R1 in typical operation, which is calculated in the CH_2 basis. As can be seen, the concentration of ethylene is balanced during this operation, and this means that the net production of ethylene can be minimized.

Figure 7. The scheme of DMTP process

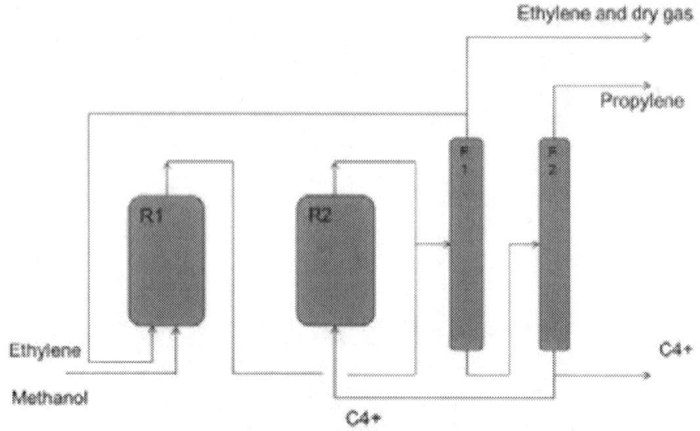

Figure 8. The concentration of ethylene at the inlet and outlet of the reactor R1 in typical operation

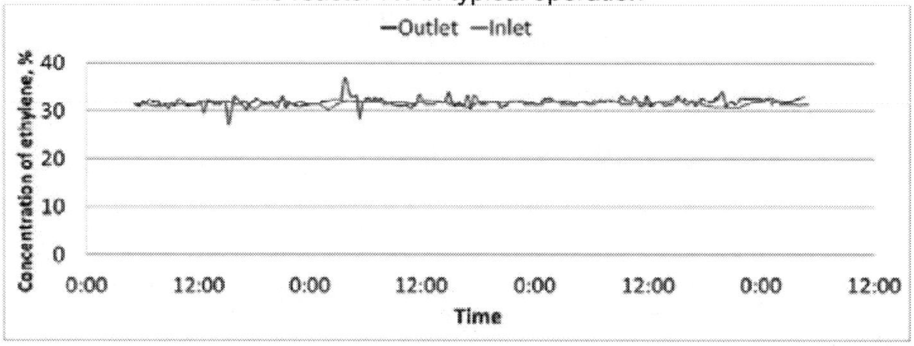

A process simulation has been carried out, and it has been shown that the overall selectivity to propylene of 75% can be reached. This technology is ready for commercialization in DICP.

5.3 Methanol-coupled-naphtha to light olefins

Naphtha steam cracking is still the main stream for light olefins production. Typically about 30% of ethylene and 15% of propylene can be obtained at a temperature of 800~900 ℃ by controlling the contact time of the reactants. Catalytic cracking of naphtha can take place at around 600°C, which is economically attractive in terms of energy consumption. But naphtha cracking reaction is still characterized by highly endothermic, and requires a continuous heat supply. A possible approach is to feed methanol to the naphtha cracking reactor. As the conversion of methanol to light olefin is exothermic, if coupled with methanol feed, the heat required by the naphtha cracking reaction can be partly compensated by heat of methanol conversion. It is known that both naphtha cracking and methanol conversion reactions could be catalyzed by solid acids, and zeolites have been proved to be most efficient catalysts, which provides a possibility to develop one zeolitic catalyst for the both reactions.

The mechanism of methanol-coupled naphtha cracking (MCNC) process is very complicated. The notable character of this coupling transformations is the presence of methanol can lower activation energy required for the alkenes cracking reactions, and promote the cracking reaction rate. This has been confirmed by the coupling conversion of n-hexane and methanol over HZSM-5 zeolite catalysts in a pulse reaction experiment in DICP (as shown in Figure 9). In addition to that, the exothermic conversion of methanol to light olefin can partly compensate the heat required by the naphtha cracking reaction.

Figure 9. Methanol-coupled n-hexane cracking reaction over HZSM-5 catalyst.

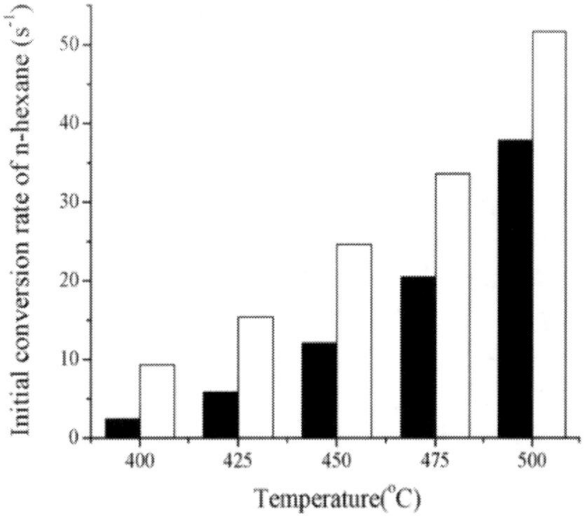

Table 3. The typical light olefins yield for different production processes.

	Naphtha Steam Cracking	Naphtha Catalytic Cracking	Methanol-coupled naphtha cracking		MTO
Reaction Temperature, °C	800-900	600-650	600-650		450-500
Feedstock, %					
Naphtha		83.33	75.00	50.00	0
Water		16.67	16.67	14.29	60
MeOH		0.00	8.33	35.71	40
Total		100.00	100.00	100.00	100.00
Product,%					
Ethylene + propylene	43.37	40.38	40.16	44.3	80.00
BTX	11.51	13.07	11.95	9.84	0

In Table 3, we show the typical yield data for four olefin production processes. The data for naphtha steam cracking was obtained from open literature. The data for naphtha catalytic cracking and methanol-coupled naphtha cracking process are based on lab-scale fluidized bed reactor experiments, which were carried out in DICP. It should be noted, however, that the yield distribution is quite rough and the attention was paid to the main components such as ethylene, propylene, and BTX. All the data are collected in a dry basis, and water is excluded in the analysis. Apparently, the advantages of the methanol-coupled naphtha cracking process include:

(a) heat required for cracking reaction is partially compensated by that released by MTO reaction;
(b) one catalyst for two coupled reaction;
(c) presence of methanol lowers the activation energy required for the alkanes cracking reactions;
(d) product including more aromatics;
(e) flexibility of feedstock.

In fact, methanol is a feedstock that can be readily obtained from either natural gas or coal. The wide availability of natural gas and coal makes methanol much cheaper than naphtha in the market. Commercialization of methanol-coupled naphtha cracking process can certainly improve the margin of olefin production plant.

6. Conclusion

This paper introduced the research and development (R&D) activities at DICP on DMTO and DMTO-II technologies, which covers the zeolite and catalyst synthesis, process development, reactor scale-up, and commercial operations. Beyond that, DICP's strategy of innovative catalyst and process research for on-purpose production of light olefins from methanol is also highlighted by two examples, i.e. the methanol-to-propylene (DMTP) and methanol-coupled naphtha cracking processes. The DMTO technologies, as well as these new processes, will certainly provide extra choices for light olefins production in the market.

7. References

[1] Tian Peng, Wei, Yingxu, Ye, Mao, and Liu, Zhongmin, "Methanol to olefins (MTO): from fundamentals to commercialization", *ACS Catalysis*, Volume 5, NO.3, 2015, pp.1922-1938

[2] Ye, Mao, Li, Hua, Zhao, Yinfeng, Zhang, Tao, and Liu, Zhongmin, "MTO processes development: the key of mesoscale studies", Advances in Chemical Engineering, Volume 47, NO.2, 2015, pp.79-335

[3] Liu, Zhongmin et al., *Methanol to olefins*. Science Press, Beijing, 2015 (inn Chinese)

[4] Jiao, Feng et al., "Selective conversion of syngas to light olefins", *Science*, Volume 351, NO. 6277, 2016, pp.1065-1068

AIChE PAPER 192b

BOOST ETHYLENE AND PROPYLENE PRODUCTION WITH AIR LIQUIDE'S MTP PROCESS AND TECHNIP'S ETHYLENE TECHNOLOGY

Jean-Francois Fournier
Supervising Process Engineer
Technip Stone & Webster Process Technology
Eric Wagner
Vice President, Technology Development
Technip Stone & Webster Process Technology
Jean-Paul Laugier
Vice President, Ethylene Product Line
Technip

Martin Gorny
Product Manager Methanol and MTP Technology
Air Liquide Global E&C Solutions
Thomas Wurzel
E & C Director, Downstream and Petrochemical Product Line
Air Liquide Global E&C Solutions
Stéphane Haag
Senior Research Scientist
Air Liquide Forschung und Entwicklung

Prepared for Presentation at the 2016 Spring National Meeting
Houston, Texas, April 11 – 14, 2016

BOOST ETHYLENE AND PROPYLENE PRODUCTION WITH AIR LIQUIDE'S MTP PROCESS AND TECHNIP'S ETHYLENE TECHNOLOGY

Jean-Francois Fournier
Supervising Process Engineer
Technip Stone & Webster Process Technology
Eric Wagner
Vice President, Technology Development
Technip Stone & Webster Process Technology
Jean-Paul Laugier
Vice President, Ethylene Product Line
Technip
Martin Gorny
Product Manager Methanol and MTP Technology
Air Liquide Global E&C Solutions
Thomas Wurzel
E & C Director, Downstream and Petrochemical Product Line
Air Liquide Global E&C Solutions
Stéphane Haag
Senior Research Scientist
Air Liquide Forschung und Entwicklung

Abstract: In recent years, the steam cracker industry has moved to lighter feedstocks. In particular, a significant number of steam crackers in the US have been revamped to switch to ethane feedstock, thereby reducing the propylene production. As a result, olefin producers must find new ways to restore the balance between the ethylene and propylene products.

Conversion of methanol to propylene via Lurgi MTP™ from Air Liquide is one of the proven technologies available for on-purpose propylene production. Up to now, Air Liquide MTP™ plants have been designed and installed as stand-alone units.

Air Liquide and Technip are currently proposing a new design approach, which combines an MTP plant and a steam cracker unit. Compared to a stand-alone MTP plant and a stand-alone ethane cracker, the integration of the two units boosts the overall ethylene and propylene production by 5 to 10%, depending of the site configuration. At the same time the integration reduces the specific CAPEX per ton of light olefin by a similar percentage. The proposed integration is flexible and can be applied to both existing steam crackers as well as grassroot plants.

BOOST ETHYLENE AND PROPYLENE PRODUCTION WITH AIR LIQUIDE'S MTP PROCESS AND TECHNIP'S ETHYLENE TECHNOLOGY

Jean-Francois Fournier and Eric Wagner, Technip Stone & Webster Process Technology
Jean-Paul Laugier, Technip
Martin Gorny, and Thomas Wurzel, Air Liquide Global E&C Solutions
Stéphane Haag, Air Liquide Forschung und Entwicklung

Introduction

In recent years, the steam cracker industry has moved to lighter feedstocks, in particular ethane, which offers a high olefin yield at a low feedstock price. In the USA, the shale gas development has boosted availability of ethane and a significant number of steam crackers in the USA have already been revamped to switch to ethane, thereby reducing the propylene production. As a result, olefin producers must find new ways to restore the balance between the ethylene and propylene products.

Conversion of methanol to propylene via Lurgi MTP™ process from Air Liquide is a well referenced technology for on-purpose propylene production. Integrated with natural gas based methanol production it offers the best in class technology, and low cash cost route to on-purpose propylene [1].

Up to now, Air Liquide MTP plants have been designed and installed as stand-alone units. Integration was only applied with up and downstream units, such as methanol and air separation units.

Air Liquide and Technip have jointly developed a new design approach, which combines an MTP unit and a Steam Cracker unit. In addition to sharing common equipment between the two plants, the MTP – Steam Cracker integration effort has also identified several synergies that boost the ethylene + propylene production using the same feedstock input as the corresponding stand-alone units.

In the following, different levels of integration are described and referenced to a grassroot concept for both Steam Cracker and MTP unit. The corresponding grassroot plants are used as a basis as they allow for simpler comparison and therefore help to illustrate better the benefits of different levels of integration.

Subsequently the potential for integration of an MTP unit as a "propylene booster" into an existing Steam Cracker is described. Since this integration is more site dependent, the effect of different levels of integration will focus on additional benefits that can be derived from the add-on MTP unit.

Assuming that the reader is quite familiar with Steam Cracking technology but not methanol to olefins processes, an introduction to the MTP process features is given upfront. A more detailed description of the Lurgi MTP™ process can be found here [2].

Lurgi MTP™ Technology

The Lurgi MTP™ process produces propylene from methanol employing a zeolite-based catalyst from Clariant and a fixed bed, multi-stage reactor system as shown in Figure 1.

Figure 1
MTP – Process Overview

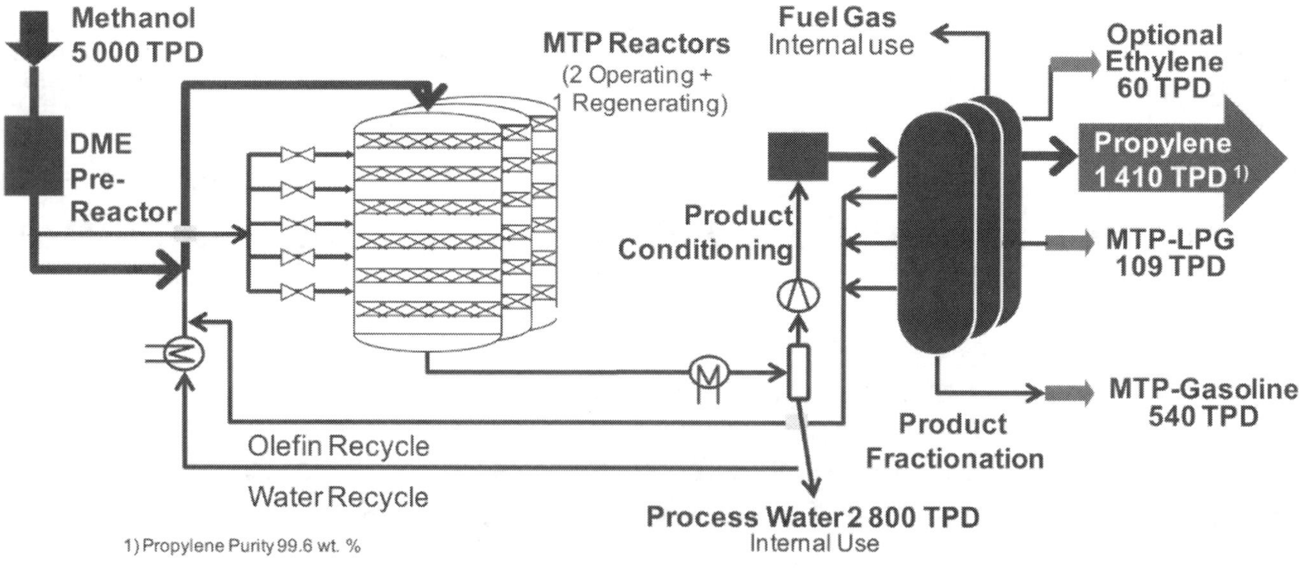

Vaporized methanol is converted to DME (Di-Methyl Ether) and water in an adiabatic DME pre-reactor on an alumina oxide catalyst according to the following equation:

2 CH_3OH (methanol) -> CH_3OCH_3 (DME) + H_2O

The catalyst, which is robust, highly active and highly selective towards DME, nearly achieves thermodynamic equilibrium for this reaction step.

One part of the DME product stream is combined with process steam and a hydrocarbon recycle from the MTP product fractionation. The combined stream is heated up further in a gas-fired heater and routed to the first catalyst bed of the MTP reactor. The major portion of the remaining DME reactor product is cooled down prior to being fed to the subsequent catalyst beds of the MTP reactor.

The MTP reactor is designed as a multi-stage adiabatic fixed bed reactor. In the MTP reactor the DME/methanol – mixture is converted to olefins on a zeolite-based catalyst according to the following main reaction:

$$n \;\; CH_3OCH_3 \;\; (DME) \qquad -> \qquad 2 \;\; C_n H_{2n} \;\; + \;\; n \;\; H_2O \qquad n = 2 \ldots 8$$

The conversion of oxygenates to olefins is exothermic and other reaction pathways result in formation of various paraffins, naphtenes and aromatics. High selectivity towards propylene requires control of the reaction conditions. Dedicated feed distribution and process conditions for the MTP reactor stages maintain similar reaction conditions optimized for maximum overall propylene yield. For this the hydrocarbon recycle is very important as it allows for conversion of olefins to propylene and therefore substantially enhances overall propylene yield. The product of the MTP reactor is an olefin-rich hydrocarbon mixture with propylene as the predominant component.

The MTP reactors are installed in a 2+1 reactor concept (2 reactors in operation and 1 reactor in regeneration mode or standby) which allows regeneration as well as the option to exchange the catalyst while production continues. Each reactor represents 50% of the plant capacity.

After a cycle of about 600 hours of operation, the MTP catalyst can be regenerated in-situ by burning off the coke on the catalyst with a nitrogen/air mixture. The regeneration is carried out at similar temperatures as the reaction itself so that temperature stress within the catalyst structure is avoided. Because of the 2+1 reactor concept, regeneration of one reactor does not disturb operation of the MTP unit.

A nitrogen purge after completion of the regeneration ensures that oxygen cannot get in contact with the MTP synthesis. This reduces operation risk and minimizes the formation of oxygenates other than methanol and DME.

The MTP reactor effluent is cooled in a heat recovery system and finally in the quench system, where the hydrocarbon product is separated from the bulk of the byproduct water. The hydrocarbons leave the quench system as overhead vapor whereas the water is condensed and sent to the Methanol Recovery Column. In this column methanol and DME are recovered and subsequently fed as unconverted fresh feed to the DME reactor.

The gas separation and compression as well as the product separation and purification follow standard techniques, quite similar to the practices used in common steam cracker processes.

The compressed hydrocarbon vapour is routed to a product purification section together with the liquid hydrocarbons. Traces of water, carbon dioxide and DME are removed by standard techniques including adsorption and extraction. The hydrocarbon product is then further processed by distillation yielding polymer-grade propylene and optionally some ethylene.

Optional ethylene production is tailored to meet requirements of downstream propylene polymerization units to be able to manufacture Co-Polymer grades. Several olefin-containing streams are sent back to the main MTP synthesis loop. To avoid accumulation of inert or undesired materials in the loop, a small purge is required for light-ends used as fuel gas. Besides the main products - propylene and ethylene - MTP-LPG and MTP-Gasoline are produced as valuable co-products.

MTP – Steam Cracker Integration: Main Constraints

The integration of Steam Cracker and MTP unit is limited by certain constraints derived from the two processes.

The MTP reactor uses a hydrocarbon recycle rich in C4 - C6 olefins. As described above, the olefins in this Hydrocarbon Recycle increase the propylene yield of the MTP unit. Therefore any stream containing C4 – C6 olefins sent to the MTP reactor will increase propylene yield. At the same time such streams should be limited in aromatics, specifically benzene, as their conversion consumes methanol and increases coking, and in acetylenics and di-olefins, which promote catalyst fouling. In addition, it must be ensured that streams sent to the MTP unit are virtually free of poisons for zeolite catalysts such as alkaline metals and sulphur.

Contamination of the Steam Cracker feeds by oxygenates shall be kept to the minimum, i.e. at ppm levels, as it can increase corrosion and fouling issues. The impact of oxygenates in a Steam Cracker have been discussed in previous papers [3, 4].

The effluent from cracking furnaces contain relatively large amounts of di-olefins and benzene. For example, in our Grassroot Study Case (ethane steam cracking), butadiene accounts for 70% of the C4s produced by the Steam Cracker and benzene for 60% of the C6s.

The MTP reactor effluent contains significant amounts of oxygenates, mostly unreacted methanol and DME. It also contains trace amounts of other oxygenates, such as ketones. In the MTP stand-alone plant, the process scheme progressively and efficiently recovers and recycles the unreacted methanol and DME. The bulk of the Oxygenates Removal is actually combined with the MTP Product Fractionation generating the Hydrocarbon Recycle.

Based on the above considerations, Air Liquide and Technip have developed the integration of the MTP unit and the Steam Cracker unit on the following principles:

- For the MTP unit:
    - Segregation of the MTP reactor effluent from the cracking furnaces effluent until the Hydrocarbon Recycle to the MTP unit is generated.
    - Enhancement of the Oxygenates Removal scheme for MTP streams sent to the Steam Cracker.
- For the Steam Cracker:
    - Segregation of the cracking furnace effluent from the MTP reactor effluent until oxygenates are removed in the MTP unit.
    - Additional treatment (such as hydrogenation or extraction) of Cracker streams sent to the MTP reactor and containing acetylenics, di-olefins or aromatics.

MTP – Steam Cracker Integration: Overall Material Balance Improvements

Air Liquide and Technip understand that each plant configuration is unique and that production objectives and associated economics can vary widely from one plant location to another, and/or from one company to another.

As a result, Air Liquide and Technip have developed several flexible integration steps that can be implemented, or not, to improve the Overall Material Balance of a combined MTP – Steam Cracker plant. A few examples of such integration steps are presented in this paper to illustrate their benefits.

For each integration step, the improvements in the Overall Material Balance are quantified against the production of two stand-alone units and at constant fresh feedstock consumption, i.e. at constant methanol and ethane feed consumption.

For a stand-alone MTP unit, propylene production is usually favored over ethylene production due to the purpose of the stand-alone MTP unit and the MTP reactions. For a stand-alone Steam Cracker, ethylene production is usually the primary target and propylene production a secondary target, in particular for gas crackers.

In the context of the MTP – Steam Cracker integration, both the ethylene and propylene productions become primary targets. Improving the Overall Material Balance therefore translates into maximizing the combined ethylene + propylene production.

As a result, the MTP unit design no longer needs to maximize propylene production over ethylene production, but can rather target the maximum light olefins production when it is combined with a Steam Cracker.

MTP – Steam Cracker Integration: Grassroot Study Case

Three sequential degrees of integration will be applied to a Grassroot Study Case which combines two separate and independent grassroot units:

- Lurgi MTP™ unit with a name plate capacity of 1,410 TPD propylene based on 5,000 TPD methanol feed.

- Gas Steam Cracker unit, producing 1,000 kTA ethylene, ethane feed, Technip SMK™ cracking furnaces, ethane and propane recycled to extinction.

The Overall Material Balance for the two stand-alone units is shown in Table 1, considering middle of run catalysts conditions and an on-stream factor is 8,400 hr / year.

Table 1
Stand-alone Steam Cracker + Stand-alone MTP
Overall Material Balance

Hydrocarbons Products	Cracker kTA	MTP kTA	Total kTA	wt%
Ethylene	1,000	26.4	1,026	50.0%
Propylene	22.5	500	523	25.4%
Fuel Gas	197	13.2	210	10.2%
Raw C4 Cut / LPG	35.8	37.2	73.0	3.6%
Pygas / Gasoline	30.1	188	218	10.6%
Fuel Oil	3.9	-	3.9	0.2%
Total Hydrocarbons	**1,289**	**765**	**2,054**	**100%**

MTP – Steam Cracker Integration: Equipment Sharing

Sharing equipment in common is driven by the overall CAPEX reduction. However, equipment sharing shall not compromise:

- the process chemistry,
- the operability of the units,
- the overall energy efficiency, i.e. with significant higher OPEX.

The treatment of the C2s and lighter can advantageously be combined between the two units. While this process section is at the heart of the Steam Cracker, in the MTP unit many equipment are required to recover small quantities of ethylene.

The C2s and lighter from the MTP Deethanizer are routed to the Steam Cracker unit upstream of the Cracker Caustic Wash. This allows in the MTP unit the elimination of all equipment related to ethylene purification such as: CO_2 Scrubber and Adsorber, Demethanizer, C2 Splitter and C2 Refrigeration. It as well eliminates all cryogenic heat exchangers.

In our Grassroot Case Study example, as per Table 1, the impact on the size of the Steam Cracker equipment is limited due to the fairly low quantity of C2s and lighter produced by the MTP unit compared with the Steam Cracker.

In the same way, the C3 Splitter system can also be combined between the two units.

For a configuration similar to our Grassroot Study Case, the Steam Cracker C3 Cut is routed directly to the MTP C3 Splitter, eliminating the need for a dedicated C3 Splitter in the Steam Cracker. As per Table 1, the impact on equipment size is limited due to the fairly low quantity of C3s produced by the Steam Cracker compared with the MTP unit.

For grassroot plants, the cooling required in the MTP unit at temperatures below cooling water can be done by a common C3 Refrigeration system shared by the two units.

For some of the steps of integration, additional equipment is required and described under the respective step.

Integration Step 1

The overall process scheme for this step is illustrated in Figure 2. In this step, the equipment sharing concepts previously developed are implemented:

- the MTP C2s and lighter are sent to the Steam Cracker upstream of the Caustic Wash,
- the Cracker C3 Cut is sent to the MTP C3 Splitter.

The propane from the MTP C3 Splitter, which now combines both the MTP propane and the Cracker propane, is mixed with the MTP C4s. This C3/C4 mixture, which has also been subjected to enhanced oxygenate recovery, is sent to the cracking furnaces.

With this configuration, the MTP C2s and light ends are recovered and fractionated in the Steam Cracker. In the end, the MTP ethane and the MTP LPG (C3/C4) are used as a supplemental feed to the cracking furnaces.

Figure 2
MTP - Steam Cracker Integration Step 1

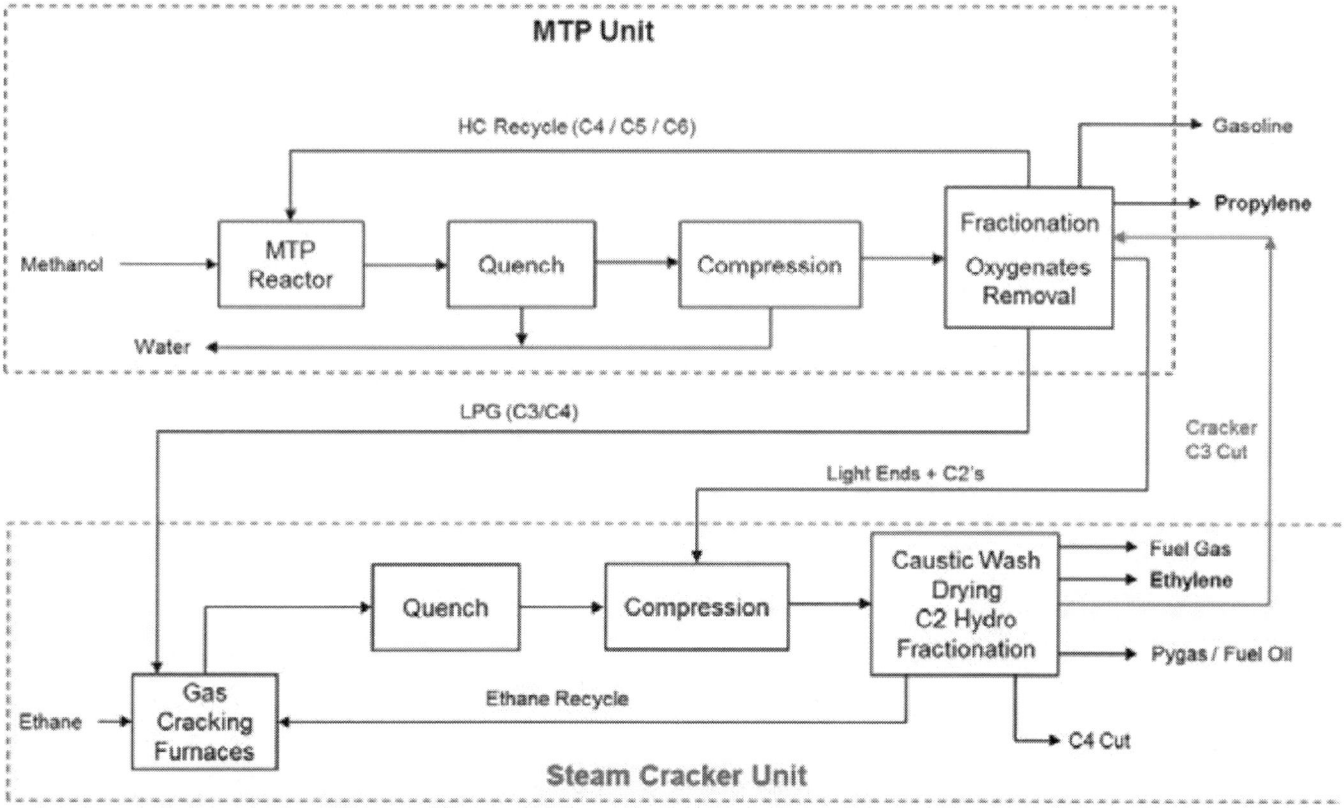

Integration Step 2

In a second integration step, the MTP Gasoline is sent to the Steam Cracker to feed a dedicated liquid cracking furnace, using Technip GK6® coil technology in this study case. Since the MTP Gasoline contains very low amounts of di-olefins, no partial or full hydrogenation of the stream is required prior to feeding it to the cracking furnace. Once-through steam cracking yields evaluated with Technip's SPYRO® software show that approximately 30% of the MTP Gasoline can be directly converted to light olefins. In a stand-alone MTP plant, the Gasoline product is normally used as low sulfur blend stock for gasoline pools.

The overall process scheme for this step is illustrated in Figure 3.

Figure 3
MTP - Steam Cracker Integration Step 2

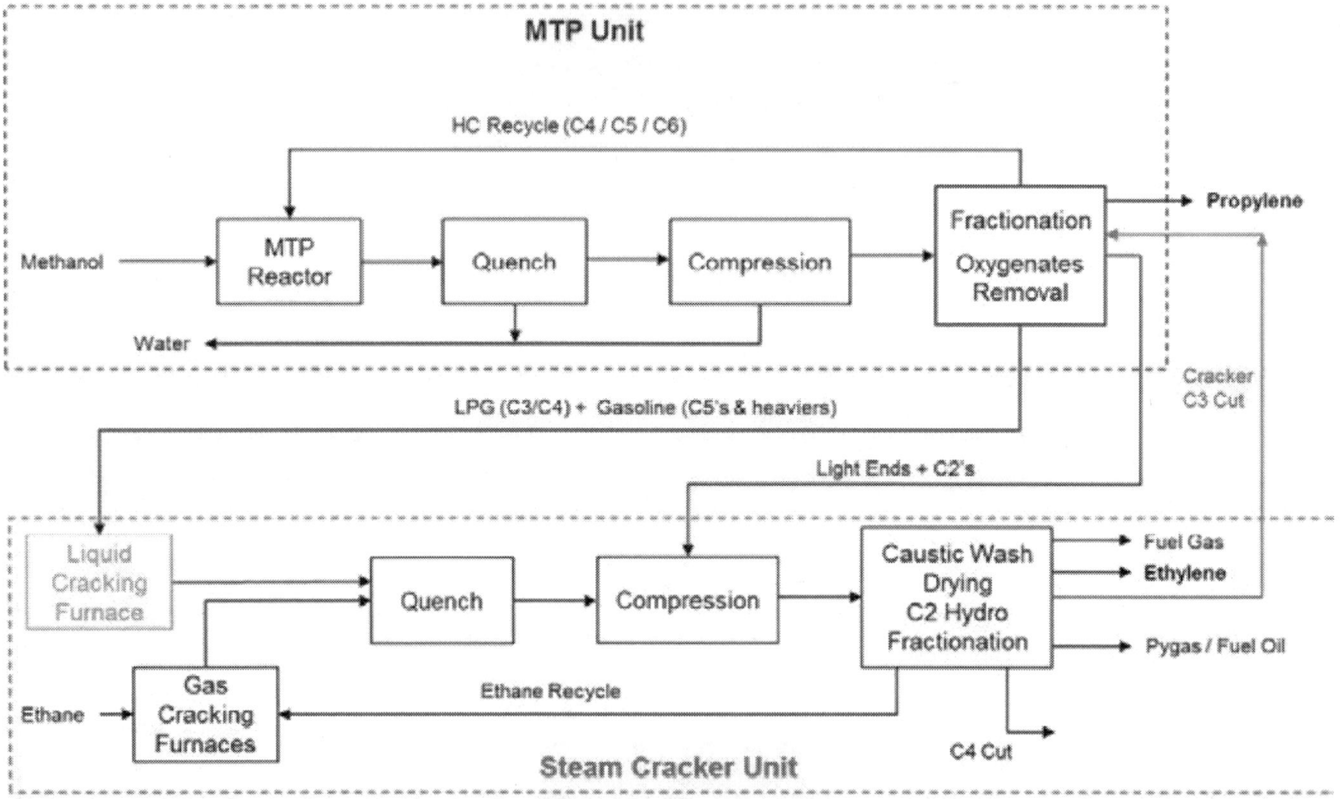

A further improvement of the gasoline conversion to light olefins could be achieved through hydrogenation of the MTP Gasoline before feeding it to the Steam Cracker. This hydrogenation step to convert olefins into paraffins would only be done to further increase the cracking yields to ethylene and propylene.

Integration Step 3

In a third integration step, the olefin-rich streams from the Steam Cracker are sent to the MTP reactor for conversion to propylene. The most obvious candidate is the C4 Cut from the Steam Cracker.

In general, conversion of the C4 Cut is in competition with other potential uses of the stream, such as extraction of butadiene or recycle of the C4 stream to the Cracker.
For the third integration step the recycle of the full C4 Cut after partial hydrogenation of the di-olefins is shown.

Any other olefin-rich stream from the Steam Cracker may be a good candidate for feed to the MTP reactor as well, as long as it does not contain di-olefins or aromatics.

The overall process scheme for this step is illustrated in Figure 4.

Figure 4
MTP - Steam Cracker Integration Step 3

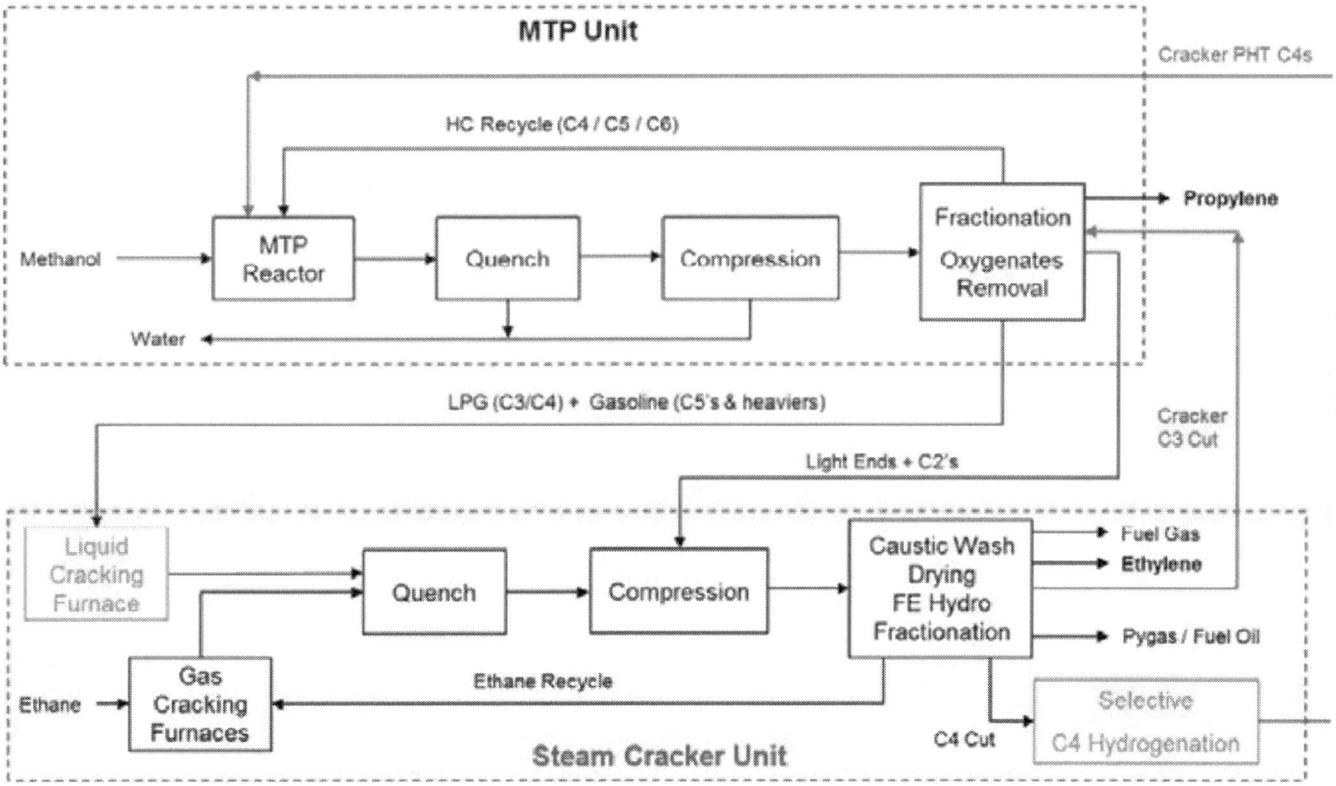

Comparison of the steps of integration and Conclusion

For the integration steps 1 and 2, the MTP by-products are converted to light olefins by pyrolysis in the Steam Cracker furnaces.

For the integration step 3, the Steam Cracker C4 Cut by-product is converted to light olefins in the MTP reactor.

Air Liquide and Technip have performed a cost analysis of each integration step for comparison with the Total Installed Cost of two stand-alone units using existing projects.

The benefits of each integration step is summarized in Table 2.

Table 2
MTP - Steam Cracker Integration
Grassroot Study Case
Integration Steps Summary

Production kTA	Stand-alone 1 + 1	Step 1	Step 2	Step 3
Ethylene	1,026	1,086	1,120	1,132
Propylene	523	507	533	556
Total Light Olefins	**1,549**	**1,593** **+ 3%**	**1,653** **+ 7%**	**1,688** **+ 9%**
Butadiene export	24.8	25.5	33.3	0.0
CAPEX	**Base**	**- 4.5%**	**- 1%**	**- 0%**
CAPEX / kTA Light Olefins	**Base**	**- 7%**	**- 7.5%**	**- 8.5%**

Compared to two stand-alone units, the first integration step leads to a slightly lower propylene production. While this may seem surprising, it is actually in line with our target of maximizing the light olefins production. It is a deliberate result from design simplifications made in the integrated MTP unit.

In the context of our Case Study based on an ethane cracker equipped with gas cracking furnaces, both the second and third integration steps require the addition of dedicated liquid cracking capabilities in the Steam Cracker to handle the MTP Gasoline.

This was selected as a "worst-case scenario" from a CAPEX stand-point. Nevertheless, the savings resulting from the equipment sharing between the two units balance the additional costs of the new liquid cracking capabilities.
The CAPEX benefits of the integration steps 2 and 3 would largely increase when combining an MTP unit with a Mixed Feed, Flexible Feed or Liquid Feed Steam Cracker already including some liquid cracking capabilities.

Sending the partially hydrogenated C4 Cut from the Steam Cracker to the MTP unit in order to increase the Propylene production can also be implemented without sending the MTP Gasoline to the Steam Cracker.

However, it shall be noted that cracking the MTP Gasoline increases the Butadiene / C4 Cut production from the Cracker, therefore it has a synergetic effect on the light olefins production when sending the treated C4 Cut back to the MTP reactor.

As mentioned previously, the Cracker C4 Cut may compete with other users. For example, a common practice is to recycle the C4 Cut to the cracking furnaces. In terms of yield to light olefins, sending the partially hydrogenated C4 Cut to the MTP unit is equivalent to recycling the fully hydrogenated C4 Cut to the cracking furnaces. With the MTP route, however, the yield to propylene is much higher than the yield to ethylene. This advantage offers an excellent opportunity to use the Cracker C4 Cut as a "propylene booster" in an integrated MTP – Steam Cracker plant.

The different steps of integration described above represent the range of options available. The optimum solution for an individual project requires all specific aspects to be analyzed.

For better overview, the impact of step 3 on the Overall Material Balance is quantified against the two stand-alone units in Table 3. The integration step 3 boosts the light olefins production (ethylene + propylene) by almost 10% while reducing the quantity of by-products by nearly 30%.

Table 3
MTP - Steam Cracker Integration
Grassroot Study Case
Overall Material Balance

Hydrocarbons Products kTA	Stand-alone Units 1 + 1	Integration Step 3
Ethylene	1,026	1,132
Propylene	523	556
Total Light Olefins	**1,549**	**1,688** **+ 9%**
Fuel Gas	210	245
Raw C4 Cut / LPG	73	0
Pygas / Gasoline	218	88
Fuel Oil	4	33
Total By-Products	**505**	**366** **- 28%**
Total Hydrocarbons	2,054	2,054

MTP – Steam Cracker Integration: Existing Plant Study Case

This Study Case has been developed to highlight the benefits of combining a new MTP unit to an actual existing Steam Cracker where the existing C3 Splitter is no longer used to its full capacity.

Such an integration has the potential for a substantially higher degree of use of common equipment as the Grassroot Study Case. The most obvious is to route the MTP C3 Cut directly to the existing Steam Cracker C3 Splitter to eliminate the need for a new dedicated C3 Splitter in the MTP unit, hence reducing CAPEX.

On the other hand the use of a common C3 Refrigerant system may be more difficult, as the existing refrigeration system may not have sufficient spare capacity. In such cases specific concepts must be developed to meet the MTP plants cooling needs (e.g. Absorption cooling).

The Study Case presented here is based on the following scenario. The existing Steam Cracker is a flexible feed unit equipped with cracking furnaces able to handle gas and liquid feeds, including naphtha. The C3 Splitter design capacity is around 250 kTA propylene. Due to a switch to lighter feedstock, the actual propylene production is down to 42 kTA while the ethylene production is 954 kTA. The raw C4 Cut is exported.

The capacity of the new MTP plant is set at 2,000 TPD methanol feed in order to fit the design capacity of the existing Cracker C3 Splitter. The MTP unit production is 200 kTA propylene and 10 kTA ethylene. For the two stand-alone plants the combined propylene production (MTP + Steam Cracker) is 242 kTA.

Since the existing Steam Cracker has already liquid cracking capabilities, the new MTP unit is combined as per the integration step 3 previously described, refer also to Figure 4. The main difference is that in this case the MTP C3 Cut is sent to the existing C3 Splitter in the steam cracker. The MTP LPG / Gasoline can be handled by a single existing liquid furnace.

No estimation of the overall investment cost is presented, as it depends of the actual spare capacities available in the cracker. However, with an existing cracker already equipped with a C3 Splitter and liquid cracking capabilities, most of the CAPEX will be spent for the new MTP unit. Air Liquide and Technip have therefore performed a cost analysis for the MTP unit alone to illustrate the potential in CAPEX reduction.

The benefits of the integration is summarized in Table 4.

Table 4
MTP – Steam Cracker Integration
Existing Plant Study Case

Production kTA	**Stand-alone Units 1+1**	**Integrated Units**
Ethylene	964	1,010 **+ 5%**
Propylene	242	269 **+ 11%**
Total Light Olefins	**1,206**	**1,279 + 6%**
Butadiene Export	28	0
MTP unit CAPEX	**Base**	**-30%**

Compared to our Grassroot Study Case, the impact of sending the Cracker C4 Cut to the MTP unit is more visible on the propylene production, due to the change in the relative sizes of the combined units.

As mentioned previously, a range of options can further improve the integration benefits. For an existing steam cracker, some of the options mentioned previously may not be applicable; however additional feedstock opportunities may be available on site.

For example, the partially hydrogenated C5 Cut from an existing pygas hydrogenation unit is typically recycled to the cracking furnaces. Instead, it can be routed as a feed to the new MTP unit to boost propylene production.

Some petrochemical sites include multiple steam crackers. The treated cracker C4 Cuts and/or C5 Cuts can be combined and send to a single new integrated MTP unit. As illustrated in Table 4, the relative size of the combined units can significantly boost the propylene production, therefore the crackers C4 / C5 by-products would in this example provide a very substantial amount of the final propylene production.

This example illustrates the strong advantages to integrate a new MTP unit as a "propylene booster" into an existing Steam Cracker through synergistic effects of exchanging streams and CAPEX reduction through use of spare capacities in existing equipment.

MTP – Steam Cracker Integration: Overall Conclusion

The cases of Grassroot Study Case show that the integration of a Steam Cracker with an MTP plant offers the potential to achieve a synergistic increase in light olefin yield of nearly 10% compared to independent systems without any increase in CAPEX. The integration scheme is flexible and can alternatively achieve a 4.5% reduction in CAPEX while still providing a 3% increase in light olefin production.

The integration of the two technologies can be done in different steps offering best adaptability to local conditions. The proposed integration offers the required flexibility to meet specific objectives and addresses the full range of Steam Cracker configurations, from ethane to naphtha crackers including mixed and flexible feed crackers.

The Existing Plant Study Case shows that integration of a new MTP plant into an existing steam cracker allows for CAPEX reduction of up to 30% for the new MTP unit and at the same time for a synergistic increase in light olefin production of 6%. This makes an MTP unit an ideal "propylene booster" for existing steam crackers.

With Technip's vast experience in Steam Cracking technologies and Air Liquide's expertise in MTP technology the integration can be tailored to best meet the needs of any specific application and its relevant markets.

References

1. Diercks Rainer(2014), "Petrochemicals The Heart of BASF's Verbund" BASF Investor Day Chemicals Segment May 22, 2014. *https://www.basf.com/documents/corp/en/investor-relations/calendar-and-publications/calendar/2014/basf_investor_day_chemicals/BASF_Investor-Day-Chemicals_Petrochemicals.pdf (accessed February 17 2016)*
2. Haag, Stéphane; Pohl, Sven; Gorny, Martin and Martin Rothaemel (2012), "Methanol to propylene: from development to commercialization" proceedings from DGMK Tagung 2012-3, p 115-122.
3. Nowowiejski, George and John A. Reid (2003), "An Overview of Oxygenates in Olefins Units in Relation to Corrosion, Fouling, Product Specifications, and Safety," proceedings from the 15th Ethylene Producers' Conference.
4. Hood, David and Steve Coleman (2007), "Minor Components, Major Issues: An Overview of Contaminants in Ethylene Plants" proceedings from the 19th Ethylene Producers' Conference.

AIChE Paper 192C

A NEW ERA IN ON-PURPOSE BUTADIENE TECHNOLOGY

Clifford Alan Maat

Senior Staff Engineer

TPC Group

John Senetar

Fellow

UOP LLC, a Honeywell Company

Christopher DiGiulio

Senior Engineer

UOP LLC, a Honeywell Company

Jeannie Blommel

Lead Design Engineer

UOP LLC, a Honeywell Company

Joseph Duff

Staff Research Chemist

TPC Group

Jillian Horn

Senior Research Scientist

TPC Group

Michael Nutt

Principal Research Engineer

TPC Group

Prepared for Presentation at the 2016 Spring National Meeting

EPC Fundamentals of Ethylene Technology Session

Houston, Texas, April 11 -14, 2016

AIChE Paper 192C

A NEW ERA IN ON-PURPOSE BUTADIENE TECHNOLOGY

Clifford Alan Maat

Senior Staff Engineer

TPC Group

John Senetar

Fellow

UOP LLC, a Honeywell Company

Christopher DiGiulio

Senior Engineer

UOP LLC, a Honeywell Company

Jeannie Blommel

Lead Design Engineer

UOP LLC, a Honeywell Company

Joseph Duff

Staff Research Chemist

TPC Group

Jillian Horn

Senior Research Scientist

TPC Group

Michael Nutt

Principal Research Engineer

TPC Group

Abstract: Butadiene is produced almost entirely as a byproduct of ethylene plants. As ethylene plants switch from naphtha feedstock to cheaper ethane, the amount of butadiene byproduct decreases. A supply shortage is anticipated. On-Purpose Butadiene will be needed to meet the demand. One way to make On-Purpose Butadiene is Oxidative Dehydrogenation of butenes, traditionally done by TPC Group (Texas Petrochemicals, Petro-Tex). TPC and UOP have jointly developed an improved process based on the TPC technology. The improvements are reviewed and the new technology compared with alternative routes for On-Purpose Butadiene.

Background

Butadiene was produced in moderate volumes from ethanol or acetylene prior to the Second World War.

Direct Dehydrogenation - early 1940's to mid 1960's

Interruption of global trade in natural rubber by the Second World War created an immediate and urgent demand for on-purpose butadiene. Most of the new production capacity used direct dehydrogenation processes, with n-butane or normal butenes as the feedstock. Further expansion of butadiene production capacity took place in the 1950's.

Oxidative Dehydrogenation - mid 1960's to late 1980's

A new process for butadiene production from butenes was introduced by Petro-Tex in 1965. Addition of oxygen to the feedstock resulted in a chemical equilibrium much more favorable for butadiene. The oxidative dehydrogenation process eventually displaced much of the direct dehydrogenation production.

Adequate Supply From Byproduct Butadiene - late 1980's to mid-2010's

A wave of construction of new ethylene plants in the late 1980's enlarged the supply of butadiene as a byproduct of ethylene manufacture. The low-cost byproduct butadiene displaced all other manufacturing routes, and large scale on-purpose butadiene production eventually stopped.

New Demand for On-Purpose Butadiene - mid-2010's...

The shale gas revolution has made low-cost ethane available to many ethylene plants in North America. As the ethylene plants have switched to feed more ethane, the amount of byproduct butadiene has decreased. On-purpose butadiene production capacity has been installed in China, and a need for new on-purpose butadiene capacity in other regions is anticipated in the near future.

Commercial History of the Oxidative Dehydrogenation Process

In 1965, Petro-Tex (predecessor of Texas Petrochemicals and TPC Group LLC) commercialized butadiene production via oxidative dehydrogenation of butenes in its TPC Group OXO-D® process [1]. The OXO-D technology provided significant savings in raw material and energy costs when compared to conventional dehydrogenation technologies, which are endothermic and equilibrium limited. The first OXO-D plant was operated by TPC Group in Houston, Texas, with an annual butadiene production capacity of 350 thousand metric tons/yr (kMTA). In the years that followed, six additional OXO-D plants were licensed with production capacities ranging from 75 to 350 kMTA. The OXO-D units successfully operated with a wide range of C4 feedstock compositions and downstream butadiene extractive distillation technologies. All of the OXO-D units had excellent safety records.

The original OXO-D plant in Houston remained in operation until 2004, when it was finally decommissioned after four decades of successful operation.

Process Chemistry/Catalysts

The reaction chemistry is summarized in Figure 1, with the heats of reaction reported per mol of 1-butene and oxygen consumed, respectively [2]. Butene is reacted with molecular oxygen to form butadiene and water, with air being the most readily-available oxygen source. The formation of 1,3-butadiene is greatly favored over 1,2-butadiene. Thus for convenience, the desired 1,3-butadiene product is herein simply referred to as butadiene. The conversion and selectivity to butadiene for the normal-butene isomers (1-butene, cis-2-butene, and trans-2-butene) are similar in the OXO-D process. The main byproducts are the combustion products CO and CO2. In this process, oxygen is consumed to completion.

FIGURE 1 – Oxidative Dehydrogenation Reactions

Stoichiometry	ΔH_{RXN} (kJ/mol_{BUTENE})	ΔH_{RXN} (kJ/mol_{OXYGEN})
C_4H_8 + ½O_2 → C_4H_6 + H_2O 1 or 2 Butenes, Oxygen, 1,3 Butadiene, Water	-131.9	-263.8
C_4H_8 + $4O_2$ → 4CO + $4H_2O$ 1 or 2 Butenes, Oxygen, Carbon Monoxide, Water	-1409.3	-352.3
C_4H_8 + $6O_2$ → $4CO_2$ + $4H_2O$ 1 or 2 Butenes, Oxygen, Carbon Dioxide, Water	-2541.3	-423.6
iC_4H_8 + $6O_2$ → $4CO_2$ + $4H_2O$ Isobutene, Oxygen, Carbon Dioxide, Water	-2524.3	-420.7

Minor byproducts include ethyl and vinyl acetylenes, formaldehyde, acetaldehyde, acrolein, methacrolein, acetone, furan, isoprene, vinylcyclohexene, and aromatics (BTX, styrene, naphthalene). Even small amounts of acetylenes and some oxygenates can cause operating problems in downstream processing units, so there is a premium on minimizing the amounts of these chemicals in the butadiene product.

Process Description

Figure 2 provides a simplified process flow diagram for the TPC/UOP OXO-D process. Butenes, steam and air are mixed and fed into a two-stage reactor system. The reactors operate adiabatically between 0.1 to 0.3 MPa(g) , inlet temperatures from 300 to 400 °C, and outlet temperatures from 500 to 600 °C. Depending on the feed composition and targeted yield, the steam to butene molar ratio for the reactors can be between 9:1 to 14:1. Additional air is introduced in between Stage 1 and Stage 2 to maximize butene conversion. Splitting the air addition between the two stages has the added benefit of limiting the exotherm in each stage, which reduces the overall steam requirement for the process. The reactor effluent is sent to a Quench Tower, which serves to condense both the dilution steam added to the feed and the additional water vapor generated via the oxidative reactions. Condensed water is recovered and vaporized to provide all of the reactor steam requirements. The utilities for steam generation are minimized by using process heat recovery (not shown). Net water produced by the process is sent to water treatment. After bulk water removal, the net product vapor is compressed and sent to an Aldehyde Scrubber to remove trace aldehydes and other oxygenates. The rich solvent from the Aldehyde Scrubber is sent to the Aldehyde Stripper to generate a lean solvent for reuse and a vent stream containing the oxygenate contaminates. The contaminant-free vapor from the Aldehyde Scrubber is sent to the Product Absorber where C4+ hydrocarbons are absorbed in a circulating solvent. The rich solvent from the Product Absorber is sent to the Degasser to remove any absorbed light species, and then sent to the Product Stripper for recovery overhead of the crude butadiene product, which is typically in the range of 55 to 65 percent butadiene. Lean solvent from the Product Stripper is recycled to the Product Absorber. Solvent regeneration (not shown) is included to remove byproducts accumulated in the solvent. The crude butadiene is then sent to an extractive distillation unit to produce high-purity butadiene.

Figure 2

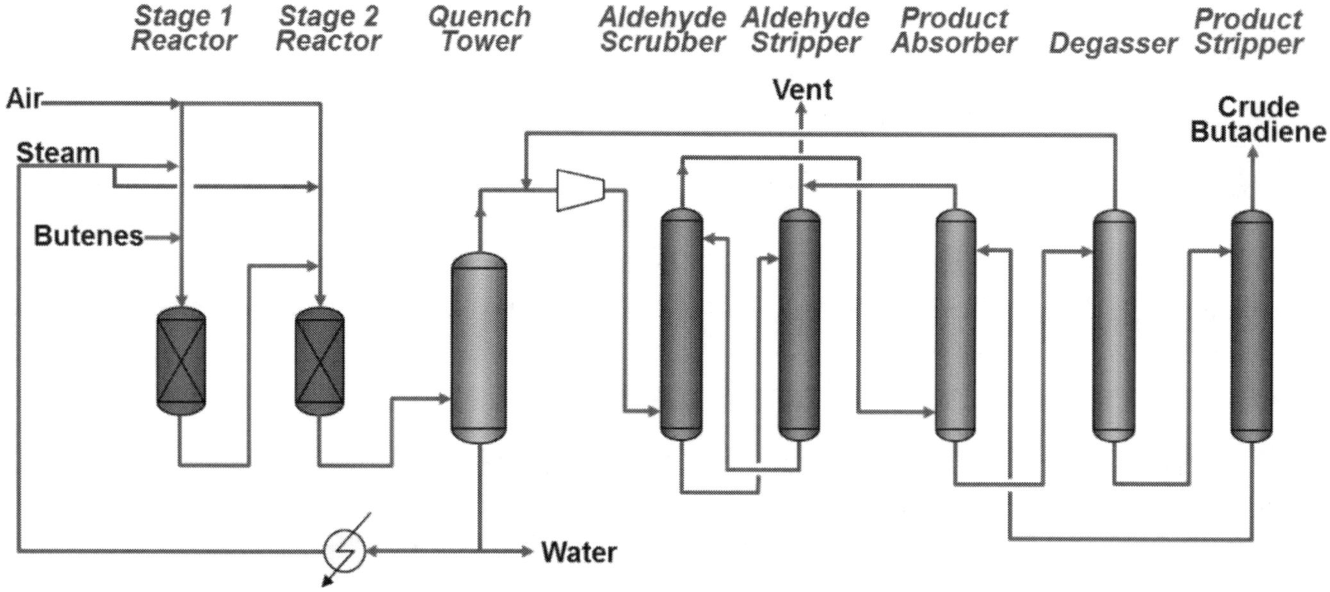

TPC/UOP OXO-D Process

Feedstocks

The unit design is flexible and can process a wide variety of feeds. Butanes pass through the reactor essentially unconverted. As indicated above, isobutene will largely convert to carbon dioxide. However, isobutene in the feed is acceptable and the suitable level is simply an economic consideration. Generally a raffinate-II with isobutene less than 2 percent is a preferred feedstock. Byproduct C4s from a methanol-to-olefins (MTO) process are an excellent feed, as MTO C4s are primarily 1- and 2-butenes with only a small amount of saturates and almost no isobutene.

Reactor Design and Operation

Conversion and Selectivity

Unlike the direct dehydrogenation processes, in which conversion is limited by the equilibrium concentrations of the products and feed, the oxidative dehydrogenation reaction proceeds essentially to completion. The conversion in each reactor can be adjusted simply by increasing or decreasing the amount of oxygen feed.

Selectivity, however, presents a practical limit to the feasible conversion per reactor pass. As incremental oxygen is added, selectivity decreases, producing relatively more CO and CO2. There is a point of maximum yield, beyond which further addition of oxygen results in lower butadiene production.

Heat and Temperature

Reactor temperature can also limit the feasible conversion per pass. Because both the main reaction and the significant side reactions to CO/CO2 are exothermic, the reactor effluent temperatures increase rapidly as the reactions progress. Typical temperature profiles are shown in Figure 3:

Figure 3

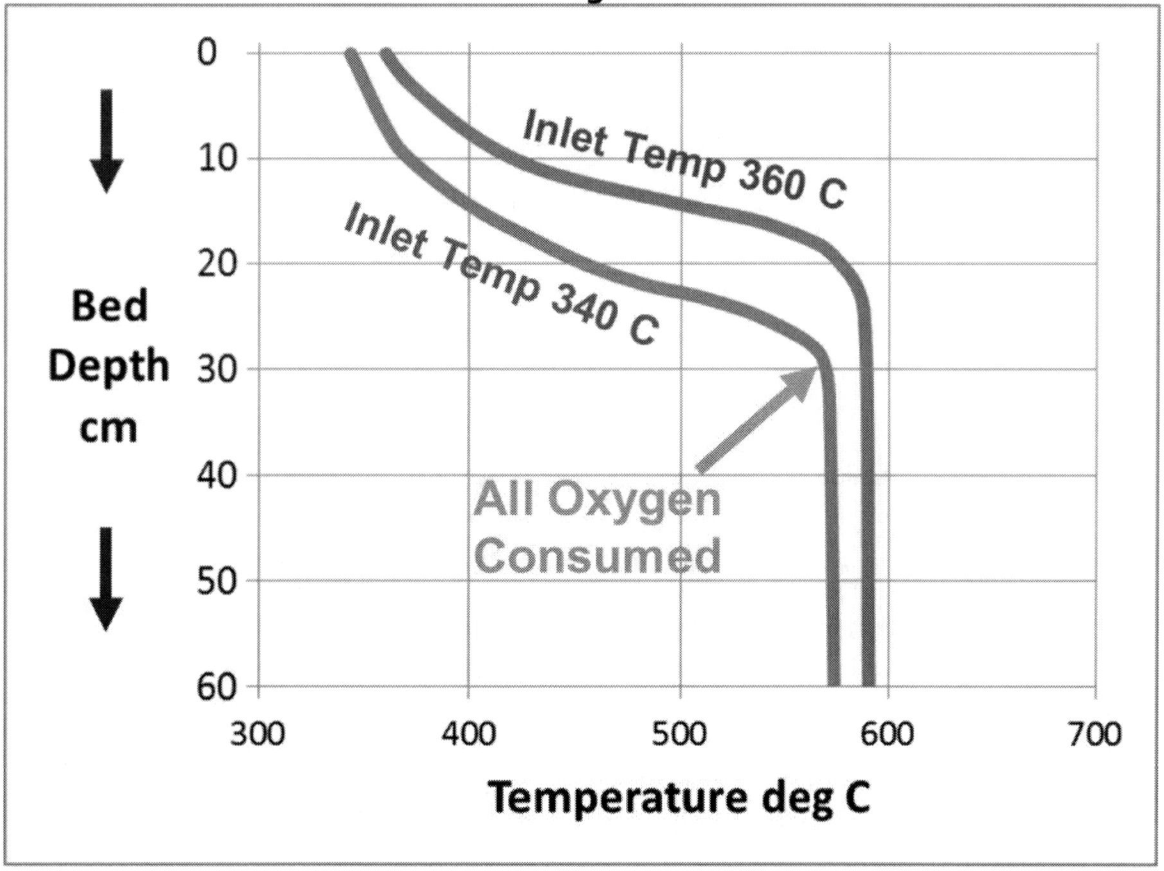

Temperature Profiles in an OXO-D Reactor

Steam feed serves partly to control the reactor outlet temperature; the steam absorbs a large part of the heat of reaction.

Single Stage versus Two Stage

The traditional commercial OXO-D process used a single stage reactor. While use of a second stage implies a loss of selectivity by feeding butadiene to the second stage, the two-stage reactor can be designed to minimize butadiene loss and, by virtue of much higher yield, result in higher overall plant efficiency. A qualitative comparison of reactor yields is shown in Figure 4:

Figure 4

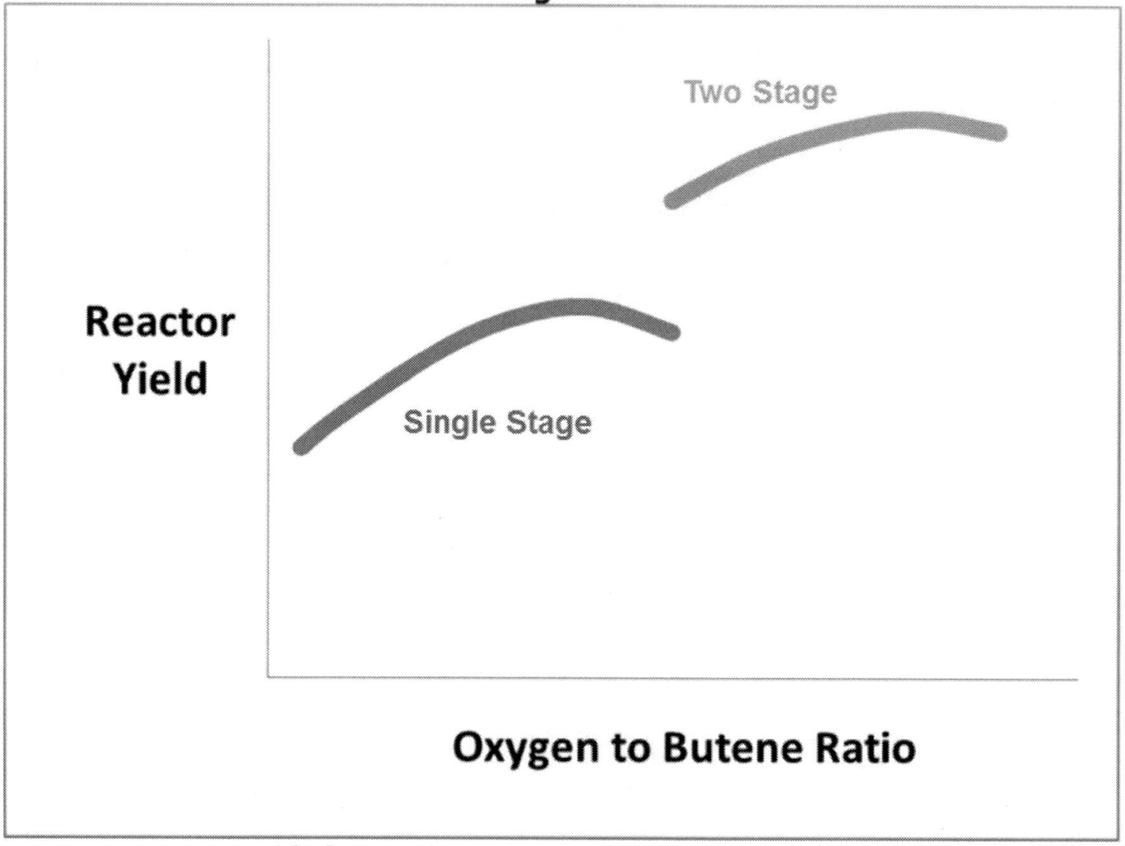

Yields for Single and Two Stage OXO-D Reactors

Byproduct Reduction

Oxygenated byproducts, as listed previously, are likely to have detrimental effects on downstream butadiene processing units. Some oxidative dehydrogenation processes, such as some with molybdenum-bismuth catalyst systems, use excess oxygen in the reactor and therefore have a propensity to form relatively large amounts of oxygenated byproducts. The OXO-D process is designed to minimize these byproducts as follows:

1. Butene is fed in excess, so all oxygen is consumed and the residence time of the oxygen is minimized.

2. The reactor catalyst system is designed such that any byproduct formation favors the relatively benign byproduct CO2. Figure 5 shows the difference between the OXO-D catalyst and an oxidative dehydrogenation catalyst designed only for high yield.

Figure 5

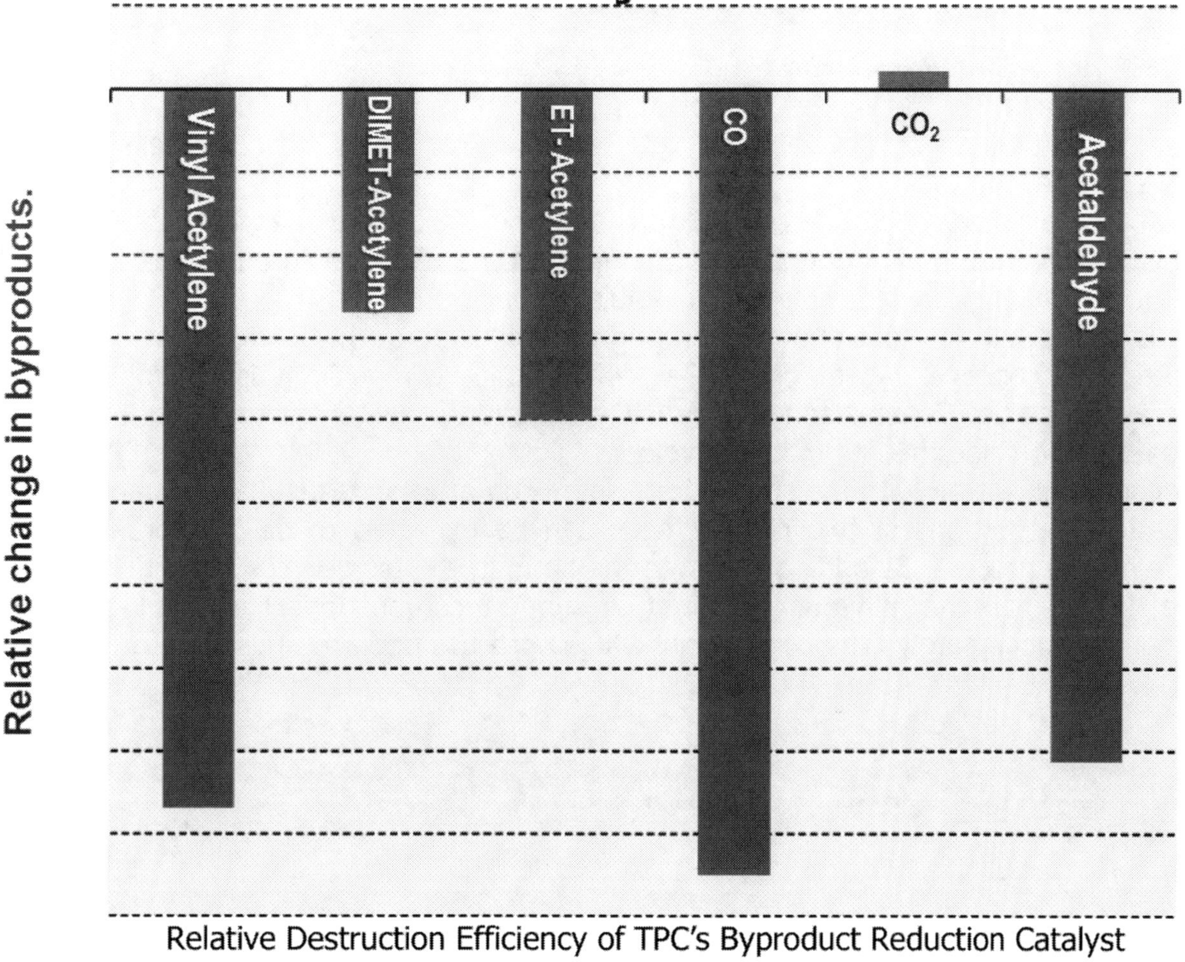

3. Oxygenated byproducts that do manage to form in the reactor are mostly removed in the Aldehyde Scrubber.

Process Safety

Reactor Temperature Control

The oxidative dehydrogenation reactions are exothermic and have a high initiation temperature (300 to 400 °C), so the reactor effluent temperature must not be allowed to get too hot. Some oxidative dehydrogenation processes and many other refining and chemical processes, hydrogenation reactors for example, also must maintain temperatures within a certain range while generating heat, and temperature control measures such as cooling sprays and emergency depressuring systems are common.

The OXO-D process temperature control is comparatively simple, because once initiated, the reactions proceed to completion. There is no need to remove heat from the catalyst bed or to try to control the temperature within the reactor. The temperature is controlled by

1. Inlet Temperature
2. Oxygen Feed
3. Steam Feed

All of these are easily instrumented and controlled upstream of the reactor.

Mixing Oxygen with Hydrocarbon

The possibility of unwanted combustion must always be considered when mixing oxygen and hydrocarbons, especially at high temperature. The OXO-D process has been operated safely as follows:

First, the air-steam-hydrocarbon mixture entering the reactor has too little oxygen to burn. This is illustrated in Figure 6, a triangular composition diagram and flammability envelope for 1-Butene/Oxygen/Nitrogen at ambient conditions (the true OXO-D flammability envelope is somewhat different due to operating conditions and other components, but Figure 6 illustrates the concept). A representative OXO-D inlet composition is the point labeled "Example Reactor Inlet Composition." All practical operating reactor inlet compositions would be in this region, outside the flammable envelope.

Second, mix the air with the steam first. No mixing process takes place instantaneously, so all possible proportions of two mixing feeds can be expected to be temporarily present during the mixing process. By adding steam to the air first, the oxygen concentration is diluted to a level lower than the minimum required to support combustion. Then, any amount of butene may be added safely. The commercial OXO-D process operated this way for over thirty years.

Figure 6

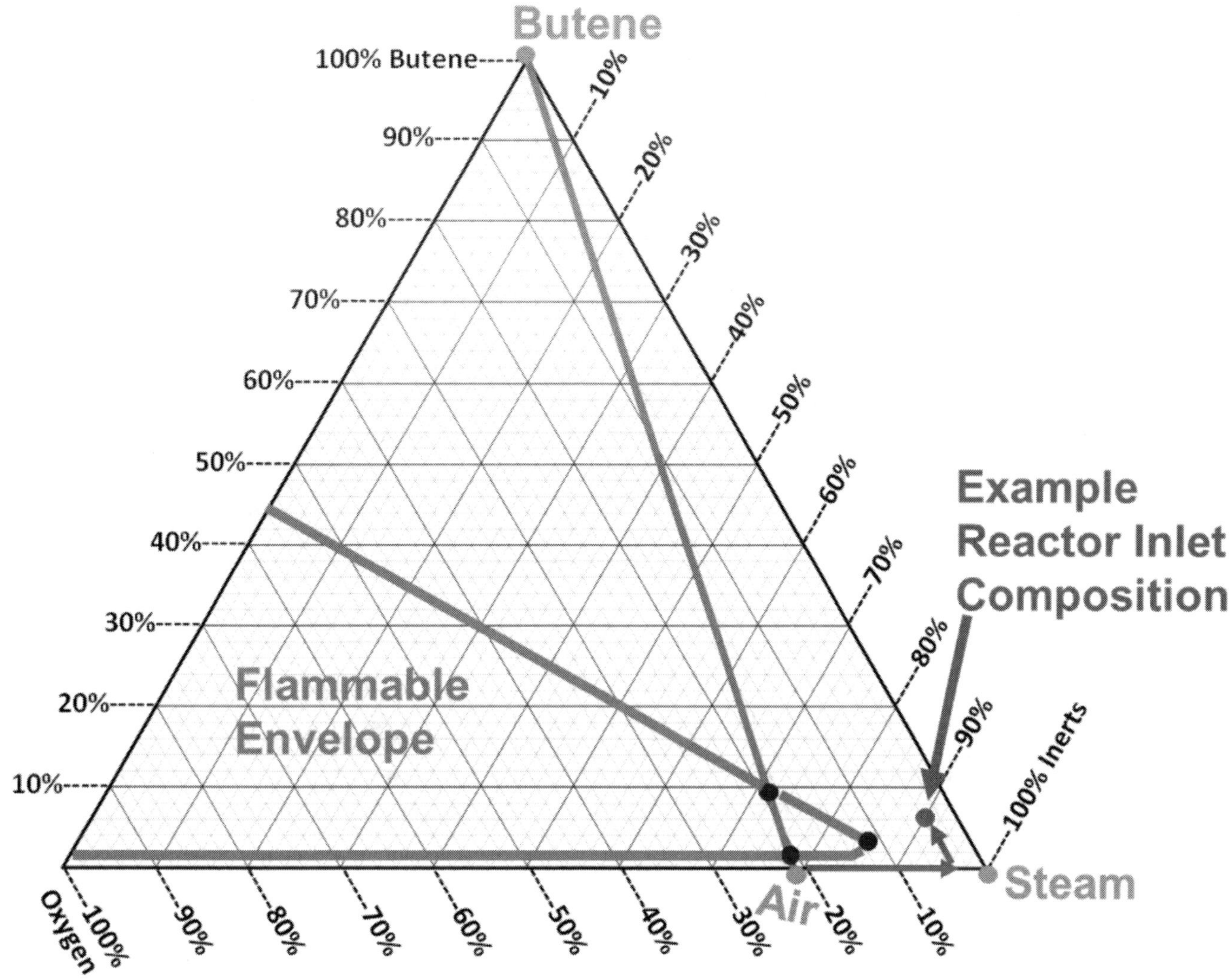

Flammability Envelope for 1-Butene/Oxygen/Nitrogen

Environmental Performance

Water Management

The steam feed to the OXO-D reactors is generated entirely from process condensate from the Quench Tower. Commercial experience from the TPC OXO-D unit as well as from the licensed plants enables proper design of process water treatment to minimize fouling and contamination problems. Since no fresh water is added to the process during normal operation, only a small purge stream of wastewater equal to the net production of water in the reactors must be treated for disposal.

Absorber Offgas

The Absorber offgas is mostly nitrogen with some CO2 and minor components. The heating value is too low for direct use as fuelgas, but it can be mixed with fuelgas and used in fired heaters or thermal oxidizers according to local environmental regulations.

Energy Efficiency

The hot reactor effluent contains high quality sensible heat that is recovered. The Quench Tower bottoms contains a large amount of heat; though it is at lower temperature and therefore more difficult to recover, opportunities should exist in most plants for heat integration. Further heat integration within the OXO-D unit has been incorporated into the TPC/UOP design.

Reliability and Campaign Length

All process plant designs involve some trade-off between capital cost and operating reliability and run-time. The commercial TPC plant operated with moderate campaign lengths. The TPC/UOP OXO-D process is designed for continuous four-year operation. Among the design features are

1. Spare Reactor – The catalyst does not last four years; consequently, one reactor can be taken offline for catalyst regeneration/replacement while full production rates are maintained.
2. Ceiling Temperatures – Streams containing large amounts of butadiene are kept cool enough to eliminate fouling problems caused by polymerization.
3. Spare Reboilers – While the reboilers in the butadiene recovery section do not foul rapidly, they cannot run continuously for four years.

If a four year run time is not needed, capital cost may be reduced by installing fewer spares and allowing a wider range of operating conditions.

Economics

Material Balance

An overall unit material balance is shown in Table 1 for a plant producing 100 kMTA of butadiene (as butadiene contained in the crude butadiene product) based on 8,000 operating hours per year. The feed in this case is assumed to be a raffinate-II feed containing 72wt% 1- and 2-butene and 0.5% isobutene. Butadiene yield is 71.6 wt% of combined 1- and 2-butenes in the fresh feed. Yields can range from 70wt% to 75wt%, depending on the feed purity and the desired operating severity.

Table 1 - Overall Unit Material Balance

	Inputs (kg/hr)
C4 Feed	24,263
1- and 2-Butene	17,469
Isobutene	121
Butanes	6,672
Air to Process	40,399
Deaerated Water	1,224
MP Steam to Process	1,631
Make-up Solvent	138
Total	67,656
	Outputs (kg/hr)
Crude Butadiene Product	21,464
Butadiene	12,500
1- and 2-Butene	2,299
Isobutene and Butanes	6,633
Other	33
Water	7,688
Vent	38,390
Heavies Purge	114
Total	67,656

Utility Consumption

Utility consumption for an OXO-D process unit producing 100 kMTA of butadiene molecules in the crude butadiene product is given in Table 2. This utility consumption does not include the utilities required for the downstream butadiene extractive distillation unit.

Table 2 – Utility Consumption

ISBL Utility Requirements		Time Averaged Consumption
Electricity (1)	kW	7,768
HP Steam (46 barg)	MT/hr	3.2
MP Steam (16 barg)	MT/hr	59.8
Boiler Feed Water	MT/hr	2.6
Steam Condensate (2)	MT/hr	(65.5)
Water Loss	MT/hr	0.10
Cooling Water Circulation	m3/hr	2,088
Net Fuel Import/Export	MM kcal/hr	5.55

1. Assumes electric compressor drivers
2. Steam condensate export to utility area

Summary

Butadiene industry conditions are changing such that new on-purpose butadiene production capacity will be needed. The commercially proven TPC OXO-D process is well suited to meet the new demand and has been further improved by TPC and UOP. Features of this advanced design include

- Safe handling and complete consumption of oxygen
- Feedstock flexibility
- High conversion and selectivity
- Minimization of acetylene and oxygenate byproducts
- Simple and robust controls
- Energy efficiency
- Reliable extended operation

References

1. Welch, L.M., Croce, L.J., and Christmann, H.F. (1978), "Butadiene Via Oxidative Dehydrogenation," *Hydrocarbon Processing,* pp. 131–136.

2. *CRC Handbook of Chemistry and Physics,* 95th Edition.

AIChE Paper Number 192d

Dow Fluidized Catalytic Dehydrogenation (FCDh)
The Future of On-Purpose Propylene Production

Matt Pretz

Sr. Research Scientist
The Dow Chemical Company

Barry Fish

Process Development Leader
The Dow Chemical Company

Lin Luo

Sr. Research Scientist
The Dow Chemical Company

Brien Stears

Associate Director
The Dow Chemical Company

Prepared for Presentation at the 2016 Spring National Meeting

Houston, TX, April 13, 2016

Dow Fluidized Catalytic Dehydrogenation (FCDh)

The Future of On-Purpose Propylene Production

Matt Pretz

Sr. Research Scientist
The Dow Chemical Company

Barry Fish

Process Development Leader
The Dow Chemical Company

Lin Luo

Sr. Research Scientist
The Dow Chemical Company

Brien Stears

Associate Director
The Dow Chemical Company

Abstract: Shale gas produced an imbalance in propylene supply and demand. This has driven the implementation of several projects using available propane dehydrogenation commercial technology. These technologies were adapted from existing processes. These adapted technologies are not tailored to efficiently manage the inherent challenges of catalyst activity, heat input, reaction equilibrium and unselective thermal reactions. Therefore, the Dow Chemical Company has developed a Catalytic Dehydrogenation Process that is designed from the standpoint of an operating company to efficiently solve the problems inherent in catalytic dehydrogenation processes.

The new technology is a circulating fluid bed technology based on existing Fluid Catalytic Cracking (FCC) technology. The Dow process utilizes a proprietary catalyst with a proprietary reactor and regenerator design to achieve 45% propane conversion at approximately 93 mol% selectivity to propylene. The simplicity of the system enables a capital savings greater than 20% versus other commercial offerings. In addition, the higher conversion at moderate pressure reduces the energy requirement per pound of propylene.

This new technology can be fully integrated with existing ethylene crackers in parallel with new or existing furnaces. In this way, a producer can increase production in an existing light hydrocarbon facility or build a new facility with exactly the amount of ethylene and propylene product that is desired based on ethane and propane feeds. Dow is currently planning a project to implement this technology in an existing cracker at commercial scale and would license the technology upon request.

Introduction

In the early 2000's, very few predicted the innovation that was going to change the landscape in the energy, refining and petrochemical sectors for the next several decades. But, innovators successfully combined hydraulic fracturing and horizontal drilling to produce unprecedented shale gas and shale oil. This innovation rapidly enabled petrochemical producers to begin producing ethylene from cheap ethane. Initially, the price of crude oil remained high which meant naphtha was rapidly phased out as the preferred feedstock where feasible in place of ethane. When the rate of naphtha feed was reduced, the co-product propylene that was produced also was reduced. This created an imbalance in the supply and demand for propylene. As this imbalance occurred, the economics for producing propylene from propane began to look very attractive. Fortunately, the Catofin® and Oleflex™ processes had been adapted from existing technologies for the production of propylene(11,12). The licensors of these technologies were able to take advantage of market conditions and sold many units to help bring the markets back into balance.

Now there is another innovation ready to change the landscape of the catalytic dehydrogenation market place. Few are predicting the change that this technology will bring to the production of hydrocarbons in the petrochemical industry. In order to understand the significance of this new innovative technology, it is imperative to understand the fundamentals and evolution of catalytic cracking process technology.

Fundamentals and Evolution of Catalytic Cracking Process Technology

Fundamentals

There are several basic fundamentals that constrain potential process solutions when cracking or dehydrogenating hydrocarbons. In this paper, cracking and dehydrogenation will be referred to collectively as "cracking". The first is that the reactions are endothermic which means heat has to be added to the process to accomplish the reaction. Secondly, the chemical equilibrium of the dehydrogenation reaction favors olefins at low pressure and high temperature. The low partial pressure preference of the dehydrogenation reaction is at odds with the cryogenic separation requirement for light hydrocarbons; therefore, the reactor pressure needs to be optimized with the entire process.

$$C_3H_8 \rightleftarrows C_3H_6 + H_2$$

Third, the high temperature requirement inherently results in the formation of coke on the catalyst which depresses catalyst activity. The resulting loss of catalyst activity requires regeneration to remove coke. Fourth, the thermal or gas phase reaction has a much lower selectivity to the desired product than the catalytic reaction, requiring the gas residence time at temperature to be minimized.

Each process that has been successful in the area of catalytic cracking of hydrocarbons has solved or managed each of these fundamental constraints. Table 1 illustrates 4 processes in this area including Fluidized Catalytic Cracking (FCC) for gasoline, Catofin® and Oleflex™ for propylene, and styrene technology.

Table 1. Fundamentals of Catalytic Cracking Process Technology

	Gasoline	Propylene	Propylene	Styrene
Process	Fluidized Catalytic Cracking (FCC)	CATOFIN® Dehydrogenation Process	OLEFLEX™ Dehydrogenation Process	
Heat Input	Catalyst[22]	Catalyst[6]	Feed[2,11]	Steam Diluent[5]
Pressure controls equilibrium	25 psia[22]	7 psia[1], [3]	16-24 psia[7]	6 psia[5]
Catalyst Activity Recovery	Burn Coke with air[22]	Burn Coke with air [1], [6]	Burn Coke off with air[2], reactivate with chlorine[3], inject sulfur & recycle H2 to inhibit coke[2], [4]	React coke off with steam and catalyst promoter[5]
Reactor Type controls residence time	Fluidized Bed[22]	Fixed Bed with cyclic operation[6]	4 Radial Adiabatic Moving Beds with inter-stage heaters [2]	2 or 3 Radial Adiabatic Fixed Beds with inter-stage reheat[5]

To put heat into the reaction, both FCC and Catofin®[12,22] primarily use the catalyst as a source of heat while Oleflex™ uses the superheated feed[11] and conventional styrene processes use a co-feed steam diluent[5]. Processes that don't require the feed to be preheated over the thermal reaction temperature have an inherent advantage in product selectivity.

To drive the reaction equilibrium, the styrene and Catofin® processes use vacuum[1,3,5]. For the styrene process, creating vacuum can be done efficiently since the diluent steam and product liquids can be condensed prior to compression of the vent gas which minimizes the compression duty. The FCC process is not as adversely impacted by higher pressure due to different chemistry; therefore, the process has tended to increase pressure over the life of the technology to reduce capital investment per ton of product. Finally, the Oleflex™ process tends to use more moderate pressures to minimize the energy and capital cost of the compression train. The Oleflex™ process does require a significant hydrogen recycle which is one reason for the moderate pressure selection[7].

The catalyst activity loss due to coke formation is mitigated primarily through burning the coke in the FCC, Catofin® and Oleflex™ processes[1,2,4,22]. The Oleflex™ process does require a relatively complex regeneration in that platinum must be redistributed with chlorine[3]. In addition, sulfur and hydrogen are intentionally injected to control coke formation[2,4]. Uniquely, the styrene process uses a novel solution by reacting the coke off of the catalyst in situ using steam and a catalyst promoter[5].

Finally, the residence time of the gas at temperature must be managed to minimize selectivity loss and that is dictated by the reactor type and heat input scheme. Both the conventional styrene process and the Oleflex™ process use radial adiabatic contacting reactors which are known for low residence times[2,5]. The Catofin® process uses fixed adiabatic reactors that operate in a cyclic mode[12]. Finally, FCC uses a circulating fluid bed and rapid gas-solid separation to control gas residence time[22].

All of these processes include designs and mitigations to manage the fundamental technical constraints, but some more than others have evolved and form the basis for the next generation of catalytic dehydrogenation processes. Technology ultimately will converge to the best overall economic solution as capitalism forces companies to find additional ways to increase profits. The question is not if .... but rather when.

Evolution

Table 2 illustrates the evolution of processes for the production of gasoline, styrene, and propylene. The gasoline or refining industry is large with over 1,200,000 kTa of gasoline produced each year[8]. With a large market, there are economic drivers to improve technologies that can reduce the cost of production. Because of this, there has been significant improvement in technologies to produce gasoline starting with the Houdry process in 1936 which utilized a batch reaction step followed by batch regeneration[21]. The Houdry process was the original technology that gave rise to the Catofin® technology[12]. Later in 1938, a moving bed technology with bucket conveyors enabled the process to transition to a continuous one[22]. This process was plagued with mechanical problems which drove the need for further innovation. Finally, in 1942, during the middle of the World War II, the Fluidized Catalytic Cracking technology (FCC) was born[22]. The initial process used acidized clay catalysts which were not as active and required bed cracking[22]. Finally, in the 1960's, zeolite catalysts were invented which had higher activity and ultimately resulted in riser cracking[22]. After the catalyst introduction, several hardware improvements followed such as the 2 stage regenerator and rapid separation riser termination devices[22]. But, the lesson here is that the technology evolved rapidly to the most economical solution which was a circulating fluid bed.

The styrene market is much smaller with only 27,500 kTa of styrene produced (2014)[9]. Therefore, the rate of innovation was slower over the years. In 1940, the United States required styrene for rubber production for World War 2; therefore, axial adiabatic reactors were rapidly built to produce styrene due to their ease of construction and speed of implementation, but the high temperatures and residence time resulted in low selectivity[23]. The heated case reactors were introduced in 1965 which essentially used a heat exchanger to improve the heat transfer rates which enabled lower operating temperature gradients and lower gas residence times[23]. The radial adiabatic reactors were introduced in 1972 which further decreased residence time and allowed very large single train plants to be built[23]. Finally, in 2005, Dow and Snamprogetti successfully demonstrated the first circulating fluidized bed process for the production of styrene without a steam diluent. Unfortunately, due to commercial decisions to exit the styrene business, the commercial implementation of this technology wasn't realized.

Finally, the propylene market is about 89,000 kTa but only 5,000 kTa of this propylene is on-purpose propylene[9]. Up until 1990, there was no on-purpose propylene production as all of the propylene was produced as an FCC byproduct or a steam cracking byproduct[11].

Therefore, there has been very little emphasis as an industry in regards to improving technology in this area. The first Catofin® system was commissioned in 1980's for C4 production and was later used for propane dehydrogenation in the 1990's[12]. In Thailand, UOP commissioned the first Oleflex™ unit in 1990 due to a unique market condition in which propane was available and propylene was needed[11]. Both of these technologies were initially developed due to unique circumstances which enabled both licensors to be ready for the unforeseen impacts of shale gas. But, the truth of technology evolution still remains. Technology will ultimately converge to the best economic solution which is a circulating fluid bed.

Table 2. Evolution of Catalytic Cracking Processes

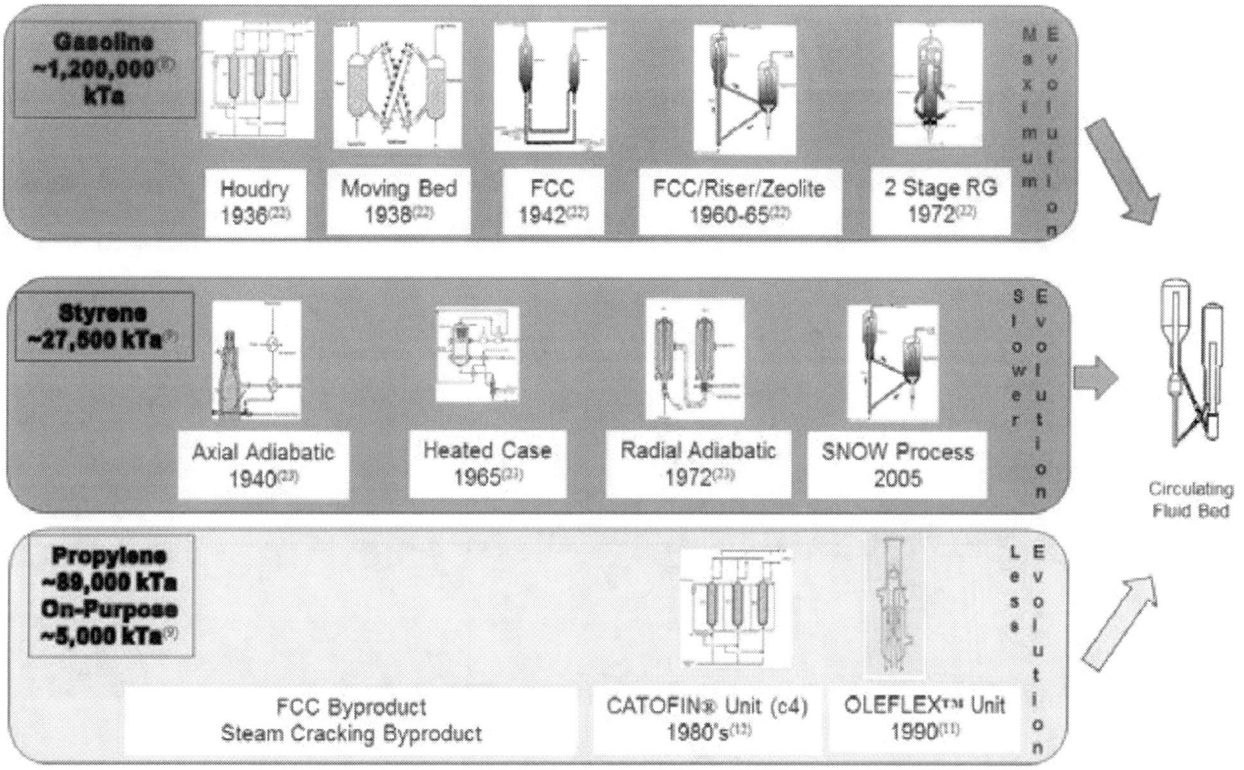

Introduction to Dow's Advanced PDH Technology

Process

The Dow Catalytic Dehydrogenation Reactor-Regenerator system is shown in Figure 1. The technology is a circulating fluid bed system with an alumina supported catalyst. Propane is introduced into a turbulent or fast fluidized reactor followed by a plug flow riser reactor which transports the catalyst to close coupled cyclones for rapid product and catalyst separation. The catalyst is then stripped with a stripping agent to recover any product entrained with the catalyst. The spent catalyst is transported to a combustor where it is reheated in a fluidized bed using external fuel for the required heat. The catalyst and flue gas then flow through a riser and a termination device. The remaining catalyst is then separated from the flue gas in cyclones and returned to a bubbling bed. The catalyst is then rejuvenated with air to recover the activity in an oxygen soak zone.

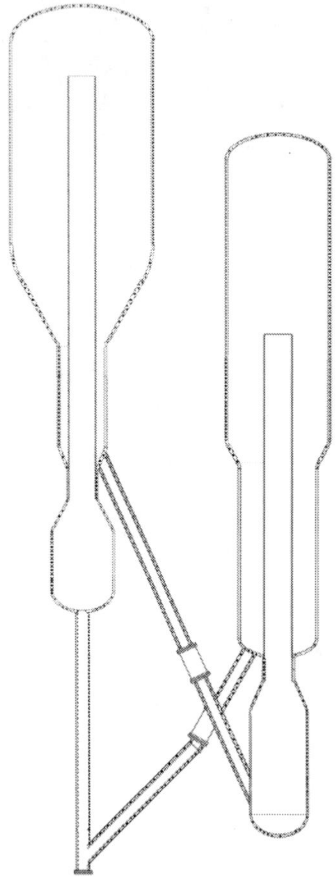

Figure 1. Dow Catalytic Dehydrogenation Reactor-Regenerator System

Catalyst

The catalyst shown in Figure 2 is produced utilizing a commercially available alumina support and impregnated with the active components which include gallium and low levels of platinum. The catalyst has been developed over many decades and has been shown to meet all the requirements of operating in a fluid bed dehydrogenation process. For example, the catalyst can perform the dehydrogenation of paraffins and alkyl aromatic compounds including: ethane, propane, butane, isobutane, ethylbenzene and others. The catalyst also promotes the combustion of supplemental fuels such as methane, hydrogen, ethane and others. As a catalyst operates in a circulating fluid bed system, it is desirable that any deactivating function that may occur in the reactor, reactor stripper or combustor be reversible. As such, the catalyst is capable of cyclic reactivation after any temporary activity loss is experienced. In addition, the catalyst can tolerate reasonable levels of impurities that are commonly found in commercially available propane streams. Finally, the catalyst has the appropriate mechanical properties to make it fluidizable (particle size and density) as well as attrition resistant.

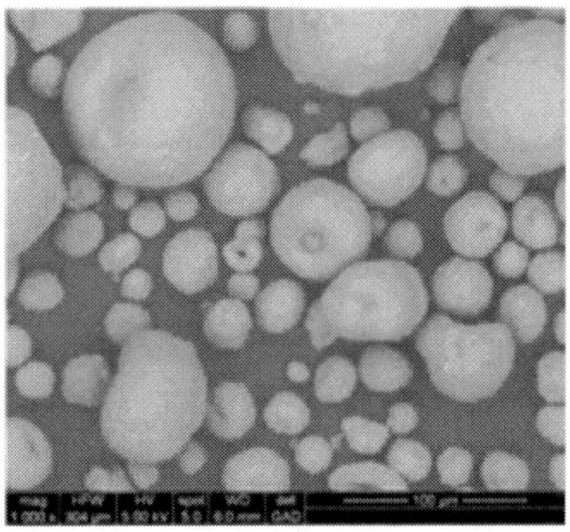

Figure 2. Dow Catalytic Dehydrogenation Catalyst

Dow Reactor Design

The reactor design (Figure 3) for the Dow circulating fluid bed process is an up flow reactor where the average velocity of both the catalyst and gas are in the upward direction. The bottom portion of the reactor operates as a turbulent or fast fluidized reactor and is followed by a conventional riser reactor. Upon completion of the catalytic dehydrogenation reaction, the catalyst and product gases are rapidly separated in a cyclonic separation device that is commonly used in Fluidized Catalytic Cracking processes. After the catalyst is removed, a stripping gas is injected in the annular space and the product gases are recovered prior to transporting the catalyst to the regenerator.

The advantage of the Dow reactor and catalyst process is that it can achieve high conversions such as 45% of the propane feed at moderate pressures. In addition, there is a minimal impact from feed impurities such as sulfur which eliminates the need for costly front-end capital investment. Finally, the selectivity to propylene is maximized due to the low gas residence time at temperature requirements and the low levels of coke that are formed on the catalyst during the reaction step.

Figure 3. Dow Reactor Design

Dow Regenerator Design

The Dow regenerator design as shown in Figure 4 consists of a lower combustor section and is followed by a catalyst removal section at the top and a catalyst reactivation section in the center. Due to the low coke generation, supplemental fuel and air are injected into the combustor which generates the heat for the exothermic dehydrogenation process. The combustor is a bubbling bed that operates with an average gas and catalyst velocity in the upward direction. The catalyst and flue gas are then transported through a riser where the catalyst is partially removed from the flue gas. The flue gas is then processed in a cyclone system which removes any remaining catalyst. The catalyst is then reactivated by a simple air treatment prior to transport back to the reaction system.

Advantages of the Dow regenerator process include the catalytic combustion of a fuel which minimizes any NOx formation versus competing alternatives which require a flame for combustion. In addition, the catalyst is continuously regenerated which eliminates the need for large high temperature valves operating at a high frequency which have been prone to failures in batch regenerator processes. In addition, there is no need for complex regeneration steps such as chlorine introduction.

Figure 4. Dow Regenerator Design

Comparison to Existing Technologies

As with any new technology, it is imperative to benchmark any new process with the current technologies to ensure a sufficient improvement in performance can be obtained. Table 3 summarizes the key differences between the Catofin®, Oleflex™, and Dow PDH process.

In reactor technology, both Catofin® and Oleflex™ require multiple reactors to process the required amount of propane to achieve PDH economies of scale. For Oleflex™, four reactors are used to achieve the target conversion due to multiple heat up steps being required[2]. The Catofin® process requires multiple reactors depending on capacity and each reactor spends a significant time operating in the regeneration step[12]. The Dow process utilizes a single reactor and operates in reaction mode continuously.

Heat introduction into the reactor is extremely important as it impacts the selectivity of the process as well as the capital intensity. The Catofin® process provides the heat by a combination of hot catalyst generated in the decoking cycle and hot propane[1,10,12]. The cyclic heat introduction by catalyst results in a high temperature initially during the start of a reaction cycle and then a low temperature just prior to regeneration. This cycle results in a non-optimum temperature for the reaction thereby reducing the overall selectivity of the process. In addition, by feeding the propane at temperatures above the thermal cracking temperatures non-selective products are formed. Alternatively, the Oleflex™ process preheats the propane and gas products before each reactor to temperatures above the thermal cracking temperature of propane[2,11]. This results in non-selective reactions due to the residence time in these sections as well as increased capital costs and NOx emissions from the fired heaters. The Dow process introduces propane below the thermal cracking temperature and utilizes the catalyst to rapidly heat up the propane to reaction temperature. The high surface area of the catalyst and rapid heat transfer is one of the most efficient ways to transfer heat to the propane vapor.

Each process utilizes a different catalyst which impacts the economics and handling requirements. For example, the Catofin® process uses a very large quantity of chromium catalyst which is relatively inexpensive[12]. The catalyst possesses environmental handling challenges which must be managed. Alternatively, the Oleflex™ process utilizes a catalyst which contains ~ 3000 ppm of platinum supported on alumina[13]. The Dow process uses an more sustainable catalyst that is significantly less expensive than the Oleflex catalyst as the platinum levels are significantly lower.

The regeneration of all the processes involves burning any coke build-up on the catalyst in either a continuous or batch process. The Oleflex™ process uses a complex cycle of coke combustion, drying, platinum re-dispersion with chlorine and oxygen, followed by a hydrogen reduction[14,15]. Both the Dow and Catofin® regeneration process are relatively simple and straight forward.

Table 3. Leading Commercial Technologies vs Dow PDH

	CATOFIN®	OLEFLEX™	Dow's Process
Reactor Design	(20) 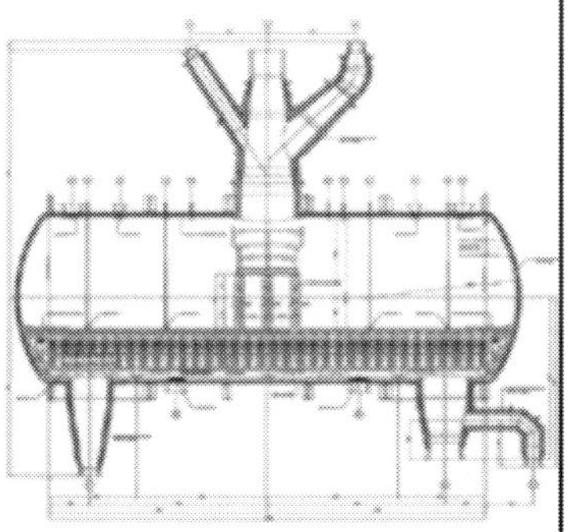	(11) 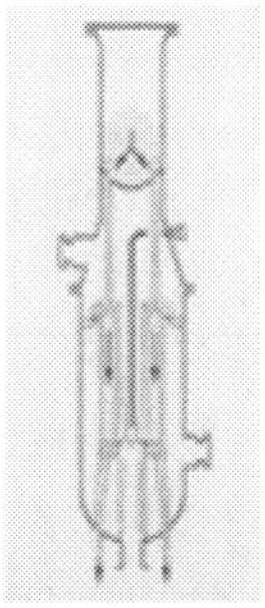	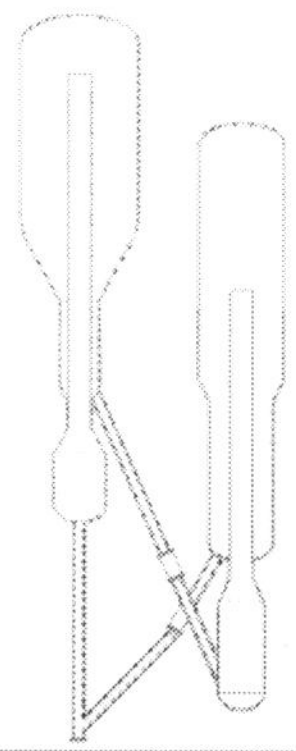
Reactor Technology	Fixed Bed Parallel Reactors [1]	4 Moving Bed Reactors 1 CCR Regenerator[2]	1 Fluidized Reactor 1 Fluidized Regenerator
Method of Heat Introduction for Reaction	Cyclic operation of reaction vs. coke combustion with air in regeneration step to reintroduce heat after reaction step cools catalyst bed[10] Propane Feed at 590-650 C[10], [1]	Direct-fired furnace heats process gas prior to each of 4 adiabatic reactors in series to 650 C[2], [11]	Heat carried to reactor via fluidized catalyst after heating by combustion in regeneration step. Propane fed at temperatures below thermal cracking
Catalyst type	Cr_2O_3 on alumina[12]	0.3 wt% Pt supported on alumina[3], [13]	Alumina with Pt and Ga
Catalyst Regeneration	Coke from reaction step burned with air.	Complex cycle of coke combustion, drying, Pt re-dispersion with chlorine and O_2 and then H_2 reduction. [14], [15]	FCC like cycling to perform combustion, decoke, and regeneration after reaction
RX Catalyst Residence Time	7-15 minutes [12]	~7,200-14,400 minutes (5-10 days) [16]	< 2 minutes
Pressure	7 psia[1], [3]	16-24 psia[7]	19-25 psia
Conversion	48 - 53%[6]	35(7) - 40% [19]	43-53%
Selectivity (mol%)	86+[6] @ XX% conv	89%(18)@ XX conv 84% @ 40% conv [19]	92-96% @ 43%-48% conv

The residence time of the catalyst in the reactor varies widely between the 3 processes and generally impacts the ability of the catalyst to withstand feed impurities and deactivation. For example, the Catofin® reactor undergoes a regeneration every 7-15 minutes which enables any deactivation that may have occurred from feed impurities to be removed each cycle[(12)]. Alternatively, the Oleflex™ process has a catalyst residence time of 7,200-14,400 minutes (5-10 days) which makes the catalyst much more susceptible to feed impurities and may require additional capital to mitigate[(16)]. The Dow PDH process has a catalyst residence time in the reactor of less than 2 minutes which makes the process more forgiving to feed impurities and potential impacts to the catalyst.

The pressure of the processes dictates the maximum potential conversion due to chemical equilibrium. For example, the Catofin® reactor operates at a low pressure of ~7 psia to drive higher conversions[(1,3)]. This increases compressor operating cost and capital. Alternatively, both the Dow PDH and Oleflex™ process operate between ~16 and ~25 psia which is slightly above atmospheric pressure[(7)]. The Oleflex™ process does require recycle hydrogen to suppress coke make due to the long catalyst residence times which additionally suppresses the maximum possible conversion[(2,4)].

Finally, the reactor conversion and selectivity ultimately dictate the economics of compression and separation sections of a commercial plant. After a thorough literature review, the conversion and selectivity for both the Catofin® and Oleflex™ processes are difficult to find reported together. The Catofin® process has published propane conversion numbers ranging from 48-53% with a separate selectivity performance of 86 mol% at an unknown conversion[(6)]. The Oleflex™ process has published propane conversions ranging from 35-40%[(7,19)]. The Oleflex™ selectivity has been reported to be 89 mol% at an unknown conversion and 84 mol% at 40% conversion in a UOP patent[(18,19)]. The Dow PDH process is expected to operate at 43-48% conversion with propane to propylene selectivities of 92-96 mol% which is significantly improved over the current commercialized technologies. The performance of the units is expected to change over time and as a function of plant scale which provides some challenges when comparing conversion and selectivity performance.

The Dow process delivers significant capital and operating cost advantages due its inherent simplicity and multi-functional nature. The capital outlay is expected to be at least 20% less than other existing commercialized technologies based on internal Dow studies. In addition, by performing the reaction at moderate pressure without hydrogen recycle, the process is expected to have 20% lower energy consumption. The low gas residence time at temperature design enables the process to achieve a 5% lower propane unit ratio than current processes. In addition, through catalytic combustion of process gas, the NOx emissions will be lower than alternative technologies. In conclusion, the technical advantages of Dow PDH technology offer significant economic advantages versus current technologies.

Dow's History of Technology Development and Status

Dow scientists identified a fluidized reaction process as the most economical for catalytic dehydrogenation more than 25 years ago. Since that time, a fleet of different lab reactors have been built and operated to test and develop improved catalysts as well as understand the chemistry of the process to enable design of a commercial unit. Fixed bed reactors, bubbling bed reactors, ACE fluidized lab reactors, attrition units, combustion units, and several others have been designed and operated. In addition, two pilot units such as a Davidson Circulating Reactor (DCR) and a large Process Demonstration Unit (PDU) were constructed to study the circulating aspects of the chemistry and better understand the fluid dynamics at high temperature. Finally, more than 3 large cold flow programs were chartered at PSRI to measure and quantify the catalyst hold-up as a function of superficial gas velocity and catalyst flux over a broad range of reactor design conditions. PSRI subject matter experts on fluidization consulted on the project and provided recommendations for multiple areas in the plant design.

In addition to the hardware and R&D programs, Fluidized Catalytic Cracking (FCC) subject matter experts with a broad range of experience were leveraged to design and scale-up the process. Many of the challenges of the development were synonymous with FCC such as catalyst circulation and catalyst retention. Dow modeling experts also generated extremely valuable scale-up tools in the area of catalyst attrition and retention, catalyst kinetics, reactor hydrodynamics, and catalyst – gas separation devices.

In conclusion, Dow brought its entire arsenal of R&D and process know-how (Figure 5) to the table as well as leveraged leading industry fluidized solids experts to bring this process to fruition. The next step in the development is the construction of a small commercial demonstration unit. Upon completion, the data and know-how generated from the commercial demonstration unit will enable the design and construction of world scale units and future technology proliferation to the industry.

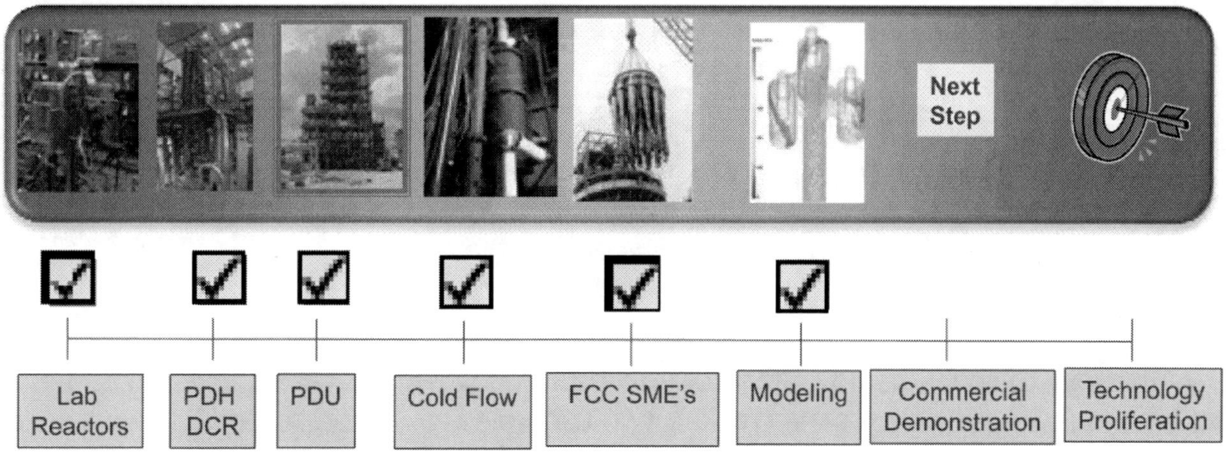

Figure 5. Dow Process Development Pathway

Commercial Application

The Fluidized Catalytic Dehydrogenation (FCDh) process is a revolutionary platform reactor technology. The potential applications for the process are numerous as shown in Figure 6. For example, in PDH, the reactor technology can plug into steam crackers that are new or be bolted on to an existing cracker. The reactors can also be used to build a new PDH unit or retrofit an existing PDH unit. Additionally, the reactors can be used to build new or retrofit existing ethylbenzene to styrene dehydrogenation facilities, butane to butene, or isobutane to isobutene plants. C4 dehydrogenation units can be integrated into refineries to upgrade butane or isobutane for use in the alkylation section to produce alkylate.

In the case of PDH plant integration with existing steam crackers, the Dow process is the only technology that can integrate economically with a cracker due to is comparable operating pressure, high propane conversion, lack of hydrogen recycle requirement and simple reactor section design. Because of the simple reactor-regenerator configuration, small plants still maintain attractive capital intensity when integrated with a steam cracker unlike competing technologies.

Figure 6. Platform Reactor Technology with Multiple Plug and Play Options

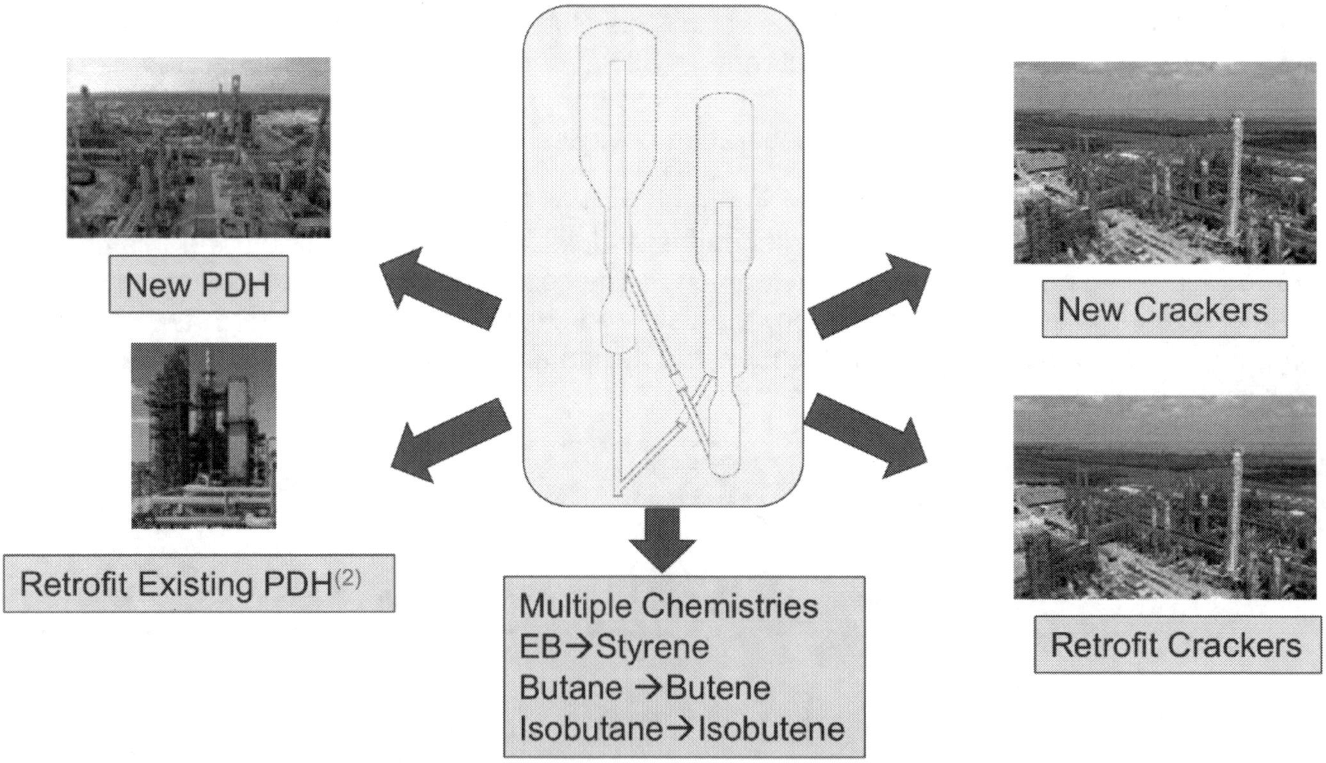

In order to show the advantage of an integrated Dow PDH reactor system and ethane cracker versus an ethane-propane cracker, two examples will be presented. The first example is an ethane-propane cracker and is shown in Figure 7. In this example, if a producer wanted to produce 1500 kTa of ethylene and 500 kTa of propylene, an EP cracker with approximately 8 furnaces could be envisioned. 2 furnaces would feed ethane and 5 would feed propane including the recycle streams. One furnace would be in decoke mode. The feedstock would be 250 kTa of ethane and 3000 kTa of propane. This cracker would produce 1250 kTa of other products.

Figure 7. 2000 kTa Ethane-Propane Cracker

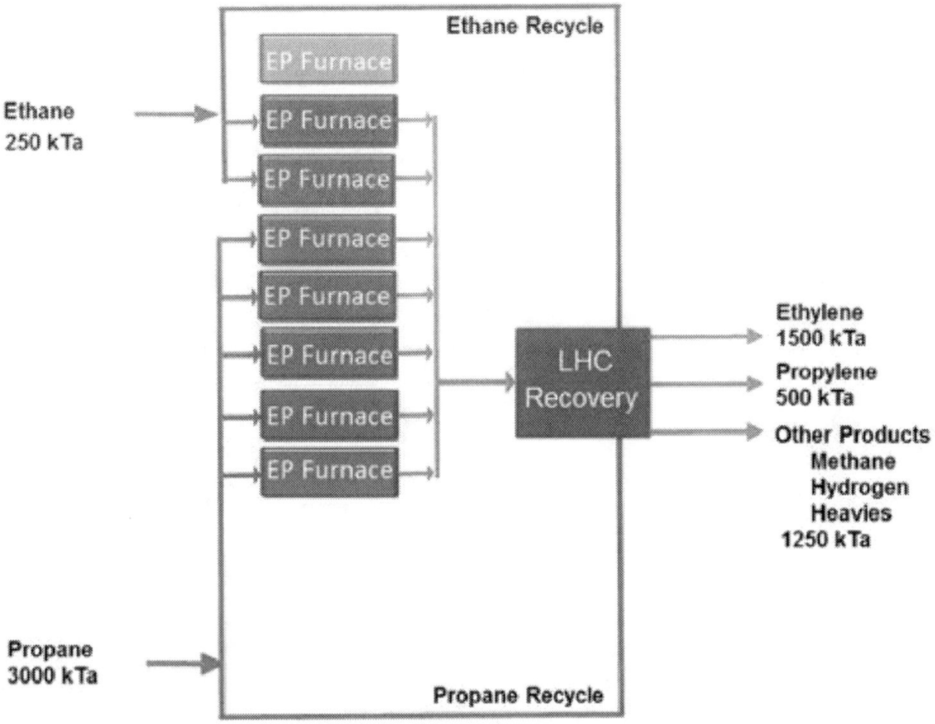

In the second example, an integrated Dow PDH reactor system with ethane furnaces is presented in Figure 8. In this example, 7 furnaces are used to crack ethane and 1 PDH reactor is used to process propane which in turn produces about 1500 kTa of ethylene and 500 kTa of propylene. In this example, 1950 kTa of ethane is required and 550 kTa of propane feedstock. There are only 500 kTa of byproducts produced. This is a ~750 kTa reduction in the amount of byproducts that are produced. The total predicted economic advantage of the integrated Dow PDH reactor and ethane cracker versus an ethane propane cracker that produces essentially the same product mix is $234 MM/yr of additional value due to the products versus the raw materials based on the average predicted price from 2016-2020[24].

Figure 8. Integrated Dow PDH Reactor System with Ethane Cracker

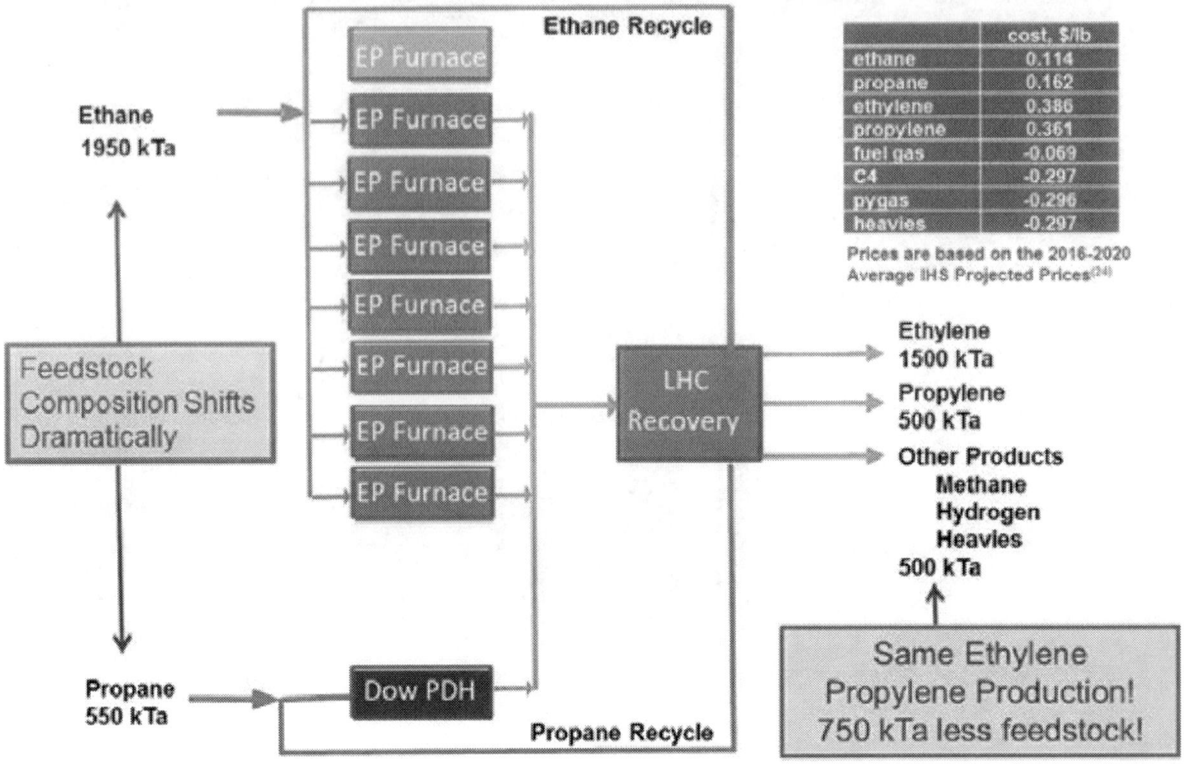

Conclusion

The Dow Catalytic Dehydrogenation reaction system is a breakthrough technology that provides significant advantage in the production of propylene from propane. Dow believes the technology represents the future of how on-purpose propylene will be produced. The process's elegant and simple design provides the most capital and energy efficient method to produce propylene from propane that is currently available. In addition, it provides for lower raw material usage and simplicity of operation versus other options.

This technology has the potential to change how hydrocarbons are produced in the chemical industry. To support this vision, Dow plans to implement a project at Dow and has plans for strategic expansion of the technology to other geographies and facilities. In addition, Dow would be willing to license the technology for applications in which catalytic dehydrogenation is the preferred solution.

References

1) Seung-Taek Seo, Wangyun Won, Kwang Soon Lee, Chansul Jung, and Seokho Lee, Repetitive control of CATOFIN process, Korean Journal of Chemical Engineering, Volume 24, Number 6 / November, 2007.

2) UOP Oleflex Process Brochure "UOP Oleflex™ Process for Propylene Production".

3) Weckhuysen, Sattler, Ruiz-Martinez, Santillan-Jimenez. Catalytic Dehydrogenation of Light Alkanes on Metals and Metal Oxides. ACS Publications. August 2014.

4) Heyse, Johnson, Mulaskey. "Dehydrogenation Processes, Equipment and Catalyst Loads Therefore." US Patent 5,406,014, April 11, 1995 to Chevron.

5) Styrene. Chen, Shiou-Shan. Raytheon Engineers and Constructors. Kirk-Othmer Encyclopedia of Chemical Technology. John Wily and Sons. Inc. December 4, 2000.

6) "CATOFIN® Dehydrogentation", Lummus Technology a CB&I Company, http://www.cbi.com/images/uploads/tech_sheets/Dehyrogenation.pdf, 2009.

7) Kogan, Herskowitz, Woerde, and van den Oosterkamp. "Article 16A: Select Dehydrogenation of Propane using an improved Dehydrogenation catalyst." 10th Ethylene Producers Conference, 1998.

8) World Crude Oil Production by Year. United States Energy Administration. 2014.

9) Global Annual Production of Styrene and Propylene Monomer. US Energy and Information Administration. 2015.

10) Gartside, Robert J., "Catalytic Hydrocarbon Dehydrogenation System with Prereaction", US Patent 6,392,113 B1, May 21, 2002 to ABB Lummus.

11) Chaiyavech, Pramote, "Commercialization of the World's First Oleflex Unit", The Journal of the Royal Institute of Thailand, Vol 27, No 3, Jul-Sept 2002.

12) Oviol, Bruns, Fridman, Merriam, Urbancic. "Mind the Tap", Hydrocarbon Engineering. Sept 2012.

13) Oleflex™ DeH-16 Catalyst Description, UOP Website, http://www.uop.com/objects/DeH16.pdf.

14) Cottrell, P. R., Fettis, M. E. "Process for the dehydrogenation of hydrocarbons", US 5,087,792 A to UOP.

15) Sechrist, P. A., Robinson, D. W., Schlueter, W. D., "*Method for controlling moisture in a catalyst regeneration process*", US 6,123,833 A to UOP.

16) Meyers, R.A., "Handbook of Petrochemicals Production Processes" McGraw-Hill Handbooks, **2005**.

17) Owen, Hartley, Mead, Belle, Zahner, John C., "Conversion of LPG Hydrocarbons into Distillate Fuels Using and Integral LPG Dehydrogenation-MOGD Process", Patent 4,675,461, June 23, 1987 to Mobil.

18) Houdek, J. Mark, Anderson, James. "On-Purpose" Propylene Technology Developments". UOP LLC. ARTC 8th Annual Meeting 4/29/2005 Kula Lumpur

19) Glover, B. K., Zarraga, J. A., Schultz, M. A., "Fluidized bed reactor with back-mixing for dehydrogenation of light paraffins", US patent application, US 2008/0161624 A1, to UOP.

20) "APPC PDH/PP Project – 8 Catofin Reactors – Jubail Saudi Arabia Engineering, Design, and Installation of Refractory Lining", Chimney and Refractory International SRL, Presentation, http://www.chimneysandrefractoriesinternational.com/attachment/APPC-SAMSUNG%20PDH-PP%20CATOFIN%20REACTORS%20PROJECT.pdf.

21) Letzsch, Warren. Hemler, Chuck. Nieskens, Mart. The 70 year History and Future of Cat Cracking. AFPM Conference: Cat Cracker Seminar. August 21-22, 2012.

22) Wilson, Joseph. Fluid Catalytic Cracking: Technology and Operation. PennWell Books. 1997.

23) Mitchell, J.E. Styrene: Its Polymers, Copolymers and Derivatives. Chapter 2: Manufacture of Styrene Monomer. Hafner Publishing Company. 1952

24) Chemical Price and Economics. IHS. January 5, 2016.

Catofin® is a registred trademark of CBI Lummus

Oleflex™ is a registered trademark of UOP

AIChE Paper Number 192E

C_2 Front-End Hydrogenation Improvements

Ernesto J Angueira

Technical Service Manager, Americas

CRI Catalyst Company LP

Pascal Coenen

Licensing Technical Manager

Shell Global Solutions International B.V.

Prepared for Presentation at the 2016 Spring National Meeting

Houston, Texas, April 12, 2016

C_2 Front-End Hydrogenation Improvements

Ernesto J Angueira

Technical Service Manager, Americas

CRI Catalyst Company LP

Pascal Coenen

Licensing Technical Manager

Shell Global Solutions International B.V.

Abstract: The presentation will address the parameters influencing selectivity and the impact over time. A side-by-side comparison of 5+ year operating data of the Shell/CRI' C_2 Front-End catalysts in both gas and liquid crackers gave new insides in excelling performance. The difference between the latest generations of catalyst looking at the key performance parameters for ethylene producers will be addressed. The selectivity comparison between the modeling (predicted selectivity) and the 5+ year real plant data (achieved selectivity) will be compared.

A business case demonstrates the value (margin gain) which has been achieved by reference plant which upgraded to the latest generation catalyst will be explained.

Ethylene continues to be one of the most important building blocks in the chemical industry; its production is a highly competitive business. The producers are continuously experiencing growing competition (e.g. low-cost feedstock producers), variation in feedstock costs, increasing energy costs and additional environmental regulations, requires plants to operate at maximum reliability and efficiency.

A few challenges in front-end hydrogenation units include: excess hydrogen, therefore a selective catalyst is very important to avoid runaways and to ensure high yield of ethylene. Another challenge for front-end hydrogenation operation is fluctuation of CO concentration in the feed. CO functions as an activity inhibitor to hydrogenation as it is adsorbed on the catalyst active sites. Temperature adjustments to CO activity inhibition and sudden CO content drop can trigger temperature runaway (Figure 1). As producers are generating hydrogenation feed with lower CO levels, depending on operation, feedstock and processes, a very stable catalyst is required.

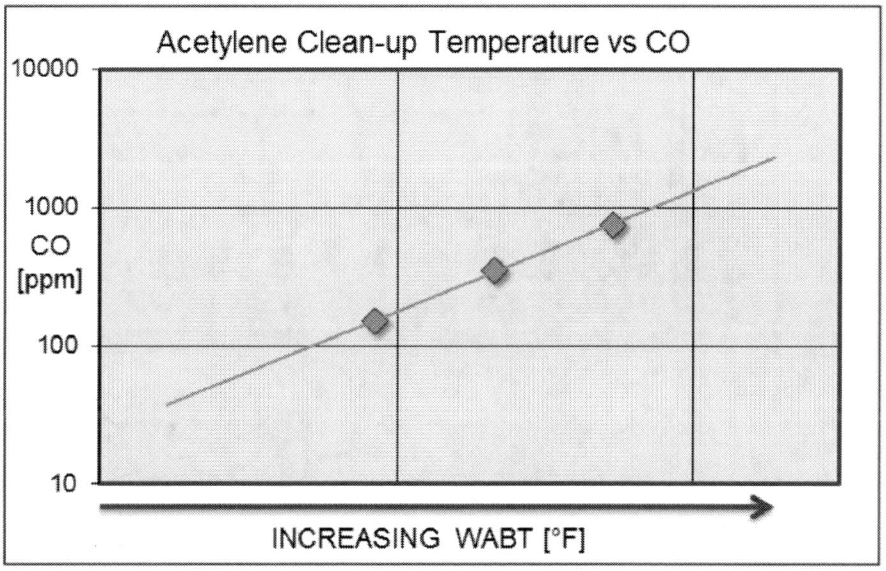

Figure 1. Temperature adjustments due to CO inhibition.

The selective hydrogenation of acetylenes is a very important unit operation within the ethylene process and catalyst selection can significantly overcome the different challenges while offering increased profitability and ease of operation to ethylene manufacturers.

This paper will discuss how new catalyst development can enhance the profitability of an ethylene plant by improvement in olefin yields, consistent delivery of on-specification product, and lower utility costs, resulting in significant improvements in overall plant operating economics.

Why new catalyst development?

Catalyst research, besides changing the formulation and palladium (Pd) impregnating techniques, are also looking into possibilities of improving catalyst performance by varying the size and shape of the catalyst as CRI developed in it KL7741B series.

One focus of catalyst shape optimization is the desired to reduce catalyst bed pressure drop which translates directly into energy savings as a result of reduced power consumption or higher throughput maintaining the same pressure drop.

A ring shape catalyst provides several advantages over the traditional tablet (or spherical) catalyst. The ring shape exposes more catalyst surface area to the process gas per catalyst volume than the tablet shaped particle. The catalyst is impregnated with Pd and since there is high geometric surface area, the Pd is better dispersed, resulting in high activity and poison resistance. Pressure drop across the catalyst bed is also reduced due to a higher void volume for the ring shape. Higher void volume results in low pressure drop, thus higher space velocity or increase in throughput (Figure 2). Lower density, therefore lower fill cost.

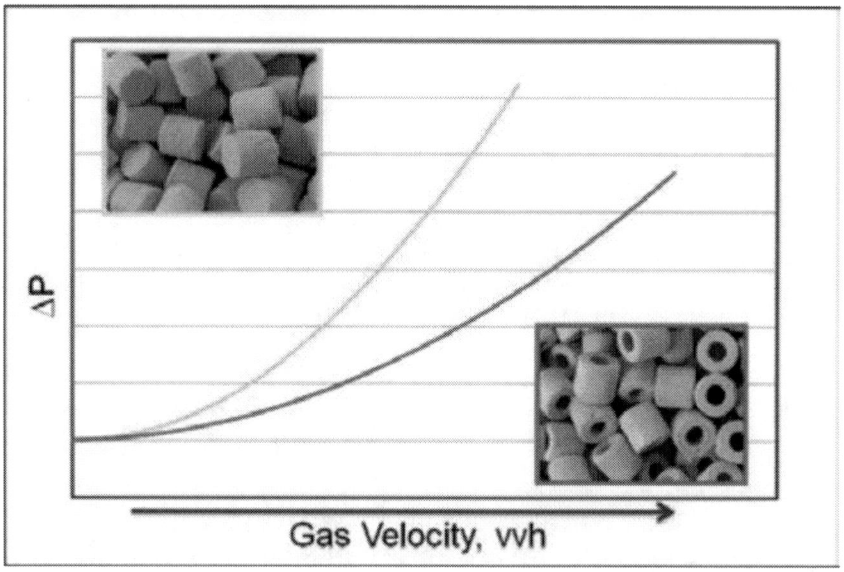

Figure 2. Pressure drop as function of GHSV.

The catalyst shape must achieve all of the above without compromising on catalyst activity or mechanical strength.

Performance

Commercial units that switched from tablets to rings shape catalyst experienced significant improvements. The ring shape catalyst features extreme stability maintaining a high level of ethylene selectivity for prolonged time as shown in Figure 3.

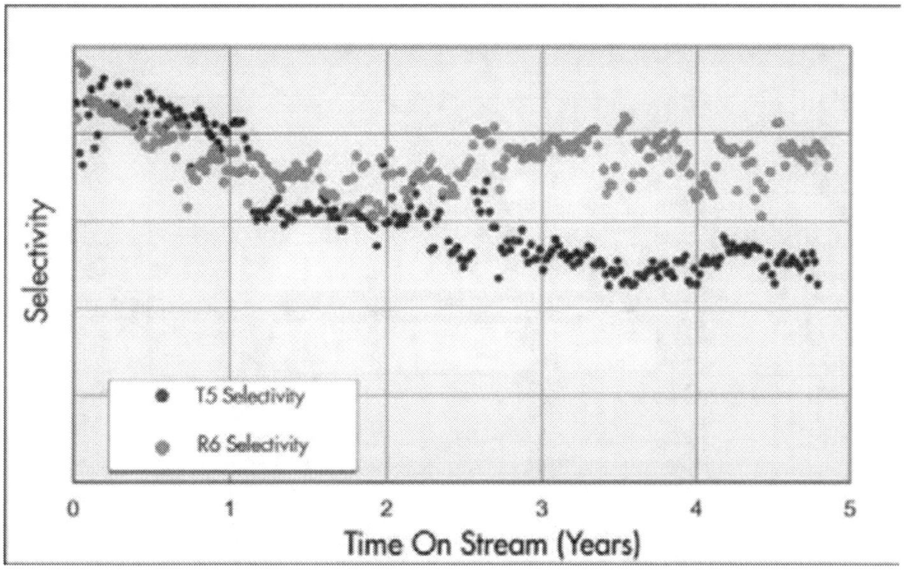

Figure 3. Rings vs Tablets shape – selectivity.

The stability of the catalyst resulted in higher selectivity during the cycle translating to extra margins therefore increased profitability.

In addition to the stability of the catalyst, its improved Pd dispersion provides a more active catalyst, therefore requiring lower operation temperature (Figure 4).

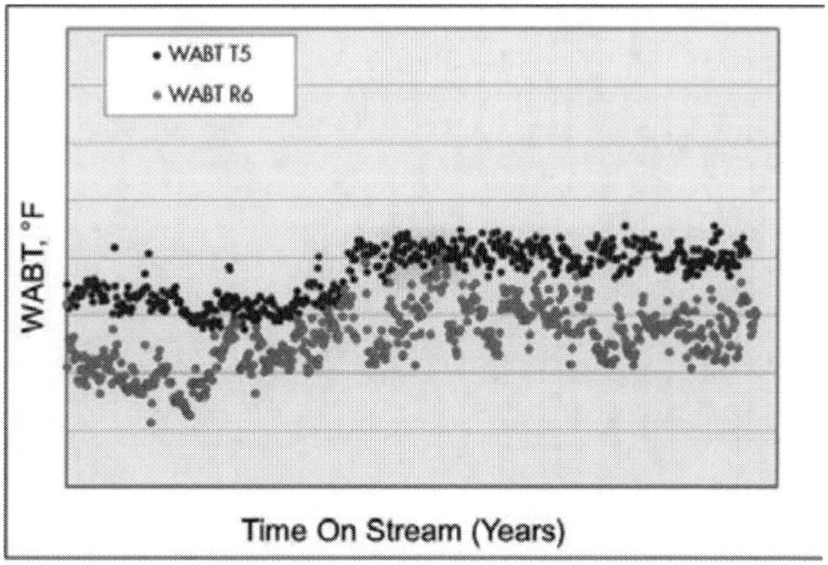

Figure 4. Rings vs Tablets shape – WABT.

KL7741B catalyst has proven to be very stable during CO swings, even under very low CO concentrations (100 ppm).

The ring shape catalyst, even with better metal dispersion exhibited robust operation during CO swings.

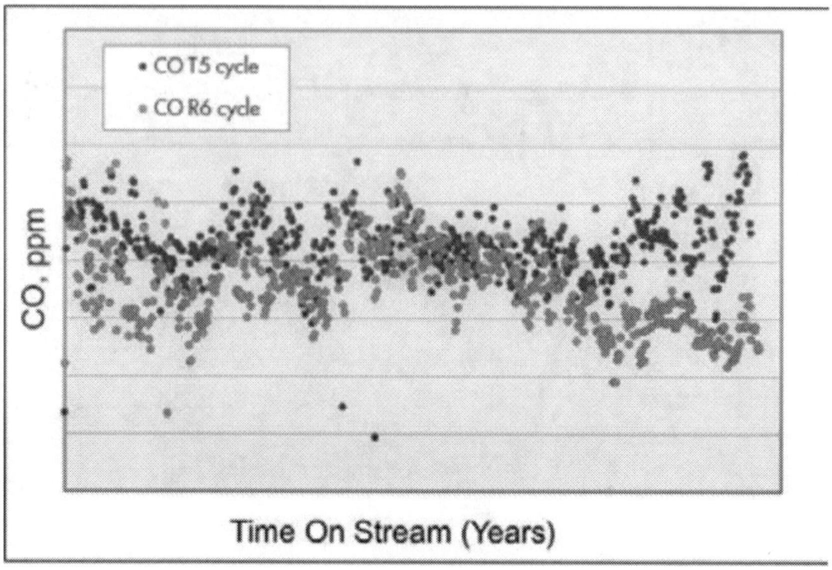

Figure 5. Rings vs Tablets shape – CO stability.

Modeling plant data

Using the data of multiple plants, an empirical equation was developed to model the aging of the catalyst.

ethylene selectivitytime = -A * (log (1 + [time]*B) – 1)

A = ethylene selectivity at SOR

B = C * D * E * constant

C = GHSV in h-1

D = "catalyst shape factor"

E = $[C_2H_2]^2$ in vol%

constant = y

Using several references gives the input to model the benefit as show in the business case.

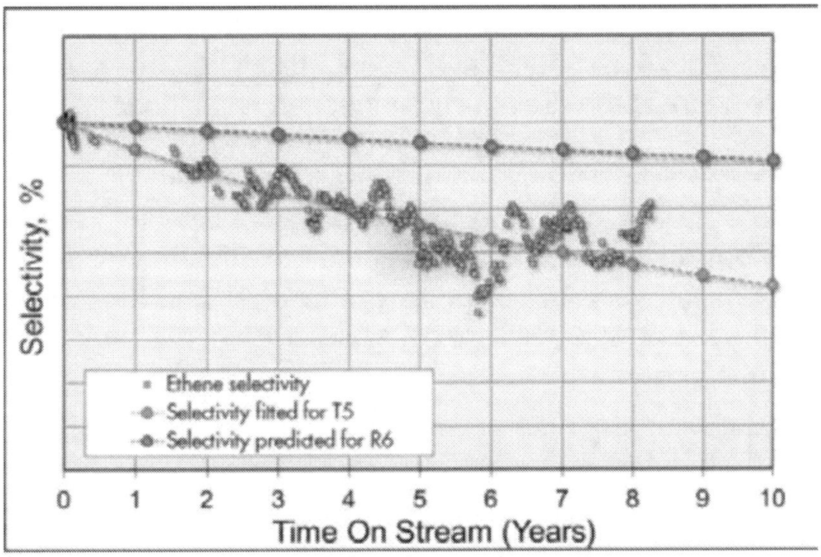

Figure 6. Predicted selectivity for rings.

Business Case Example

Based on the commercial data from several plants operating the tablet and ring shaped catalyst of CRI a great improvement (average over 5 year period) in ethylene yield has been observed. For this world scale steam cracker with the assumptions listed below leads to 2 million $ per year increase in margin as shown in Figure 7.

Steam Cracker - Gas		
Plant Capacity	kTA	World Scale
Ethylene margin	$/ton	400
On-stream factor	hr/yr	8,500

Yearly Benefit over 5 year cycle

Benefits		
Increased direct Ethylene Yield	$/yr	> 1,100,000
Increased benefit (more fresh feed instead of recycle)	$/yr	> 900,000
Estimated Total Margin Improvement	**$/yr**	**> 2,000,000**

Figure 7. Predicted selectivity for rings.

Conclusion

Ethylene producers expect the acetylene selective hydrogenation unit to operate in a safe and reliable manner. Due to high competitive market efficiency and profitability are key and therefore catalyst selection is critical.

Catalyst manufacturers continue the development of front-end hydrogenation catalysts, improving the performance with each new generation. Catalyst shape demonstrated to be an effective solution and KL7741B rings have been introduced to markets and commercially proven to successfully address the previously mentioned challenges.

The ring shape catalyst demonstrated enhanced stability, better selectivity, improved activity, and therefore lower operating temperature. Lower pressure drop resulted in economic benefits, from lower energy consumption to higher unit throughput. The largest benefit is achieved from the very low aging of the catalyst and high selectivity achieved over the cycle as shown in the example for a world scale cracker the benefit can be in excess of 2 million US$ per annum.

192F

FULL FURNACE SIMULATIONS AND OPTIMIZATION WITH COILSIM1D

Alexander Vervust
Laboratory for Chemical Technology
Ghent University

Ismaël Amghizar
Laboratory for Chemical Technology
Ghent University

Andrés E. Muñoz G.
R&D Manager
AVGI

Kevin M. Van Geem
Laboratory for Chemical Technology
Ghent University

Guy B. Marin
Laboratory for Chemical Technology
Ghent University

Prepared for Presentation at the 2016 Spring National Meeting
Houston, Texas, April, 13, 2016

FULL FURNACE SIMULATIONS AND OPTIMIZATION WITH COILSIM1D

Alexander Vervust

Laboratory for Chemical Technology

Ghent University

Ismaël Amghizar

Laboratory for Chemical Technology

Ghent University

Andrés E. Muñoz G.

R&D Manager

AVGI

Kevin M. Van Geem

Laboratory for Chemical Technology

Ghent University

Guy B. Marin

Laboratory for Chemical Technology

Ghent University

Abstract: COILSIM1D is a steam cracking simulation package that has been developed as the result of several decades of independent research and modelling activities at the Laboratory for Chemical Technology (LCT) of Ghent University. Using a friendly graphical user interface, it combines a detailed fundamental kinetic model with a one-dimensional reactor model, which provides high flexibility regarding reactor geometries, feeds and simulation strategies. Simulations for a very broad range of complex feeds up to (hydrotreated) VGO can be easily performed thanks to SIMCO, a molecular reconstruction module incorporated in the software based on commercial indices of the feed. Over time, and based on own and client initiatives, various other features have been incorporated into COILSIM1D, such as TLE and run length simulations, or a shooting method that allows to perform simulations using a variety of severity indicators as boundary conditions at the reactor outlet. Recently, the kinetic model has been expanded and fine-tuned based on data from a large ethylene producer, including oxygenates; a firebox and convection section model has been incorporated, to allow full cracker simulations (i.e. convection, firebox, reactor and TLE altogether) in a unique package, and with short simulation times. COILSIM1D has also been linked with Aspen and other process simulators. Illustrative simulations have been carried out to show the operation of these new features, which effectively take COILSIM1D to a new level, transforming it into a whole-furnace simulation and optimization package.

Introduction

COILSIM1D is a comprehensive steam cracking simulation package, developed as the natural result of several decades of independent research and modelling by the Laboratory for Chemical Technology (LCT) of Ghent University. AVGI, a spin-off company from Ghent University, recently acquired all technical and commercial rights to COILSIM1D. With this, further and faster development of the software can be carried out, whilst interacting on a business-to-business structure with clients spread over four continents, allowing for customization of the package, tailored to the specific needs of the client.

The first two AVGI-driven developments of COILSIM1D have been carried out within a framework of collaborative work with LCT, and comprise the expansion and fine-tuning of the kinetic network used to describe the chemistry of the process, as well as the incorporation of furnace simulation capabilities that allow users to carry out full cracker simulations in one software package, with an even higher level of accuracy at the reactor level.

COILSIM1D has been traditionally developed as a reactor simulator aimed at the prediction of yields at the reactor outlet, combining a fundamental single-event kinetic network with a one-dimensional plug-flow reactor model. Over time, and based on own insights and permanent interaction with the increasing number of licensees, new functionalities have been gradually added to the package, that are made available to all clients having annually renewable license agreements by means of regular updates. Clients using evaluation licenses also receive the most updated version of COILSIM1D.

All these functionalities have conferred large flexibility to COILSIM1D, making it able to handle a very broad range of reactor geometries, feeds and operating conditions. A feedstock reconstruction method (SIMCO) is integrated in the package, which allows to reconstruct the molecular composition of light mixtures to HVGO feeds, based on simple and easily available commercial indices of the feed. Also a module to carry out TLE simulations is incorporated, allowing to account for the effect of this equipment on the process yields, effluent temperature, and heat recovery.

Different simulation approaches can be applied with COILSIM1D, depending on the information that the user has available: heat-flux or temperature profiles along the reactor can be defined; outlet conditions based on frequently used severity indices can also be specified, such as the Coil Outlet Temperature (COT), the conversion of a given species, the ratio of propylene to ethylene in the effluent (P/E), or that of methane to propylene (M/P). Additionally, dedicated routines for yield maximization, sensitivity analysis and run length estimation are incorporated in COILSIM1D, enabling users to straightforwardly optimize their process while taking multiple targets into account.

Following the market's urge to count on a single tool that is able to accurately simulate steam crackers as a unit (i.e. firebox, convection section, reactor and TLE in one package), AVGI and LCT took on the challenge of taking COILSIM1D to the next level, successfully transforming it into a whole cracker simulation package.

This paper presents an overview of the main features of COILSIM1D, paying particular attention to the most recent developments regarding the kinetic network, as well as the firebox and convection section models. To demonstrate the operation of these new features and their harmonious integration in the already existing package, illustrative simulations combining the new and already existing models are presented. The user-friendliness of COILSIM1D and its ability to easily interact with external software packages are also discussed.

The COILSIM1D Simulation Package

SIMCO: Feedstock Reconstruction from Common Commercial Indices

To be able to take advantage of the flexibility and accuracy that a fundamental kinetic model can offer, knowledge of the detailed molecular composition of the used feed is crucial and will affect the accuracy of the obtained simulation results. [1] For liquid fractions, which can be made up of several thousands of different components, detailed feedstock compositions are not straightforward to determine; not even for naphtha fractions. Existing analytical state-of-the-art techniques such as GCxGC, can be used to determine the detailed feedstock composition, but they are costly and can be very time consuming. In industry, feed composition can change significantly on a daily or even hourly basis and hence these analyses are not feasible to perform.

A feasible alternative are feedstock reconstruction methods, that allow to compute a detailed molecular composition of the feedstock based on common commercial indices, e.g. the average molecular weight, information regarding boiling points, the specific density, the global PIONA weight fractions, etc. The ease to reconstruct different feeds makes SIMCO a powerful tool for rapid and accurate feed reconstruction. This numerical tool comes with a user friendly GUI, depicted in Figure 1, in which the different input parameters are filled in by the user in the units of choice.

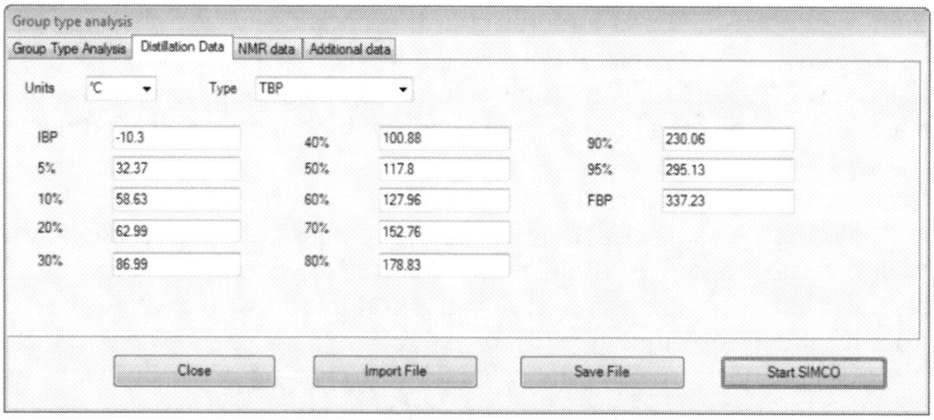

Figure 1: The GUI of SIMCO, where different parameters can be filled in as input for reconstruction of a feed.

Commercial indices can be determined by means of relatively fast and low cost standardized analytical procedures. In this respect, SIMCO, the feedstock reconstruction method available within the COILSIM1D package, is a powerful tool for industry to simulate the complex molecular composition as input for a fundamental kinetic model. SIMCO uses two methodologies, one based on the Maximisation of the Shannon Entropy (MSE) [2, 3], and another based on an artificial neural network (ANN). [4] MSE is the most general method and can be used for virtually every hydrocarbon feed. ANN, on the other hand, has shown to be more accurate for naphthas, provided that their characteristics fall within the application range of the network's training set. [4] The most suited method to reconstruct the feedstock is automatically selected by the software, offering the highest reliability on the reconstructed feed and the subsequent simulated product spectra. This enables users to take optimal advantage of feedstock flexibility. The feed reconstruction method also incorporates sulfur components. The reconstructed feed can be directly imported into a reactor simulation.

Reactor Definition

COILSIM1D incorporates an extensive built-in library that includes the reactor configurations most commonly used in industrial steam crackers. In this way only the length, internal diameter and wall thickness of each section have to be provided as input by the user. Since the reactor model of COILSIM1D can be applied to any type of tubular reactor, virtually any other reactor geometry that is not included in the built-in library can be easily implemented using the 'New coil geometry' tool in the GUI, as shown in Figure 2. The user can also choose from an assortment of reactor materials. Fins or other 3D geometries (e.g. straight, riffled, spiral) are also supported and accounted for. User defined reactor geometries can be saved and reused later in future simulations.

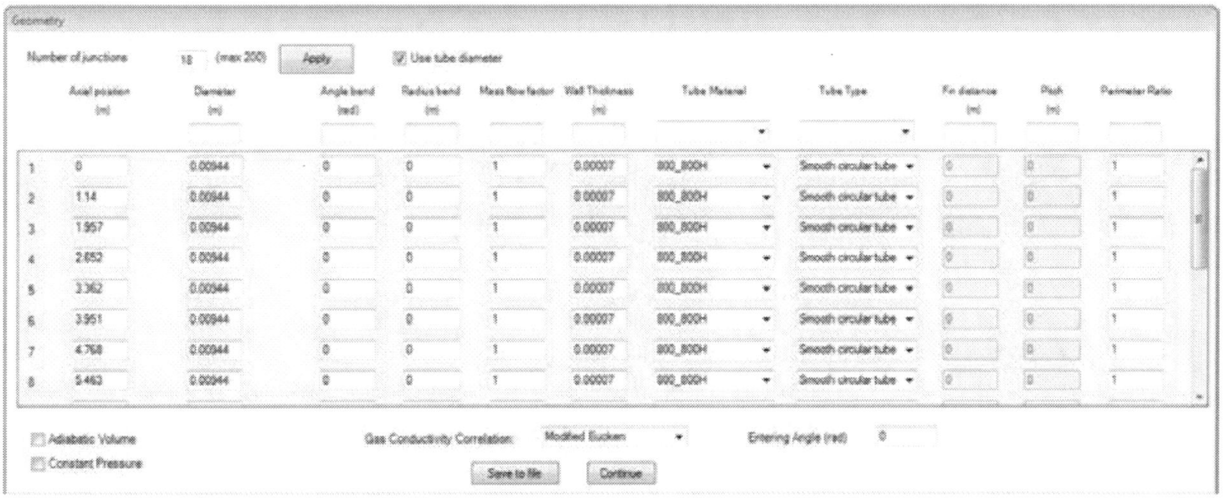

Figure 2. New coil geometry tool included in the GUI of COILSIM1D.

Transfer Line Exchanger

To explicitly account for the effect of the rapid quenching of the process gas leaving the reactor, COILSIM1D incorporates a TLE simulation model. [5] It takes into account the heat transfer from the process gas side to the steam/water mixture used for cooling. The high heat content of the process gas is usually used to produce high pressure steam and hence accurate simulations of this element are highly important for integral simulations of the furnace as a whole.

Different simulation strategies are possible with COILSIM1D, using either a specific water mass flow rate or a specific steam quality as boundary conditions. The model also allows the simulation of a full thermosiphon, which can be used to estimate the water recirculation rate as well as the steam production.

The TLE model is designed specifically to simulate double piped TLEs, like the one shown in Figure 3, which are the most common configuration in current commercial furnaces. In this type of TLE, the cracked gas flows through the inner tube and the water/steam mixture flows through the outer tube. Since the temperature of the cracked gas is still considerably high, the secondary reactions that can take place in the inner tube of the TLE are explicitly accounted for.

This module also includes a dedicated model to account for fouling in the TLE, caused mainly by the condensation of the heavy hydrocarbons that are often present in the reactor effluent. [6-8] Combined with the coke deposition models for gaseous and liquid feeds used in the reactor section, this feature provides the user with a very handy tool for the estimation of run lengths.

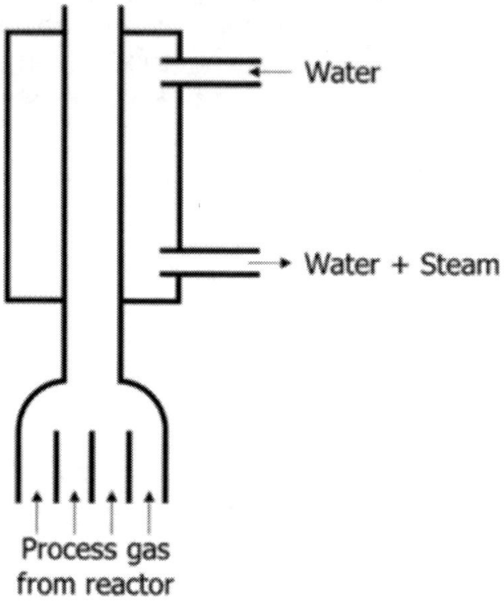

Figure 3. Schematic of a double piped TLE used to quench the effluent of reactor coils.

Run Length Simulations

Steam cracking is inevitably associated with the deposition of coke on the inner wall of the reactor, which largely affects the economics of the process. [9] Since decoking operations translate into loss of production capacity, accurate prediction of the run length of a furnace is highly interesting for the optimization of cracking furnaces. COILSIM1D offers the possibility to perform run length simulations that account for the effect of coke depositing inside the reactor over the course of a run, until the operational limitations are reached.

To accurately predict the formation of coke, two different models are incorporated: one for gaseous feeds [10] and one for liquid feeds [11]. Both account for the heterogeneous non-catalytic or so-called asymptotic coking regime, which is dominant over most of the run length. [12, 13] The effect of the brief catalytic coking regime can be accounted for in COILSIM1D, by giving the model an estimate of the expected duration of the catalytic coking regime.

The coking rate is computed using the selected model, and the increase in coke thickness during each specific time step is calculated and added to the already deposited layer, until reaching the maximum TMT or maximum pressure drop, as defined by the user in the GUI. An illustrative example of simulation results is shown in Figure 4, where the calculated TMT (accounting for faster catalytic coking at the beginning of the run) is plotted as a function of time. It is clear that the slope of the TMT rapidly drops to a constant value as a consequence of the shift of the dominant coke deposition mechanism.

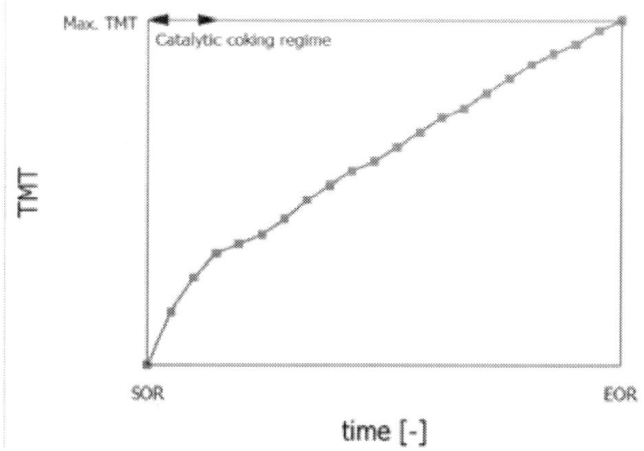

Figure 4: TMT as a function of time for a simulation case that accounts for a faster catalytic coking regime at the beginning of the run.

New Features of COILSIM1D

Kinetic Model for Steam Cracking of Hydrocarbons

The kinetic model of COILSIM1D is assuredly the most important feature of this software package. COILSIM1D comprises a very thorough mathematical model that consists of two parts: (i) A single event kinetic model called CRACKSIM, [1] which accurately describes the chemistry taking place in the reactor; and (ii) a one dimensional tubular reactor model describing the transport phenomena. The model equations are valid for any type of tubular reactor, allowing COILSIM1D to be used for a broad range of coil configurations. Both parts complement each other and help to structure a mathematical description of the problem to simulate. A dedicated fast and robust solving algorithm is applied to solve the mathematical problem. The required simulation time is usually less than 1 second allowing to simulate within a broad window of conditions and feeds in a limited time.

The accuracy and robustness that are characteristic of CRACKSIM are, to a great extent, the result of the fundamental approach that it uses, i.e. describing each of the occurring reactions as a single step, instead of simply accounting for global reactions. The corresponding kinetic and thermodynamic parameters are calculated using consistent and dedicated high level ab-initio calculations. [14-19] It is therefore no surprise that COILSIM1D incorporates the broadest and most accurate reaction network for steam cracking of hydrocarbons.

The kinetic model has been recently extended, to incorporate also common oxygenate contaminant species whose thermochemical decomposition yields products such as methanol and aldehydes, known for leading to several downstream process difficulties such as gum formation and off-spec products.

Currently more than 720 different molecular species and 43 radical species are incorporated, leading to a global network of more than 300,000 reactions. The model has been validated using an extensive set of data points originating from industrial furnaces, as well from the pilot plant and bench scale set steam cracking setups of LCT. The validation dataset includes a broad assortment of feeds, ranging from ethane to complex and heavy mixtures such as gasoil, waxes and even crude oil. Excellent agreement is observed between the measured and predicted yields of the main and minor products, [1, 20] as shown in the parity plots presented in Figure 5.

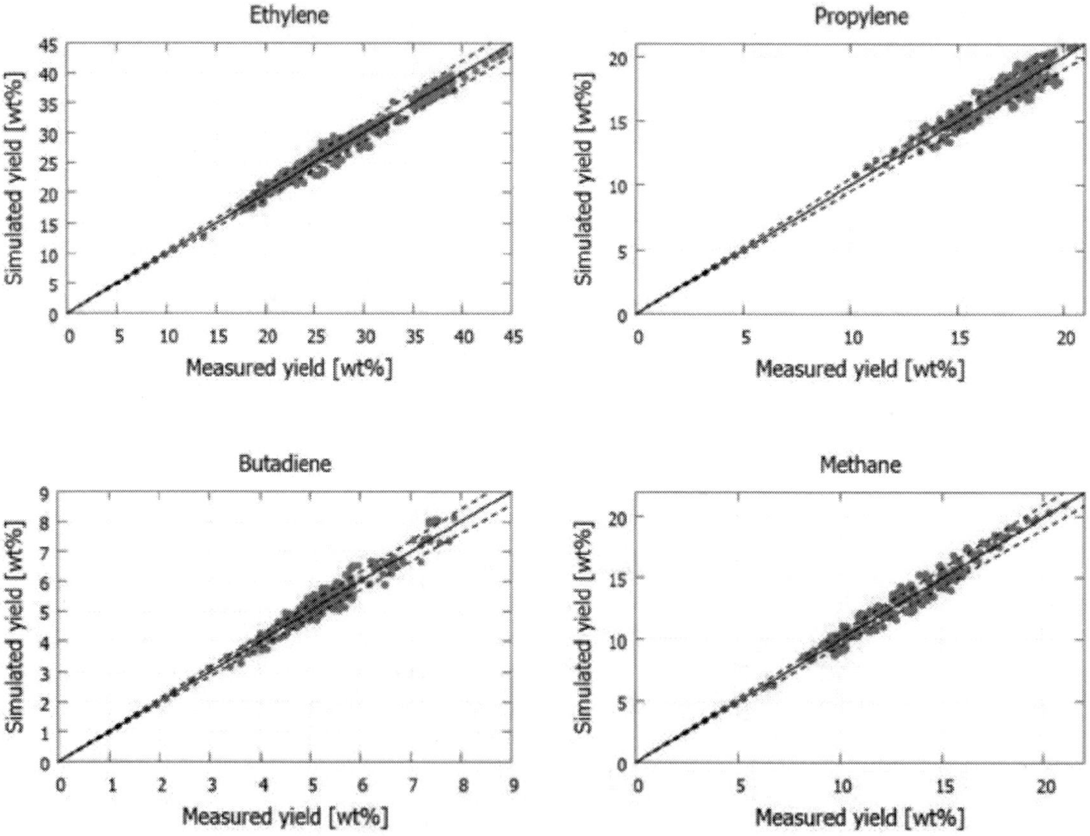

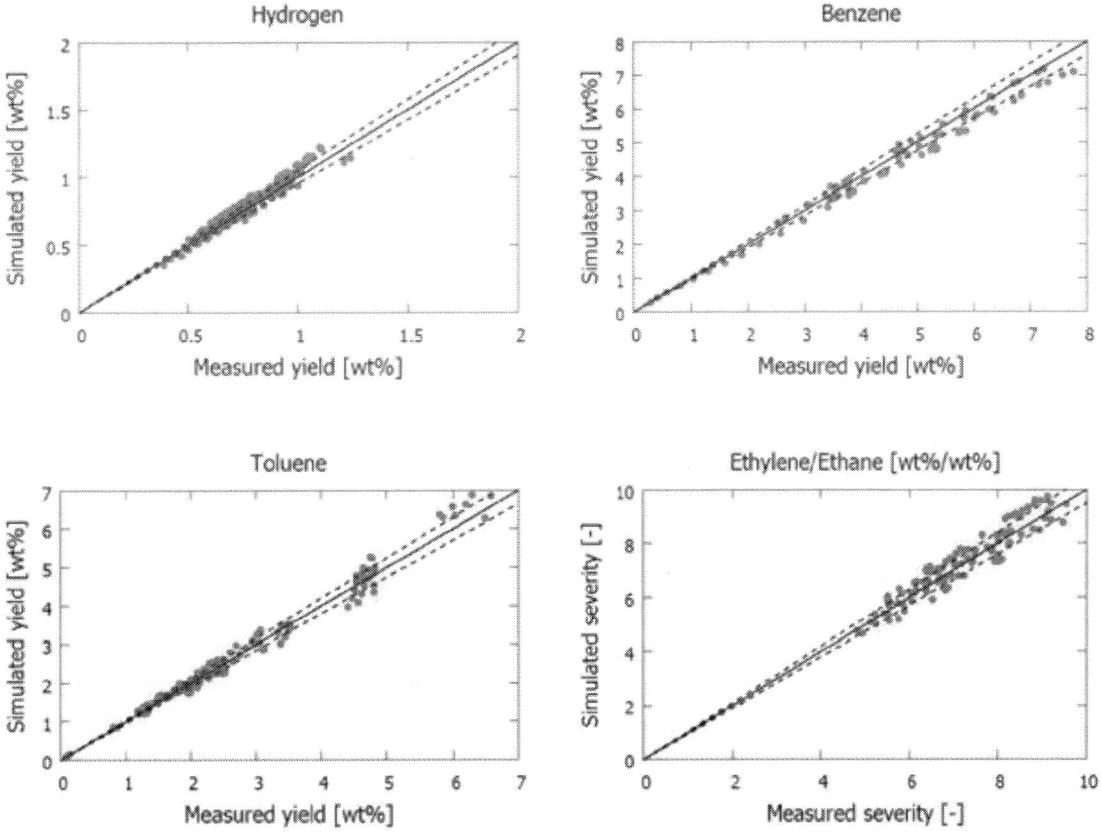

Figure 5: Parity plots comparing measured and simulated yields of various cracking products. Dashed lines indicate ±5% deviations.

Convection Section

The convection section is a new feature of COILSIM1D, and it is part of the furnace module of the GUI, which comprises two complementary models for simulations of firebox and convection section. The convection section simulations help to estimate different process parameters of interest such as flue gas temperature, utility steam production and total thermal power exchange based on limited user input.

When feeding liquids, evaporation also takes place in the convection section, which is an important aspect to consider for design and optimization, especially when dealing with heavy feeds that might not be completely evaporated and thus lead to fouling, in particular in the lower banks. Explicitly taking evaporation into account can help to locate hot spots and hence prevent extensive fouling of the convection banks. To meet this challenge, the a dedicated equilibrium based evaporation model has been developed.

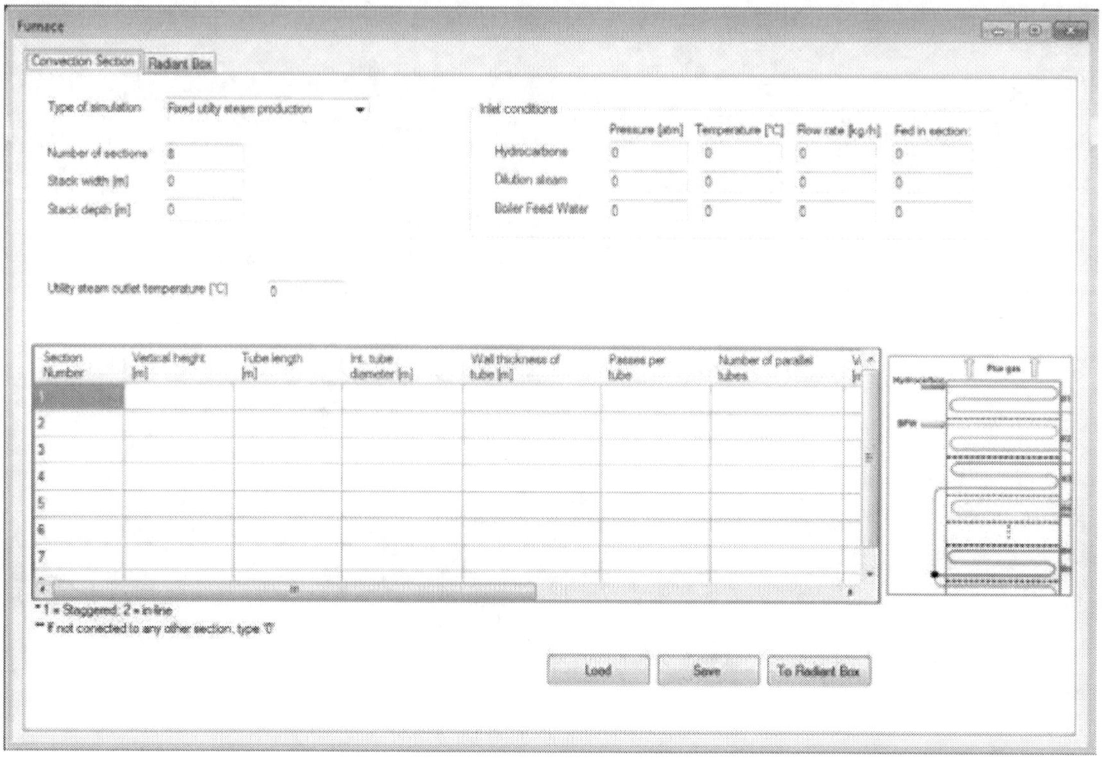

Figure 6: GUI of the convection section simulation module of COILISM1D.

The convection section is modelled as a finite number of tube banks in-series. As can be seen in Figure 6, the user can define the number of banks and the number of parallel tubes on each bank, which can make multiple passes. The feed entering each bank can be either a stream entering the convection section for the first time, a stream originating from another bank, or a mixture thereof. Furthermore, different tube arrangements (in-line or staggered) are supported, as well as different configurations (counter and concurrent flow), as shown in Figure 7. Additionally, different tube materials, with characteristic roughness and thermal properties can be chosen. The model can also account for the usage of finned tubes. In this way, virtually any type of convection section configuration can be easily simulated with COILSIM1D.

The model describes the duty pick-up of each bank from the hot flue gas to the process stream, based on different users input and process conditions (e.g. throughputs, composition, temperatures), parameters from of the firebox (thermal efficiency, combustion fuel rate, bridge wall temperature) and TLE (heat duty, steam flow rate). The tool contains an extensive library of pressure- and temperature-dependent physical properties for all the molecular species available in COILSIM1D.

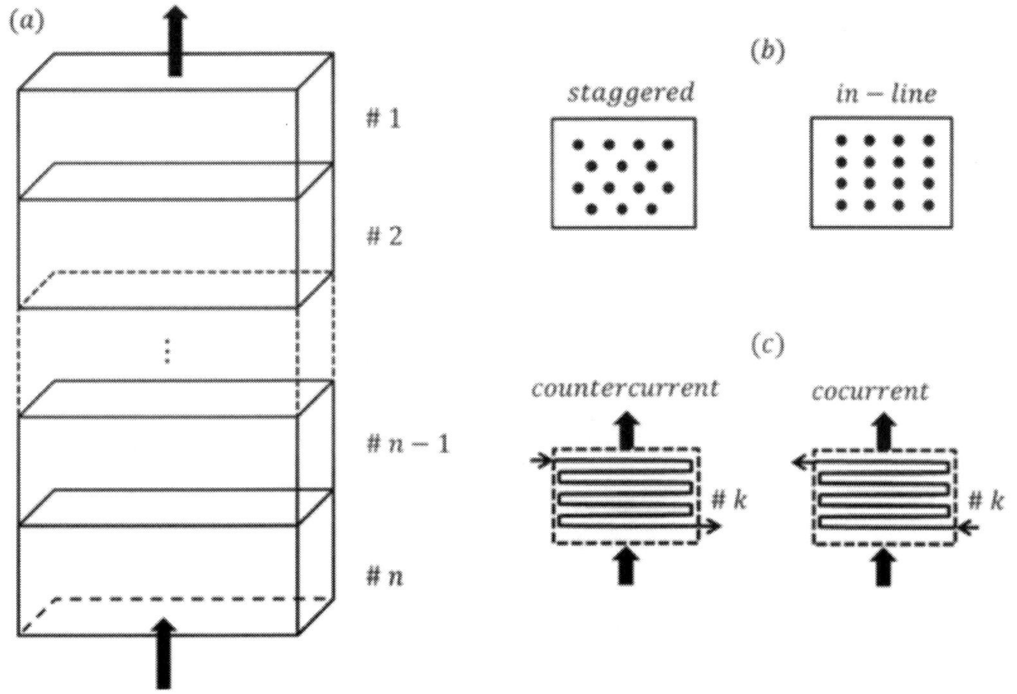

Figure 7. Schematic representation of the convection section in COILSIM1D: (a) tube banks in series; (b) supported tube arrangements; (c) supported flow pattern arrangements.

To illustrate the use of the convection section module, a demonstration case has been simulated for a typical arrangement, presented in Figure 8. The geometry of the convection section (i.e. number of passes, tube lengths, number of parallel tubes and pitch lengths) was defined via the GUI, as well as the mass flows of the input streams and the number of the bank in which they enter the convection section. Eight heat exchanging tube banks and six different inlet streams are defined. The naphtha is fed to the first bank, then mixed with dilution steam and taken to the third bank. The mixture is further heated in banks 3 and 4. The heated mixture is then mixed with superheated dilution steam (outlet stream from bank 7), and conducted to the lowest bank. The second bank is used to heat water at high pressure. High pressure steam is overheated in bank 5 and then mixed with desuperheating water. The resulting stream is then heated up in bank 6.

The hydrocarbon feed was reconstructed using SIMCO, based on a set of common commercial indices presented in Table 1. The reconstructed naphtha consists of 72 different components ranging from C_4 to C_{11}.

The calculated Coil Inlet Temperature (CIT) is 578.3°C and the pressure before the Venturi nozzles is 3.8 bara. These values are typical for a naphtha cracker, [21] and can be subsequently used as input for the reactor model to determine the product distribution.

Table 1. Commercial indices used for reconstructing the naphtha feed with SIMCO

Commercial indices of the naphtha feed	
Specific density [kg/m³]	708.8
ASTM D86 [°C]	
IBP	39
50%	99
FBP	165
PIONA [wt%]	
Paraffins	36.5
Iso-paraffins	32.8
Olefins	0.0
Naphtenes	21.4
Aromatics	9.20

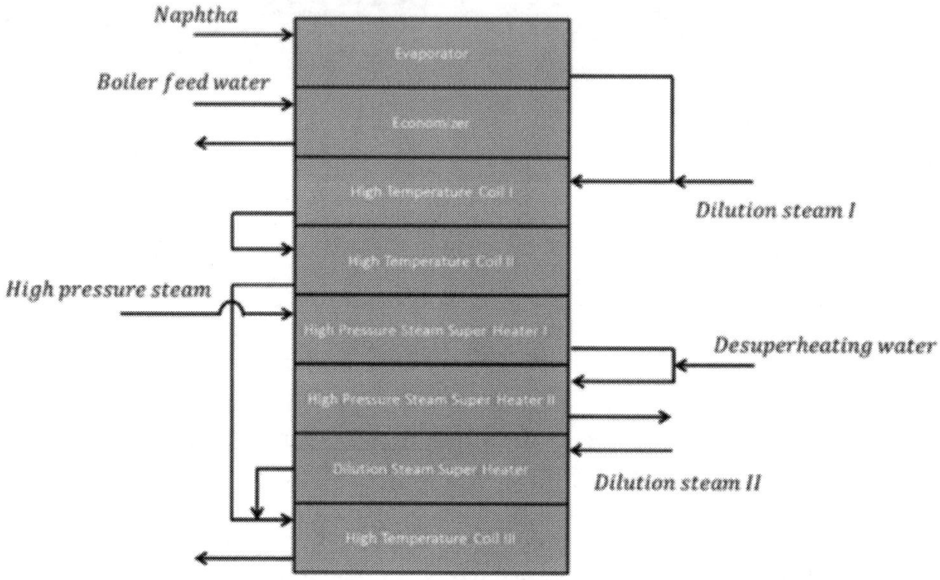

Figure 8. Schematic representation of the simulated convection section.

The calculated temperature profiles of the flue gas and process streams are shown in Figure 9. The flue gas leaves the stack at a temperature of 283°C, whilst the bridge wall temperature is estimated to be 1195°C. The vapor fraction is estimated to be 98.39% at a temperature of 301.4°C, indicating that the feed is almost completely evaporated in the first bank. It is worth noticing that, according to the ASTM D86 boiling points given in Table 1, the feed should be completely evaporated at such a high temperature, but due to the substantially high pressure in the evaporator (6 bar), the boiling points of the species are significantly higher than under atmospheric pressure. This shows the importance of accounting for pressure dependence when simulating the phase transition from liquid to vapor.

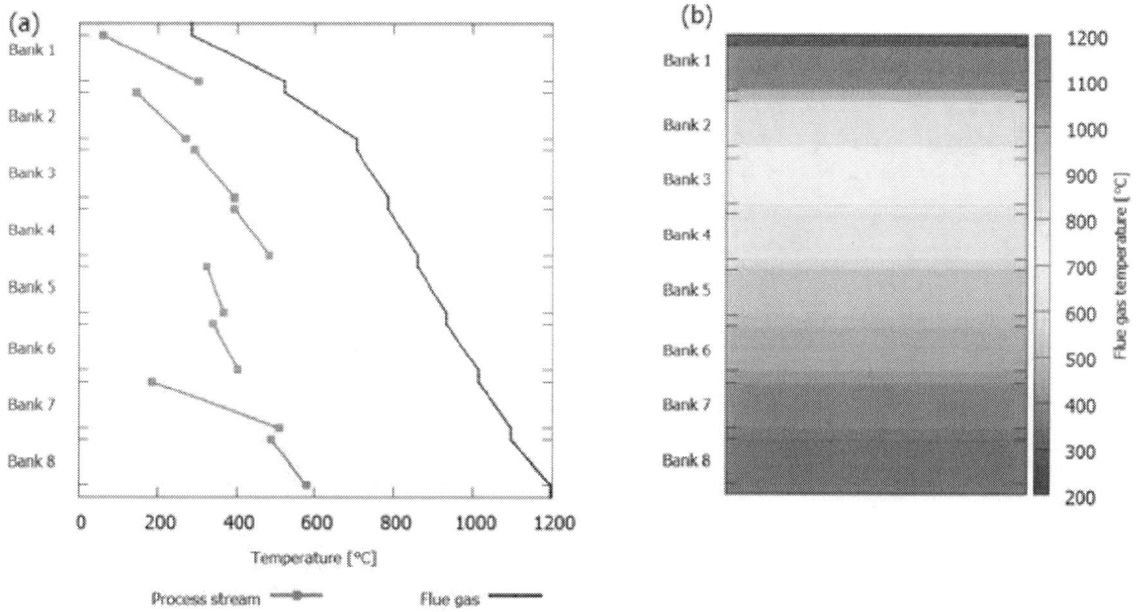

Figure 9. (a) Temperature profile of the process streams for all banks, (b) flue gas temperature profile along the convection section

Firebox Simulation Model

With the implementation of a firebox model, COILSIM1D is able to estimate additional process parameters of interest, such as the flue gas temperature profile inside the firebox, the bridge-wall temperature, the furnace efficiency and the heat flux profile to the reactor coils. Using the GUI of COILSIM1D, shown in Figure 10, the user can easily provide the geometry and operating conditions of the firebox, required as input for the simulation. The user is able to specify the furnace geometry, the type of burners (i.e. floor and/or wall burners) and, in the case of wall burners, the heights at which they are located. Also the operating conditions, such as the inlet temperatures and mass flow rates of fuel and oxidizer (e.g. air) per burner, and the operating pressure, are to be specified. This flexibility in user input guarantees the applicability of the model for a broad range of cracking furnaces.

The firebox model has been coupled to the reactor model, enabling the estimation of process yields, COT, COP, etc. of cracking furnaces, based on the geometry and inlet conditions of the furnace and tubular reactors, specified by the user.

An illustrative simulation of the firebox, reactor and TLE has been carried out with COILSIM1D for a Kellogg Millisecond furnace, shown in Figure 11 (left). The dimensions and operating conditions of the reactor and TLE are presented in Table 2. The simulation parameters are straightforwardly input by means of the GUI of COILSIM1D, as illustrated in Figure 10 for the geometry and operating conditions of the firebox. 53.9 kg/h of fuel (CH_4 + 5% H_2) are supplied to each long flame burner and burnt using air at ambient conditions. The software converts the three-dimensional furnace dimensions to their one-dimensional equivalent and discretizes the furnace along its height, as illustrated by the dotted lines in Figure 11 (right).

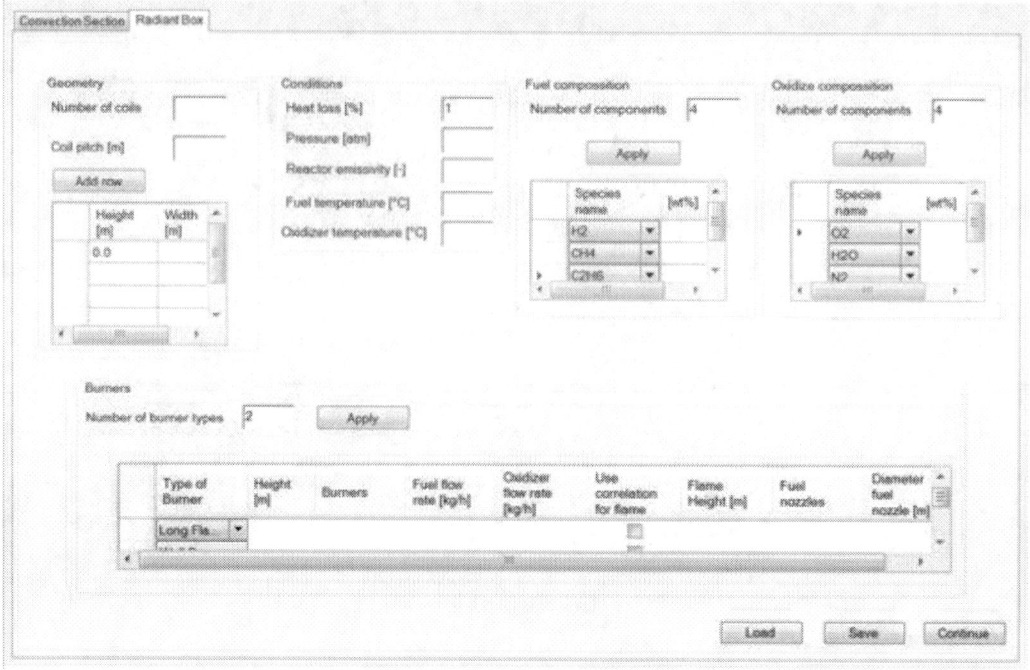

Figure 10. GUI of the firebox simulation module of COILISM1D.

Table 2. Dimensions and operating conditions of the reactor and TLE of the Kellogg Millisecond furnace

Reactor dimensions		**Feed composition**	[wt%]
Length [m]	10.556	C_3H_8	92.25
Internal diameter [m]	0.0348	C_2H_6	1.05
		C_3H_6	3.02
Operating conditions		i-C_4H_8	0.33
Feed mass flow rate [kg/h]	118.5	1-C_4H_8	0.27
Steam dilution [kg/kg]	0.3265	2-C_4H_8	0.08
Coil inlet temperature [°C]	630.5	i-C_4H_{10}	1.93
Coil inlet pressure [bara]	2.35	n-C_4H_{10}	0.45
		C_5^+	0.62
TLE dimensions			
Length [m]	9.369		
Diameter inner tube [m]	0.067		
Diameter outer tube [m]	0.119		
Number of joining coils [-]	4		

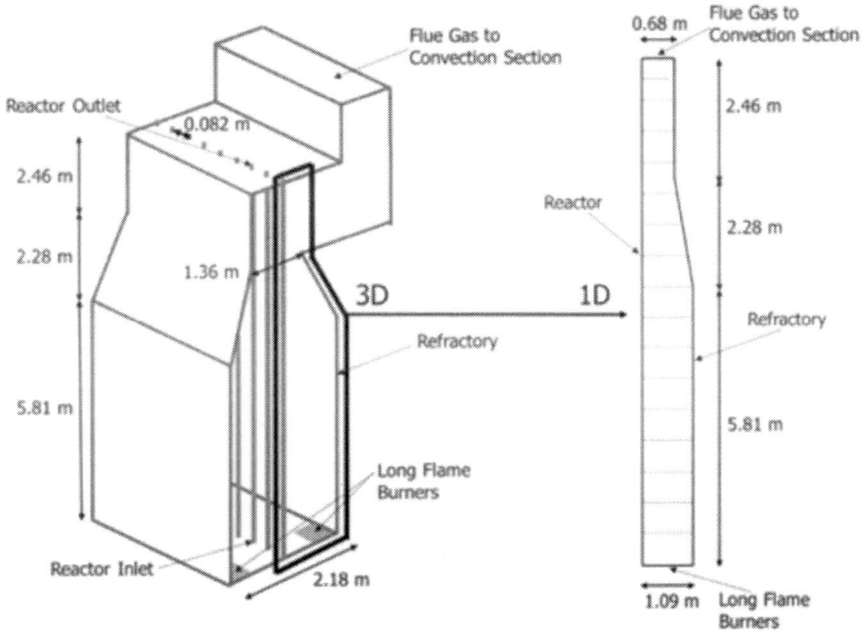

Figure 11. Three-dimensional geometry of the radiant section of a Kellogg Millisecond furnace (left) and its equivalent one-dimensional geometry used for the COILSIM1D simulation (right).

Table 3 shows the results of the COILSIM1D simulation, together with industrially measured values. The simulated values agree well with the available industrial data. The simulation results demonstrate not only the accuracy with which COILSIM1D predicts the process yields, but also its ability to estimate other parameters, such as the COT, furnace efficiency and bridge-wall temperature. Since the GUI facilitates setting up the simulation and considering that the computational time is less than one minute on one core of an Intel® Core™ i7-5600U CPU @ 2.60 GHz processor, this simulation case illustrates the ease and speed with which steam crackers can be simulated using COILSIM1D.

Table 3. COILSIM1D-simulated and industrial [22] data for the Kellogg Millisecond furnace.

	Simulated	**Industrial***
Product Yields [wt%]		
H_2	1.5	1.6
CH_4	19.9	19.3
C_2H_2	0.7	0.8
C_2H_4	35.1	35.1
C_2H_6	2.7	2.9
C_3H_4 (MA)	0.3	0.7
C_3H_4 (PD)	0.1	
C_3H_6	17.0	17.0
C_3H_8	15.8	15.7
1,3-C_4H_6	1.8	1.9
1-C_4H_8	0.8	1.2

2-C_4H_8	0.4	
i-C_4H_8	0.3	0.2
i-C_4H_{10}	0.2	0.1
n-C_4H_{10}	0.1	0.1
Other properties		
Reactor coil outlet temperature [°C]	878.9	
Temperature at exit TLE [°C]	445.9	435
Furnace efficiency [%]	43.2	
Bridge-wall temperature [°C]	1138	

*Industrial data are measured downstream from the transfer line exchanger (TLE).

Input/Results Files and Interaction with External Software

COILSIM1D is a text-based simulation tool. The user friendly GUI allows the user to easily set up a simulation case step by step. Operating parameters (and their units), simulation strategies and geometrical input can be typed in manually by the user, or can be loaded from previous simulation cases.

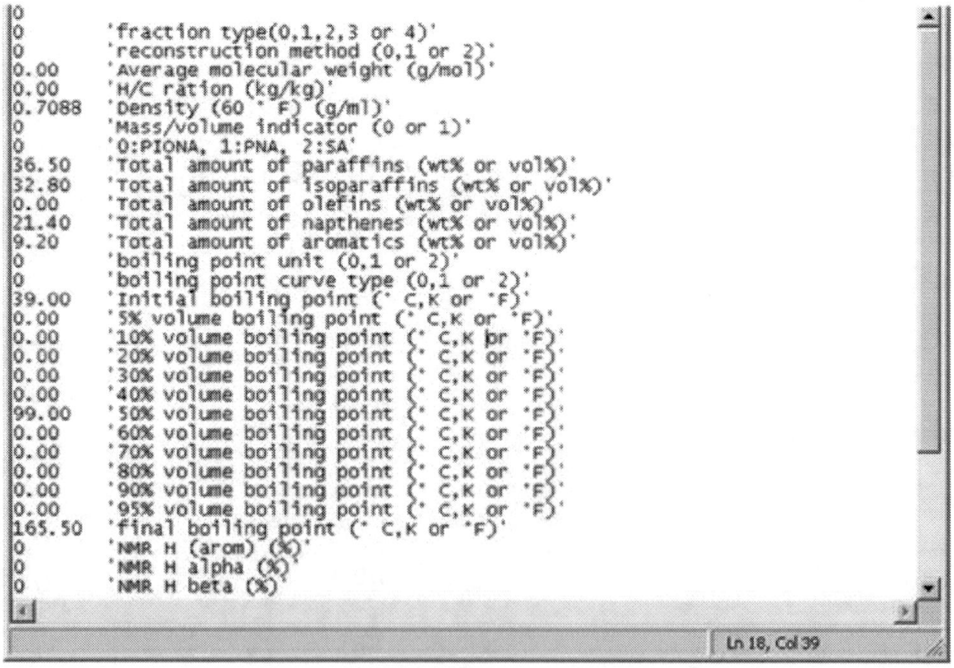

Figure 12: Example of an input file used for feedstock reconstruction with SIMCO.

All the input values are then used to create well-organized input files that are stored into project folders specifically created for each simulation. These input files are simple, self-explanatory, and have a fixed format. An example of an input file used for feedstock reconstruction is shown in Figure 12. The user manual of COILSIM1D explains the purpose and structure of all the input and output files.

The GUI incorporates a tool for quick visualization of the main results of a simulation, as shown in Figure 13. Furthermore, the GUI also displays the results (yields, temperatures, ...) in the form of graphs, which facilitates interpretation of the simulated results. This feature is illustrated in Figure 14. The files with all the simulation results are also stored in the project folder, and can be imported into spreadsheets for further analysis or post-processing.

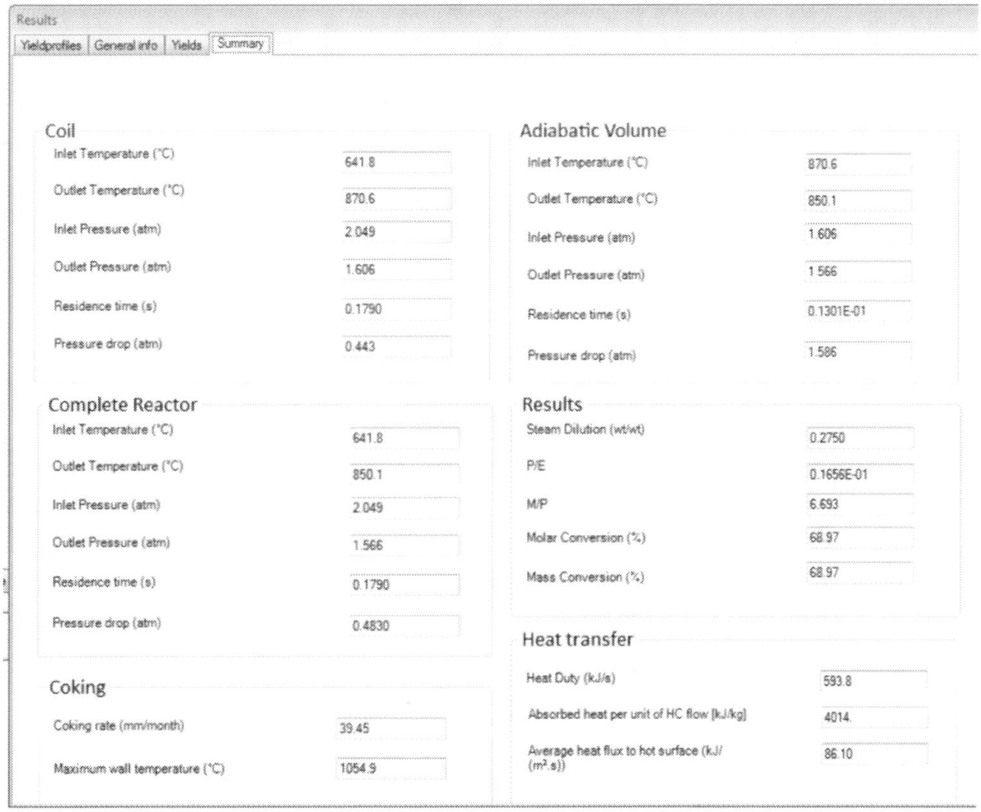

Figure 13: Results tab of the GUI, showing the main results of a simulation

Since the input/output files have a fixed format, they can be easily created and modified without using the GUI, thus allowing users to make use of scripts in the programming language of their preference to take care of these tasks. COILSIM1D includes also an executable file, which allows to carry out simulations bypassing the GUI. This enables users to write scripts for automated creation and execution of large batches of simulations without the need of constant human input.

The ability of COILSIM1D to work using text input/output files has the additional benefit of enabling easy interaction with external software packages, e.g. Aspen Plus®, HYSYS® or gPROMS. This way, whole plant simulations can be made taking into account the constraints of all the units involved. The interaction is simple and based on an interfacing script that reversibly converts the species and process conditions representation of both software tools. Integration with unit operation software packages or in-house optimization tools has been successfully performed with several licensees, considering their specific requirements. Integration of COILSIM1D in real time optimization has also been successfully carried out in a demonstration unit in Asia, showing great potential for industrial scale implementation.

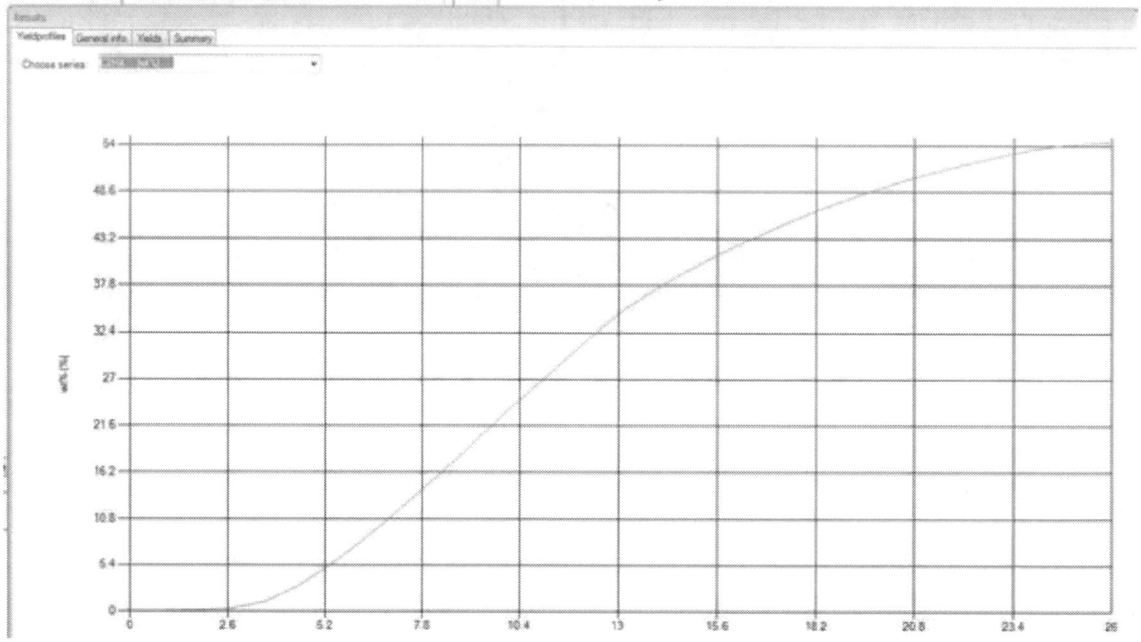

Figure 14: Graphic representation of yield profiles in COILSIM1D

Conclusions

COILSIM1D is a comprehensive simulation package for steam cracking processes. Its main strength lies in its fundamental kinetic model, the broadest currently available. Various features and simulation strategies have been developed and implemented in COILSIM1D: SIMCO, a molecular feedstock reconstruction method; a module to simulate the TLE; a module to estimate reactor run lengths using dedicated coking models; and a shooting method, that allows to carry out simulations using severity-related indices as boundary conditions. To take COILSIM1D to the next level and transform it into a whole furnace simulation package, new features have been incorporated and the kinetic model has been fine-tuned and further extended to account for oxygenate species.

A model to simulate the convection section of a steam cracker has been developed. It allows the user to evaluate the impact of a broad range of different process conditions and feeds in the process. Despite requiring a limited amount of input data, this model helps to easily evaluate the effect of a broad range of different process conditions and feeds. This is a very helpful tool for design and optimization of new and existing configurations. The model accurately accounts for the evaporation of the feed and can be used to prevent fouling in the convection section, which has significant technologic and economic impact on the process. An example simulation was defined and executed yielding satisfactory results.

Another new feature is the firebox model, which allows to reliably determine heat flux profiles for the reactor, and various other parameters of interest, such as the bridge wall temperature and thermal efficiencies. The model can be applied to a variety of firebox geometries, burner arrangements and fuel compositions. It has been successfully used to simulate a propane Millisecond reactor with its corresponding TLE, showing the accuracy and ease of using COILSIM1D when dealing with simulations that explicitly link various aspects of the process altogether.

COILSIM1D comes with a user-friendly GUI, which makes it extremely easy to create and execute all types of simulations. The input and output files of the simulations are stored in Excel readable files, that can be easily imported into spreadsheets for further analysis and processing. This has the additional advantage of enabling easy interaction of COILSIM1D with user-created scripts for large batches of simulations, or to couple COILSIM1D to external software, e.g. to carry out whole-plant simulation and optimization.

The combination of all these new and already existing features enables users to perform accurate simulations of the whole furnace with limited number of inputs, reasonable computational effort and unprecedented accuracy, all with one click.

References

1. Van Geem, K. M.; Reyniers, M. F.; Marin, G. B., Challenges of modeling steam cracking of heavy feedstocks. *Oil & Gas Science and Technology-Revue D Ifp Energies Nouvelles* 2008, 63, (1), 79-94.
2. Hudebine, D.; Verstraete, J. J., Reconstruction of petroleum feedstocks by entropy maximization. Application to FCC gasolines. *Oil & Gas Science and Technology–Revue d'IFP Energies nouvelles* 2011, 66, (3), 437-460.
3. Van Geem, K. M.; Hudebine, D.; Reyniers, M. F.; Wahl, F.; Verstraete, J. J.; Marin, G. B., Molecular reconstruction of naphtha steam cracking feedstocks based on commercial indices. *Computers & Chemical Engineering* 2007, 31, (9), 1020-1034.
4. Pyl, S. P.; Van Geem, K. M.; Reyniers, M. F.; Marin, G. B., Molecular reconstruction of complex hydrocarbon mixtures: An application of principal component analysis. *AIChE journal* 2010, 56, (12), 3174-3188.
5. Dijkmans, T. Steam cracking: from feedstock analysis to plant optimization. Ghent University, 2014.
6. Lahaye, J.; Badie, P.; Ducret, J., Mechanism of carbon formation during steamcracking of hydrocarbons. *Carbon* 1977, 15, (2), 87-93.
7. Albright, L. F.; Marek, J. C., Mechanistic model for formation of coke in pyrolysis units producing ethylene. *Industrial & Engineering Chemistry Research* 1988, 27, (5), 755-759.
8. Kopinke, F. D.; Bach, G.; Zimmermann, G., New results about the mechanism of TLE fouling in steam crackers. *Journal of Analytical and Applied Pyrolysis* 1993, 27, (1), 45-55.
9. Muñoz Gandarillas, A. E.; Van Geem, K. M.; Reyniers, M. F.; Marin, G. B., Coking resistance of specialized coil materials during steam cracking of sulfur-free naphtha. *Industrial & Engineering Chemistry Research* 2014, 53, (35), 13644-13655.
10. Plehiers, P. M.; Reyniers, G. C.; Froment, G. F., Simulation of the run length of an ethane cracking Furnace. *Industrial & Engineering Chemistry Research* 1990, 29, (4), 636-641.
11. Reyniers, G. C.; Froment, G. F.; Kopinke, F.-D.; Zimmermann, G., Coke formation in the thermal cracking of hydrocarbons. 4. Modeling of coke formation in naphtha cracking. *Industrial & engineering chemistry research* 1994, 33, (11), 2584-2590.
12. Reyniers, M.-F. S.; Froment, G. F., Influence of metal surface and sulfur addition on coke deposition in the thermal cracking of hydrocarbons. *Industrial & engineering chemistry research* 1995, 34, (3), 773-785.
13. Muñoz Gandarillas, A. E.; Van Geem, K. M.; Reyniers, M. F.; Marin, G. B., Influence of the reactor material composition on coke formation during ethane steam cracking. *Industrial & Engineering Chemistry Research* 2014, 53, (15), 6358-6371.
14. Sabbe, M. K.; De Vleeschouwer, F.; Reyniers, M.-F.; Waroquier, M.; Marin, G. B., First principles based group additive values for the gas phase standard entropy and heat capacity of hydrocarbons and hydrocarbon radicals. *The Journal of Physical Chemistry A* 2008, 112, (47), 12235-12251.
15. Sabbe, M. K.; Reyniers, M. F.; Van Speybroeck, V.; Waroquier, M.; Marin, G. B., Carbon-Centered Radical Addition and β-Scission Reactions: Modeling of Activation Energies and Pre-exponential Factors. *ChemPhysChem* 2008, 9, (1), 124-140.

16. Sabbe, M. K.; Reyniers, M. F.; Waroquier, M.; Marin, G. B., Hydrogen Radical Additions to Unsaturated Hydrocarbons and the Reverse β-Scission Reactions: Modeling of Activation Energies and Pre-Exponential Factors. *ChemPhysChem* 2010, 11, (1), 195-210.
17. Sabbe, M. K.; Vandeputte, A. G.; Reyniers, M.-F.; Waroquier, M.; Marin, G. B., Modeling the influence of resonance stabilization on the kinetics of hydrogen abstractions. *Physical Chemistry Chemical Physics* 2010, 12, (6), 1278-1298.
18. Saeys, M.; Reyniers, M. F.; Marin, G. B.; Van Speybroeck, V.; Waroquier, M., Ab initio group contribution method for activation energies for radical additions. *AIChE journal* 2004, 50, (2), 426-444.
19. Saeys, M.; Reyniers, M. F.; Van Speybroeck, V.; Waroquier, M.; Marin, G. B., Ab initio group contribution method for activation energies of hydrogen abstraction reactions. *ChemPhysChem* 2006, 7, (1), 188-199.
20. Van Geem, K. M.; Reyniers, M.-F.; Marin, G. B. In *Taking optimal advantage of feedstock flexibility with COILSIM1D*, AIChE Spring Meeting: Ethylene producers conference, New Orleans, LA, USA, 2008; New Orleans, LA, USA, 2008.
21. Zimmermann, H.; Walzl, R., Ethylene. *Ullmann's Encyclopedia of Industrial Chemistry* 2009.
22. Heynderickx, G. J. Modellering en Simulatie van Huidige en Nieuwe Technologieën voor de Thermische Kraking van Koolwaterstoffen. University Ghent, Ghent, 1993.

443239 Acetylene Hydrogenation Tutorial

Abstract:

Acetylene Hydrogenation Tutorial

Author: Edgar Mohundro –

Acetylene hydrogenation is a very important step in most olefins plant operations along the process flow path of achieving on-specification ethylene and propylene production. The tight limit on the allowable C_2 acetylene contaminant content levels driven by polymer producers necessitates close scrutiny and evaluation of the operation and performance of acetylene converter systems.

The focus of the Acetylene Hydrogenation Tutorial will be upon the know-how and experience base for operating acetylene hydrogenation units. The intent of the Tutorial is to convey, from the Operator's perspective, the essentials and insights that an Operator should know and understand when it comes to successful acetylene converter operation.

The Tutorial will:

- Touch upon a brief overview of history and background of the technology.
- Review the various process flow schemes that an Operator may encounter.
- Overview on chemistry that is involved with the hydrogenation process.
- Discuss the factors related to various steam-cracker hydrocarbon feedstocks and the various contaminants and poisons.
- Discuss the reactor operating conditions and control parameters.
- Discuss monitoring the system performance.
- Comment upon catalyst regeneration and activation considerations.
- Aspects of safety, the element of which an Operator must always be mindful.

- Present key start-up and shutdown considerations for acetylene convertor systems.

ACETYLENE HYDROGENATION WITH OLEMAX®207 FOR BACK END UNITS

Yu Liu[1], *Lin Luo*[1], *Marcos Martinez*[1], *Cristina Martinez*[2], *Denise B. Cooper*[3], *Brian Heasley*[3] *and Mingyong Sun*[4]

(1) Olefins, Aromatics, and Alternatives R&D, The Dow Chemical Company, Freeport, TX, (2) LHC, Freeport, TX, The Dow Chemical Company, (3) Clariant Corporation, Houston, TX, (3) Clariant Corporation, Louisville, KY

Abstract

Selective acetylene hydrogenation is widely used to reduce acetylene in the ethylene stream to meet the specifications of polymer grade ethylene. For a backend selective hydrogenation system, frequent catalyst regeneration is typically required due to relatively fast catalyst fouling. The new generation backend catalyst, OleMax®207 supplied by Clariant, has reduced green oil formation and consequently led to much longer cycle time. OleMax®207 was first implemented in one of Dow's units in Freeport, TX in 2007. In this presentation, Dow's experience of using this catalyst will be shared. OleMax®207 has been functioning well for the past eight years and the catalyst can provide sufficient activity for 3-6 year increments between regeneration. An example of bringing the reactor system on-line efficiently after a poisoning event will also be discussed.

Introduction

An acetylene hydrogenation unit is normally used to reduce acetylene in the ethylene stream from an ethane cracker in order to meet the specifications of polymer grade ethylene. There are four types of acetylene selective hydrogenation units, front-end de-ethanizer (FEDE), front-end de-propanizer (FEDP), back-end (BE), and raw gas, depending on the configuration of the cracker system. In this presentation, we focus on BE acetylene removal. Back-end acetylene reactor systems are characterized by their relatively short runtimes (typical 8 months) mainly due to catalyst deactivation through green-oil formation[i]. The deactivation is reflected in plant operation by an increase in operation temperature and a significant decline in selectivity through the catalyst run cycle. At the end of each run, catalyst regeneration is required in general. This type of catalyst, typically, has a much shorter lifetime than the front end systems (>10 years) where much less green oil is produced due to the large excess of H_2 in the feed stream. As shown in Figure 1, a typical configuration of a BE unit, the products from the cracker go through both the de-methanizer (to remove offgas: hydrogen, carbon monoxide and methane) and de-ethanizer (to remove heavier than C_2 hydrocarbons). The feed going to the reactor contains only the C_2 cut (overhead products from de-ethanizer: acetylene, ethylene and ethane). H_2 has to be fed to the reactor separately to mix with the C_2 feed to remove the acetylene through hydrogenation. Depending upon the acetylene concentration in the feed, the reactor can be a single bed, two beds in parallel or two beds in series. The plant in Texas has three reactors in the BE system, with two online in parallel and one in regeneration or standby position.

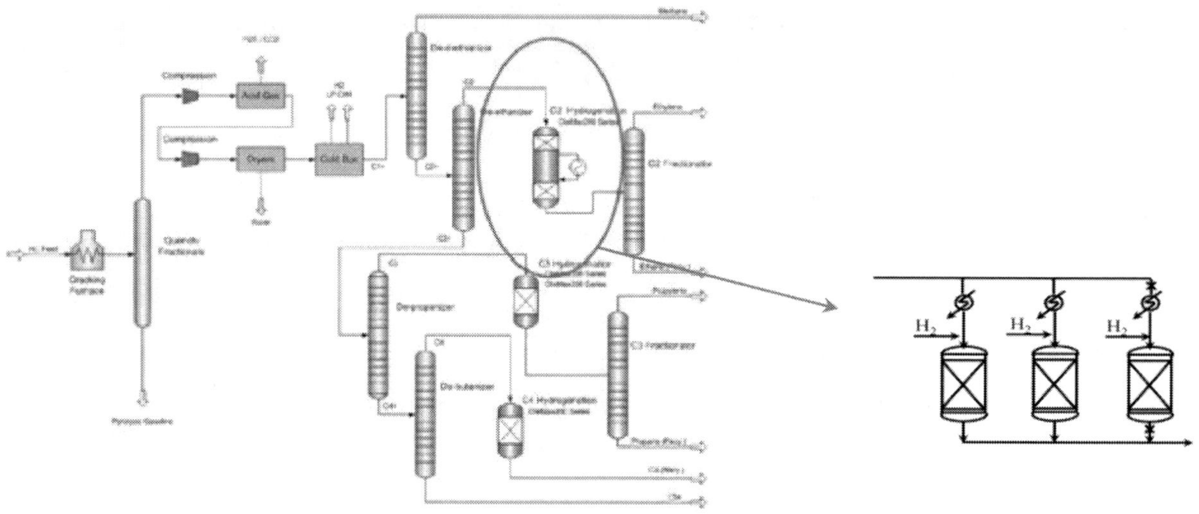

Figure 1 Schematic flow diagrams of the BE reactor system

Since the implementation of OleMax®207 catalyst in Dow's Texas plant, significant improvement has been achieved in the unit performance in terms of ethylene selectivity, hydrogen consumption, and run cycle length. With the previous generation of catalyst, the average selectivity was below zero and catalyst regeneration was required every 8-18 months. Currently, the reactors are in operation with ethylene selectivities maintaining at a significantly high value. For the two currently operating reactors (R60A and R60B) in this system, OleMax®207 catalyst has been running for nearly 3 and 6 years, respectively, since the last catalyst regeneration. This paper documents an overview of the performance in this BE unit and one example of the catalyst activity/selectivity quick recovery from a H_2S contamination.

Performance Comparison of Previous Catalyst with OleMex®207

OleMax®207 was implemented in one of Dow's Texas plants in 2007. This was the first commercial implementation of this catalyst. Prior to February of 2007, a Clariant's previous generation catalyst was used in the plant. As shown in Figure 2, a much higher H_2 to C_2H_2 ratio had to be applied to achieve the cleanup of acetylene with poor selectivity to ethylene when the previous catalyst was used. A significant increase in the selectivity to ethylene was achieved at lower H_2 to C_2H_2 ratio after the catalyst was swapped to OleMax®207.

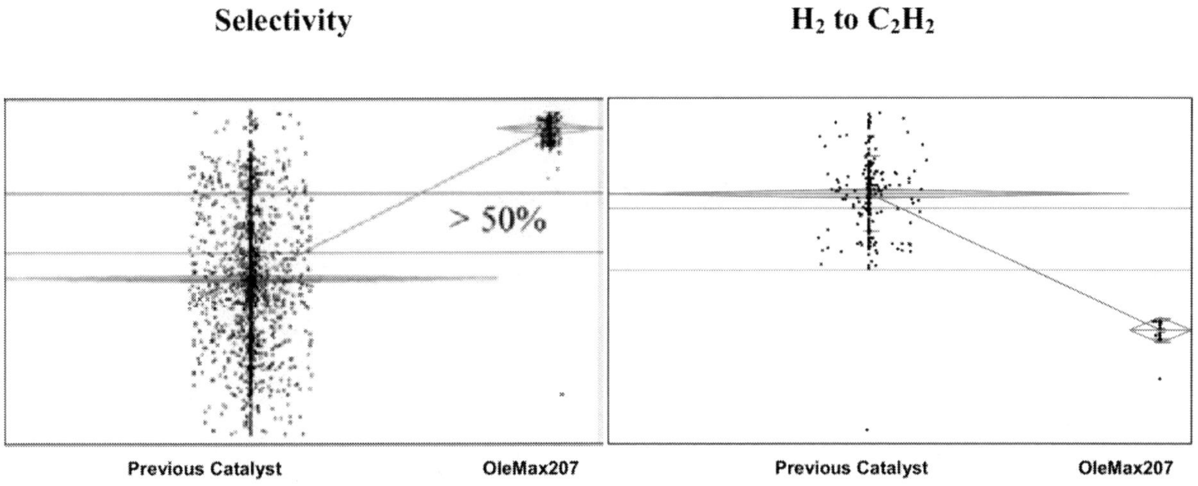

Figure 2 Comparison of the selectivities (left) and H_2/C_2H_2 ratios (right) of the previous catalyst and OleMax207 in Dow's Texas plant

Overview of the Performance of OleMax®207

Figure 3 illustrates the trend in selectivities of OleMax®207 in R60 A and R60B since 2013.

Typical in BE operation, ethylene selectivities decrease in each cycle due to "green oil" formation which blocks the active sites on the catalyst[ii]. For R60A and R60B, the ethylene selectivities were maintained at a high value at the start of the run and dropped gradually during the operation. After the regeneration, the selectivities of R60A and R60B were recovered, reaching the same starting value of the previous cycle.

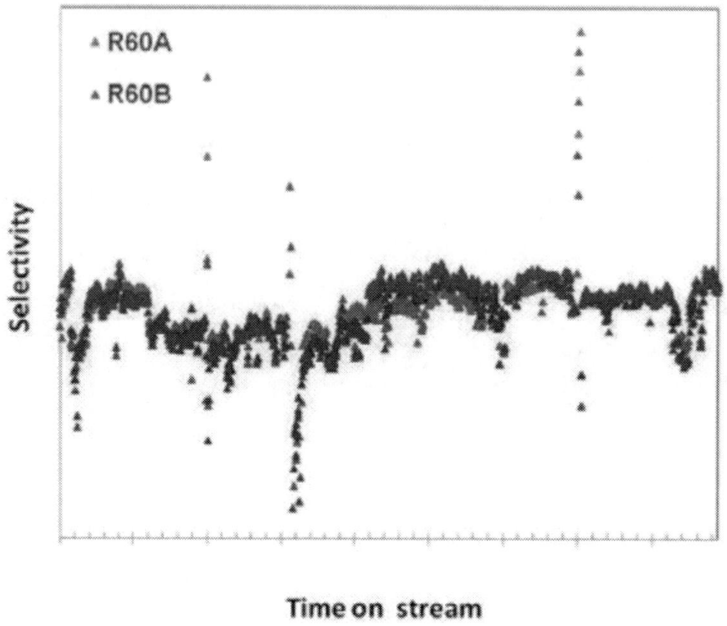

Figure 3 Selectivity of OleMax207 in last three years

So far, OleMax®207 in R60A and R60B have gone through two and one regeneration cycles, respectively. As shown in Figure 3, with the exception of a few plant events (including a short duration turn around), the selectivities of both R60A and R60B have maintained at fairly constant similar values and followed the same trend since 2013. For BE, H_2/C_2H_2 ratio has a significant impact on the selectivity. A stable H_2/C_2H_2 ratio during the operation not only allows steady (high) selectivity but also improves the operational in plant. As illustrated in Figure 4, the H_2/C_2H_2 ratios in both R60A and R60B have been controlled at nearly a constant value since 2013. There has been a slightly increase (~4 °C) in the inlet temperature of both reactors since 2013 due to the aging of the catalyst. By optimizing both inlet temperature and H_2/C_2H_2 ratio[i], around 70% of the temperature rise has been achieved in the first half of the catalyst bed and the hydrogen concentration in the outlet is within the target desired range.

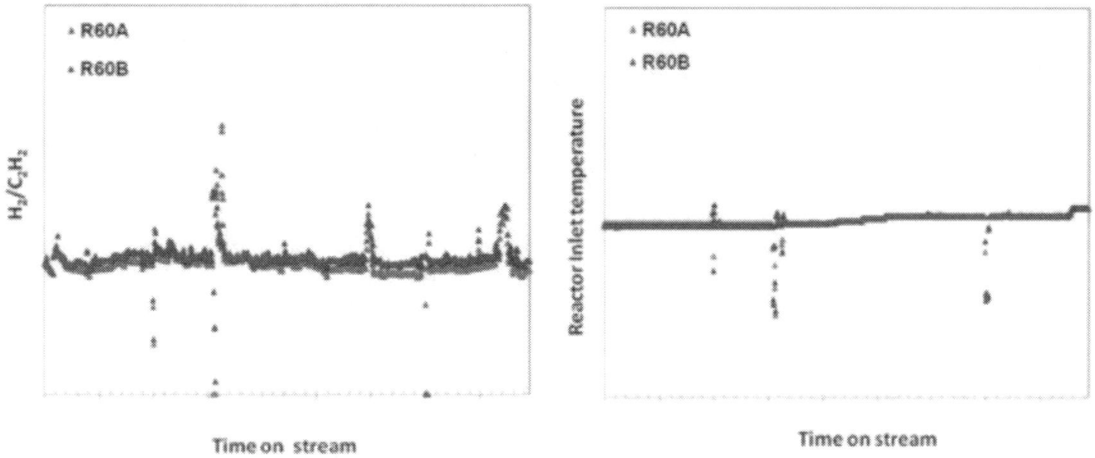

Figure 4 H_2/C_2H_2 ratio (left) and reactor inlet temperature (right) in last three years

Restoration of the H_2S contaminated OleMax®207 in a Plant Event

In March of 2014, an unknown amount of H_2S was accidently leaked into the both reactors during a plant event. As shown in Figure 5, the initial attempt at start-up of the reactor was not successful. There was no temperature rise in the bed after both inlet temperature and H_2/C_2H_2 ratio reached the target conditions of the start of run. After discussion with a Clariant representative, the Dow team took the following actions: purged both reactors with the plant off gas at 95 to 110 F inlet temperature until no H_2S could be detected in the outlet stream. A bed temperate rise was observed quickly once the inlet temperature and H_2/C_2H_2 ratio reached the target conditions of the start of run. The performance of the OleMax®207 has not been impacted by this event. The selectivities of both reactors are currently maintained at significant high values.

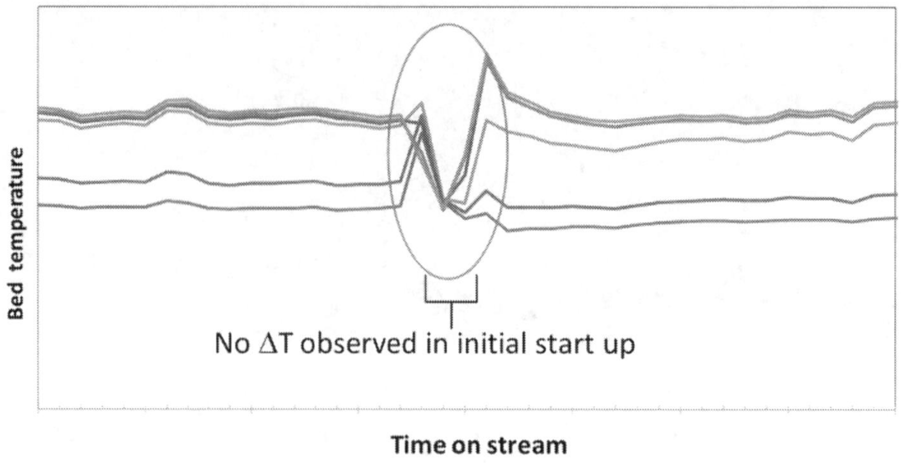

Figure 5 Temperature profile of R60A bed before and after the removal of H_2S

Summary

Since the implementation of OleMax®207 in one of Dow's TX plants, lower H_2/C_2H_2 ratio, much higher selectivity and reliability of operation have been achieved relative to the previously used catalyst. The long catalyst cycle time observed in R60A and R60B reactors far exceeds the expectation. The improved performance is believed to be resulted from the improvement in catalyst technology which allows for less green oil formation as supported by skid rig results.

Acknowledgements

The authors are grateful to Howard Clark, Greg Barbay, Swetha Akula and Mias Allison for their contributions to this work.

---

[i] C2 Tail-End Acetylene Converters, Technical Bulletin, Clariant.

[ii] Calvin H.Bartholomew, Mechanism of Catalyst Deactivation, Appl. Catal. 212(2001)17

443473 Molecular Sieve Cracked Gas Dryer Tutorial

Abstract:

This tutorial provides an overview of adsorption technology. It will cover the following topics: adsorption phenomenon, structure and chemistry of adsorbents, equilibrium and dynamic characteristics of an adsorbent bed, regeneration by thermal swing, operation of an adsorption unit, and an overview of commonly encountered issues. This tutorial is focused on adsorption applications in ethylene production plants.

AIChE Paper Number

Global Ethylene Operating Rates – What Is It Really Telling Me?

Stephen Lewandowski
Global Business Lead – Olefins
IHS

Prepared for Presentation at the 2016 Spring National Meeting
Houston, Texas, April 14-16, 2016

Global Ethylene Operating Rates – What Is It Really Telling Me?

Stephen Lewandowski

Global Business Lead - Olefins

IHS

Abstract: Ethylene operating rates of steam crackers are used to indicate the fundamentals of the ethylene market, with ethylene producers primarily concerned with adequate surplus ethylene production capacity to manage supply during global upsets. Operating rates have been trending higher over the last 6 to 7 years indicating the market is getting tighter on ethylene supply. As we review the historical trend, we also need to be aware of the key assumptions driving operating rates for the next 3 to 5 years. How much capacity are we adding versus demand growth? Is demand growth real and what is driving this value? What is the feed slate and will a shift to lighter feeds cause producers to de-rate? How do on-purpose units, Methanol-To-Olefins and Coal-To-Olefins, fit into the mix? Do plastic recycle trends matter? Does Polymer substitution incentive grow? What about coal chemistry to ethylene derivatives direct, bypassing ethylene production altogether? A change in any one of these assumptions could have a dramatic impact on the ability of the system to supply ethylene to meet market needs.

Presenting ethylene from an industry consultant's perspective, one is obliged to take a 30,000 foot view to reflect on the many different stakeholders' impacts and interest. We have to harmonize an energy view, against a global economic view, against an industrial view, and finally keeping in mind the financial view as many US dollar equivalents are spent in this capital intensive industry. Determining the ideal region to investment has been and remains the main focus of all the operators in such a competitive environment. The answer to this becomes more complicated as the industry ebbs and flows from privately owned to government owned/controlled industrial development. Each ownership group has different drivers for investment.

Ethylene is and has been the bellwether of the health of the global petrochemical markets as it is the largest by volume basic chemical used in the manufacture of many durable and non-durable goods. So it is said, "...how ethylene goes so does the fate of many other basic chemicals and the petrochemical industry as a whole."

The diagram below shows the connections between energy and economy on the steam cracker operations that is linked through: cost structures (driven by energy), operating rates (driven by the economy and demand), and then ultimately to margins (which encourage re-investment in sustaining capex, debottlenecks, and green/brown field investments in/on steam cracking units.) The connections and links at times are not so obvious, and even at times counterintuitive. The big question facing petrochemical producers today is: what are the opportunities on the steam cracker front as a consequence of all the other dynamics happening around it?

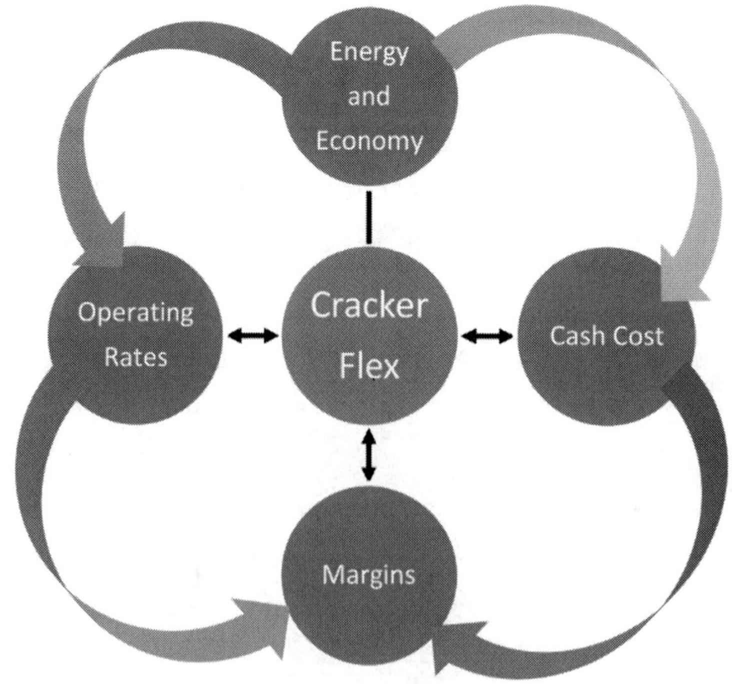

ECONOMY

The global economic forecast is the best place to begin. The health of the global economy and the individual countries will ultimately drive demand for petrochemicals. Consumption growth in each region is tied to GDP growth and as regions move from different stages on the economic development continuum, the consumption will change. The current population of a country as well as its population growth can be linked to its economic development, which impacts total tons of petrochemicals consumed in the key metric of ***consumption per capital****. The world has a large population in the developing stages that, with even modest increases in consumption per capital, will have enormous impact in investments in many new petrochemicals assets which will be required to satisfy that appetite. Basic standard of living improvements converted to petrochemical use can be seen first in non-durable good growth like packaging, later followed by durable good demand growth such as appliance and automobiles, as the country moves along the development continuum. To drive durable good demand, disposable income has to be available, with less spent on what it takes "to live" and more spent on making living even more comfortable. Disposable income is linked in many regions directly to the cost of energy as many developing countries import energy so as the cost of energy falls, the money available for other spending is liberated. The current outlook for the global economy, in spite of the perceived slowdown of the Chinese economy, is stable GDP growth in the range of 3%.*

ENERGY - CRUDE Oil

Crude oil demand on a global basis continues to grow. This is directly linked to economic development. So as demand for oil grows, two key issues immerge in the energy / oil patch:

- *a) Depletion of existing production wells - Wells have a finite life and decline rates are a fact of life in the industry. A rapid decline rate is prominent in shale oil / shale gas geology which is prominent in the US.*
- *b) Year on year demand growth – How much is this growth and can we keep pace.*

To cover the oil growth needs we are dependent on two tranches of oil supply:

- *a) Do the fields discovered but under assessment have the oil production levels to justify investments in production to cover some of the depletion and new oil discovery needs?*
- *b) Can we find new plays and develop them to cover demand growth?*

Many of the current fields that are under assessment and other fields that have yet to find oil, are by definition in less than obvious locations, meaning that oil produced from these fields will likely come to market at a higher cost. Why is this oil supply critical? It is critical because with oil production comes associated gas (usually). This associated gas is typically a mix of natural gas and Natural Gas Liquids (NGL). What becomes challenging is the fact that the amount and quality of the associated gas can vary from field to field and play to play. In certain North America (NAM) plays, exceptionally high volumes of associated gas can be produced. It is this rich associated gas which is adding the surplus NGL production in NAM and is why US shale oil/gas is so important for global NGL balancing and US NGL pricing.

NATURAL GAS (NG)

On the NG front, NAM continues to surprise the world with an increasing level of proven NG reserves. For the petrochemical space, the keys are: a) where the play exists and b) how wet (content of NGL's) exist in the gas play. So the shale oil / gas renaissance has converted NAM from what a few short years ago was expected to be an importer of NG to a major liquefied natural gas (LNG) exporter. In fact, the US is now what appears to be NG/LNG outlet constrained. We have pushed back NG imports from Canada and have increased exports to Mexico. Even with these changes in trade patterns, NAM continues to have a surplus of NG. Investments are being made to export LNG out of the US and the US has the economics (low cost production) to support this high level of LNG exports. The US could push enough LNG to the market to satisfy global demand on its own. The constraint holding the US from this position is that Canada has very similar geology to the US and is looking to do the same with its NG excess. Australia and many other regions of the world are on a similar LNG export push with stranded domestic gas reserves. Economic incentive/return on investment to move NG to LNG exports is reduced in today's lower crude environment with alternative energy sources to the potential importers coming in at a lower cost, but in the longer run, as discussed in the crude oil section, we anticipate a ramp up in oil prices, which will bring stranded NG and hence LNG exports back into vogue.

NATURAL GAS LIQUIDS (NGL)

NGL's (ethane, propane, butane, and natural gasoline) have three primary sources of production: a) via associated gas from crude production, b) via non-associated gas from natural gas production, and c) from crude oil refineries. The key message though, is that from all of these sources, NGL's are not the primary product, but rather a co-product of the production of other key materials: a) stabilized crude oil, b) NG, and c) transportation fuels from refining. This makes NGL's a fatal product / necessary evil of making the primary products: oil, gas, transportation fuels. So with supply dependent on another product the demand has to be found to balance out all of the NGL production. Pricing of the NGL's barrel looks to alternative outlets for each of the NGL components. The producer has to clear its production in the market place so they will price it to a level that keeps the primary operations: oil, gas, and transportation fuels profitable, which normally means displacing some other energy source with NGL.

In the case of North America, all the outlets for the NGL's have been saturated. The US is now or soon to be exporting ethane, propane, butane and natural gasoline to the global market place. Just a few short years ago the US was short ethane, balanced to short on propane/butane and short natural gasoline. The rest of the world (Europe, Asia, etc) is not experiencing such a production windfall, but rather are somewhat a benefactor from extra supply to the market by acting as the marginal/alternative outlets as described above. But to incent these alternatives, pricing comes into play, as a net backed price of the NGL in the US as it relates to the clearing price in the destination market.

ETHYLENE ECONOMICS

The most common metric to compare ethylene production economics is cash cost. This analysis can be done regionally and for varying feedstock. Each region that has ethylene production has a dominate feed as well. In NAM and the Middle East (MDE) this is ethane. In Europe and Asia it is predominantly Naphtha. But each region has some flexibility on feed types. In China, in addition to steam cracker based ethylene production, they also have added new chemistry via methanol and are indifferent to whether that methanol comes for indigenous coal or based on methanol imported from methane chemistry, which also produces methanol. Cost of ethylene production by feed type and chemistry is important both in low overall ethylene production operating rates and high operating rates as the marginal cost feed on a cumulative cost curve will normally have influence on the price of ethylene.

As a consequence of the NAM shale oil / gas boon, the preferred feedstock for ethylene production changed in the US from liquids (propane/butane, naphtha and heavier) to gas (ethane) over the past five to six years. In fact during the mid-2000's cracker capacity was being shut down in the US from lack of competitiveness. Over the last three years, US crackers have been extremely advantaged via the oversupplied low price ethane, advantaged to levels near that of the MDE. But unfortunately for US operators this advantaged has been reduced both by lower energy prices (crude oil) and a changing supply / demand picture for ethane with new export terminals coming on stream and new cracking capacity soon starting up.

The MDE will retain its low cost advantaged position even with new Kingdom of Saudi Arabia pricing for NG and NGL's. Naphtha fed crackers in the rest of the world are experiencing some new found hope with the drop in crude and the ethane based crackers having lost 50% plus of their advantage. But unfortunately for the naphtha crackers, they still sit on the higher end of the cost curve. This is largely due to high volumes of co products made via liquid feeds and these coproducts are pricing at a much lower premium to the naphtha in the current market.

Cash cost for other sources of ethylene, even though a smaller percentage of global supply, become more competitive in the longer run. As long as coal in China and stranded NG (which produces methanol) remain inexpensive, coal to olefins and integrated gas to olefins (via methanol) can be very competitive on cost. For those China assets buying merchant methanol to operate their methanol to olefins (MTO) plants, the economics become a bit less obvious with other more profitable outlets available to methanol producers. The issue for these other sources of ethylene even with lower cash cost is the investment cost and margin needed to cover the large capital outlay. Cost is not the only metric when comparing the various technologies coming into the market place, returns on investment become important too.

ETHYLENE FUNDAMENTALS

Demand for ethylene continues on a steady climb at a multiple to GDP at just over 1.0 times GDP. The ethylene derivative mix is dominated by polyethylene at over 60%. Ethylene oxide / ethylene glycol (MEG), Polyvinyl Chloride (PVC) and Styrene come in at a distance 2nd, 3rd, and 4th all around 10% of the ethylene demand with another ten derivatives making up the balance. Supply continues to be sourced mainly from steam crackers with the work horse feeds being naphtha and ethane. LPG's and heavier feeds do fit into the mix, but ethane and naphtha lead the way. Non steam cracker production capacity and supply growth continues, but this on purpose remains a small piece of the global ethylene supply pie and in the longer run we expect to peak at 6% of global ethylene supply.

Year on year capacity additions during the next two years, per our forecast, outpaces demand growth. It is important to recognize that capacity is not supply. History has shown that sustainable operating rates on steam crackers does not exceed 89% on a global basis, so it would be expected 10% more capacity needs to be added over time to stay ahead of demand growth. In the 5 year time horizon IHS demand forecast is strong according to our economic forecast as discussed above, and in this five year window demand begins to outpace production capacity additions, which could stress the global systems capability to supply demand needs.

Side Bar: THE US WAVE - A quick focus on US ethylene capacity post the shale renaissance estimates 15 MMTA of new ethylene production capacity in the queue via debottlenecks and green field grass root projects between 2014 and 2020. The grass roots crackers are under various stages of construction. A second wave of an additional 15 MMTA has been announced post 2020. These projects remain under study as a few investment drivers need to be assessed: a) capital cost, b) ethane balance going forward in new crude price environment, c) global demand growth, and lastly d) how much feed flex should I design into my new cracker. It is unlikely the US supply of ethane will be sufficient to allow all of the second wave to happen, but maybe 2/3rds of it could find ethane supply. The large number of announcements does indicate a belief more ethane could be made available.

To summarize the fundamental picture over the next five years:

a) *Based on IHS economic forecasts demand should be strong, no global recession.*
b) *Global ethylene capacity under construction to be started up over the next five years may not be sufficient to satisfy demand.*
c) *Timing of startup of assets under construction is not certain.*
d) *Operating rates dip in 2018 based on an optimistic construction schedule, but climb to levels at or near the 2015 peak*

The operating rate story is the most critical issue facing ethylene supply globally. At the high operating rates seen in 2015 for both Asia and Europe, disruptions to supply caused a spike in ethylene pricing and near record margins in both of those regions. The message is that any demand gain or supply disruption over the next five years will push the global ethylene supply system into a volatile market at a minimum and possibly into an infeasible situation.

ETHYLENE IS NOT AS SIMPLE AS IT ONCE WAS PORTRAYED

With a base case view defined, the question at hand is this: in a high global operating regime, what are the supply/demand dynamics working at the fringes? The world of ethylene works at the margin, and in a world of high operating rates small shifts can mean major impacts to margins. How the cracker fits into this mix is more important than ever. The figure below depicts the supply/demand forces pressing on the ethylene balance. A short discussion of each pressure point will show how changes could impact ethylene margins.

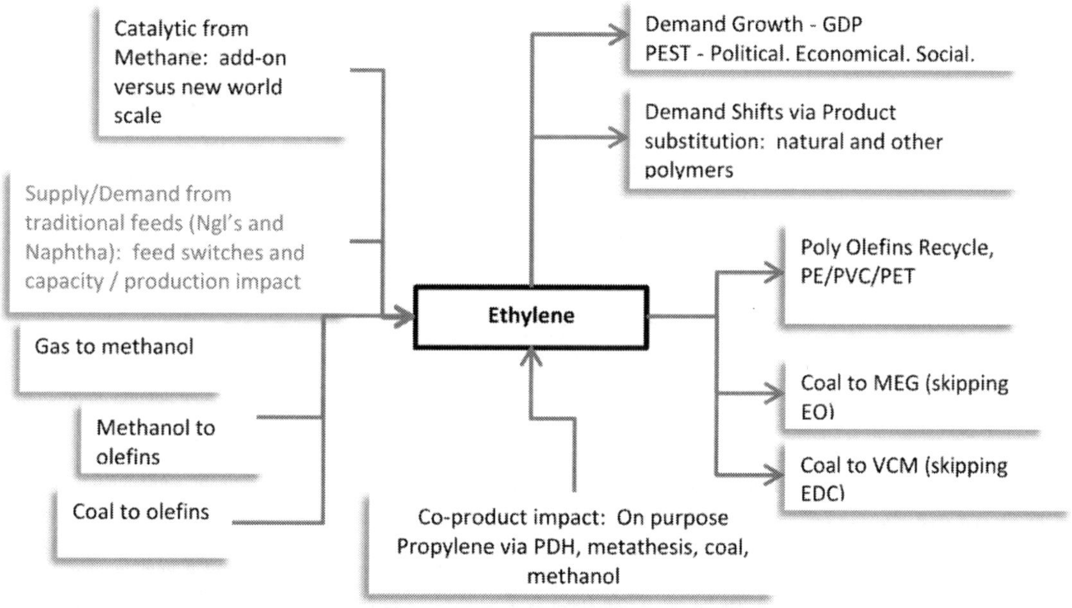

FEED Shifts

With dramatic shifts in the NGL supply picture as discussed earlier, price sensitive demand of NGL's into steam crackers becomes a critical assessment. The trend will be for NGL's to price lower to encourage steam crackers to shift from naphtha to a lower cost feed. This switch from naphtha to NGL does not come without a cost. Operating a cracker with different feed will generate a different set of single pass and ultimate yields. Recycles within the unit from feed switching become a major constraint unless investments in the hardware are made. The level of these investments depends on the desired level of the switch to NGL feed. In many cases, the investment becomes cost prohibitive. So, switching cracker feed slate, with limited investment, ultimately results in a de-rate of the ethylene production capacity.

This de-rate phenomena was demonstrated when many US crackers were driven to run lower cost ethane at the expense of naphtha, the US crackers de-rated ethylene production capacity but still managed to increase revenue. Over time the operators began spending money to eliminate some constraints, but the investment for conversion takes conviction that the investment in flex will pay out with the trade-off that some conversion may not be reversible in case feed cost trends shift back. This de-rate is also being experienced in Europe as more LPG is being processed at the expense of Naphtha. The NGL producers now have Asian crackers in their sites as an additional outlet. Asia has a high percentage of ethylene production from naphtha fed crackers. This makes this a logical offtake to be studied by NGL producers. The Asian market could experience some ethylene production de-rating as was the case in the US and Europe if they choose to switch feeds with limited investment.

Ethylene cash cost on flex crackers feeding propane and butane (LPG) are keys to be watched with the glut of LPG molecules coming to the market. The annual average cash costs are not as good as ethane in NAM or naphtha in WEP/Asia. But propane and butane have high levels of seasonal pricing with high winter values and low summer values. This makes them ideal for swing or price sensitive feed to steam crackers. But shifting feed to crackers with no or limited debottleneck investment would de-rate crackers already pushing their production capability and will cause a tighter ethylene market. In the feed switching scenario, operating rate peak levels have to be reassessed as this metric is a moving target in a feed switching world.

Coal and Methanol
Coal / methanol economics also play an interesting role over the next few years. The current forecast for methanol price and MTO derivative pricing pushes the downstream integrated MTO assets into the red. With these assets being owned and operated by non-traditional Chinese petrochemical producers, this negative profit operating mode may not be acceptable and slowdowns/shutdowns could come into play. The lost ethylene production would have to come from some other source as demand is not eliminated, the fallback source of ethylene is the steam crackers. But, we already have the steam crackers pushing the limit, meaning a tighter ethylene market once again.

Co-Product Propylene
The largest yielding co-product from liquid crackers is propylene. This market is making a transition into more of an on-purpose production asset base from what was a co-product market with supply coming from: a) steam crackers (which make ethylene) and b) refinery fluid catalytic crackers (which were designed to make gasoline).

On purpose units such as: a) Propane dehydrogenation units (PDH) using low price propane, b) CTO/MTO and methanol to propylene (MTP), and c) metathesis units converting ethylene and butylenes to propylene, are growing to around 30 percent of the global propylene supply capacity. Propylene's pricing will have a high level of volatility relative to the ethylene price with all the new capacity and all the seasonality in propane pricing. The shift in propylene values impact the cash cost on steam crackers which could change dramatically month to month the advantage of running LPG versus naphtha and the operating mode of crackers.

Recycle Polyethylene (PE)

A very key impact on ethylene demand is the use of recycle PE. Over the last 10 years as absolute PE price increased, the level of recycle has increased from 0 to 5 MMTA, all in China. Two key issues are changing the recycle usage:

a) *pricing of prime/virgin PE resin coming down with energy, the consumers are shifting back to the prime material.*
b) *policy changes in China may impact its availability as existing recycle plants are required to move from community/residential areas to industrial zones which are more capable of handling truck traffic, water treatment, odors, etc. with Low margins on recycle operations today and little economic incentive to move these operations, recycle may have to convert to prime resin*

Key point being, prime resin needs ethylene, ethylene is already tight, adding 1-2% more demand to an already stressed supply will definitely cause price increases on ethylene.

Shift from Traditional ethylene based Chemistry

Two molecules are in the sights of Coal chemistry: a) the production of Vinyl Chloride monomer (VCM) used to make PVC, and b) mono-ethylene Glycol (MEG) to make polyester. MEG from coal continues in the scale up and commercializing mode but, continues slowly due to technical challenges, so at the moment not much of a threat, but something to be watched.

Coal based VCM production is China's main route to PVC, but this chemistry has issues of scale, cost and pollution. In a lower energy market, the cost competitiveness of the VCM from coal chemistry may shift to ethylene chemistry. This shift will have the effect of more ethylene demand. In an already tight ethylene market, the story continues, where do I get the ethylene from?

SUMMARY.

Crude, NG, and NGL's are positioned to be well supplied globally. This over supply will change pricing dynamics as NGL's are fatal products from production of other key materials. The pricing dynamics are critical as the one key outlet for NGL's is steam cracking. We have seen many forces pressing upon the ethylene balance. The message is clear that the steam cracker may be drawn upon more and more to close ethylene suuply/demand balances. Without some short term constraint removal ethylene prices are poised to move up strongly. For those in this space, on the radar should be cost/benefit analysis on feed switching in the market of NGL advantage. Longer term the world has 5.5-6.0 MMTA of ethylene demand growth meaning 6.0—6.6 MMTA of new ethylene production capacity. Is the world thinking along these lines?

#199b

Feed & Byproduct Flexibility for New and Existing Crackers

Carmen A. Fornarotto
Process Manager
Lummus Technology Inc., a CB&I Company

Stephen J. Stanley
VP – Base Olefins
Lummus Technology Inc., a CB&I Company

Prepared for Presentation at the 2016 Spring National Meeting
Houston, TX, April 14, 2016

Feed & Byproduct Flexibility for New and Existing Crackers

Carmen A. Fornarotto
Process Manager
Lummus Technology Inc., a CB&I Company

Stephen J. Stanley
VP – Base Olefins
Lummus Technology Inc., a CB&I Company

Abstract: The shale gas revolution has significantly increased the percentage of ethylene produced from ethane feed while reducing the amount produced from heavier feeds, particularly naphtha. However, with falling oil prices, naphtha feed prices have dropped and the operating margin of naphtha feed units has improved. In addition, in some regions the availability of ethane feed has dropped and the conversion of ethane feed plants to naphtha feed units is being studied. Many new plants are being designed for a range of feeds from ethane through naphtha and are identified as mixed feed crackers. To address these issues the paper will examine the design considerations of several cases including adding ethane flexibility to an existing liquid feed plant and naphtha flexibility to an existing ethane feed plant. In addition, adding propane and butane flexibility to an ethane cracker will also be examined. In addition to presenting material and energy balance information, the paper will also present the impact on various plant systems feed flexibility has and examine the modifications required to achieve various levels of feed flexibility. Finally, the paper will examine the impact on the byproducts slates the various feed combination has and will examine alternate technologies that can help meet the byproduct demand of existing downstream units in the complex.

Ethylene Plant Feedstocks

The availability of low cost ethane in the United States due to shale gas in recent years has significantly increased the amount of ethylene produced from ethane. This increase in ethylene production from ethane has seen the amount produced from heavier feeds reduced. In the past 10 years, the amount of ethylene produced from ethane in the U.S. has increased by almost 33%, while the amount produced from heavier feeds such as naphtha and LPG has decrease by 34%. With many new ethylene projects recently awarded in the U.S. utilizing ethane feed this trend will continue as these new ethane crackers are brought on-stream.

With readily available low cost ethane, U.S. ethylene plants have been shifting from liquid feedstock to ethane even though there are large supplies of low cost propane and butane also available. Ethylene manufacturing costs for North American LPG crackers were lower than for North American ethane crackers in 2015 as seen in Figure 1. Furthermore, naphtha feed prices have recently dropped, due to falling oil prices, which has improved the operating margin for naphtha feed plants.

Figure 1: Global Ethylene Production Cost and Supply

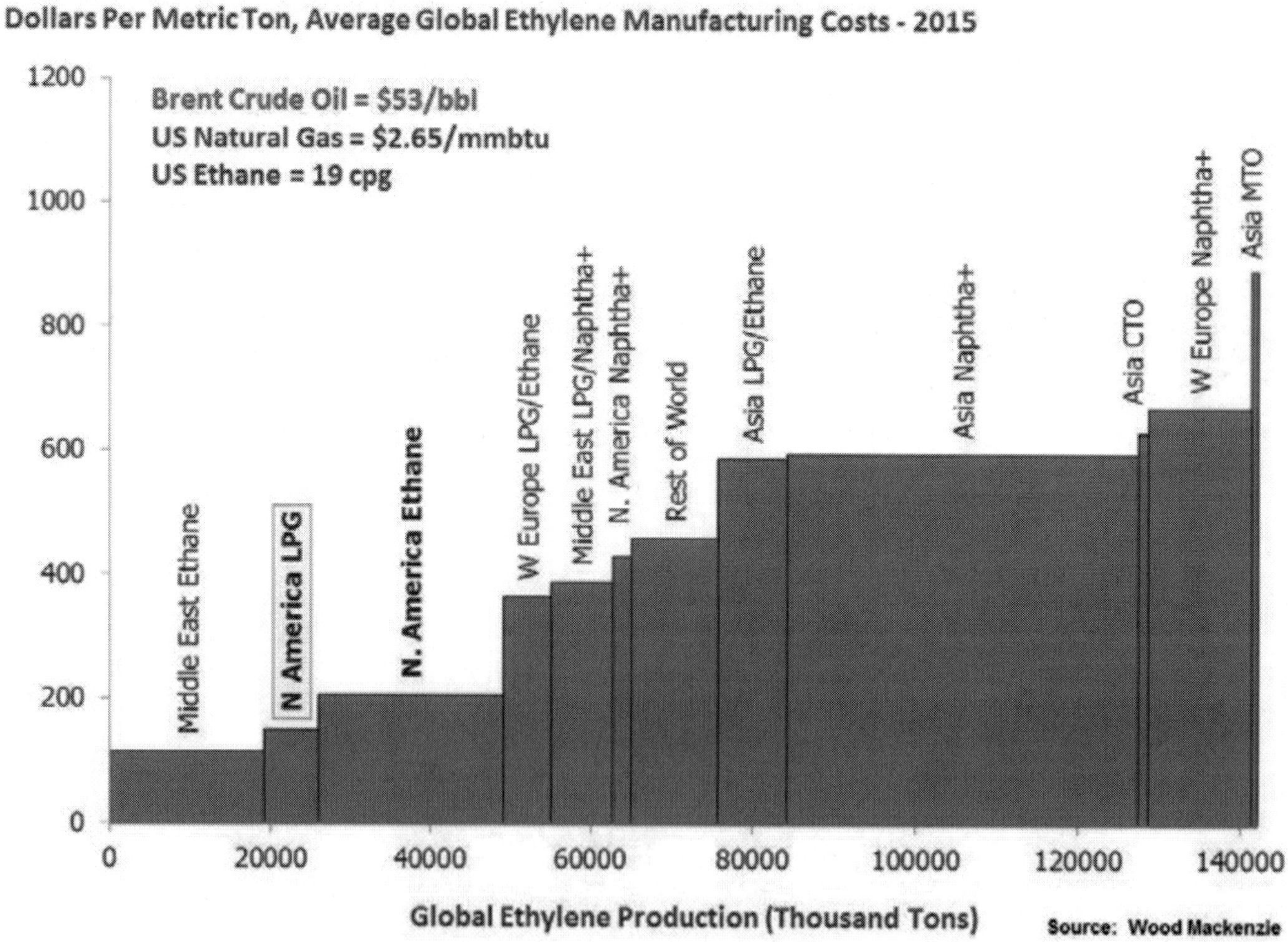

Table 1 shows typical material balances for 1,000 kta ethylene plants utilizing naphtha, n-butane, propane, and ethane as feedstock and compares the product margin for each feed using feed and product prices from January 2016. Table 1 illustrates that the product margin for naphtha feed plants would be higher than for an ethane feed plant using recent prices. Even with the very low price of ethane, the correspondingly low price of ethylene and the lack of byproducts produced have affected the product margin for ethane crackers. The byproducts produced in the n-butane and propane feed cases lead to higher overall product margins than for the ethane feed plant. Naphtha cracking, which produces the most byproducts, while showing higher product margins than ethane cracking is still behind propane and butane cracking because of the cheaper feedstock pricing for propane and butane.

Table 1: Product margins for various feeds using January 2016 USGC spot prices

	Naphtha	n-Butane	Propane	Ethane
FEEDS	**kta**	**kta**	**kta**	**kta**
Naphtha	3,331	0	0	0
Butane	0	2,360	0	0
Propane	0	0	2,220	0
Ethane	0	0	0	1,265
TOTAL	3,331	2,360	2,220	1,265
PRODUCTS	**kta**	**kta**	**kta**	**kta**
Hydrogen	28	25	35	64
Methane Rich Offgas	456	528	584	127
PG Ethylene	1,000	1,000	1,000	1,000
PG Propylene	617	392	311	18
Mixed C4's (w/o BD)	279	160	27	8
Potential Butadiene	164	96	93	23
Raw PyGas (w/o Bz)	592	102	97	17
Potential Benzene	127	50	63	7
PGO	23	2	4	0
PFO	45	4	7	1
TOTAL	3,331	2,360	2,220	1,265
Product Margin, MM$/yr	**526**	**574**	**604**	**508**

Figure 2 illustrates how feedstock prices have changed over the last five years. If feed and product pricing from other periods in time were used, the results would be very different than those shown in Table 1. In 2013 for example, ethane cracking showed the best product margins even with the lack of byproducts produced due to the low cost of ethane coupled with the high price for ethylene. Since feed and product prices are constantly changing, having a plant with feedstock flexibility is the best way to take advantage of unpredictable market fluctuations.

Figure 2: Ethylene plant feedstock prices over the last 5 years

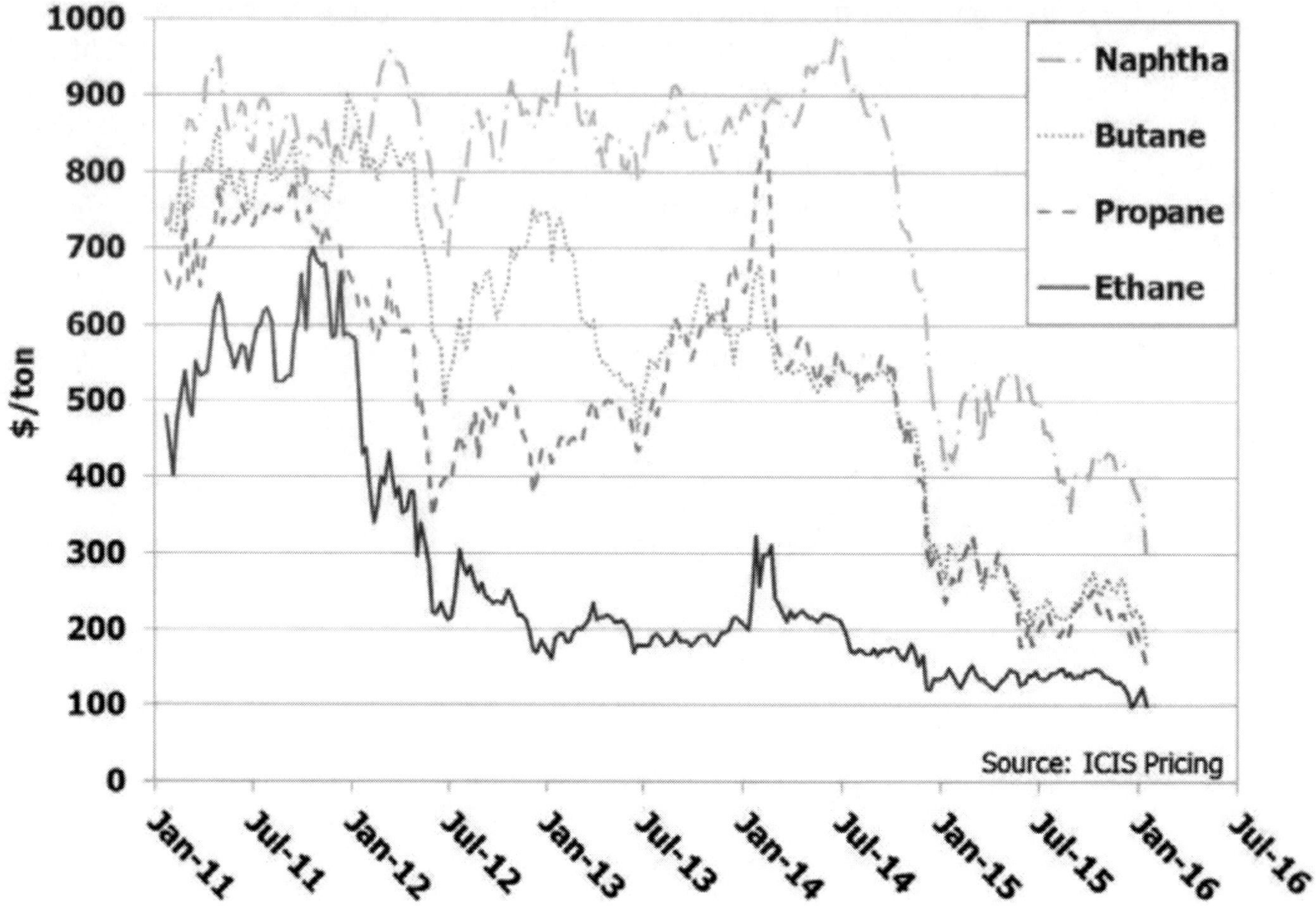

Feed Flexibility Case Studies

The remainder of this paper will discuss three separate case studies from actual plants that were looked at to add feed flexibility. The first case study will cover the addition of ethane flexibility to a liquid feed plant. The second case study will look at the opposite, adding naphtha flexibility to an ethane feed plant. Finally, the last case study will look at adding propane and butane flexibility to an ethane feed plant. The cases discussed in this paper are specific to the actual plants that were studied, and while there are some similarities between plants designed for the same feedstocks, the modifications required depend on the actual equipment installed and the level of flexibility that is to be added.

Adding Ethane Flexibility to a Liquid Feed Plant

Readily available low cost ethane has made it very attractive to add ethane flexibility to existing liquid feed plants. The modifications required to add ethane flexibility to a naphtha plant will be discussed based on CB&I's experience from a recent study. Table 1.1 shows the design case material balance for the 1190 kta naphtha cracker along with the material balance for adding ethane flexibility. The plant that was studied was expanded once from an original design capacity of 950 kta of ethylene and includes other units in order to produce a butadiene product, an isobutene/isobutane product, and a benzene product. As can be seen from the material balances, adding ethane flexibility reduces the total plant throughput and lowers the production of byproducts, notably propylene, butadiene, and benzene. However, even with the lower byproduct production, when using USGC average prices from 2015 as shown in Table 1.1, the case with added ethane flexibility shows a higher product margin than the 100% naphtha feed case due to the low cost of ethane.

Table 1.1: Overall material balances for adding ethane flexibility to a naphtha feed plant

	Naphtha	Naphtha Add Ethane
FEEDS	**kta**	**kta**
Naphtha	3,543	1,945
Ethane	0	682
External Feed(s)	46	46
TOTAL	3,589	2,672
PRODUCTS	**kta**	**kta**
Hydrogen	25	52
Methane Rich Offgas	584	388
PG Ethylene	1,190	1,190
PG Propylene	572	338
Butadiene	180	128
Mixed C4's	164	82
C5 Product	137	78
Benzene	245	140
C8-C9+ Product	303	174
PFO	189	103
TOTAL	3,589	2,672
RECYCLES	**kta**	**kta**
Ethane	224	476
Propane	27	19
C4's	29	14
C6/C7	117	70
Product Margin, MM$/yr	**370**	**621**

After expansion, the ethylene plant had 11 cracking heaters, with 1 heater designed to process the ethane/propane recycle and the remaining 10 heaters designed to process naphtha feed. One of the 10 naphtha heaters is also capable of processing the recycle ethane/propane during decoke of the recycle heater. For the feed flexibility case, 4 of the cracking heaters originally designed to crack naphtha require modifications in order to crack the fresh ethane and the increased ethane recycle. In addition to the heater modifications, new secondary TLEs will be added to the modified heaters, a new ethane feed vaporizer is added for the fresh ethane feed, and a new ethane preheater is added for the fresh ethane and additional recycle ethane.

The lower steam to oil ratio for ethane cracking leads to an overall reduction in the dilution steam requirement for the ethylene plant and therefore the flexibility case does not require any modifications to the dilution steam system. The cracking of ethane also reduces the total heater effluent and lowers the fuel oil production, resulting in lower loads in the gasoline fractionator, PGO stripper, and PFO stripper, along with less duty required to be removed via the quench oil system. No modifications are necessary for these systems as a result of adding ethane flexibility. Similarly, no modifications are required for the quench tower as the loads are also reduced, and with the lower dilution steam requirements the existing quench water settler and process water stripper do not require any modifications.

Due to the lower molecular weight of the charge gas resulting from the added ethane flexibility, the loads in the charge gas compressor increase compared to the expansion design case. Since the charge gas compressor was already revamped for the expansion, it was not possible to modify the compressor any further in order to meet the process requirements for the ethane flexibility case. Therefore, a booster compressor was added to increase the 1st stage suction pressure and allow the existing charge gas compressor to be reused. The booster compressor was the preferred solution as the alternative options were a footprint replacement of the entire compressor or to operate the compressor at a higher speed with essentially no margin on the rated turbine speed and a low margin on the rated turbine power. No additional changes are required for other equipment in the charge gas compressor area (aftercoolers, suction drums, gasoline stripper, caustic tower, condensate stripper, and charge gas dryers) as the process requirements for the ethane flexibility case fall within the design limits of the equipment with sufficient margin.

For the chilling train and demethanizer, due to the radical change in the hydrogen to methane ratio in the charge gas resulting from the added ethane flexibility, a small parallel cold box and hydrogen methane separator drums are added and some of the internals in the demethanizer are replaced. In addition, the methane refrigerant compressor is revamped with new internals and a new motor. Parallel exchangers are added to handle the increased ethane recycle that gets vaporized and superheated in the chilling train. No changes to the methanator and hydrogen dryer systems are required as the excess hydrogen that is produced when adding ethane flexibility is exported to OSBL upstream of these systems.

The loads in the deethanizer increase as a result of the added ethane flexibility, but the high capacity trays that were installed during the expansion have sufficient margin to handle the flexibility case and no changes are required. The deethanizer reflux drum is however modified with a high capacity mist eliminator, the reflux pump impeller is replaced, and the parallel reboiler added for the expansion is replaced with a larger exchanger. For the acetylene converter, the feed heater is replaced with a larger exchanger, additional catalyst is added to the reactor beds, and the mist eliminator in the green oil KO drum is replaced. Due to the increased ethane recycle, the feed purity to the ethylene fractionator is reduced. In order to handle the resulting increased loads in the ethylene fractionator, an alternate feed location is utilized, the high capacity trays that were added for the expansion are replaced with even higher capacity trays, and parallel exchangers are added for the condenser and both side reboilers.

As shown in Table 1.1, the products and recycle streams from the back-end of the plant are all reduced for the ethane flexibility case. The propylene product goes down to approximately 60% and the propane recycle is reduced to 70% of the expansion rates. All of the C4 and C5 products and recycles are also reduced from the expansion rates. These lower product and recycle rates lead to reduced loads in the depropanizer, MAPD converter, propylene fractionator, and debutanizer systems. These systems will operate at turndown conditions and therefore no modifications are required.

For the expansion, the ethylene and propylene refrigerant compressors were not modified and a new parallel enhanced binary refrigeration system was added to handle the expansion refrigeration requirements. For the ethane flexibility case, with the added duty required for the ethylene fractionator condenser the loads in the propylene refrigerant compressor increase by 5-10% for each stage. Since the propylene refrigerant compressor was not previously modified for the expansion, the increased loads for the feed flexibility case can be met by revamping the internals of the compressor. The loads in the ethylene refrigerant system and the enhanced binary refrigerant system do not increase from the expansion case and no modifications are required for either system.

The modifications required for the ethylene plant expansion were much more intensive than those required for adding ethane feed flexibility. The results of this study show that ethane feed flexibility could have been added when the plant was expanded for a relatively minor incremental cost to the expansion cost. The major modification for feed flexibility are modifying 4 cracking heaters to process ethane, adding a booster compressor for the charge gas compressor, adding a small parallel cold box to handle the increased hydrogen production, revamping the methane and propylene refrigerant compressors, and retraying the ethylene fractionator.

The modifications for adding ethane feed flexibility largely overlap with other modifications done for the plant expansion. For example, instead of modifying 4 heaters to process ethane for feed flexibility, the 2 new heaters added for the expansion could have been designed for both naphtha and ethane feed, then only 2 additional heaters would require modification for ethane feed. Similarly, the charge gas compressor and turbine were revamped for the expansion; if ethane flexibility had been considered at the time, instead of just revamping the turbine, it could have replaced to ensure adequate margins were available to handle the flexibility case and a booster compressor would not have been required. The new small cold box and revamp of the methane and propylene compressors could have been avoided by designing the enhanced binary refrigeration system that was added for the expansion with ethane feed flexibility as an alternate design case. Finally, instead of retraying the ethylene fractionator twice, once for the expansion and a second time for feed flexibility, the tower could have been modified the first time to ensure operation for both cases.

Adding Naphtha Flexibility to an Ethane Feed Plant

In some regions of the world the availability of ethane feed has dropped and the conversion of ethane feed plants to naphtha feed plants is being studied. CB&I recently completed a study that looked at converting an ethane feed plant to 100% naphtha feed while maintaining the ability to crack ethane in the future if ethane supply were to become available again. After the study was completed an alternate case with less naphtha flexibility was looked at to determine the optimum revamp approach. Ultimately, because of the future unavailability of ethane feedstock, the revamp viability was judged in comparison to a new grassroots plant to crack naphtha.

The plant that was studied for this effort was previously expanded twice. The plant was originally designed to produce 335 kta of ethylene and after the second expansion the ethylene capacity was increased to 461 kta. The second expansion was based on primarily ethane (85 wt%) with the remainder C3/C4 LPG as the fresh feeds to the cracking heaters. The goal of the study was to determine the impact of converting the ethylene plant to 100% light naphtha feed while maintaining the same ethylene production from the second expansion without ethane and propane recycle. Table 2.1 shows the overall material balances for the second expansion along with the balances for 100% naphtha and 100% ethane that were used as the basis for the study and the alternate case with less naphtha flexibility.

Table 2.1: Overall material balances for adding naphtha flexibility to an ethane feed plant

	2nd Expansion Design	Case 1 100% Naphtha	Case 2 100% Ethane	Alternate Case 50% Ethane with LPG
FEEDS, KTA				
Ethane	586		610	290
LPG	111			111
Light Naphtha		1532		612
External Feed(s)	17	50	17	50
TOTAL	**714**	**1582**	**627**	**1063**
PRODUCTS, KTA				
Fuel Gas	142	289	105	214
PG Ethylene	461	461	461	461
PG Propylene	34	244	15	139
Ethane Product		76		
Propane Product		28		
Mixed C4 Product	36	155	22	98
C5+ Product	41		24	
C6-C8 Product		220		106
C9+ Product		34		15
Pyrolysis Fuel Oil		76		29
TOTAL	**714**	**1582**	**627**	**1063**
RECYCLES, KTA				
Ethane	212		271	146
Propane	10		5	28
C5's		53		26

The material balances show that the feed to the plant goes up considerably when cracking naphtha compared to ethane and LPG feeds due to the much higher amounts of by-products formed when cracking naphtha. Some major modifications are required in order to convert the plant to 100% naphtha feed and enable recovery all of the by-products produced. A summary of the major modification required are discussed below. Figure 2.1 shows a block flow diagram illustrating the modifications required; blue boxes represent sections of the plant that do not require any modifications, yellow boxes represent modified systems, green boxes represent new systems, and the orange box represents a system that is partially new and partially modified.

Figure 2.1: Block flow diagram for adding naphtha flexibility to ethane feed plant

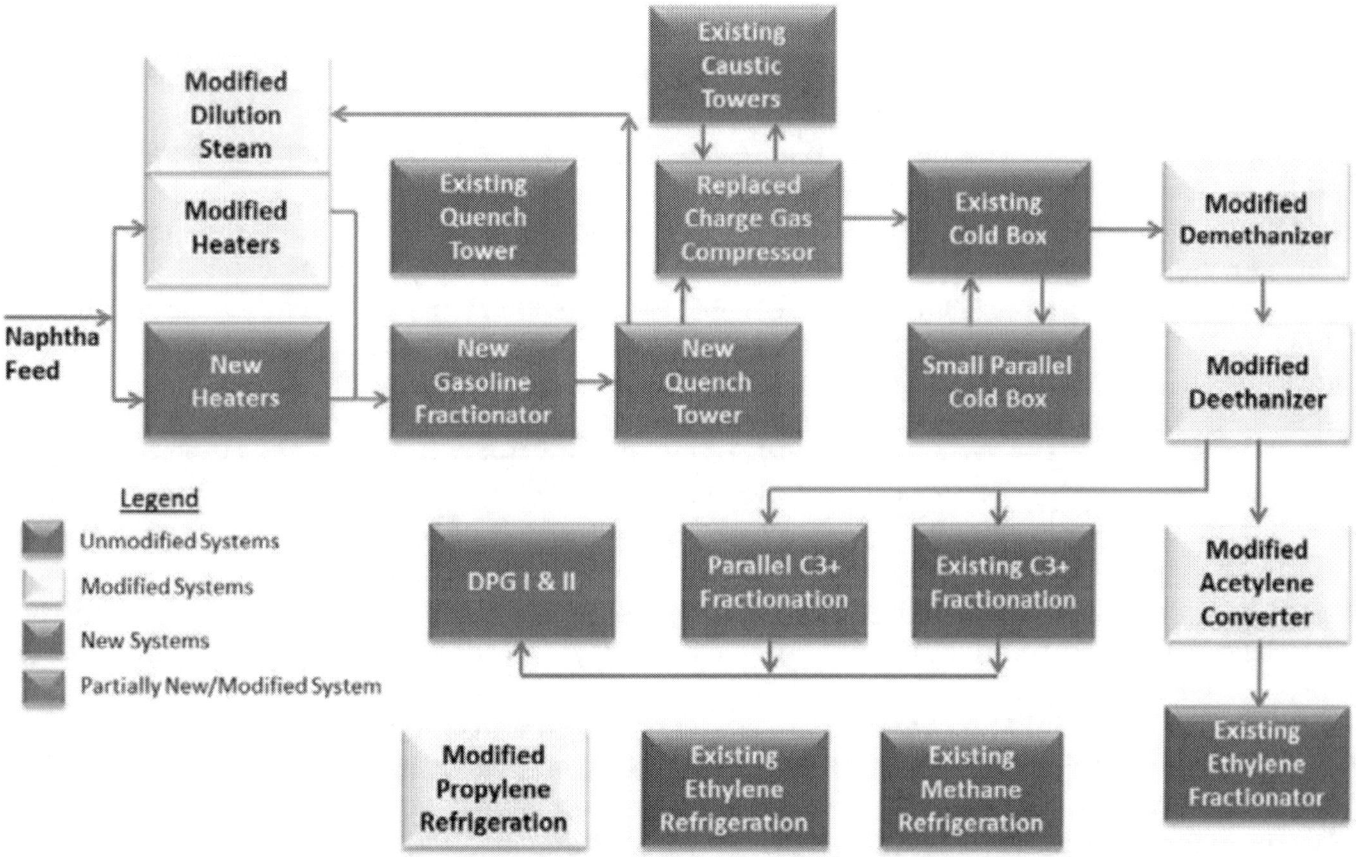

The existing cracking heaters will be revamped and two new heaters, identical to the revamped heaters, will be added to achieve the plant capacity required while cracking the light naphtha feedstock. The existing radiant coils are replaced with more selective radiant coils which are designed to fit in the existing firebox with the same end-to-end dimension as the existing radiant coils. In addition, the primary TLEs, burners, and ID fans will need to be replaced in the existing heaters. The secondary TLEs in the existing heaters are taken out of service and bypassed during liquid cracking. The convection section of the existing heaters is completely replaced with new coil services that fit within the existing structure to preserve the same steam drum platform and drum location. The new convection sections are designed to enable both naphtha and ethane cracking by designing convection tubes that are only required for naphtha operation to run dry for extended periods (years) when the plant is switched to ethane feed.

Cracking of naphtha feedstock requires the addition of a gasoline fractionator with quench oil system and a pyrolysis fuel oil stripper to cool the cracked gases and separate out the pyrolysis fuel oil product. The existing quench tower cannot be reused as it would only be able to handle approximately 60% of the new loads, therefore a new larger quench tower is added due to the change in feedstock to the plant and the increase in dilution steam required. A process water stripper will also be added to remove any acid gases and volatile hydrocarbons from the dilution steam that is condensed in the new quench tower. For the dilution steam drum, the existing mist eliminator is replaced with a high capacity mist eliminator and additional dilution steam generators utilizing quench oil and MP steam are added to in order to generate the additional dilution steam required.

The existing charge gas compressor will need to be replaced on the existing footprint in order to compress the quench tower overhead to the required pressure. The charge gas compressor drums require various levels of modifications and/or replacement depending on how much the vapor loads and/or amount of condensed liquid in each drum increases. Most of the exchangers in the charge gas compressor area will need replacement or parallel shells added to supply the additional duty required. Because of the heavier feedstock and the additional gasoline produced, a gasoline stripper is added to remove light materials from the net heavy gasoline condensed in the quench tower and from the medium gasoline condensed in the charge gas compression section. The existing condensate stripper can handle only about 35% of the new loads; therefore, a new condensate stripper must be added. The benzene wash tower, which is located upstream of the charge gas dryers, requires new high capacity internals. The existing caustic towers and charge gas dryers do not require any modifications.

After the charge gas compression area, the addition of a new small cold box and hydrogen methane separator drum are required to handle the increased amount of methane rich fuel gas and to purify the hydrogen feed to the existing PSA units. For the demethanizer, the top section of the tower needs to be replaced with a larger diameter to handle the higher loads resulting from the increased amounts of methane from naphtha cracking. Due to the increase in loads in the deethanizer, the existing trays will be replaced with higher capacity trays. A parallel condenser and side reboiler are also required for the deethanizer. For the acetylene converter, due to the increase in acetylene content in the feed to the reactor, the existing system will be converted to have three beds in series with intercooling between the beds; this involves modifying the existing reactors and adding one additional reactor vessel. Since the feed to the ethylene fractionator is richer in ethylene for naphtha cracking compared to ethane cracking, no modifications are necessary for the ethylene fractionator.

The increase in C3 and heavier components that result from cracking naphtha feed requires the addition of parallel C3 and C4 processing equipment (depropanizers, MAPD converter, propylene fractionators, and debutanizer). Since it would be necessary to install new tower systems even after modifying the existing C3 system, the existing systems will not be modified and the new parallel C3 system will process anything the existing unmodified C3 system cannot handle.

The existing propylene refrigeration compressor will need to be modified within the existing casings in order to provide the refrigeration requirements when converting the plant to naphtha feed. Additional refrigeration users from new parallel systems will be added to the propylene refrigeration system as required. The existing ethylene refrigeration system does not require any modifications to handle the loads for naphtha cracking operation.

A new two-stage pyrolysis gasoline hydrogenation unit (DPG-1 and DPG-2) is added as part of the study. C5's are separated out after DPG-1 and recycled to the cracking heaters. DPG-2 produces a C6-C8 product suitable for aromatics extraction and a C9-180°C product.

Figure 2.2: Cost breakdown for adding naphtha flexibility to an ethane feed plant

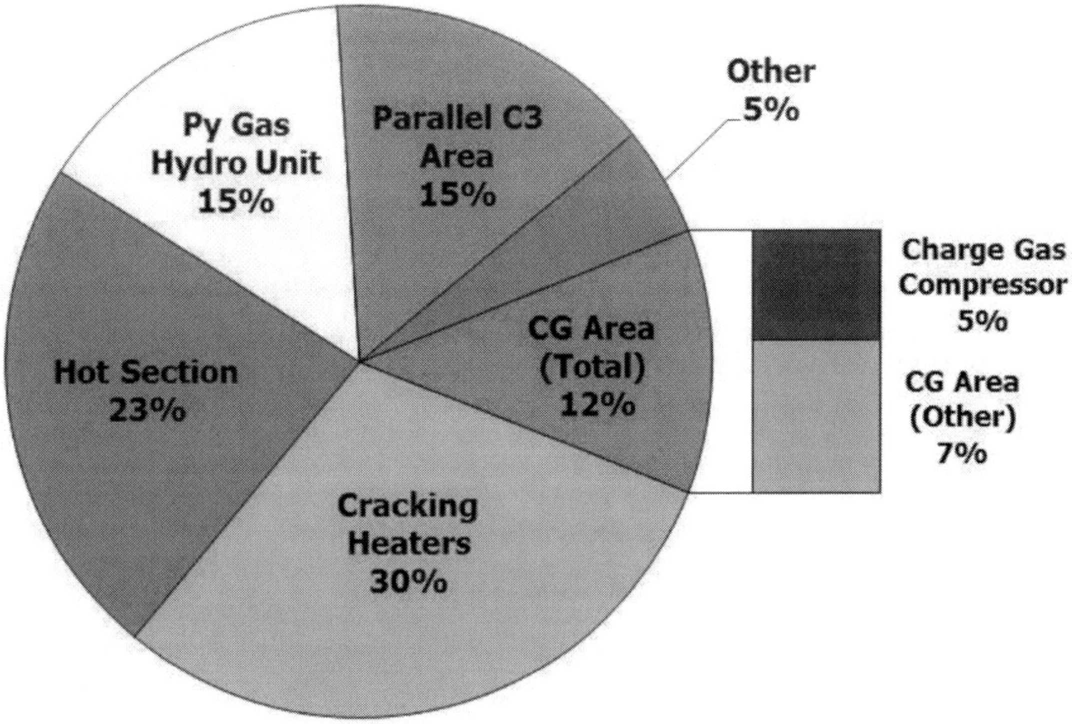

Figure 2.2 shows the breakdown of costs for the revamp for each area of the plant. The largest contributor to the cost is the new and modified cracking heaters, which make up 30% of the total cost for the revamp. The new hot section towers (gasoline fractionator, quench tower, pyrolysis fuel oil stripper, and process water stripper) and associated equipment make up the next largest portion of the revamp cost at about 23% of the total. The two-stage pyrolysis gasoline hydrogenation unit and the parallel C3 system each make up approximately 15% of the total revamp cost. The modifications to the charge gas compressor area make up 12% of the total cost, with the replacement charge gas compressor 5% by itself, and the other modifications and new equipment required making up remainder of the charge gas area cost. All of the other modifications (demethanizer, deethanizer, acetylene converter, etc.) make up the final 5% of the revamp cost.

Compared to a new grassroots naphtha cracker and looking at the ISBL costs, the revamp to 100% naphtha feed of the existing ethane feed plant came out to be 53% of the total installed cost (TIC) of the grassroots plant. The ISBL TIC for both the revamp and the grassroots plant were developed on a similar basis and included estimates for working in a tight plot space for both and working in an operating plant for the revamp.

An alternate revamp case with less naphtha feed flexibility was also examined. The alternate case considered 50% of the ethane feed and the same C3/C4 LPG feed rate as the second expansion. The material balance for the alternate case is presented in Table 2.1. This case was examined as a way to avoid modifications to the existing cracking heaters which would have been done while the plant was operating in order to avoid prolonged downtime. For this option, instead of adding two new cracking heaters and modifying all of the existing heaters, four new heaters would be added to crack the naphtha feed with the remaining unmodified heaters used to crack the fresh ethane and LPG along with the recycle ethane and propane. For the heater cost alone, the alternate case results in TIC savings of 23% compared to the 100% naphtha conversion heater costs or 7% of the total revamp cost.

As shown in the material balance, propylene production is significantly reduced for the alternate revamp case. This reduction eliminates the need for the new parallel C3 system as the propylene produced can be processed in the existing equipment with some modifications, saving 12% of the revamp TIC. The feed to the pyrolysis gasoline hydrogenation unit is reduced by about 50% for the alternate case, resulting in a much smaller unit, which saves 5% of the revamp TIC. Finally due to the reduction in the total throughput for the plant, other new systems are reduced in size and modifications to existing systems are less severe, which results in a reduction of approximately 2% of the revamp TIC. Overall, the alternate scenario reduces the TIC of the revamp by 26%, making the alternate revamp case TIC approximately 40% of the cost of a new grassroots naphtha cracker.

Adding Propane and Butane Flexibility to an Ethane Feed Plant

The final feed flexibility case study this paper will cover is for a study that was done in parallel to the basic engineering for a grassroots 1000 kta ethylene plant based on 100% ethane feed. The goal of the study was to determine the maximum amount of propane and butane that could be processed in the ethane feed plant with minimal changes to the design. Since this study was conducted in parallel to the basic engineering of the ethane feed plant, equipment evaluation was based on preliminary process equipment datasheets and proposal equipment data. A block flow diagram showing the ethane feed plant configuration that was examined for this study is shown in Figure 3.1.

Figure 3.1: Block flow diagram for ethane feed plant configuration

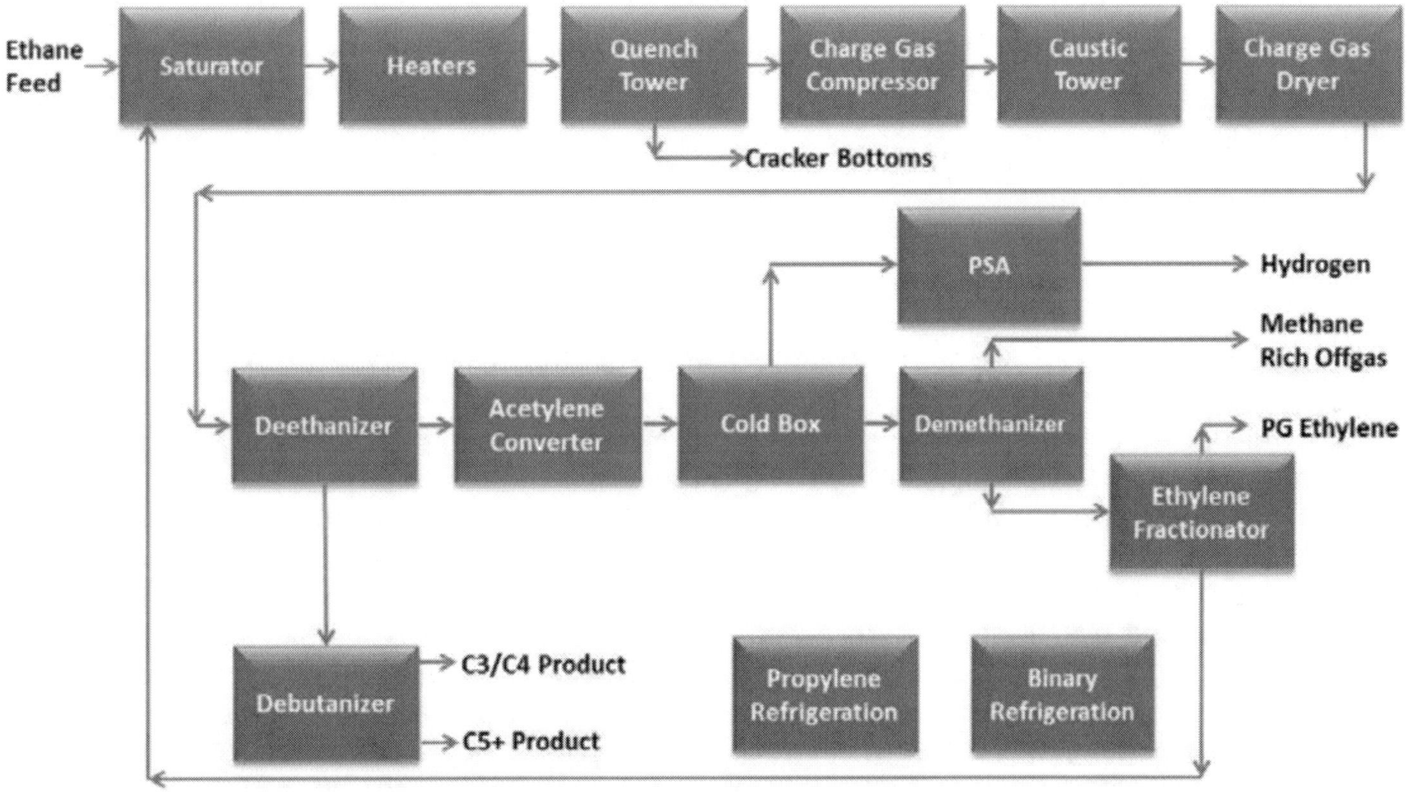

The study evaluated the maximum amount of propane that could be processed without modification to the refrigeration compressors, this was determined to be 15 wt% of the fresh feed. Table 3.1 shows the overall material balances for the 100% ethane feed case along with the propane and butane flexibility cases. In order to co-crack propane in the cracking heaters with ethane, a new propane vaporizer system has to be added and the ethane feed saturator system needs to be modified to provide the additional heat required to obtain the desired steam to oil ratio for the ethane-propane mixture. Since propane is co-cracked with ethane no modifications to the cracking heaters are required for adding propane flexibility. In addition, with the amount of propane flexibility limited to 15 wt%, no modifications are required for the quench tower and quench water system, charge gas compression system, acid gas removal system, spent caustic system, or the charge gas dryers.

Table 3.1: Overall material balances for adding propane and butane flexibility to an ethane feed plant

	100% Ethane	85 wt% Ethane 15 wt% Propane	80 wt% Ethane 20 wt% Butane
FEEDS, KTA			
Ethane	1307	1198	1160
Propane		211	
Butane			289
TOTAL	**1307**	**1409**	**1449**
PRODUCTS, KTA			
Hydrogen Product	66	61	56
Methane Rich Offgas	141	188	195
Polymer Grade Ethylene	1000	1000	1000
Refinery Grade Propylene			78
C3/C4 Product	66	110	
C4 Product			65
C5+ Product	32	45	50
Cracker Bottoms	2	4	5
TOTAL	**1307**	**1409**	**1449**

For the deethanizer, in order to limit the impact to the refrigeration system, additional trays are added to the stripping section of the tower in order to maintain the same tower bottoms specification as the 100% ethane cracking case. The flow and acetylene concentration in the deethanizer overhead both go up slightly for the propane flexibility case, but the increases are minor enough that no modifications are required for the acetylene converter system. The additional methane content and the resulting increased duty required for the deethanizer overhead condenser, demethanizer feed chilling, and demethanizer overhead condenser results in a bottleneck for the binary refrigeration system. To mitigate this bottleneck, the ethylene losses in the raw hydrogen offgas stream are increased and in order to maintain the overall ethylene recovery of the plant, the ethylene in the ethane recycle is correspondingly reduced. Since the ethylene fractionator is relatively under loaded for the propane flexibility case, this reduction of ethylene in the ethane recycle is possible without requiring any changes to the ethylene fractionator column design. The flows and duties in the cold box show some significant variations compared to the 100% ethane feed case, therefore a new design case should also be evaluated by the cold box vendor to ensure the cold box can operate with the require propane feed flexibility. The cost impact for adding additional surface area to the required sections of the cold box in order to cover the propane flexibility case is minimal relative to the cost of the entire cold box.

The PSA capacity is limited by the purge gas compressor and with the increase in methane the required purge gas rate goes up. Modifications to the PSA unit are avoided by bypassing a portion of the PSA feed to fuel. The loads in the demethanizer are significantly higher for the propane flexibility case, but since the tower has a wide range of operability, the tower size for the 100% ethane case is adequate. The cost impact for changing distributors and internals, along with any changes in packing type are minimal. The feed purity to the ethylene fractionator is higher for the propane flexibility case due to the decrease in ethane recycle. The higher feed purity results in lower loads in the column along with lower condensing and reboiling duties. As previously mentioned, some of this extra capacity in the ethylene fractionator design is used up to help unload the binary refrigeration system. The debutanizer system will need to be upgraded to handle the increased amounts of C3 and heavier material resulting from the additional propane cracking and essentially all of the equipment in the debutanizer system is made larger.

The propane flexibility case was evaluated with the maximum amount of propane that could be processed without modification to the refrigeration compressors. With the changes noted above no changes to the propylene and binary refrigeration systems are required for processing 15 wt% propane feed. With this amount of propane, the propylene refrigerant compressor is at the limit in the 3rd and 4th stage while the binary refrigerant compressor still has some margin.

The basis for evaluating butane flexibility was to use the maximum amount of butane that could be processed to fill up one heater. Since, unlike propane, butane is not co-cracked with ethane, the butane flexibility had to be evaluated in discreet amounts instead of determining the limit without modification to the refrigeration compressors. The modifications required for one heater on butane, were determined for all the equipment from the 100% ethane feed case. In addition, the equipment required to produce refinery-grade (RG) propylene were identified. As shown in Table 3.2, fresh butane equal to 20 wt% of the total fresh feed can be processed in one heater.

In order to crack butane, two heaters need to be modified so that when the heater normally cracking butane is being decoked, butane can be cracked in another heater. Modifications to the heaters include a new lower feed preheat coil in the convection section, new feed piping and block valves, and new dilution steam control valves. In order to maintain the total ethylene production, the ethane feed rate to the other heaters increases by 4%. All of the heaters therefore require upgrades to the ID fans, burners, SCR catalyst, and convection section tube supports. The dilution steam system is designed for recycle C3/C4 cracking with two heaters on high steam standby (HSS) for the 100% ethane feed case and is therefore adequate for one heater on butane with only one heater on HSS. Similar to the propane flexibility case, a new butane vaporizer system is added for butane flexibility. Since the ethane feed is reduced and dilution steam for the butane heater is added via the dilution steam system, no modifications to the saturator are required for butane flexibility.

A new external quench water settler is added for butane flexibility. The bottom of the quench tower has insufficient hold-up to meet the required residence times to ensure proper separation of the hydrocarbon, process water, and quench water. In addition, to mitigate fouling when cracking butane, debutanizer bottoms will be mixed with the feed to the settler and provisions for wash gasoline injection to the top of the quench tower will be added. The new settler is designed to allow operation with up to 20 wt% butane, above this a gasoline fractionator would be required as the fouling and emulsion issues would be too great to handle in a settler alone. For the other systems upstream of the deethanizer (charge gas compression system, acid gas removal system, spent caustic system, and the charge gas dryers) no modifications are required for the additional butane flexibility.

For the butane flexibility case, the deethanizer bottoms is sent to the new depropanizer and MAPD converter system to produce refinery grade propylene. This allows the deethanizer bottom specification to be relaxed in comparison to the 100% ethane feed case while still meeting the refinery grade propylene product specification. The deethanizer column therefore does not require any modifications for the added butane flexibility. The capacity of the deethanizer reflux pump will be increased to limit the propylene in the overhead. Similar to the propane flexibility case, no modifications are required for the acetylene converter system as the increases in flow and acetylene concentration are minor for the butane flexibility case.

Also similar to the propane flexibility case, the duty required for the deethanizer overhead condenser, demethanizer feed chilling, and demethanizer overhead condenser increase for the butane flexibility case. The same approach is taken for the butane flexibility case, where the ethylene losses in the raw hydrogen offgas stream are increased while the ethylene in the ethane recycle is correspondingly reduced to maintain the overall ethylene recovery of the plant while minimizing the impact to the binary refrigeration system. A new design case is also required for the cold box to handle the variations in flows and duties. The additional design case has minimal cost impact relative to the cost of the entire cold box.

The added butane flexibility results in similar impacts to the PSA unit, demethanizer, and ethylene fractionator as the propane flexibility case. Modifications to the PSA unit are avoided by bypassing a portion of the PSA feed to fuel. The loads in the demethanizer are significantly higher but can be processed in the tower with some changes to the tower internals, which have minimal cost impact. For the butane flexibility case, the demethanizer reflux pump capacity will be increased to supply the additional reflux required. No changes are required to the ethylene fractionator design as the feed purity to the tower is increased for the butane flexibility case.

A depropanizer and MAPD converter system are added for the butane flexibility case. The deethanizer bottoms is sent to the depropanizer which utilizes -6°C propylene refrigerant in the condenser and desuperheated LP steam in the reboiler. The net overhead from the depropanizer is sent to the MAPD converter system to selectively hydrogenate the MAPD and produce the refinery grade propylene product. Vent gas from the MAPD effluent separator is sent back to the charge gas compressor, but does not have any impact to the charge gas compression system. The bottoms from the depropanizer is sent to the debutanizer. For butane flexibility, the debutanizer needs to be significantly larger than for the 100% ethane feed case and essentially all of the equipment in the debutanizer system is changed.

The addition of the depropanizer condenser significantly increases the required capacity of the 3rd and 4th stages of the propylene refrigerant compressor and therefore requires the compressor and turbine to be modified from the 100% ethane feed case design. In addition to the new user, the 3rd stage suction drum requires a mist eliminator to handle the increased vapor flow. For the binary refrigeration compressor, while shifting some of the ethylene losses from the ethane recycle to the raw hydrogen offgas helped mitigate the increase, it was not sufficient to keep all stages of the compressor within the maximum capacity of the 100% ethane feed case. The binary refrigerant compressor therefore needs to be modified to handle the increased capacity in the 1st stage for the butane flexibility case.

The amount of propane flexibility can be increased if the modifications identified for adding butane flexibility are considered for propane flexibility. Propane flexibility can be increased to 25 wt% of propane by making the modifications required for the butane flexibility case and in addition making modifications to the feed saturator as required for propane flexibility. No modifications to the deethanizer are required for the higher amount of propane flexibility because of the depropanizer that was added for butane flexibility which allows the bottom specification of the deethanizer to be relaxed.

The cost of adding propane and butane flexibility are minor in relation to the cost of the entire ethylene plant. Adding propane flexibility equal to 15 wt% of the fresh feed and making the modifications described above for the propane flexibility case increases the cost of the ethylene plant by approximately 1%. For adding 20 wt% butane flexibility with the ability to also have 25 wt% propane flexibility the cost of the ethylene plant increases by approximately 5%. Adding feed flexibility before the plant is built for a relatively small cost allows the use of alternate feedstocks in the future in order to maximize product margin depending on feed and product prices.

AIChE Paper Number

Cracker Feedstock Flexibility from Naphtha to Ethane: Design Criteria and Concepts for Quench Water and Quench Oil System

Thomas Emmert

Manager Process Engineering Petrochemical Plants

Linde Engineering North America

Sam Antonas

Olefins Lead Process Engineer & Process Technologist

Qenos Pty. Ltd

Gunther Kracker

General Manager Process Engineering Petrochemical Plants

The Linde Group, Engineering Division

Prepared for Presentation at the 2016 Spring National Meeting

Houston, Texas, April 10-14, 2016

Cracker Feedstock Flexibility from Naphtha to Ethane: Design Criteria and Concepts for Quench Water and Quench Oil System

Thomas Emmert
Manager Process Engineering Petrochemical Plants
Linde Engineering North America
Sam Antonas
Olefins Lead Process Engineer & Process Technologist
Qenos Pty. Ltd
Gunther Kracker
General Manager Process Engineering Petrochemical Plants
The Linde Group, Engineering Division

Abstract: Driven by the US shale gas boom and associated ethane supply and pricing dynamics, cracking ethane has become much more attractive than liquid feedstocks in the last years. Owners of liquid crackers have correspondingly minimized liquid feedstocks or even fully replaced them with gaseous feedstocks. However, the recent decline of global crude oil prices has shown how unpredictable the market is. The key to success, therefore, is to maximize cracker feedstock flexibility.

Nearly twenty years ago, in 1996, an expansion and ethane conversion project for the Qenos Naphtha/LPG Cracker was executed by Linde Engineering. The plant originally was designed for 50% Naphtha and 50% LPG Cracking in 1983. After the conversion project, the cracker was then able to crack the full range of feedstocks: from Naphtha/LPG to 100% ethane.

This paper briefly describes challenges in process engineering when converting an olefin cracker from liquid to gaseous feedstocks. Process unit concepts and associated modifications are explained in more detail, particularly for the hot section containing the quench oil and quench water systems.

In order to stably operate the quench oil system given the lack of liquid cracking products when processing gaseous feedstocks, a hydrocarbon stream must be imported with similar boiling range and aromaticity. This concept has been optimized in the past two decades to the maximum extent possible. In the same way, prediction methods and design criteria have been developed further and validated based on long term operational experience, as this paper will elaborate.

1. Introduction

Converting an ethylene plant from liquid to gaseous feedstock cracking presents challenges to the operation of the hot section of the plant. As a basic rule, lighter (gaseous) feedstocks will yield lighter products while heavier (liquid) feedstocks tend to yield a greater percentage of heavies, particularly C5's and above. A typical cracking product distribution for both feedstock cases is shown in Figure 1.

This difference in product ratios of gas cracking to that of the original liquid cracking case has a profound impact on the plant operation. As a result, several downstream process units and equipment must be modified in order to ensure optimum and reliable performance. Although this paper will focus primarily on the modification to the hot section containing the quench oil, quench water and cracked gas compression system, several other sections, including the precooling and drying section, C2/C3 separation, and C2 splitter must be modified to perform optimally.

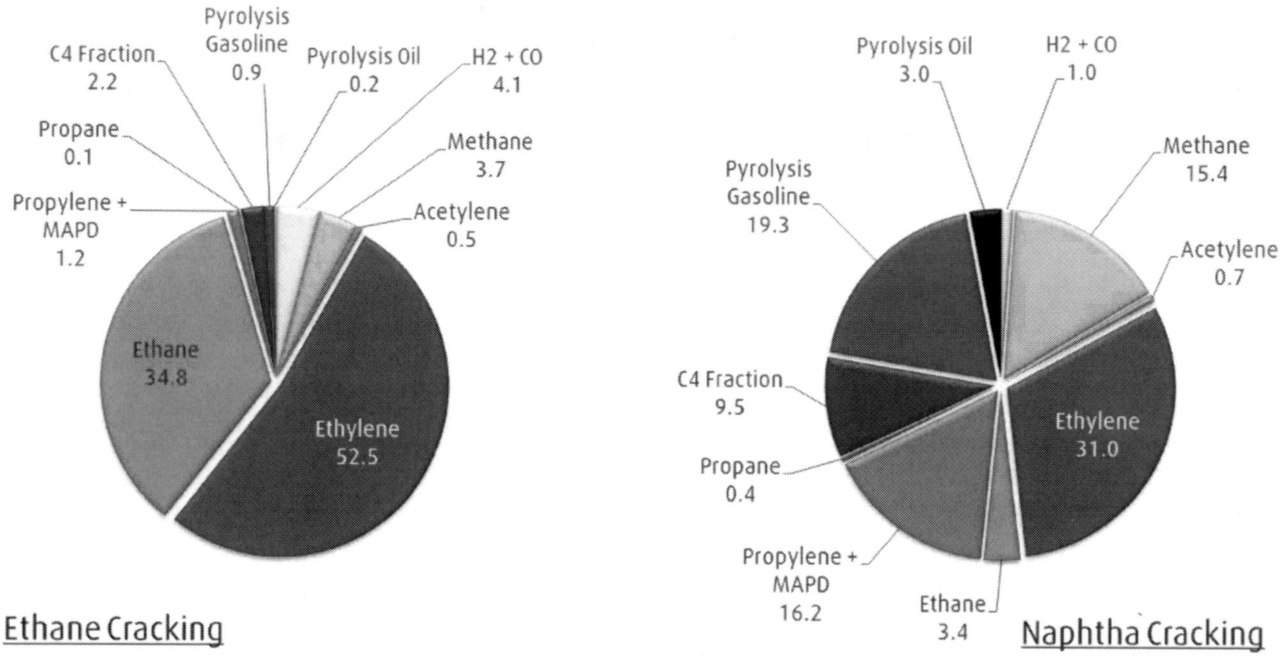

Figure 1: Once through furnace yields [wt%].

When cracking liquid feedstock, a quench oil column, followed by a quench water column, is used to cool the cracked gas to almost ambient conditions. However, utilizing gaseous feedstock, insufficient heavy oil is available in the cracked products to stably operate the quench oil system.

There are two strategies for hot section revamp when changing from liquid to gas cracking.

The first option is to convert the quench oil column to a quench water column. Depending on the existing topology and quench oil column height, various concepts are possible [1]. Following the Linde quench water tower design philosophy, the quench oil column should contain a separate saturator section, where the cracked gas is cooled and saturated by a quench water pump around. Heavy hydrocarbons and coke need to be removed from the column bottom. If the height of the existing column allows, a second separate quench water cycle can be installed to cool the cracked gas down to an intermediate temperature. This keeps heavy hydrocarbons in the warm zone of the column and prevents fouling in the colder downstream quench water system. This scenario presents its own unique design and construction challenges, particularly within an already confined or congested plant area. Additionally, it removes the flexibility to revert back to liquid feedstock. However the operational costs are expected to be low for this concept.

2) The second option is to import hydrocarbon streams with appropriate boiling point curve and aromaticity to compensate the lack of fuel oil make. Potential streams are available when the petrochemical complex is integrated with a refinery. Such a solution preserves the flexibility of an ethylene plant to return to liquid cracking and offers the lowest capital solution. Since the operational costs for this import stream are high, it is important to minimize the import stream. Thanks to long term experience of Qenos operating a Linde naphtha cracker under gas cracking, strategies have been developed to minimize import of hydrocarbons as described below.

2. Reference Plant History

Originally built by Linde in 1983 to crack 50%/50% Naphtha/LPG, the reference plant underwent modification in 1996 (Linde) to crack 100% ethane feedstock. The plant operated by Qenos has a capacity of 250 kta and typically cracks a combination of 95% ethane and 5% LPG, with an ethane conversion of 68%. The plant has enjoyed successful operation on this feed slate, with a run of seven to ten years between turnarounds.

Figure 2 represents the process scheme of the reference plant hot section. The furnace effluent is cooled by quench oil injection and routed to the quench oil column. In the quench oil column, the cracked gas is cooled by a quench oil pump around. Heat is mainly removed from the quench oil by generating LP steam and by preheating feedstock. The quench oil column overhead is further cooled in a quench water tower, where most of the dilution steam is condensed. Gasoline is decanted from process water in the gasoline/water separator. Gasoline is recycled back to the quench oil column as top reflux to remove oil components from the cracked gas. The balance is sent to B.L. via a gasoline stripper.

In order to maintain flexibility in the system to revert back to liquid cracking, the quench oil system has been retained in its original operation. To make up for the lack of heavy product produced from ethane cracking and to maintain the system inventory in the quench oil pump around loop, diesel and gasoline are imported to the system, as shown in Figure 2. During ethane cracking, the reference plant maintains direct quench oil injection to the furnace effluent because no tertiary quench exchangers are available. Further details about the conversion project are provided in [2].

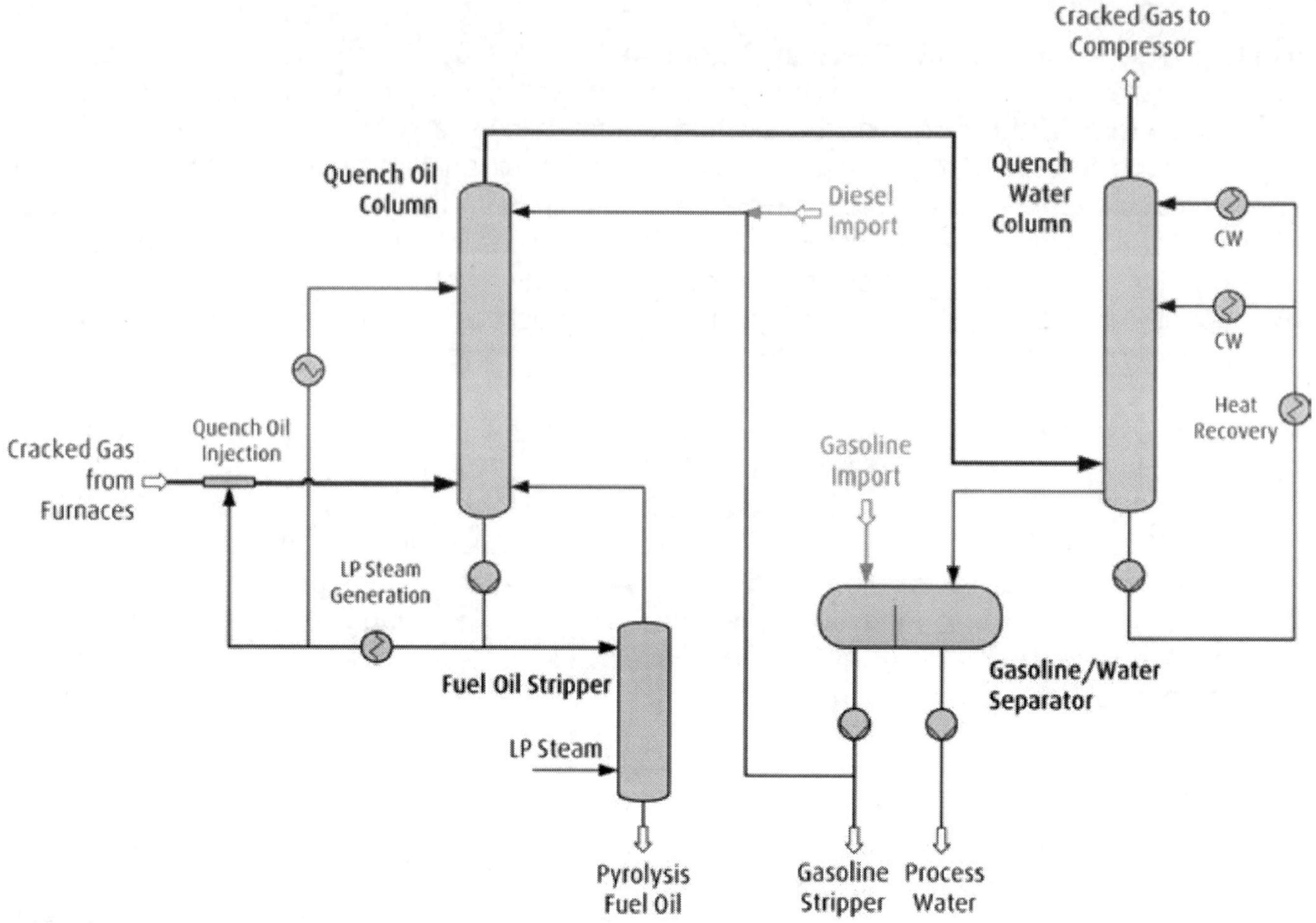

Figure 2: Process flow diagram of the reference plant hot section.

3. Hot Section Fouling Mechanisms

The control of viscosity and fouling in the quench oil loops is crucial for stable operation of cracker hot section. Fouling potentially results in line plugging and reduced equipment availability but also seriously affects heat transfer and has a negative impact on energy efficiency.

The dominant mechanisms that contribute to fouling in a quench oil system are:

- **Thermal polymerization of reactive monomers (e.g. indene, styrene)**
  In general, chemical reaction rates have an exponential dependency on temperature. The residence time influences the amount of polymer formed, the extent of reaction and the molecular weight of the resulting polymer. The parameters concentration (of fouling precursors), temperature and residence time are the main drivers for hot section fouling and need to be carefully addressed when converting from liquids to gas cracking.

- **Concentration of heavy components (e.g. Polycyclic Aromatic Hydrocarbons PAH)**
  Heavy components mainly consist of PAH. PAHs are formed from dienes and vinyl aromatics via Diels Alder mechanisms and/or free radical polymerization. Both their formation and their agglomeration contribute to an increase of quench oil viscosity.

- **Precipitation of heavy aromatics (e.g. naphthalene)**
  Naphthalene has a melting point of 80°C [3]. Depending on the concentration, naphthalene solidifies and precipitates when the process temperature drops below the pour point temperature. This typically contributes to fouling at cold surfaces in the quench water system (cold heat exchanger surface area, lines without heat tracing).

4. Operating Parameters and Design Criteria

Thanks to long term operational experience from the reference plant and other past projects, design concepts and prediction tools were validated and further developed over the past years [4].
In the following, selected operating parameters and design criteria are discussed. Figure 3 summarizes areas of concern when operating a liquid cracker hot section under gas cracking conditions.

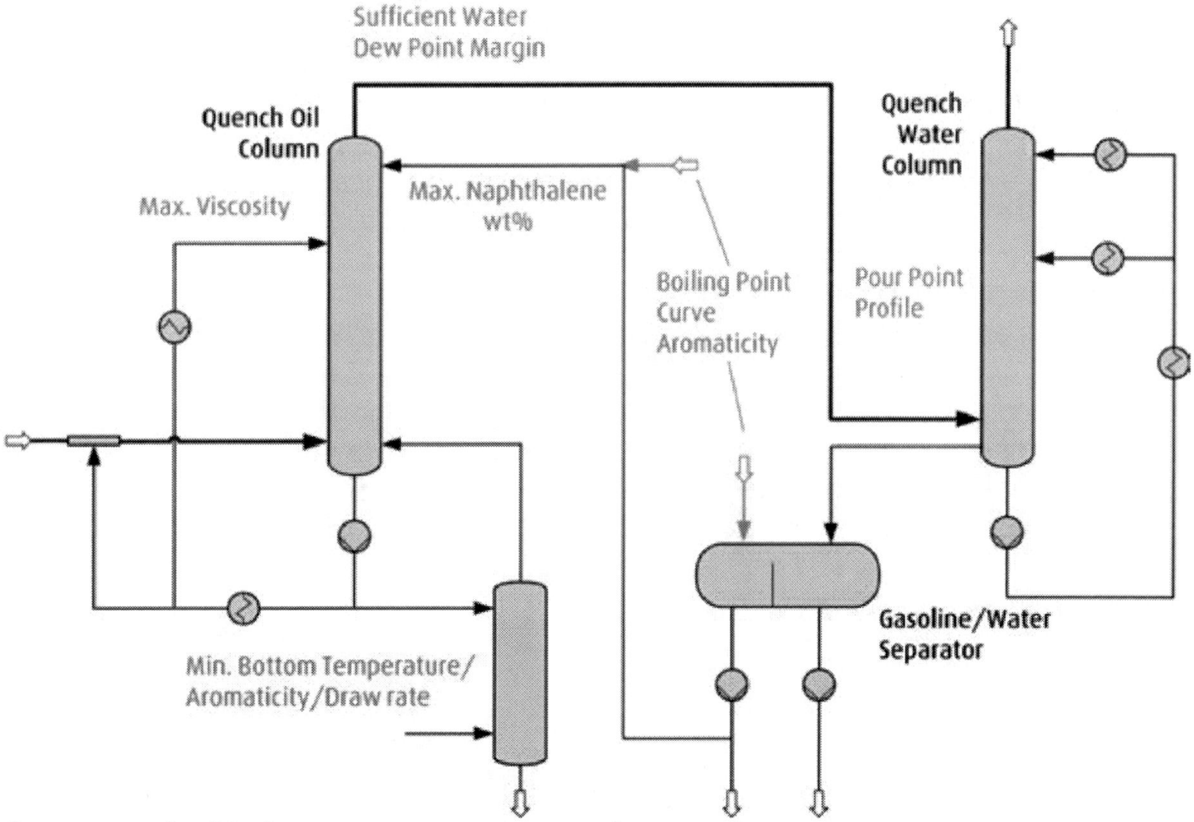

Figure 3: Summary of critical process parameters and areas of concern.

Import Hydrocarbon Stream Specification

The composition of the imported hydrocarbon stream is a critical variable in the design of the system. Proper operation of the quench oil column mandates that the import hydrocarbon stream be compatible with respect to aromaticity and boiling point curve. Incompatible hydrocarbons make the system inoperable due to fouling and increased viscosity. Light Cycle Oil (LCO) and Heavy Cycle Oil (HCO) from refineries are good candidates for import. Due to its more compatible boiling point curve HCO is the preferred choice. Typical amount of oil import is 1-5% of the feed rate which depends on boiling range, aromatic content and residence time of fouling precursors.

The reference plant imports diesel to operate the cracker hot section. Additionally, to improve cracker OPEX and control aromatics in quench oil column, pyrolysis gasoline is imported. By injecting gasoline, the diesel injection can be cut by half.

Quench Oil Tower Temperature

As discussed above, the most important parameters that govern fouling are temperature, composition, and residence time [5]. To avoid accumulation of heavy material in the quench oil column, a minimum purge of heavy material via the fuel oil product must be ensured.

The goal is to operate the quench oil column at a temperature as low as possible to minimize the polymerization rate and maximize the bottom product purge rate, while still keeping an adequate temperature margin above the dew point of water at the column overhead.

Residence time

Time influences the extent of reaction and amount of fouling material being formed and accumulating in the system. Because the fuel oil make within the cracking furnaces is significantly lower for gaseous compared to liquid feedstock, the lifetime of fouling precursors and material within the system is much longer, requiring many more cycles through the quench oil system before it can be removed. The hydrocarbon stream import increases the draw-off of heavies from the quench oil circuit and thus reduces the residence time of fouling precursors in the system. Therefore the minimum import rate is determined by the fouling precursor residence time.

To further reduce the residence time, and thus the import hydrocarbon stream, the quench oil hold up volumes in the system must be minimized. This can result in modification of column internals and minimization of equipment items in operation (e.g. stripping columns, quench oil surge drums, operation of only 1 quench oil cycle). In a revamp situation, this can be combined with technological upgrade of the existing equipment.

The residence times of heavy components within the quench oil system have been calculated for the reference plant and are used to determine the hydrocarbon make up rate for cracker feedstock conversion projects.

Fouling Precursor Concentration

Monomers like styrene and indene are fouling precursors. To minimize the fouling risk, concentration and residence time needs to be minimized by varying the quench oil column overhead temperature and gasoline reflux flow. The import of gasoline allows to reduce residence time and to reduce the import of heavy hydrocarbon stream.

Naphthalene Concentration

The concentration of naphthalene at various points within the system is a constraint of particular concern, as it could solidify and foul cold surface areas of heat exchangers (e.g. quench water coolers). Consequently, the naphthalene concentration must be carefully controlled in order to avoid exceeding the precipitation limit in the gasoline phase. The fouling risk can be assessed by calculating pour point of the hydrocarbon phase such as proposed by Schmigalle et al. [7]

Quench Water Tower Pour Point Profiles

It is important that an adequate margin be maintained between the process temperature and pour point temperature of the liquid hydrocarbon phase in the quench water tower. This margin prevents solidification of fouling components in the quench water tower and in vital exchangers, which is mainly driven by naphthalene and heavy hydrocarbon components.

Viscosity of Quench Oil Loop

A further constraint is the viscosity of the quench oil loop. A column bottom temperature that is too high enhances the formation of heavy hydrocarbon components. Volatile light components are reduced in the bottom quench oil, thus increasing the viscosity of the pump around and the fuel oil product. Aromatic components act as solvents for gums and oligomers and help to keep the viscosity below the operability limit. Therefore, the minimum concentration of aromatic components in the quench oil column bottom is an important constraint when adjusting the column temperature and the hydrocarbon import flow rate of this stream in a conversion project.

Gasoline/Water Separation

The density difference between the gasoline phase and water phase is the driving force for gasoline/water separation. For gas cracking, the density of gasoline condensed in the hot section tends to approach the water density, challenging the gasoline/water separation.
Gasoline/water separation needs to be carefully assessed during a conversion project. The gasoline density needs to be adjusted by smart and flexible condensate rerouting, while carefully avoiding recycling and concentrating fouling precursors in the system [4].

5. Conclusion

In conclusion, the current trend of ethylene producers adding flexibility between liquid and gaseous feedstock presents many challenges, particularly in the area of the quench oil and quench water system. The concept presented in this paper maintains the quench oil system in operation by the use of import hydrocarbon streams. The previously discussed process concepts and design criteria are based on the long term operational experience and feedback from a reference plant. The described approach offers the reward of low capital expenditure because the plant can be operated with minor modifications. However the import stream needs to be carefully selected and minimized to reduce operational costs and to ensure the success of the feedstock conversion project.

6. References

1. Emmert, Thomas, Kracker, Gunther, van Helmond, Paul (2014) "Cracker Feedstock Changes: From Naphtha to Ethane Cracking" in *AIChE Southwest Process Technology Conference,* Galveston, Texas.
2. Beattie, Peter, Antonas, Sam, Bairlein, Werner (1999) "Converting an LPG/Naphtha Cracking to 100% Ethane Feedstock On-line, Almost" in *Ethylene Producers' Conference, AIChE,* Houston, Texas.
3. Collin, Gerd, Höke, Hartmut and Greim, Helmut (2012), "Naphthalene and Hydronaphthalenes", in *Ullmann's Encyclopedia of Industrial Chemistry*, Vol. 23, pp. 661-670, Wiley-WCH Verlag GmbH & Co. KGaA, Weinheim.
4. Zellhuber, Mathieu, Andre, Matthias, Emmert, Thomas and Kracker, Gunther (2016) "Simulation-Based Assessment and Mitigation of Fouling in Water Quench and Charge Gas Compressor Sections of Gas Crackers" ", in *Ethylene Producers Conference, AIChE Spring National Meeting*, Houston, Texas.
5. Vachon, Jérôme, Hancock, Jessica M. and Robson, Ian (2015), "A chemist's perspective on organic fouling in ethylene operations: update", *Hydrocarbon Processing*, October, pp. 49-53.
6. Bernard, André, Fahiminia, Feridoun and Kim, Grace (2013), "Challenges of Primary Fractionators on NGL Feed" in *Ethylene Producers Conference*, *AIChE Spring National Meeting*, San Antonio, Texas.
7. Schmigalle, Holger and Tolergard, Christer (2010), "Fouling Mitigation of Quench Tower System", in *Ethylene Producers Conference*, *AIChE Spring National Meeting*, San Antonio, Texas.

16AIChE
Spring Meeting, Houston, TX

Ethane Cracking in Existing Liquid Crackers - Effects on CGC, Cold Fractionation and Refrigeration

Muhammad Imran
Lead Process Engineer
Technip Stone & Webster Process Technology

Thomas Pickett
VP of Ethylene Technology
Technip Stone & Webster Process Technology

Prepared for Presentation at
American Institute of Chemical Engineers
2016 Spring Meeting
11th Global Congress on Process Safety
Houston, Texas
April 10-14, 2016

UNPUBLISHED

Ethane Cracking in Existing Liquid Crackers - Effects on CGC, Cold Fractionation and Refrigeration

Muhammad Imran
Technip Stone & Webster Process Technology

Thomas Pickett
Technip Stone & Webster Process Technology

Keywords: cracking, compression, flexibility, ethane, refrigeration;

Abstract

Plant feedstock cost represents the largest single expense for ethylene plant operators. With the availability of low cost ethane feedstock, it may be advantageous to replace a certain portion of the liquid feedstock with ethane feed. Ethane cracking in a liquid cracker results in better ethylene yield; however, there are certain challenges associated with this shift. The amount of ethane which can be cracked in a liquid cracker depends on many factors including:

- Plant configuration
- Desired capacity increase
- Available margins in the existing equipment
- Pipeline logistics/plant infrastructure

Shifting from liquid feed to lighter feeds such as ethane changes plant traffic patterns, decreasing plant loads for systems that handle C3+ components but potentially burdening systems that handle the C2 and lighter fractions. The extent of change in traffic patterns and resulting challenges depends on the portion of the liquid feed replaced by the ethane feed. In this paper, the discussion will be focused on the effects of ethane feed cracking on cracked gas compression, cold fractionation and refrigeration areas. Key discussion items are:

- Reduction in cracked gas molecular weight and resulting impact on the cracked gas compressor (CGC) performance
- Shift in hydrogen and methane yields with traffic pattern changes in cold fractionation areas; effects on hydrogen and residue production, chilling train pressure drop, expander performance and ethylene loss
- Performance of the C2 Splitter and handling of increased ethane throughput and ethane recycle
- Performance of the Ethylene Refrigeration (C2R) and Propylene Refrigeration (C3R) compressors
- Potential refrigeration recovery from the fresh ethane feed.

1. Introduction

Since 2010, advances in the recovery of gas from shale reserves in the United States has resulted in a shift in worldwide ethylene plant feedstock economics. Strong ethane supplies have positioned the US to become the most competitive and low cost ethylene producer globally. Since 2014, declining crude oil prices have partly eroded this competitive advantage due to the narrowed price gap between naphtha and ethane. With new ethylene plants coming online in near future, the US will be well positioned to take full advantage of cheaper ethane feedstock once crude oil prices stabilize to reasonable levels.

In addition to grass-root ethylene plants, many ethylene producers cracking liquid feedstocks such as naphtha or gas oil in their existing liquid crackers have replaced a certain portion of the liquid feedstock with ethane feed. The shift in ethane feedstock has increased from 50% in 2005 to 75% in 2014 (source: Wood Mackenzie[1]). Some international companies are also evaluating the economics for importing US based ethane into their existing liquid crackers. Ethylene producers undergoing modifications in their units to accommodate lighter feeds, but who have maintained flexibility to enable the plant to return to liquid feed cracking, are in a better position to face the crude oil market volatility.

This paper describes typical challenges associated with processing ethane in a liquid cracker. A feasibility study with break point analysis is a good starting point for evaluating the amount of the ethane that can be economically processed.

A typical liquid cracker produces many co-products such as hydrogen, fuel gas, propylene, butadiene, pyrolysis gasoline, fuel oil, C4's and C5's. With ethane cracking, there is a significant shift in other products produced. Ethane cracking yields for ethylene are higher as compared to cracking liquid feeds. However being a lighter feedstock its cracking yields are significantly lower toward the heavier products. This product shift can have an impact on local markets and can cause supply/demand changes, impacting prices especially for the heavier co-products. Other on-purpose technologies can be used to offset shortage of heavier products due to the shift to ethane feed cracking. For example, propane dehydrogenation (PDH) and methanol-to-propylene (MTP) technologies are being used to offset the global propylene shortage. High severity catalytic cracking technologies are likewise becoming more attractive to combat the propylene shortfall. Contribution of on-purpose technologies for global propylene production has been increased such that propylene supplies were derived from less than 4% of the total in 2005 to 15% in 2014 and are expected to increase to 25% by end of this decade (source: Wood Mckenzie[1]).

Producers that are already cracking ethane and/or propane and butane feeds have inherent advantages since their plants require minimal or no modifications. However, plants which were configured for heavier liquid feedstocks such as naphtha and gas oil have to reconsider the impact to their units when processing ethane feeds. The extent of challenges and modifications depend on the proportion of the liquid feed replaced by ethane. The plant configuration (demethanizer, deethanizer or depropanizer tower first) can present unique challenges for ethane feed flexibility. Depending on the configuration, furnace effluent, plant traffic, flow rate and composition will vary in each section of the plant. Various sections of the ethylene plant therefore can be adequate or limiting depending on the original plant design and process configuration.

Some ethylene producers target to increase plant capacity in parallel to achieving maximum ethane cracking flexibility. Capacity increases will generally require additional capital spending, especially if the design margins on existing equipment are already tight. Ethane pricing and ethane cracking economics can fluctuate with time. It is therefore important to consider maintaining flexibility to shift back to liquid cracking when market conditions dictate.

2. Ethane cracking and product distribution

Shifting from heavier to lighter feeds substantially changes plant traffic patterns, decreasing plant loads for systems handling C3+ components, but burdening systems handling the C2 and lighter fractions.

Figures 1, 2 and 3 depict the shift in product generated with the gradually shifting of feedstock from 100% naphtha to 100% ethane. The amount of ethane that can be cracked in a particular naphtha or gas oil liquid cracker depends on factors such as ethylene production targets, available design margins (compressors, equipment, piping, etc.), plant configurations, compressor performance curves, quench system design, etc. Evaluation must be determined on case-by-case basis. It is to be noted that product yields are strongly dependent on cracking severity and ethane conversion per pass. Actual product yields will be different depending on cracking conditions on a particular plant. Figures 1, 2 and 3, therefore are included to explain the effect of increasing ethane cracking on product distribution and should not be used for design work.

2.1 Ethylene yield

As shown in Figure 1, ethane cracking results in higher ethylene yields with minimum production of by-products. Typical once-through ethylene yield with ethane cracking is greater than 50% (depending on ethane once-through % conversion). Considering recycle cracking of ethane to extinction, an ultimate yield of approaching 80% ethylene is achieved. Naphtha cracking results in a yield of 25% to 36% ethylene depending on paraffin content and cracking severity.

Due to higher ethylene yields, ethane cracking requires a significantly lower fresh feed rate (tons of fresh ethane per ton of ethylene) as compared to other feedstocks. Table 1 shows tons of feedstock required per ton of ethylene produced.

Table 1-Feedstock (T/T of Ethylene)

Feedstock	Ton Feed/Ton of Ethylene
Ethane	1.25
Propane	2.3
Naphtha	3.0
Gas Oil	3.85

Shale gas recovery in the US has resulted in reduced ethane pricing due to large surpluses. The combination of relatively lower ethane prices and better ethylene yields may result in improved feed/product margins by substituting a certain portion of the liquid feed with ethane. However, the current crude oil price drop has resulted in availability of cheaper liquid feedstock and improved economics for liquid cracking. The crude oil market is highly volatile at present and ethylene producers who have implemented ethane feed flexibility are in a better position to operate their plants for maximum possible revenues by being able to select the most economic feed slate. It is therefore worthwhile to consider implementing ethane feed flexibility to mitigate the market volatility.

2.2 Ethane and propylene yields

Figure 1 also shows yield patterns for ethane and propylene based on varying ethane and naphtha feed mixtures. There is a significant increase from 5 wt% ethane when cracking 100% naphtha, and 35 wt% ethane when cracking 100% ethane. Actual ethane yield will depend on how much ethane cracking can be achieved in a given liquid cracker. Propylene production drops significantly when shifting to maximum ethane cracking. As Figure 1 shows, propylene yield reduces from 14 wt% to only 2 wt% when shifting from 100% naphtha to 100% ethane cracking.

Figure 1-Products Shift- Ethylene and Propylene

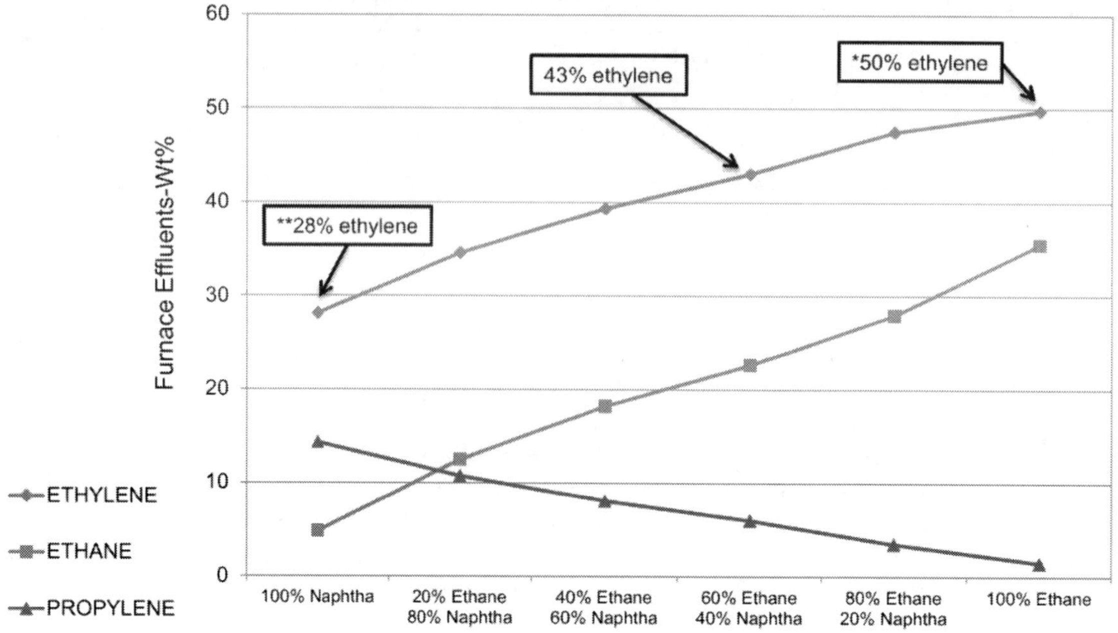

***Ethylene yield will higher depending on ethane % conversion**

****Ethylene yield will be higher depending on saturates % and cracking severity**

4

2.3 Heavier products yields

Besides propylene, shifts to crack more ethane can impact the quantity of other heavier products produced. As shown in Figure 2, 1,3 butadiene (13BUTD), butenes, C5-200, BTX and fuel oil production is significantly reduced. Due to a very low production of heavies at the higher ethane feed slate, in some cases it becomes more challenging to separate C3 and heavier products. Since plants designed for liquid cracking already have C3 and heavier fractionation systems installed, many operators may decide to keep these systems in operation by a combination of external make ups and internal recycles.

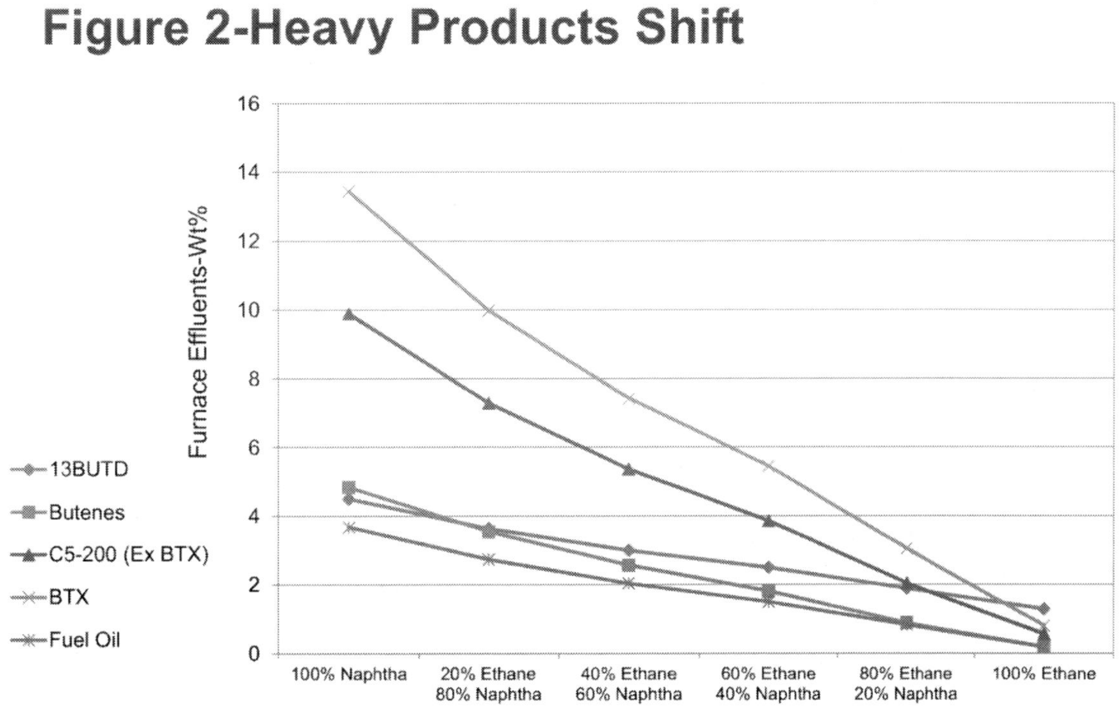

2.4 *Hydrogen and methane yields*

Figure 3 shows the shift in hydrogen and methane production with ethane cracking. Hydrogen production increases from 1 wt% to 4 wt% in the cracking effluent and methane production drops from 13 wt% to 5 wt%. A higher hydrogen and lower methane content lowers the molecular weight of cracked gas which impacts CGC performance. A higher hydrogen content also results in high pressure drop in systems handling separation of hydrogen and methane from C2 and heavier components.

Figure 3-Light Products Shift

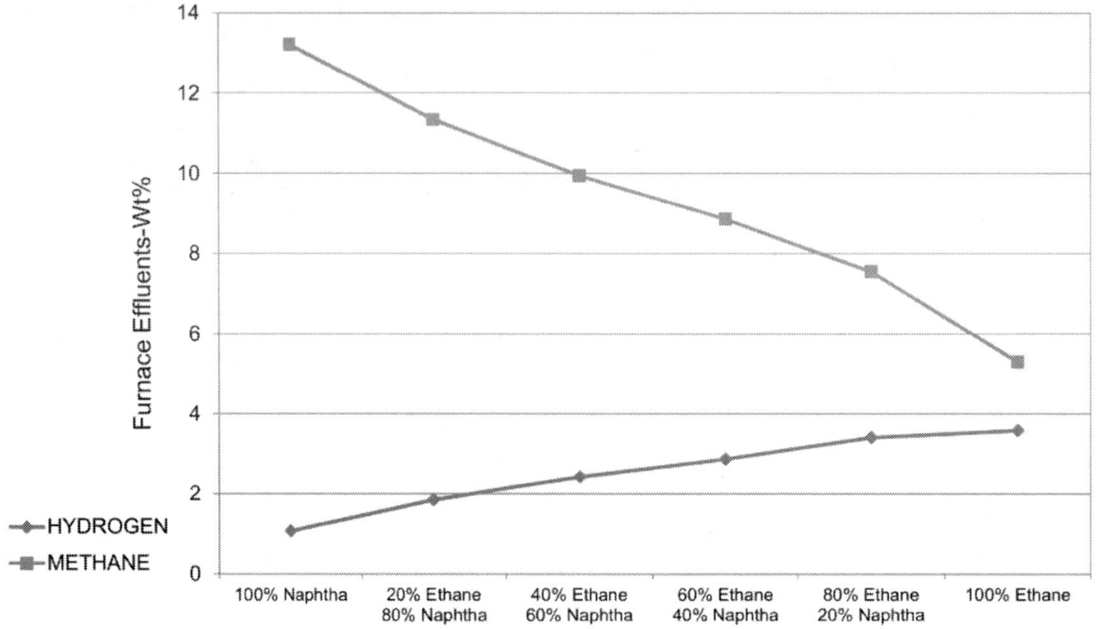

3. Ethane cracking and impact on recovery section

In this section, the discussion will mainly focus on how ethane cracking impacts the recovery (excluding quench) section of a liquid cracking plant. Furnaces are also excluded from this discussion.

3.1 Cracked gas system

CGC performance needs a thorough evaluation. With a higher ethane feed diet, molecular weight reduces from ~30 to ~19 with a possible increase in volumetric flow rate. One of the important parameters to consider when shifting from liquid cracking to ethane cracking is polytropic head. Polytropic head is inversely proportional to molecular weight. As shown in the equation 1 below, a lower molecular weight gas will require a higher polytropic head for the same compression ratio.

Equation 1

$$H_P = \frac{1545}{MW} Z_{AVG} T_1 \left(\frac{n}{n-1}\right)\left[\left(\frac{P_2}{P_1}\right)^{\left(\frac{n-1}{n}\right)} - 1\right]$$

Where:

H_P	Polytropic Head (ft)
MW	Molecular Weight
Z_{AVG}	Average Compressibility
T_1	Suction Temperature, Deg R
n	Compression Coefficient
P_1	Suction Pressure, psia
P_2	Discharge Pressure, psia

Also as shown in the equation 2 below, compressor horsepower is directly proportional to mass flow rate and polytropic head

Equation 2

$$SHP = \frac{(m) H_P}{[(\eta_P)3300]1.02}$$

Where:

SHP	Shaft Horsepower
m	Mass Flow Rate
H_P	Polytropic Head
η_P	Polytropic Efficiency
1.02	2% gear losses

From the equations above, it is clear that for the same mass flow rate and compression ratio, a lighter cracked gas from cracking ethane will require a higher shaft horsepower as compared to heavier cracked gas. However, as Figure 1 shows, ethane cracking results in higher ethylene yield and therefore a lower mass flow of cracked gas will be required to produce the same quantity of ethylene as compared to liquid cracking.

An important variable to overcome potentially higher horsepower requirements is to increase the first stage suction pressure of the CGC. As is clear from Equation 1, an increase in compressor suction pressure will reduce compression ratio and polytropic head requirements. On the negative side, a higher first stage suction pressure results in higher radiant coil outlet pressure which degrades the ethylene yield and potentially lowers ethylene production. The yield degradation for higher coil outlet pressures when cracking ethane is not as significant as the degradation when cracking heavier feeds such as naphtha. This must be considered when evaluating to shifts to lighter feed cracking and determining the optimum furnace coil outlet pressure.

A higher coil outlet pressure also lowers volumetric flow rate which helps to avoid potential casing or suction nozzle limitations. If feed flexibility is the only target with no increase in ethylene production, it is sometimes possible that a CGC driver designed for liquid cracking can work for ethane cracking effluents. In some cases, especially where both feed flexibility and ethylene production increase are targets, it may not be possible to operate by just increasing first stage suction pressure. Possible solutions for this situation are:

- Adding a booster stage,
- Adding a parallel stage with dedicated driver
- Decoupling compressor casings and adding a new driver for one or more of the stages
- Adding wheels on stages requiring higher polytropic head

Another important consideration for the CGC is to confirm that a sufficient operating margin is maintained between the operating point and surge limit. This operating margin tends to be narrow due to higher operating speed and polytropic head required for lighter cracked gas operation. Again a combination of increased suction pressure and a lower mass flow required for lighter cracked gas operation can help to stay away from the surge limit. Compressor performance curves need to be studied carefully by use of a process simulator to confirm compressor performance for lighter cracked gas effluent. Ultimately, compressor manufacturers should confirm predicted machinery performance.

Figure 4: Cracked Gas Compressor

3.2 Chilling train

As shown in Figure 3, switching from naphtha cracking to ethane cracking results in a hydrogen production increase from ~1 wt% to ~4 wt% and a lower methane production from ~13 wt% to ~5 wt%. Higher hydrogen production may result in a larger volumetric flow to the chilling train and an increased pressure drop. If hydrogen is being recovered in the original liquid cracking scheme with no slip stream to the expander, there will be an increase in hydrogen flow to the hydrogen recovery system with potential hydraulic and exchanger surface area limitations. Replacement of piping and exchangers in this section may help to lower pressure drop. In case of no modifications, this additional pressure drop has to be compensated by additional discharge pressure from the CGC. This may not be possible due to potential limitations in CGC if the design was based on heavier cracked gas.

Ethylene losses from the demethanizer overhead will also increase due to lower methane to hydrogen ratio. Modifications can be done in the overhead system to reduce ethylene losses. Details of these modifications are outside of scope of this paper. All of these modifications should be flexible enough to allow reverting back to naphtha or gas oil cracking, if required.

Another important consideration is the hydrogen content in the residue gas and fuel gas. Ethane cracking results in lower molecular weight residue gas due to the increase in hydrogen content. Since residue gas is used as a regeneration gas for dryers, there is a potential of increased pressure drop in regeneration gas system. It is however possible to reduce regeneration gas flow and therefore reduce pressure drop due to improved (higher) specific heat of the lighter regeneration gas. Cracking furnace burners are specified for a specific hydrogen content. Fuel gas composition should be checked for increased hydrogen content while considering natural gas import requirements. Modifications in the hydrogen recovery area may be required to limit the hydrogen content in residue gas. In extreme case, furnace burner modifications or replacement may be required. With no modifications or replacement, existing furnace burners can potentially set a limit on how much ethane can be cracked in a liquid cracker. Alternatively, exploring the idea of adding or increasing hydrogen recovery to the plant via the addition of a pressure swing adsorption (PSA) or methanation system may reset fuel gas composition to acceptable levels for the burners.

3.3 Expander/recompressor

A lower methane and C3+ production may reduce demethanizer warmer level feeds but may increase colder level feeds. Also there will be reduced flow to the expander from the demethanizer overhead due to lower methane content in the cracked gas. A reduced flow to the expander will lower expander efficiency. A combined effect of lower efficiency and a reduced flow is a reduction in refrigeration available to the chilling train from residue gas expansion. The expander feed system can be modified to increase the load and recover more refrigeration. Expander internals however may need to be replaced to make it suitable for the new conditions. An additional stage for the expander/recompressor machine may be justified.

Figure 5: Simplified Sketch of Expander/Recompressor

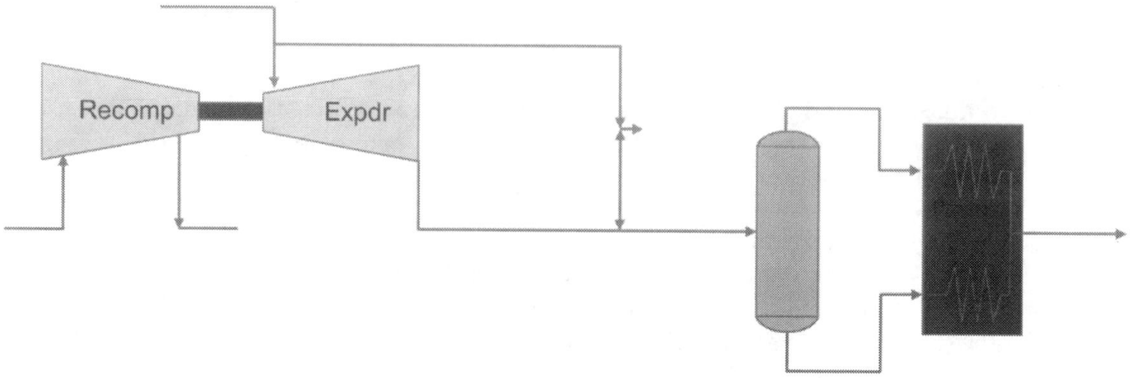

3.4 Ethylene splitter, C2 hydrogenation and deethanizer systems

Shifting from naphtha or gas oil cracking to ethane cracking results in a higher ethane content and a lower purity feed to the ethylene splitter tower. This requires a higher reflux rate for the same production capacity. If additional capacity is another objective, the required tower internal traffic will be further increased. Existing ethylene tower trays need to be examined for extra capacity. Possible solutions to overcome tray hydraulic limitations are replacing existing trays with high capacity trays or adding another tower to supplement fractionation capacity. Deethanizer and C2 hydrogenation reactors systems also need to be evaluated for hydraulic constraints and any possible negative impact due to possible higher flow rates.

3.5 Ethylene and Propylene Refrigeration Compressors

Depending on composition, available credits and mass flow rate, a higher ethylene refrigeration duty may be required in the chilling train to accommodate ethane cracking. Revamping the ethylene refrigeration compressor may be required in case it is operating close to its design limit. The scope of revamp can be extensive if the objective is to increase plant capacity in addition to adding feed flexibility. The ethylene refrigeration compressor may need a higher capacity and horsepower to enable increased ethane cracking.

Figure 6: Ethylene Compressor

The propylene refrigeration load will increase on two major users when cracking more ethane. First, as explained above, the ethylene splitter will require comparatively higher reflux rates when cracking ethane compared to naphtha or gas oil cracking. For non-heat pumped systems, a higher reflux rate will require a higher duty on ethylene splitter condenser, which is generally the largest user of -35°F propylene refrigerant. For low pressure heat pumped systems there will be an increased load on the heat pumped compressor. Secondly, the ethylene refrigeration condenser duty will increase due to higher demand of ethylene refrigeration. The ethylene refrigeration condenser is typically the second largest user of -35°F propylene refrigerant.

On the positive side, there are often two important refrigeration credits to propylene refrigeration when cracking ethane. These credits can be used to reduce the load on the propylene refrigeration compressor. First, there is an increased recycle ethane production from ~5 wt% to ~35 wt% when shifting totally from naphtha to ethane cracking. Recycle ethane can potentially be used for cold level (-35°F C3R) refrigeration credit. A higher recycle ethane refrigeration credit can help to significantly lower propylene refrigeration compressor horsepower. Second, refrigeration credits are often available from the fresh ethane feed which can be flashed to furnace feed pressure to achieve additional credit at -35°F. Ethane feed however may need to be dried before flashing to lower pressure and temperature to avoid hydrate formation. If so, additional investment for dryers and regeneration systems will be required. A higher duty on the ethylene splitter condenser and ethylene refrigeration condenser therefore may be balanced out by these additional credits when shifting to more ethane cracking. The propylene refrigeration compressor must be checked and confirmed by the compressor manufacturer on a case-by-case basis.

4. Conclusion

Ethylene producers with the ability to quickly react to fluctuating energy and feedstock prices are better prepared to compete in today's global petrochemicals market. While current energy and feedstock supply economics may favor ethane feed cracking, plant owners must rigorously examine operational and capacity constraints before transitioning to maximize ethane feeds. Detailed feasibility studies need to be undertaken to evaluate the impact of maximizing ethane cracking on plants originally designed for naphtha or gas oil cracking. A well-planned feasibility study can properly identify plant modifications and capital costs, as well as other challenges, potentially helping to improve production margins by reducing the feedstock cost, a major expense to ethylene plant owners. Each ethylene plant is unique and needs a dedicated investigation to determine suitability for feedstock transition.

5. References

[1] Steve Zinger, Vikki Medley, and Anne Keller, "The New World Order in the Olefins Industry - A Result of Developments in Advantaged Feedstocks and Technology", AIChE 2015 Spring National Meeting, Austin, Texas, April 2015

[2] Hallee, L. P., Zdonik, S. B., Green, E. J., Manufacturing Ethylene, The Petroleum Publishing Company, The Oil & Gas Journal, Copyrighted 1966, 1967, 1968, 1969, and 1970

[3] Radjen Krishnasing, Gabriele Mariotti, and Kara Byrne, "Turboexpander-Compressor Technology for Ethylene Plants", AIChE 2010 Spring National Meeting, San Antonio, Texas, March 2010

[4] GPSA, Section 13, "Compressors and Expanders"

16AIChE
Spring Meeting, Houston, TX

AICHE PAPER: 199e
CONVERSION OF A FLEXIBLE CRACKER TO MOSTLY ETHANE FEED

David Gent
NOVA Chemicals Corporation
P.O.Box 5006
Red Deer, Alberta, Canada T4N 6A1

Prepared for Presentation at the 2016 Spring National Meeting
17th Annual Ethylene Producers' Conference
Houston TX, Apr 10-14th
Session TA010 – Ethylene Plant Technology

UNPUBLISHED

AICHE PAPER: 199e
CONVERSION OF A FLEXIBLE CRACKER TO MOSTLY ETHANE FEED

David Gent

NOVA Chemicals Corporation
Engineering & Projects
P.O. Box 5006
Red Deer, AB, T4N 6A1

Phone: +1 (403) 754-4620
Mobile: +1 (403) 352-1586
Email: David.Gent@novachem.com

Abstract

Recent NGL feedstock opportunities that have arisen from shale gas development has led to the goal of processing large volumes of ethane feed in flexible feed crackers. This type of processing can result in some unique challenges for plants that have traditionally processed significant quantities of liquid feedstocks.

This paper will discuss the conversion of a flexible liquids and gas fed cracker to all gas feed with the primary feed constituent being ethane. An overview of the process systems that are most affected will be provided together with a review of some of the most significant issues and how they can be addressed. The design approach to some specific areas of modification will be discussed and some of the operating results will be described. Future challenges for phased expansion of ethylene production and full conversion to ethane will be briefly outlined.

Introduction

Plant History

This plant was originally designed in the 1970's as a naphtha cracker which included a crude oil distillation unit that produced the cracking feedstock. During the 1980's the plant was retrofit to allow processing of significant quantities of Natural Gas Liquid (NGL) feed. The 1990's saw capacity expansions and further increases in the use of NGL feeds. A major expansion was implemented in 2005 and this included conversion of some of the Furnaces to a lower residence time design as well as replacement of all of the major compressors and their steam turbine drivers.

Plant Capacity

The plant nameplate capacity at the start of the ethane feed revamp project was 1.8 billion pounds per year of ethylene when using a variety of feedstock diets. The principal feeds included naphtha, gas oil, butanes, propane and ethane recycle.

Plant Configuration

There are 14 Cracking Furnaces that include three different types. Quenching is provided by direct oil injection to the pyrolysis effluent followed by a Gasoline Fractionator and a Quench Water column. The processing sequence can be described as a high pressure Demethanizer-first with tail-end Acetylene Converters. Coproduct processing includes gas-phase methyl acetylene/propadiene and liquid-phase C4-acetylenes removal by hydrogenation. Chemical-grade propylene is produced from a C3 Splitter. The C5-plus product is processed in a Dripolene Pyrolysis Gasoline (DPG) unit where the dienes are hydrogenated and the aromatics are extracted for sales.

The steam system is supported by three Auxiliary Boilers that raise 1500 psig steam. There are also three fired Steam Superheaters that superheat the saturated 1500 psig steam that is produced by the 6 natural draft SRT3 Cracking Furnaces.

Objective

The processing objective for the plant revamp was to maintain the ethylene production capacity while changing the feedstock cracking profile to 7 Furnaces on ethane feed (and recycle) with 2 to 3 Furnaces cracking butane or propane.

Early Analysis

Equipment Comparison

The initial studies were started in 2010 and these considered a range of ethane feed processing that was equivalent to as few as 4 and as many as 7 Cracking Furnaces. One early observation was that the size of many items of equipment, including all three major compressors, was equivalent to, or better than, the same equipment in one of NOVA's ethane crackers that has a similar ethylene capacity. This perspective provided some confidence that a significant increase in ethane feed would be possible.

EQUIPMENT - VESSELS	Flexible Feed Cracker				Ethane Feed Cracker			
	D	L	Trays	Design Pressure	D	L	Trays	Design Pressure
	ft	ft	#	psig	ft	ft	#	psig
Gas Frac	27.0	72.5	Packed	50	-	-	-	-
Quench (bottom)	42.0		Settler	50	25.0	58.0		40
Quench (top)	24.5	41.8	Packed	50	15.0	26.0	21	40
CG Dryers	12.0	26.5	Media	590	11.3	32.5	Media	600
Demethanizer Stripper	9.5	107.3	Packed	487	8.5	79.0	31	500
Demethanizer (bott.)	11.5	73.3	59	487	9.0	106.3	47	500
Demethanizer (mid)	9.0	29.5	16	487	-	-	-	-
Demethanizer (top)	6.5	27.8	12	487	5.5	33.0	18	500
Deethanizer	9.5	131.3	58	408	9.5	104.0	51	440
Acetylene Converters	*10.0*	79.5	3 beds	408	11.0	40.0	1 bed	440
C2 Splitter	*13.0*	222.2	123	300	18.0	234.8	125	300

Table 1a: Equipment Comparison with Ethane Cracker

Table 1a shows a comparison of the size of the key process vessels and, as highlighted by the diameters that are in green, most of the equipment is either larger or the same size as that in the Ethane Cracker. The exceptions are the Acetylene Converters and the C2 Splitter which are highlighted in red.

In the case of the C2 Splitter, even though the diameter is significantly smaller, the capacity is much closer to that required because it contains very high capacity trays, whereas the one for the Ethane Cracker contains conventional 4-pass valve trays. Analysis of the capacity of this column for the revamp case indicated that it might achieve 90 to 95% of the required throughput once the ethane recycle was increased.

While the Acetylene Converter is smaller, it is a 3-bed system that has ample catalyst so the smaller size became mostly a potential hydraulic constraint that was not critical at this capacity.

EQUIPMENT - COMPRESSORS	Flexible Feed		Ethane Feed	
	Frame	Power	Frame	Power
		Hp		Hp
Charge Gas Compressor Casing 1	88MD2-2	13,805	88M6I	14,848
Charge Gas Compressor Casing 2	70M6I	25,542	60M8I	12,287
Charge Gas Compressor Casing 3	60M8I	23,496	46M9-8	6,148
C3 Refrigeration	88M8-6	39,073	88M 8-6	30,910
C2 Refrigeration	38M9-7	9,726	38M9-8	5,228
Turbo Expander 1		843		845
Turbo Expander 2		NA		799

Table 1b: Equipment Comparison with Ethane Cracker

As shown in Table 1b, the comparison of compressor frame sizes and design power usage suggested that in all cases the equipment was probably going to be large enough to accommodate the all-ethane Cracker loads. Although the configuration of the compressors in terms of suction and discharge nozzle locations and sizes was not going to be ideal, it was thought to be workable without modification for the initial feed revamp. When the analysis moved on to consider capacity expansion it was found that some of these external configuration differences were more limiting than the aerodynamic components inside the machines as those could typically be changed-out to support expansion cases of up to 30% and even approaching 50%.

Change in Processing Loads

The changes in product distribution that occur when each feedstock type is cracked are well understood by the industry although they are subject to plant-specific operating conditions such as cracking severity and pressure (Table 2). However, when considering a flexible feed cracker there can be a fairly wide range of product capacities depending on the feedstock combinations that are used, the cracking conditions and what equipment constraints are present in the various product processing systems.

	FEEDSTOCK			
Product Yields (weight Basis)	**Ethane**	**Propane**	**n-Butane**	**Naphtha**
Cracking Severity	**High**	**Medium**	**High**	**Medium**
Hydrogen-Rich Gas	5.9%	1.8%	1.2%	1.0%
Methane-Rich Gas	7.0%	26.8%	22.4%	16.9%
Ethylene	80.9%	46.2%	40.6%	38.7%
Propylene	1.9%	16.9%	16.6%	15.5%
Butadiene	1.8%	1.8%	3.5%	4.8%
Butylenes /Butanes	0.8%	1.3%	6.7%	5.1%
Pyrolysis Gasoline:	1.6%	4.8%	7.2%	15.6%
Benzene	*0.8%*	*1.8%*	*2.6%*	*4.4%*
Toluene	*0.1%*	*0.5%*	*1.1%*	*1.7%*
C8 Aromatics	*0.0%*	*0.0%*	*0.4%*	*2.2%*
Other	*0.7%*	*2.5%*	*2.9%*	*7.3%*
Fuel Oil	0.0%	0.5%	1.7%	2.5%
Total	**100.0%**	**100.0%**	**100.0%**	**100.0%**
Hydrocarbon Load	48.0%	84.0%	95.0%	100.0%
C3-plus Products	7.8%	29.6%	41.2%	56.5%
C4-plus Products	5.8%	12.7%	24.6%	41.1%
Butadiene in C4+	30.6%	14.2%	14.4%	11.6%

Table 2: Typical Cracker Yields by Feedstock

In the case of the plant in question there are essentially three Furnace residence time types and a wide range of product processing is included. The plant had been modified over many years from an original naphtha cracking basis towards cracking gas oils on the heavy end to cracking NGL's on the lighter end of feed diet. Some ethane cracking had been included in the most recent re-design but this was limited to one to two furnaces of feed most of which was ethane recycle.

Table 3 demonstrates the changes in total hydrocarbon flow and product flow distribution as the proportion of ethane cracking is increased from the base design. Three cases are shown:

1. Base Design
   The starting point for the revamp
   Feeds included gas oil, naphtha, butane, propane and ethane

2. Proposed
   The proposed feeds are ethane, propane and butane
   All liquids (gas oil and naphtha) cracking is eliminated

3. 100% Ethane
   All of the feed is ethane

The table expresses the change in flow at the suction of the Cracked Gas Compressor (CGC) in terms of flow factors, where:

Mass flow factor = mass flow for Case / mass flow at Base design
Molar flow factor= molar flow for Case / molar flow at Base design

Ethylene production is held constant across these cases but the total flowrate and the component flowrates change based on the feedstock mix in each case. The ethane feed change for each case is reflected as the percentage of cracked gas hydrocarbon flow.

While not initially being considered as a real case, the 100% ethane case has been used to define a book-end for the potential range of feedstock change. This can be helpful when considering ultimate equipment limitations and planning for the future.

It can be seen that at constant ethylene production the total mass flow through the plant is reduced by 30% or so as the feed diet is moved towards a full ethane case. However, the molar rate is only reduced by 8% and this is indicative that the volume being processed will be similar. On the other hand, as the feed gas is split to the various component product systems, there are substantial shifts in processing loads both to the high side and the low side.

The total C2 load increases by 36% as the proportion of ethane feed is increased toward 100% and this is responsible for pushing the limits of the C2 Splitter capacity. Acetylene removal requirements are reduced to half or less but the increase in C2 flow to the Converter pushes toward a hydraulic limitation.

Although there is the expected large increase in the mass of hydrogen production this is mostly offset by a reduction in methane make with the resulting tail-gas molar flow actually being 10% less than the base design.

The production of C3, C4 and aromatics is reduced to 30 to 40% of the base and this change needs to be considered in terms of the turndown of the associated processing systems.

Total Flow Factors (dry basis) at Cracked Gas Compressor (CGC) Suction	**Base Design**	**Proposed Case**	**100% Ethane Case**
Case Ethane [feed + recycle] rate as % of total CGC mass flow	**12%**	**63%**	**97%**
CGC total mass flow divided by Base Design mass flow	1.00	0.76	0.70
CGC total molar flow divided by Base Design molar flow	1.00	0.92	0.92
Component Flow Factors at CGC Suction	**Base**	**Proposed**	**100% Ethane**
Ethylene mass flow factor	1.00	1.00	1.00
Ethane mass flow factor	1.00	2.31	2.99
Total C2 mass flow factor	1.00	1.24	1.36
Hydrogen mass flow factor	1.00	1.53	1.82
Methane mass flow factor	1.00	0.46	0.22
Tail-gas (H2 + CH4) mass flow factor	1.00	0.55	0.35
Tail-gas (H2 + CH4) molar flow factor	1.00	0.90	0.89
Acetylene mass flow factor	1.00	0.53	0.40
Propylene mass flow factor	1.00	0.35	0.07
C4's mass flow factor	1.00	0.36	0.14
Butadiene mass flow factor	1.00	0.35	0.21
BTX mass flow factor	1.00	0.40	0.23

Table 3: Changes in Relative Flows at CGC Suction

While these changes in the production profile are mostly as expected, the specific processing loads that result have to be assessed against the capabilities of equipment that in many instances was designed for a totally different case. In the end it is unlikely that many naphtha crackers were designed with 100% ethane feed processing in mind. When thermal processing requirements are considered, then constraints in the supporting refrigeration systems become evident and these have to be addressed as they can frequently bottleneck the unit.

Technical Issues

Ethane Feed Contaminants

The ethane feedstock for this revamp was going to be supplied from a new shale gas source that was not even in production at the start of the engineering work. While a typical commercial specification was developed for a pipeline ethane feed it was not clear whether this would be met during initial operation. In general, the level of contaminants that have been seen since the feed was commissioned have met the specification. Water and methanol have been the most significant contaminants and some excursions involving these components during early pipeline and cavern operation was expected.

It was also anticipated that the plant would see some accumulation of Naturally Occurring Radioactive Materials (NORM). Training and analysis programs as well as procedures for handling NORM-contaminated items were available from NOVA's other ethane crackers. These were implemented as part of the change management processes for this project. The extent of equipment contamination with NORM will be better understood at the next plant maintenance turnaround.

Some mercury compounds were detected when analyzing downstream samples during plant troubleshooting. However, it was found that the mercury was not originating from the new ethane feed and it is likely that this is residual material from some of the liquids cracking history. No elemental mercury has been found and previous inspections of the Cold Box equipment showed no signs of mercury or mercury attack.

Ethane Feed System

While this plant already had some ethane feed and recycle vaporization equipment, this capability was substantially less than would be needed for the proposed quantity of ethane feed and this meant that a new ethane feed system was required.

The new feed system provided an opportunity to offload increases in refrigeration load by vaporizing the ethane at temperatures similar to those of the propylene refrigerant. The ethane feed has to be let-down in pressure from the pipeline (800 psig) prior to feeding the Cracking Furnaces. Depending on the operating pressure, vaporization at temperatures as low as -40 °C (100 psig) can be achieved and this allows the ethane to be used to replace refrigerant services at the first suction level of the Propylene Refrigeration Compressor (PRC). Examples of the services that can be considered are:

- Ethylene Refrigerant Compressor (ERC) Condenser
- C2 Splitter Condenser
- Chilling Train Condenser
- Ethylene Refrigerant Subcooler
- Propylene Refrigerant Subcooler

There are operability pros and cons to each of these potential uses and the ERC Condenser is generally preferred because it is typically more flexible and avoids direct interaction with operation of the process. Depending on the proportion of ethane feed, it is possible that the entire ERC discharge vapour flow can be condensed against an ethane feed coolant. In this case propylene refrigerant becomes a backup medium that is mostly used for startup of the ERC and possibly for control of the discharge pressure.

One consideration in the use of ethane feed as refrigerant is the amount of vapour flash that occurs when the feed pressure is let-down from pipeline pressure to the Cracking Furnace supply pressure. This flash can be as much as 30% and some designs incorporate a feed flash drum to disengage the vapour prior to feeding the residual liquid to the vaporizers that will be used to provide the refrigeration services. If the design doesn't include a flash drum then the vaporizer inlet piping and internal vapour disengagement area must accommodate the additional flash vapour.

Even if the type of integration that is discussed above is implemented it is advisable to have a back-up vaporizer based on a hot utility so that vaporization of the ethane feed can be assured should process-driven heating be interrupted due to upsets or trips during startup and shutdown. There may be a choice of utility and even low-grade heat streams might be considered. Typically the heating is provided by low pressure steam with alternatives being to use the steam directly or via another medium such as methanol or glycol. If direct steam heating is used then the design should incorporate features that will prevent freezing of the resulting condensate. Specialized heat exchanger designs are available for this purpose. NOVA has both indirect systems using hot methanol and direct systems using steam in its ethane crackers.

Figure 1A illustrates the use of low pressure ethane vaporization to provide the refrigerant that is used to cool the Ethylene Refrigeration Compressor Condenser. In this case, the pipeline ethane feed is cooled against the cold ethylene product from the C2 Splitter prior to being let down in pressure. This pre-cooling reduces the amount of flash vapour from around 30% to 16%. The scheme shows the potential use of a feed flash drum to reduce the amount of vapour going to the ethane vaporization kettle to around 4%. The alternative is to send 20% flash vapour to the kettle and eliminate the drum. Vaporizing the ethane at the indicated pressure of around 100 psig produces a refrigerant that is equivalent to the lowest temperature level of propylene refrigerant and this has the effect of reducing the suction flow to the Propylene Refrigeration Compressor (PRC). It was found that for 100% ethane feed cases the use of propylene refrigerant to condense the ERC could be eliminated for normal operation. This effectively breaks the energy cascade between the ERC and the PRC. The lower coolant temperature also allows the ERC to be condensed at a lower discharge pressure and consequently results in a power saving.

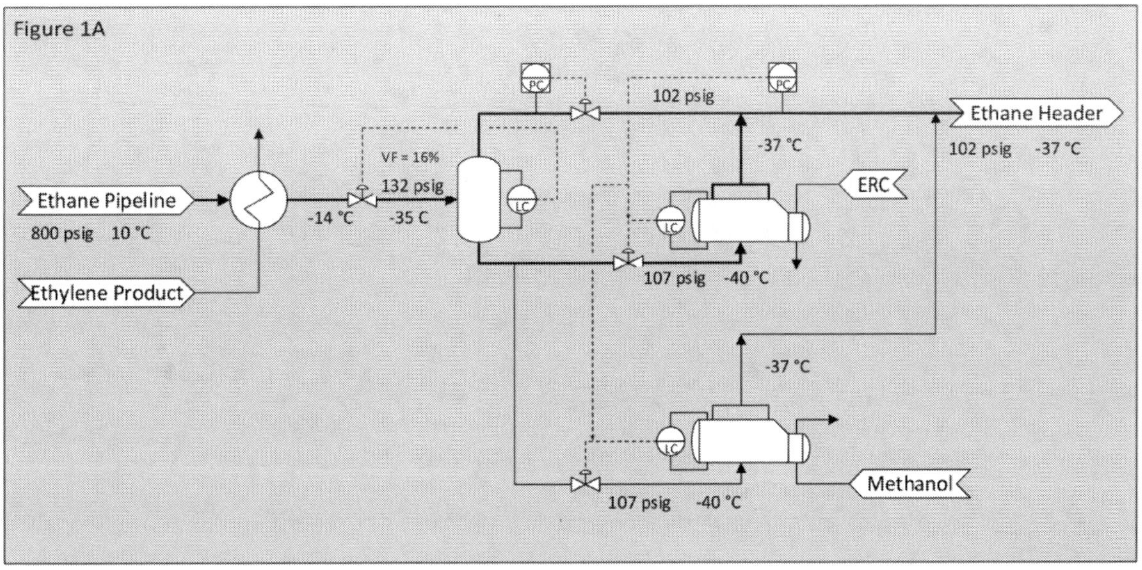

Figure 1A: Low Pressure Ethane Feed Vaporization with Flash Drum

This feed scheme also illustrates the potential incorporation of a low pressure auxiliary vaporizer that uses hot methanol as the heating medium.

Figure 1B shows a similar integration between the ethane feed vaporization and the ERC condenser but at a warmer temperature. In this case the ethane feed header is operated at around 50 psi higher pressure. While this approach also results in a significant reduction in propylene refrigerant used to condense the ERC, the economization takes place at the second suction flow to the PRC. This results in lower overall power savings and, more importantly, it does not unload the first stage of the PRC which is frequently found to be a capacity constraint in the cold section of an ethane cracker.

This figure also shows the use of a steam-heated auxiliary vaporizer. The exchanger has ethane feed flow on the tubeside and is designed to both vaporize and superheat the feed. As mentioned previously, the design must incorporate features that avoid exposure of the steam condensate to freezing temperatures. An alternative high pressure operating condition of 390 psig is also shown. The use of this back-pressure control allows the vaporization to take place at temperatures above the hydrate point. This mode can be employed to deal with a wet feedstock scenario, however it does require this specialized heat exchanger to be designed for higher pressure and this contrasts with the methanol unit above which can be designed at the downstream header pressure.

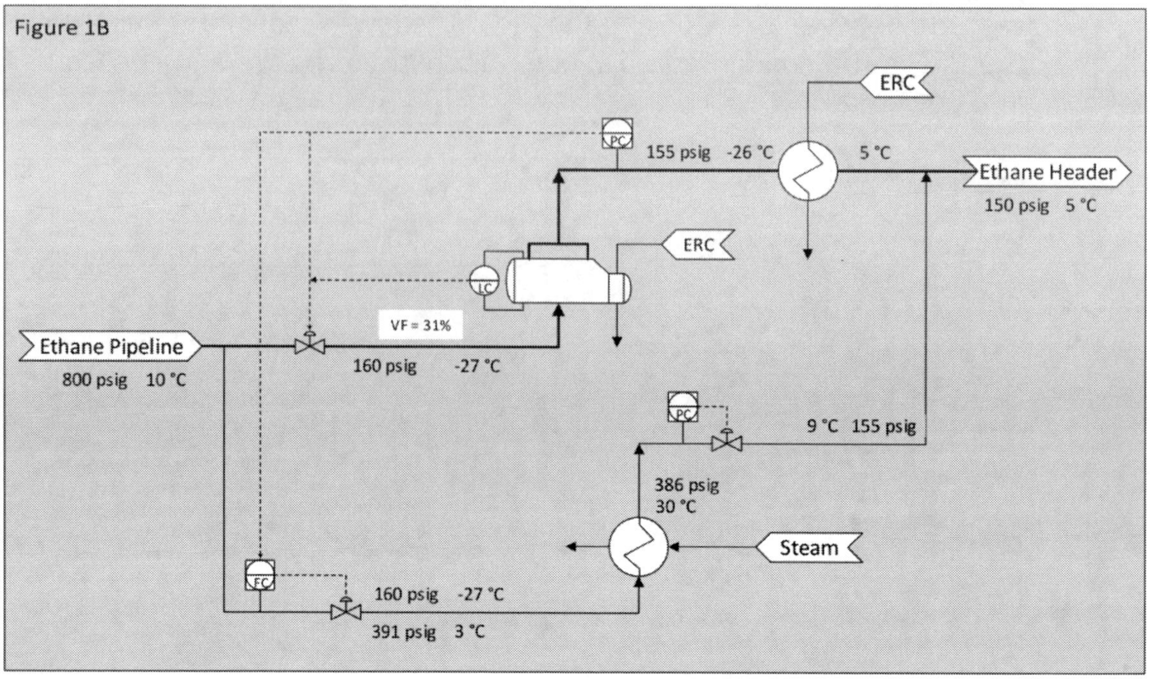

Figure 1B: Medium Pressure Ethane Feed Vaporization

This specific project did not include ISBL drying of the ethane feed because it was expected that there would ultimately be sufficient drying upstream of the plant. However, this approach did introduce some risk during the commissioning and initial operation phases. The risk was mitigated by starting the new feed system through an auxiliary steam-heated vaporizer. The vaporizer has been designed to operate at a higher pressure so that ethane feed can be vaporized and superheated prior to letdown into the feed header. This avoids the potential for any operation at low temperature with wet feed. Once the pipeline material is analyzed to be dry the feed system can be switched over to the refrigeration recovery mode.

The cracking of a predominantly ethane feed results in a substantial amount of ethane recycle being recovered at the C2 Splitter bottoms. This recycle needs to be vaporized into the ethane feed header in order to return to the Furnaces for cracking. In this plant the ethane recycle vaporization had been incorporated into an ERC discharge condensing service. As the amount of ethane cracking is increased it becomes impossible to fully use both the ethane feed and recycle vaporization services at the ERC condensers. This means that one or more of the other points of integration listed earlier must be considered. The chilling train is a typical location where all of the available ethane recycle duty can be consumed efficiently and it also allows for reasonable operability and control features. Another potential point of integration is to use the recycle (or feed) as coolant for the C2 Splitter Condensers.

Cracking Furnaces

The plant has three different fundamental Furnace types;

- 7 SRT3 Furnaces; 4-pass medium residence time coils
- 4 GK6 Furnaces; 2-pass short residence time coils
- 3 Millisecond Furnaces; 1-pass short residence time tubes

Six of the SRT Furnaces are based on natural draft to supply the combustion air and they generate saturated steam. The seventh SRT Furnace is induced draft and it generates superheated steam. All of the other Furnaces have induced draft fans and generate superheated steam.

Not all of the Furnaces were connected to NGL feed supplies and prior to this project the maximum number of Furnaces on NGL had been 8 of the 14 available. This operation was typically in conjunction with a minimum of 3 naphtha Furnaces.

The revamp project required 9 to 10 operating Furnaces in order to meet capacity with seven of these being fed with ethane plus associated ethane recycle. Prior to the project ethane cracking had only been demonstrated on the SRT-type Furnaces and that had mostly been limited to operating one Furnace on ethane recycle. Of particular concern was that there had been no prior experience with cracking NGL feeds in the Millisecond Furnaces. The scope of this project required these Furnaces to operate on NGL but it was decided to initially limit them to cracking the propane or butane portion of the proposed feed-slate.

Another aspect of the new ethane cracking was going to be a significant increase in the hydrogen content of the fuel gas. This meant that burner designs had to be re-evaluated and the results indicated that the burners on the 7 SRT Furnaces needed to be replaced and those on the 2 MSF's needed to be modified. The SRT Furnace has both raw gas hearth burners and air pre-mixed wall burners. The NOx target for the new burners was 65 ppmv (at 3% O_2) and this meant that a lower NOx design was required.

Installation of the new burners triggered a regulatory review by the local standards authority and by this point in time they were insisting on a burner safety management system. The intent of the BMS was to prove the fuel trip and isolation capability as well as proving that burners were lit while the firebox temperature remained below the auto-ignition temperature of the fuel. NOVA Chemicals presented a paper on the installation of a suitable system at the 2013 Ethylene Producers Conference (reference 2).

Furnace Effluent Quenching

The original Furnace design included quenching the effluent in a primary quench heat exchanger (PQE or TLE) followed by direct oil injection at a quench fitting. This resulted in a 'wet' transfer line operation with the hot oil being recovered in the bottom section of the Gasoline Fractionator. The quench oil was indirectly cooled by the evaporation of process water to make dilution steam (Figure 2).

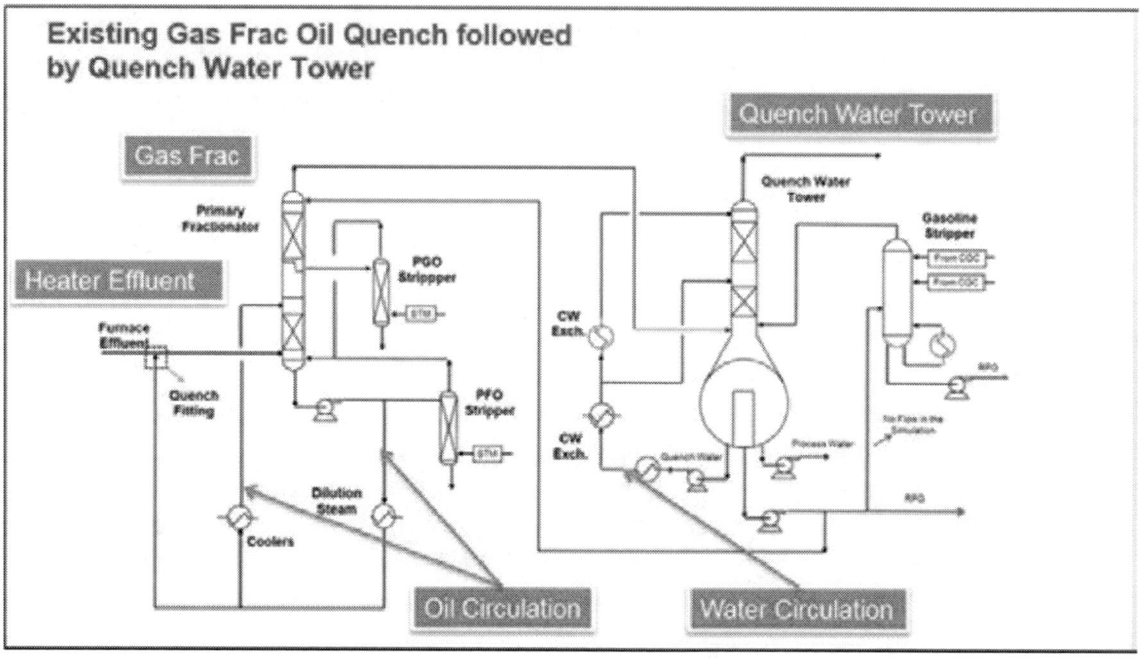

Figure 2: Gasoline Fractionator & Quench Water Tower

Cracking mostly ethane results in a substantial reduction in the make of quench oil components. The chart in Figure 3 illustrates the reduced make of light, middle and heavy fuel oil components when cracking ethane-propane feed mix compared to cracking a propane/ butane/ naphtha diet. This lack of oil make-up means that there is a real concern about the operability of the hot quench oil circuits. The increased residence time and reactivity at high temperatures was expected to create a high viscosity quench oil. NOVA Chemicals presented a paper on this subject during the 2013 Ethylene Producers Conference (reference 1). It describes an experimental process whereby a proxy Pyrolysis Fuel Oil stream taken from an E/P Cracker was used as test material to understand the impact of temperature on reactivity and viscosity. Figure 4 illustrates the Fouling Batch Mix Reactor (FBMR) that was used for these experiments. The work did indicate that an acceptable operation might be achieved if the oil temperature could be kept below 160 °C. This approach was also supported by another ethylene plant owner who had operated the Gasoline Fractionator lower oil loop below this temperature when running a mostly-NGL feed diet. In that case the furnace effluent cooling involved multiple heat exchangers with no direct quench oil injection and therefore a 'dry' transfer line. This was felt to be an important factor because it results in a much lower circulation of quench oil (30% of the NOVA rate) with the consequent reduction in reaction residence time.

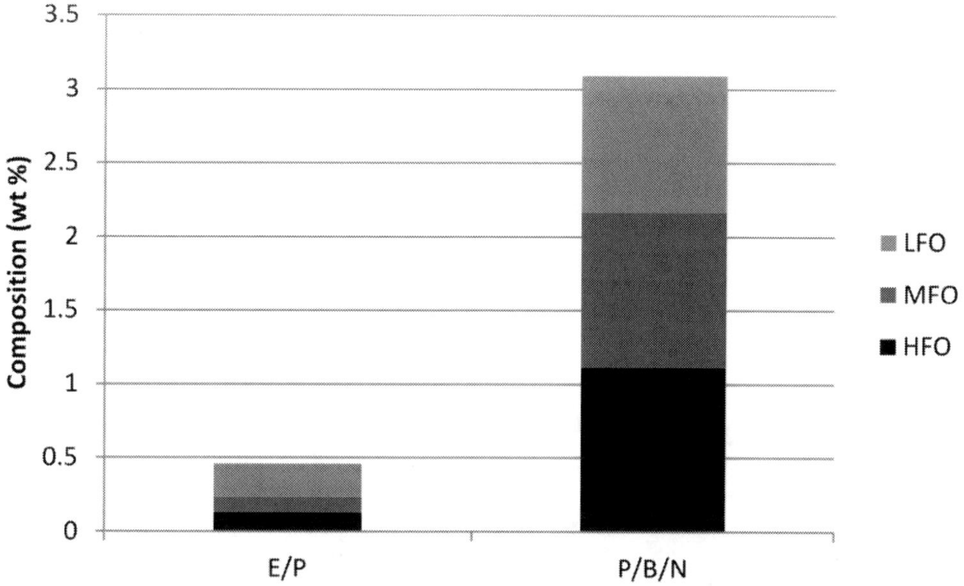

Figure 3: Reduction in Oil Components E/P vs P/B/N Feeds

Figure 4: Fouling Batch Mixed Reactor

A known industry approach to controlling the quench oil viscosity is the use of a so-called 'flux oil'. Good flux oils are identified by their aromatic content and the suitability of their boiling range for operation in the quench oil loop. The selected flux oil needs to be compatible with the quench oil in order to avoid asphaltene precipitation. Refinery streams that have suitable properties are typically Light or Heavy Cycle Oils that are produced in Fluid Catalytic Cracker operations. However, their preferred disposition towards diesel production was seen as a limitation to their economic availability at the time of this project. Another potential source of dilution oil were streams that are available from the operation of the DPG unit. These streams contain heavier components from the processing of various aromatic streams and they were already being recycled to the Gasoline Fractionator.

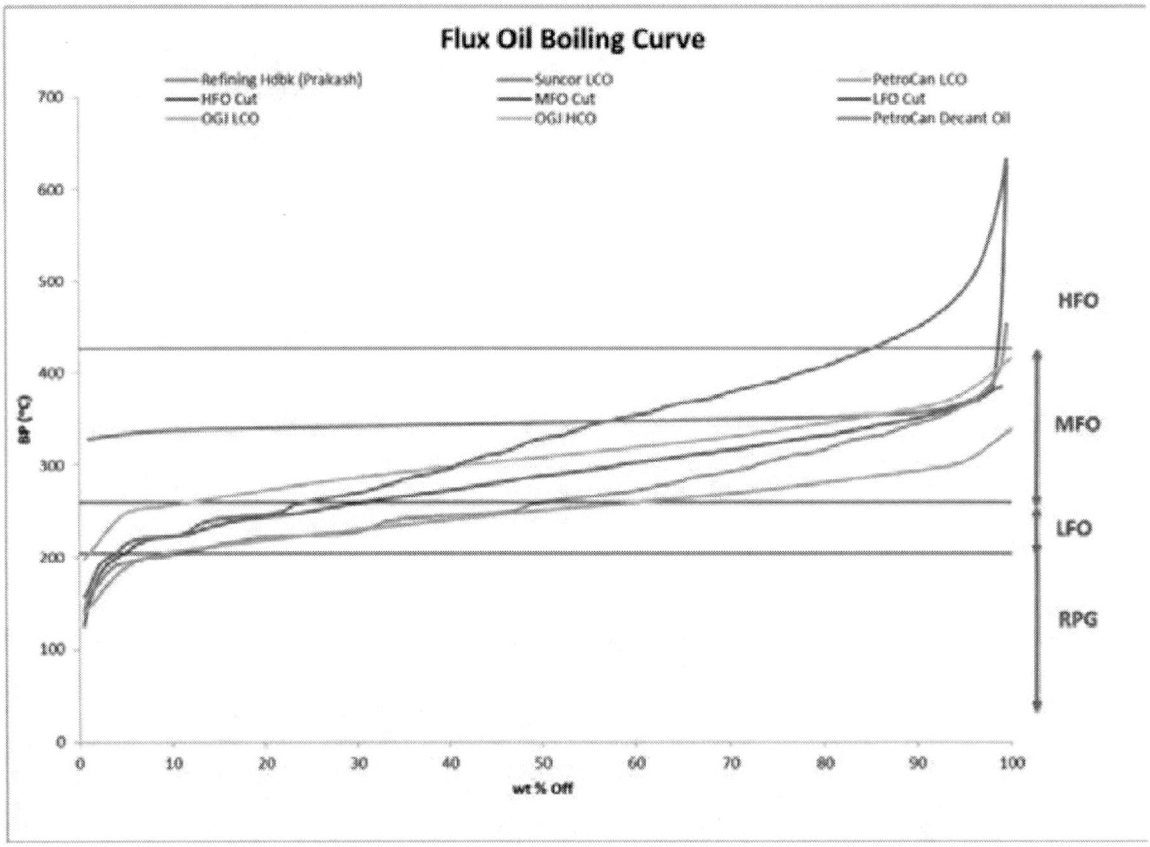

Figure 5: Boiling Curves for Potential Flux Oils

The boiling ranges for some typical refinery cycle oils are illustrated in Figure 5. These oils can be used to maintain components that boil in the Middle Fuel Oil (MFO) range in the Gasoline Fractionator.

One alternative to provide cooling of the Furnace effluent is to install an additional Transfer Line Exchanger (TLE). This configuration is fairly typical for a gas cracker and, depending on the cooling medium, the exchanger can be used to cool the effluent to temperatures as low as 170 °C. One consideration is that going too low on the effluent temperature can result in the condensation of heavy components when cracking Butane. This can lead to fouling of secondary effluent heat exchangers and, based on this concern, the effluent temperature should be kept above 200 °C. This plant has a further restriction in that the main transfer line piping must be operated below 200 °C due to thermal stress limitations. Hence, even with an additional TLE operating with a cracked gas outlet temperature of 200 °C, there is still a requirement for further cooling in order to keep the piping stresses at an acceptable level. This implied the continued need for some direct quench downstream of a secondary heat exchanger.

Given the concerns about using an oil quench with the high ethane feed diet, the possibility of using water as a direct quench medium was considered. This approach can be used whether or not a secondary indirect heat exchanger is installed. The difference would simply be in the amount of injection water that is required. The goal is to inject sufficient water to cool the effluent to the desired temperature by evaporation and to avoid adding any excess free water.

In summary, the following alternatives were considered:

a. *Continue with direct oil quench and maintain 'hot' operation*

The benefit of this approach is that the temperature is kept high enough that dilution steam can be generated from the recovered heat. The down-side is that the oil make-up from the ethane cracking is too low to provide sufficient replenishment and the oil will degrade over time leading to high viscosity, fouling, and ultimately to pluggage.

Flux oil could be introduced to mitigate the risk, however there may be limited availability of a suitable oil and the cost might be prohibitive depending on the amount used.

Chemical suppliers do provide additives that will counter the quench oil reactivity. These additives can also include dispersants to ensure flow. The effectiveness of this approach could be studied by empirical trials on an operating system once ethane feed is introduced into the plant diet, however NOVA was not in a position to engage in this type of ethane trial prior to actually having to process large volumes of ethane feed, so it wasn't viewed as an opportunity in this case.

b. Install a Secondary Transfer Line Exchanger

The benefit of installing a second cooler is the elimination of the direct oil quench (or a substantial reduction of it)

c. Use of a Water Quench

Direct water injection can be used either in conjunction with installing a second quench cooler or as a replacement for the oil quench.

While any of the approaches that have been described above could probably be made to work, NOVA's goal was to establish as much certainty as possible with the outcome. Based on this, the following scheme was selected:

- Addition of cooling to the quench oil loop to operate the temperature at 160 °C or less (i.e. a cool oil quench).
- Replacement of the effluent transfer line direct oil quench with a direct water quench to cool the effluent below the 200 °C transfer piping stress limit.

Once the decision to discontinue use of the direct oil quenches was made the next decision was whether to incorporate a second TLE to re-capture this energy. This decision is essentially an economic one that is based on the forecast of fuel costs weighed against the capital cost to install these heat exchangers, consequently, the provision of secondary quench exchangers was viewed as optional project scope. These have not been installed on 11 of the 13 furnaces for the base project but they were installed on the two Millisecond Furnaces. In this case the Primary Quench Exchangers (PQE's) have a relatively small capacity and the effluent temperature at their outlets is above 500 °C. This temperature is hot enough to allow the production of Very High Pressure steam (VHP – 1500 psig) and it was deemed to be a prudent use of capital to provide Secondary Quench Exchangers (SQE's) to generate additional steam at this pressure. A direct water quench was still required downstream of these SQE's in order to reduce the piping temperature below the stress limitation.

Figure 6 illustrates a cooling curve for the total pyrolysis effluent from several furnaces. On leaving the cracking coil the primary cooling (73% of total) occurs in the first TLE (or PQE – Millisecond Furnaces). The effluent is cooled to around 380 °C and the heat that is recovered is used to produce VHP steam. In this plant, the primary TLE effluent is then direct-quenched to the inlet temperature (160 °C) of the Gasoline Fractionator without any further heat recovery. If instead a secondary TLE (or SQE) is installed to operate in the cooling range from 380 °C to 200 °C then a further 22% heat recovery is possible resulting in a total recovery of 95%. Figure 6 also shows the heat recovery constraints that could occur depending on the temperature level of the coolant that is selected (e.g. HP steam versus MP steam vaporizing temperature).

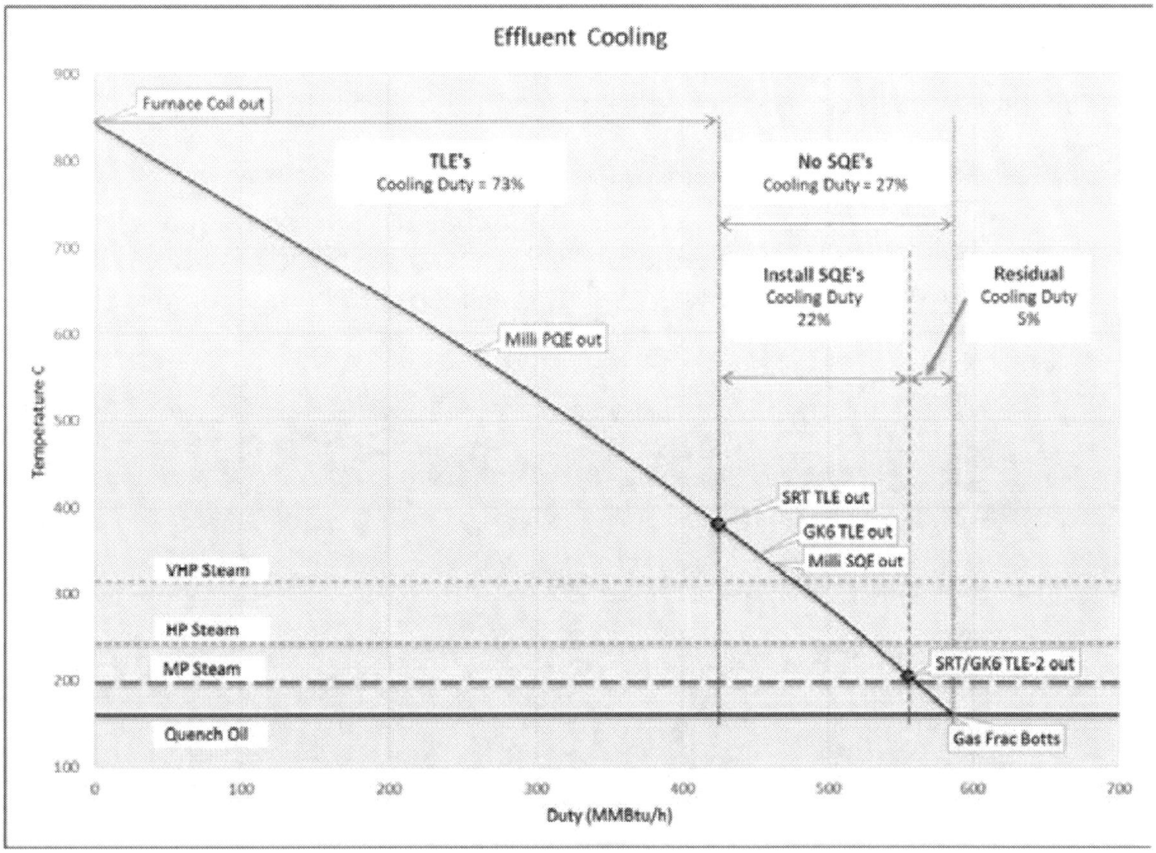

Figure 6: Heat Recovery vs Temperature of Pyrolysis Effluent

The inlet temperature to secondary heat exchangers on the SRT and GK6 Furnaces would be in the range of 300 to 400 °C depending on the capacity and condition of the primary TLE's. These effluents could be cooled by:

- Boiler Feed Water pre-heat
- Medium Pressure (MP) steam production
- Furnace feed preheat

Of these approaches BFW pre-heat (Figure 7) is probably the most typical in the industry and the heat exchanger design for this is relatively simple. However, in the case of the SRT Furnaces, where BFW is supplied at 180 °C, it was found that significant steaming of up to 18% could occur in the BFW Economizer in the convection section. Avoiding this might be viewed as a limitation to the amount of heat recovery that could be achieved.

The medium pressure steam production alternative results in fairly direct replacement of the heat recovery that was being captured in the quench oil. In this case the incremental medium pressure steam that is produced is effectively used to provide the heat input that is required to raise dilution steam from process water. It is more capital intensive than the BFW option because typically a kettle-type heat exchanger will be used. This involves more instrumentation and a local medium pressure header is required to transport the steam to the users.

Using the additional Furnace feed preheat approach was found to lead to high crossover temperatures at the outlet of the convection section mixed-feed preheat banks. This alternative was viewed as being less desirable because the crossover temperature was already considered to be high on the SRT Furnaces.

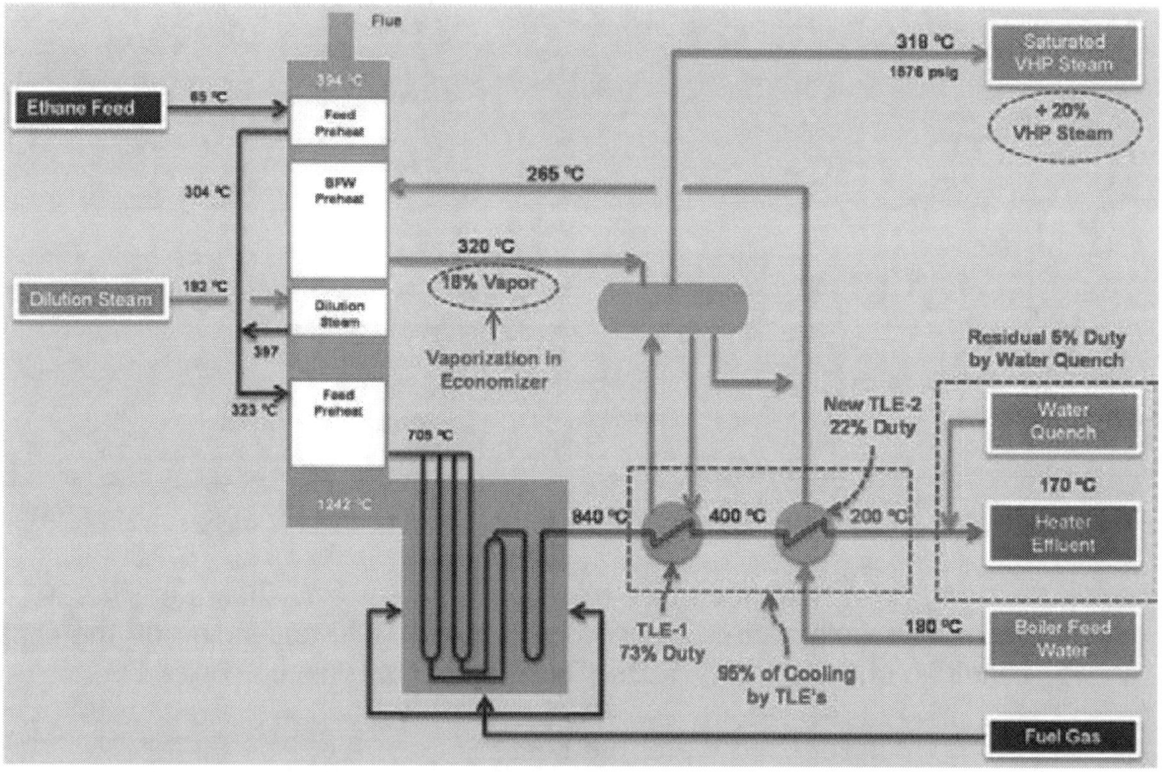

Figure 7: Addition of a Secondary TLE to Pre-heat Boiler Feed Water

Cracked Gas Compression

The Cracked Gas Compressor (CGC) had been replaced during the 2005 capacity expansion and it was clear from the outset of the ethane feed revamp that this compressor would have sufficient capacity and power for the planned operating case, so no changes were required. Fouling of the compressor had historically been mitigated by the injection of a C9 to 200 °C wash-oil that is generated in the DPG unit. This had proved to be a very good wash-oil with a reasonable heavy aromatic content. It was expected that the aromatic quality of this internal stream would deteriorate with the move toward ethane cracking and this has proved to be the case. Alternative wash oils are under consideration, although fouling has not been a significant issue so far.

As further increases in ethane feed and plant ethylene capacity are considered the CGC capability will come more into question. The initial approach to increasing throughput could be to allow the suction pressure to increase, thus reducing the head and power requirements across the machine. This approach may ultimately be limited by the design pressures of the inter-stage equipment and it is not unusual for the operating pressures to approach the set-points of the pressure relief devices. At this point consideration can be given to re-rating the design pressures or to using alternative safety devices such as pilot-operated relief valves in order to maximize the operating pressure without simmering the safety valves.

It is useful to understand the limitations of the CGC string as it is often a constraint to ultimate unit capacity. In the case of this particular machinery, it is the approach of the steam turbine operating speed to a second critical speed that results in an overall speed limitation for operation of the compressor string. The compressor wheels are able to operate at higher speeds than the turbine speed limit and the turbine itself has sufficient power to drive the compressor at higher speed and throughput. This suggests that the first upgrade to the CGC could involve modifications to the steam turbine in order to remove the speed limitation. On the other hand, replacement of the 4th and 5th stage compressor wheels would also allow the string to operate at a lower speed as well as removing the highest inter-stage pressure limitations. All of these considerations and opportunities will come into play if the plant capacity is expanded.

Operation of the CGC for the ethane revamp has been satisfactory with the main issue being high discharge pressures due to some downstream fouling. Once the impact of the fouling was removed, the CGC was able to support full capacity operation of the plant at record production rates.

Cold Boxes

One of the unique features of the plant is that it has two Cold Boxes that are configured for parallel operation. The second Box is smaller and it was added as part of the 2005 expansion project. It was intended to offload the original Cold Box by overcoming both thermal and hydraulic constraints. Not all of the tail-gas, cracked-gas and refrigerant passes are duplicated between the two Cold Boxes. One of the challenges of the ethane feed revamp was to establish what the combined capacity of these Cold Boxes would be given that the tail-gas passages would contain significantly more hydrogen where previously those flows were predominantly methane. Furthermore, this was made more difficult by the absence of some design information for the original Cold Box due to its age. A plate-fin heat exchanger Consultant was employed to provide the evaluation and it was determined that this equipment should be able to handle the revised loads although there was some doubt about the effectiveness of the refrigerant thermosiphon loops. In general, Cold Box operation for the ethane revamp has been acceptable but limitations to the available cooling have been seen and this has been partly driven by fouling and performance constraints in the upstream chilling equipment.

A move toward more ethane feed and expansion of the plant capacity will create the need to provide more Cold Box capacity. It is envisaged that the current two systems will be replaced with a single new Cold Box that will be configured more appropriately for the gas feedstocks.

Turbo Expander

A new Turbo Expander had been installed during the previous 2005 expansion project but the change in tail-gas composition toward much more hydrogen meant that it would be required to handle gas with a lower molecular weight. This could be accomplished by changing the wheels in the expander and the re-compressor. One consideration with this type of change was what type of testing could be achieved prior to installation. A physical performance test would have required the provision of (or construction of) a suitable housing that the new rotating element could be installed in. The alternative was to run a computer-simulated test to validate performance, mechanical stresses etc. NOVA opted for the simulated test and the results have proved to be acceptable both in the virtual test and in actual operation.

Ethane Crackers typically have two turbo expanders that are configured in series operation. This allows a larger overall enthalpy drop to be taken which results in more efficient cooling and better ethylene recovery from the tail-gas that is exported to fuel. In this case, while a second expander set can be added, the current expander may also still need to be modified to accommodate a capacity increase that is based on more ethane feed.

Deethanizer

The top section of the Deethanizer was going to see more load due to the increase in unconverted ethane. On the other hand, the lower section load would be reduced as the production of C3-plus coproducts is much less with ethane cracking.

One aspect of the reduction in C3-plus production is the change in composition in the bottom section of the Deethanizer. The concentration of fouling dienes increases whereas the concentration aromatic solvents is reduced. The net result is a potentially more fouling regime depending on the operating temperature. Fouling can be mitigated by including a C3 recycle from the Depropanizer overhead. This has the effect of diluting the fouling components and lowering the Deethanizer bottoms temperature. The addition of suitable anti-foulant chemicals can also be used to limit fouling.

Another concern for the lower section of the Deethanizer in this plant was that very high capacity trays had been installed as part of the 2005 expansion. This tray technology employs a suspended downcomer design which is sealed dynamically and a certain amount liquid flow is required to accomplish this, so the reduced C3-plus flow needed to be considered. NOVA had experienced a loss of this type of seal on similar trays in a Depropanizer in the past (reference 4). In that case tray fouling was believed to have influenced the seal problem and it was projected that some fouling might occur on these Deethanizer trays. The introduction of an external mixed-C3 feed upstream of the Deethanizer meant that relatively high loading of the bottom section trays could be maintained. However, a move toward more ethane cracking and the inclusion of more feed pre-heat suggests that these trays may become under-loaded in the future.

Loadings in the bottom section of the column were acceptable for most actual operation at the intended ethane feed flow although there were circumstances where some signs of flooding were seen when processing higher rates of external C3 feed. This was overcome by re-routing some of the external C3 feed directly to the Depropanizer. The C3 recycle was minimized during this type of operation as the external material provided the same benefit and internal recycle could unnecessarily limit the capacity. Some very high reboiler loads occurred during the higher C3 processing cases and it was evident that some additional feed pre-heating could be beneficial. Lowering the reboiler duty would also be expected to have a positive impact on fouling. When later opportunities to expand the use of ethane feed and the capacity of the plant were being considered it became imperative to incorporate more feed heating in order to offload the reboiler and the tray loads in the bottom of the Deethanizer. Feed heating is fairly typical for Deethanizers in ethane cracking service.

The higher C2 content of the Deethanizer feed lowers the operating temperature in the mid-section of the column and this may need to be checked against the minimum design metal temperature (MDMT) of the shell material in this zone. Some older Deethanizer designs have been known to split the design temperature of the column between the upper and lower sections with the upper section being made of impact-tested carbon steel and the lower section being untested carbon steel, even though the plate material used for the column shell is essentially the same.

Acetylene Converters

As mentioned earlier, the existing Acetylene Converters were deemed to be adequate for the revamp case. The feed acetylene concentration was going to be reduced from 1.5 mol% to 0.6 mol% and the flowrate through the reaction beds was essentially the same as for the pre-revamp design.

There were already some hydraulic constraints through the Converters and their associated heat exchange equipment and this meant that the pressure differential that is required for flow between the Deethanizer and the C2 Splitter was going to be tight. This became a more critical issue as further expansion of the plant was evaluated and several steps were considered in order to reduce pressure drop. These have ranged from eliminating some significant pressure losses at the Deethanizer overhead by replacing the reflux drum and reducing the number of tube passes in the condenser, to considering parallel operation of the Converters. There are two stacks of three-bed Acetylene Converters so it is possible to consider splitting the feed and sending half of it to each of two beds in the stack thus creating a parallel system. The third bed becomes redundant but consideration can be given to using that as the outlet Dryer subject to pressure drop. This arrangement allows a significant increase in hydraulic capacity between the Deethanizer and the C2 Splitter. None of these types of changes have been implemented as they were not required for the basic feed revamp project.

C2 Splitter Capacity

The C2 Splitter was clearly a distillation column that would receive a significant increase in load due to the additional unconverted ethane. The 2005 project had included installation of what were considered to be the maximum capacity trays at that time. Furthermore, the Splitter achieved some its capacity by drawing a dilute ethylene product in addition to a normal polymer-grade product draw. The increase in ethane in the feed meant that the dilution of the lower grade product would increase to the point that it would not really provide a meaningful capacity increment due to the need to 'spike' it with polymer grade material in order to produce an acceptable specification.

Preliminary evaluation of the tray capacity constraints suggested that the C2 Splitter would probably limit operation to something like 92% of the required production. A number of alternative approaches were considered and these ranged from new experimental trays to significant changes in the processing scheme and ultimately to the installation of a second C2 Splitter system in parallel. These considerations are discussed in a paper by NOVA Chemicals that is being presented in this conference (reference 3). The final decision was to install a second C2 Splitter and this was driven by the potential to more fully utilize the spare Cracking Furnace capability in the future by increasing the processing capacity of the rest of the plant. The design and sizing of this new distillation column also gave some consideration to longer term capacity expansion as well as the incorporation of even greater amounts of ethane feed.

Propylene Refrigeration

A new Propylene Refrigeration Compressor (PRC) had been installed in the 2005 expansion project and was typically operating on the low end of its volumetric and power capabilities. This was one reason that the ethane feedstock revamp had looked reasonably achievable. It can be expected that the PRC will pick-up some significant additional loads from chilling and condensing the lighter cracked-gas as well as the additional condensing load on the C2 Splitter due to the large quantity of ethane recycle. The load increase can be offset by using the vaporization of the ethane feed and recycle as an additional source of refrigerant at temperature levels that are equivalent to the propylene refrigerant. This was discussed in the section on ethane feed vaporization.

Once revised PRC loads were established for the revamp case they were checked against the Manufacturer's capacity performance curves. This review indicated that the compressor could operate at the required capacity without any modifications and that the steam turbine driver had sufficient power margin. Indeed, when further capacity upgrades were being reviewed, it appeared that the compressor would run at its maximum continuous operating speed before it would run out of driver power. Similar revamp projects may not have this amount of excess compressor capacity and as further ethylene capacity increases for this particular plant were evaluated the PRC became one of the critical constraints.

One key observation about the design of the compressor was that casing nozzle and internal diaphragm design placed a limitation on the opportunity to economize the machine by removing propylene vapour at lower pressure levels. The PRC provides an indirect heat pump across the C2 Splitter by compressing vapour from the condenser so that it can be condensed in the reboiler. The propylene refrigerant flows involved for ethane cracking are quite large and significant power savings can be achieved if the reboiler is fed with propylene vapour that is extracted from the compressor at an intermediate pressure level. A suitably designed PRC can allow the extraction of most, if not all, of the required reboiler flow at a much lower pressure level. This economization is one factor that can be weighed into a decision on whether to replace the PRC or to re-rate it when a capacity increase is required.

Figure 8A illustrates the extraction of propylene refrigerant vapour at an intermediate compression stage. This is then fed to the C2 Splitter column bottoms reboiler where it is condensed and the resultant liquid effectively supplies the cooling at the column condenser after letdown to the suction pressure of the PRC.

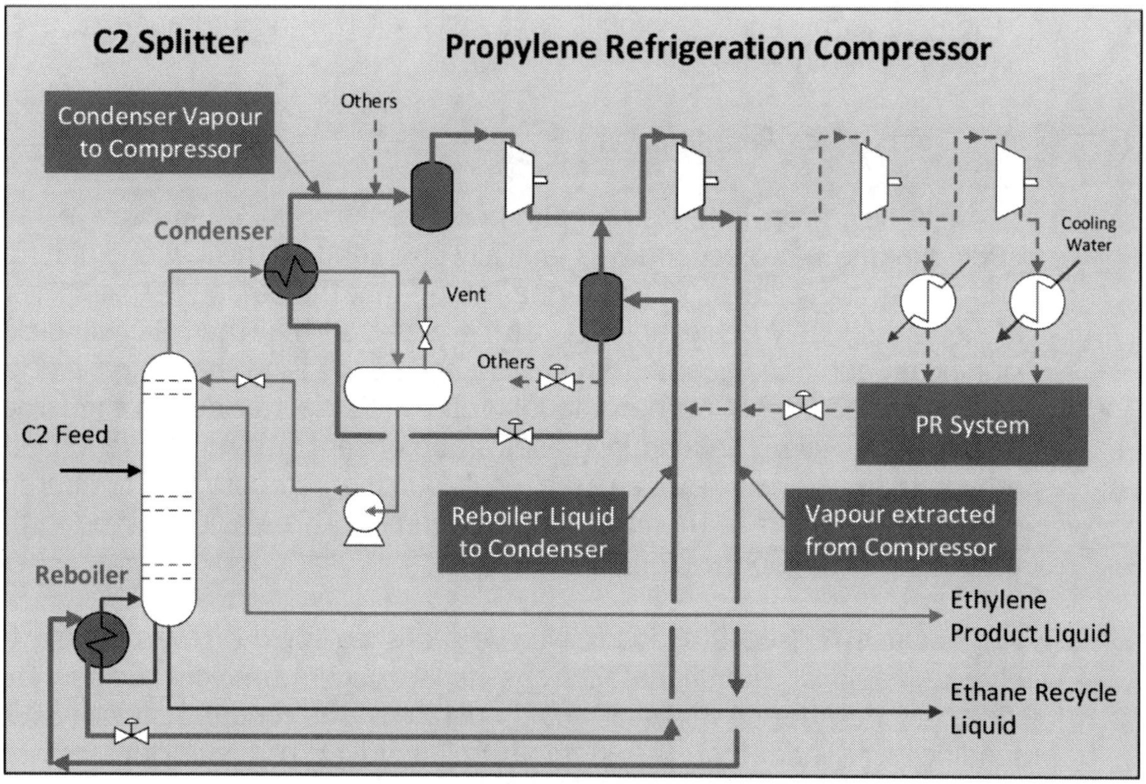

Figure 8A: PRC Heat Pumping of C2 Splitter – vapour from extraction

Figure 8B shows the route taken by the refrigerant vapour when the low pressure extraction cannot be taken or the flowrate is constrained. Make-up vapour to supply the reboiler duty must be taken from the discharge of the PRC and this increases the power demand for operation of the C2 Splitter.

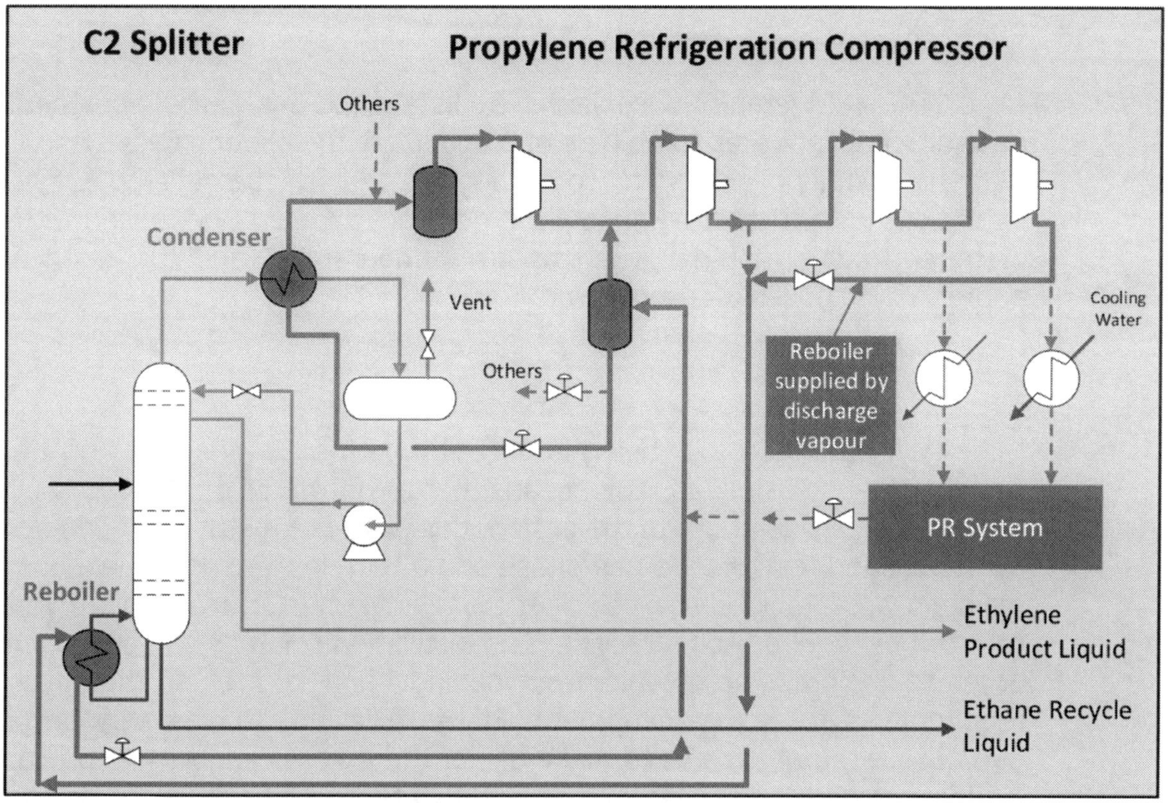

Figure 8B: PRC Heat Pumping of C2 Splitter – vapour from final discharge

Ethylene Refrigeration

The Ethylene Refrigeration Compressor (ERC) had also been replaced in 2005. However, because the plant capacity had not been fully exploited on the intended feedstock it tended to operate close to its surge control margin. While marginally higher, the projected loads for the ethane revamp feed case did not really have a significant impact on this compressor although this projection did rest somewhat on the estimated Chilling Train, Cold Box and Turbo Expander performance.

In practice a power restriction has been experienced on the ERC when reaching the intended ethylene production and ethane feed rate. This has been caused by two factors; one being lower than expected Chilling Train and Cold Box performance and the other being a large pressure drop in a refrigerant Subcooler.

As indicated in the Cold Box section, the performance of the parallel system was difficult to predict and it appears that the ER thermosiphon chillers in the newer Cold Box are under-performing which results in a higher load on the suction of the compressor. In addition, the presence of fouling at the warmer end of the Chilling

Train has shifted the cooling load downstream onto the ethylene refrigerant exchangers.

The refrigerant Subcooler is a platefin core-in-kettle type and the high pressure drop indicates some type of pluggage of the platefin passages. It is evident that this condition existed to some degree prior to the revamp project but the nature of the restriction is unknown. It does not appear to be fouling with a hydrate or other soluble material when at low temperatures and the restriction is not appreciably reduced after warm-up.

Coproducts

With the feedstock being largely ethane the C3-plus areas of the plant are significantly under-loaded. This means that turndown of those processing systems becomes a consideration. In the case of this plant the methyl acetylene/propadiene hydrogenation reactors are gas-phase and a recycle was recommended in order to maintain acceptable space velocities. Similarly, the C4 acetylenes hydrogenation reactor has required some recycle. The distillation columns can be over-refluxed in order to maintain acceptable tray loadings and control of the minimum flow requirements for some pumps should be reviewed. It may also be possible to purchase external streams to make use of the excess capability to process C3-plus olefin products. Some external C3 and C4 streams are being processed in this plant and these do help with some of the turndown issues.

The pyrolysis gasoline production is substantially reduced when cracking ethane and the lack of this as feedstock to associated Pygas processing and aromatic extraction units (collectively known as DPG) can result in the need to operate these units in a campaign mode. External oil and aromatic cuts may be purchased to supplement the internally produced feedstock and gain additional revenue.

Utilities

The key utilities that can be impacted by the feedstock change are the steam and fuel systems.

In the case of the fuel system the composition can change significantly due to the large increase in hydrogen production, although this will depend on whether is hydrogen is going to be recovered as a product. Depending on the original feed slate, there may also be a substantial loss of liquid fuels. The net result may be a need to change from liquid to gas fuel and to modify (or replace) the burner types where the gas is fueled. In the case of this plant, there was a need to convert the steam Boilers to gas firing in order to avoid the purchase of liquid fuels. There was also a requirement to replace or modify some burners at the Cracking Furnaces to accommodate the increased hydrogen in the fuel. When considering these impacts, it is possible to strategize whether some combustion equipment should operate exclusively on natural gas while other equipment uses the plant tail-gas as fuel. Each site will have its own specific situations that need to be addressed by balancing operability against additional cost. In this plant, some control of the blending of imported natural gas with the plant-produced fuel gas allows some hydrogen to be diverted for combustion in the Boilers with the rest going to the Cracking Furnaces.

Changes in steam production and use need to be considered. The steam balance will be different for each site but there will typically be a reduction in steam make at the Cracking Furnaces if liquid feedstocks are reduced in favour of ethane. Steam consumption will probably be reduced with lower power demands at the compressors although this is dependent on whether capacity increases are included in the revamp. It would not be unusual for the steam turbine power demand to be kept the same by channeling it into more production. There can also be a significant swing in steam demand if the associated Pygas unit is operated in campaign mode. One potentially very significant factor in the steam balance is the increase in medium pressure steam requirements to make dilution steam if the oil quench is operated at lower temperatures. As discussed earlier, this impact can be mitigated or eliminated if secondary TLE's are installed to recover this heat and make more steam.

Flare System

The flare system will see lower molecular weight cracked gas during major relief scenarios as well as at start-up. Relief capacities and flare hydraulics should be checked for the shift in composition and rates that is associated with ethane cracking. Flare system design for this plant had undergone a significant review during the 2005 expansion and this involved the use of both steady-state and dynamic models of the relief loads. In general, the loading was lower for the ethane revamp case. High integrity trip systems may need to be considered as a means of eliminating flows to the flare if capacity upgrades are to be included with the feedstock revamp.

Project Issues

The schedule for the revamp project had some challenges. After the decision to proceed with the seven ethane Furnace case, basic engineering started in early 2011. This meant that a limited amount of design information was going to be available prior to a plant maintenance turnaround that was already scheduled for September of that year. The challenge was to define and engineer over 100 tie-in points that could be implemented during that turnaround so that all of the necessary changes could be brought online while the plant was in operation.

A majority of the additions and modifications were completed while the plant was in operation. The most significant additions included a new parallel C2 Splitter and an ethane feed vaporization and superheating system. Conversion of parts of the dilution steam system from oil to steam heating was also a significant area of modification.

E/P-mix feed cracking was started in early 2012 and the first ethane was introduced in the third quarter of 2013. From that point forward the various equipment transitions continued toward operation of the new parallel C2 Splitter which was brought onstream in mid-2014. Full ethane feed capability was achieved in the third quarter of 2014 and has now been online for 18-months or so.

Figure 9: Erection of New C2 Splitter

Results

The overall result of this revamp is that the plant is now able to operate on the new high ethane feed diet. Nameplate ethylene production at design ethane feed rates is being achieved consistently and ethane rates up to 110% of design have been demonstrated.

The plant was successfully transitioned on-line through the significant milestones of initially ceasing gas-oil cracking and then gradually removing all naphtha feed. The transition period involved moving to an external ethane/propane-mix feedstock before finally bringing in the new ethane supply.

The new ethane feed system has been demonstrated at both high and low pressure with the high pressure auxiliary vaporizer mode proving to be a useful way of dealing with some moisture issues that occurred during early use of the new ethane pipeline.

All of the Cracking Furnaces (except the Milliseconds) have achieved target ethane capacity with acceptable run-lengths and yields. The Millisecond Furnaces have been used to crack propane and butane feeds. Start-up of furnaces with the new burner management systems has been achieved and these systems are operating reliably. The new furnace effluent water quench fittings were successfully commissioned and oil quenching has been stopped.

Operation of the Gasoline Fractionator with a colder quench-oil circuit proved to be somewhat better than expected with little or no high viscosity excursions being experienced. Quench water quality did deteriorate making oil-water separation more difficult and this is being countered by the use of emulsion-breaking chemical addition. Over the longer term, a new quench water treatment system will be added to the plant and this is expected to further improve oil-water separation. Moving to the cooler oil circulation involved converting several Dilution Steam Generators to use medium pressure steam as the heat input and these conversions were made during the transition period.

Apart from commissioning the new ethane feed and recycle vaporizers in conjunction with the Ethylene Refrigeration Compressor, the main transitions in the cold section of the plant involved bringing the new parallel C2 Splitter online as well as the modified Turbo Expander. Most of these changes were made after liquid cracking had been stopped.

Some challenges presented themselves during the course of transitioning various systems. The most notable were a high pressure drop at the front-end of the chilling train and pluggage of some of the Furnace burner tips.

The pressure drop across the first propylene-cooled heat exchanger in the chilling train increased to the point where it became a constraint to plant operating rates. Figure 10a illustrates the heat exchangers at the warm end of the chilling train prior to the first knock-out of liquid to feed the Demethanizer. The propylene-cooled chiller is a plate-fin core-in-kettle type unit and, while some improvement was seen by injecting methanol, it was usually short-lived. Sampling upstream indicated the presence of heavy oil which, while free-flowing at that point, was likely setting-up at the lower temperatures inside the plate-fin passages. The pluggage was somewhat unexpected as this heat exchanger had been operating successfully for 8 years on high NGL diets. It is believed that oil originates from the Cracked Gas Dryers although some of the content was also identified as carryover of wash oil components from the Cracked Gas Compressor. The use of other solvents such as toluene was tried and a partial bypass was installed but only 10-15% of the pressure drop could be recovered. In the end, a parallel shell and tube heat exchanger was installed (Figure 10b) and this allowed full plant rates to be achieved albeit with less than the intended level of cooling at this location.

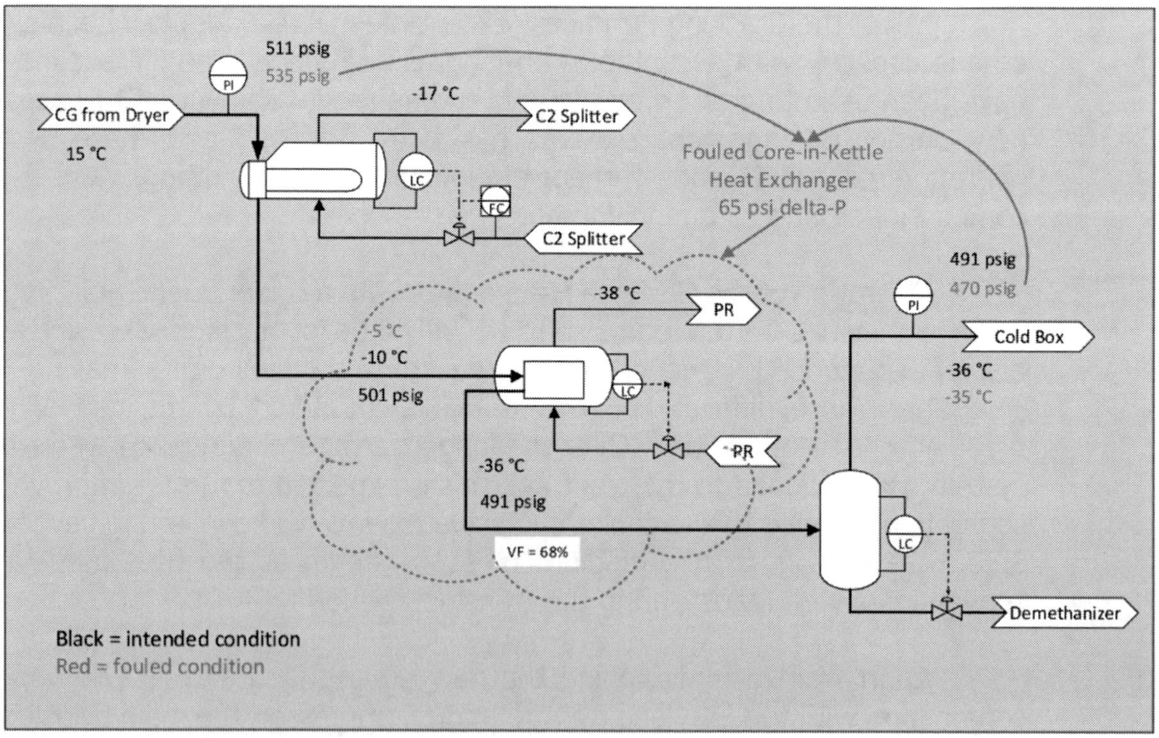

Figure 10a: Fouled Core-in-Kettle Exchanger in Chilling Train

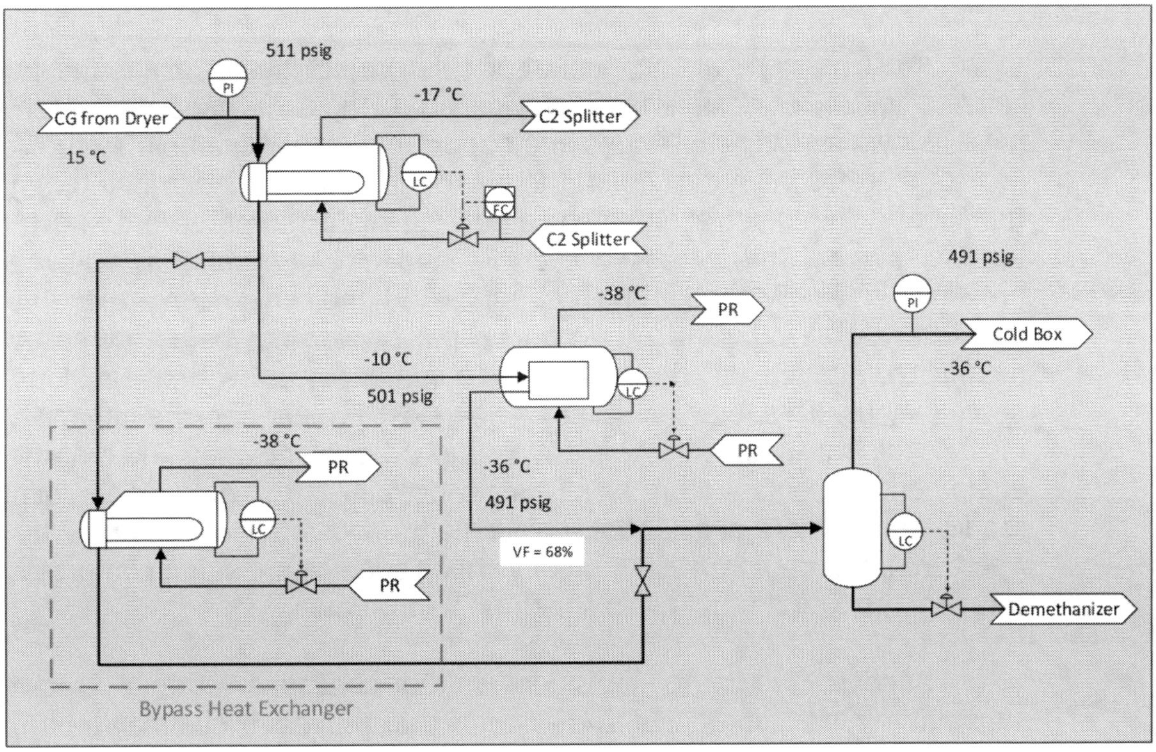

Figure 10b: Bypass Heat Exchanger in Chilling Train

Oil accumulated at the new burner tips resulting in some pluggage. The source of the oil was not totally clear as some analysis of the fuel gas indicated the presence lube oil components but there also appeared to be oligomer components in the upstream Tail-gas after it had been used for Dryer regeneration. A number of operating changes have alleviated this problem. These included modifications to the Dryer regeneration process and shutdown of a fuel gas compressor. Longer term, there is a plan to install some high efficiency coalescers on the fuel gas supply.

Conclusions

While in general the results of this project demonstrate the potential to convert liquids, or flexible feed, crackers to a mostly ethane diet, each plant will have its own specific challenges and opportunities. For example, in the case of the project that has been described, the plant had recently undergone a major retrofit that had included replacement of all of the major compressors and their steam turbine drivers. The increased capacity and efficiency of this newer equipment proved to be a very good foundation for the ethane conversion project. At the same time the overall ethylene production capacity was not being increased and this meant that much of the existing equipment was within previous capacity specifications. This led to a relatively low cost project and the capability to make most of the transition with the plant online. Other plants may find that their original equipment presents a more significant constraint and if greater production is also required then more substantial replacements and modifications may be required. Apart from being more costly, modifications such as changing compressor internal components would require a planned unit shut down.

Another factor to consider is that because ethane cracking yields substantially less coproducts, the systems that process those may be turned-down significantly. Some items, such as reactors, may need to be operated in recycle mode and that could require additional piping and controls. Ancillary processing units may need to be operated in campaign mode which could influence storage and logistics requirements.

When considering this type of conversion, it is also important not to forget that significant utility and infrastructure changes may need to be included. These considerations can be very specific to the type of operating site in which the Cracker is located. In this particular plant the loss of liquid fuel streams resulted in the need to retrofit the Boilers to a fully gaseous fuel operation and this turned-out to be a challenge both in terms of the design and the need to take the equipment offline. It also required a new external natural gas supply line.

Further analysis of the equipment capability in this plant suggests that conversion to a 100% ethane feed capability is feasible with changes that would be limited mostly to the feed system and the processing area between the Deethanizer and the C2 Splitter. Beyond that, ethylene capacity increases in the range of 20-50% could be economically achieved if consideration is given to re-wheeling the refrigeration compressors and adding, or upgrading, Cracking Furnace capacity.

If further feed and capacity upgrades are being considered for future projects it will be useful to have a good understanding of the ultimate limitations of key equipment when processing ethane feed as well as what modifications or replacements are available. This will allow the development of a sound overall strategy for the size and timing of the potential upgrades.

References

1. Challenges of Primary Fractionators on NGL Feed Systems – Paper 92a
   A. Bernard; F. Fahiminia; G. Kim
   2013 AIChE Ethylene Producers Conference, San Antonio, TX

2. Burner Management System Upgrade as a Result of Cracking Furnace Firing System Replacement – Paper 136c
   M. de Boer
   2013 AIChE Ethylene Producers Conference, San Antonio, TX

3. Design, Operation and Startup Strategy of a New Parallel C2 Splitter
   A. Bernard
   2016 AIChE Ethylene Producers Conference, Houston, TX

4. Loss of Downcomer Seal on a Depropanizer During Plant Startup
   A. Bernard
   2011 AIChE Ethylene Producers Conference, Chicago, IL

AUTHOR INDEX

AUTHOR INDEX